APPLIED MULTIVARIATE ANALYSIS

APPLIED MULTIVARIATE ANALYSIS

JOHN E. OVERALL, Ph.D.
Research Professor
Department of Neurology and Psychiatry
University of Texas Medical Branch

C. JAMES KLETT, Ph.D.
Director
Central Neuropsychiatric Research Laboratory
The Veterans Administration

ROBERT E. KRIEGER PUBLISHING COMPANY
MALABAR, FLORIDA
1983

Original Edition 1972
Reprint Edition 1983

Printed and Published by
ROBERT E. KRIEGER PUBLISHING COMPANY, INC.
KRIEGER DRIVE
MALABAR, FLORIDA 32950

Printed in the United States of America

Library of Congress Cataloging in Publication Data

Overall, John Ernest, 1929-
Applied multivariate analysis.

Reprint. Originally published: New York : McGraw-Hill, 1972. (McGraw-Hill series in psychology)
Includes bibliographies and index.
1. Psychometrics. 2. Multivariate analysis.
I. Klett, C. James, 1926- . II. Title.
III. Series: McGraw-Hill series in psychology.
BF39.083 1983 519.5'35 81-20944
ISBN 0-89874-325-7 AACR2

FOR LAUREN

CONTENTS

PREFACE

This book is written for the advanced student and researcher in the social and behavioral sciences. Although examples used to illustrate various methods are drawn from a single source to facilitate comparisons, it is anticipated that students in education, psychology, sociology, and related fields will recognize the broad applicability of the methods discussed. The examples may be of particular interest to students of psychology, psychiatry, and sociology since they are based upon extensive collections of data concerned with psychopathology and its relationships to social and demographic background variables. Because procedures for the classification of psychiatric patients and assignment of patients to homogeneous groups have not been systematically presented elsewhere, researchers in clinical psychiatry should find of interest not only the results but also the methods used to obtain them.

Experimental research tends to focus on a single dependent variable, and precision is gained by holding constant or systematically controlling extrinsic sources of variance. In psychiatry, clinical psychology, sociology, and education, it often is not possible to control experimentally for important sources of variance, and the responses that are observed tend to be multidimensional. Quantitative methodology for clinical research often requires substitution of statistical controls for experimental controls—recognizing that good research cannot be based upon poor data. Careful measurement and classification are essential, but they are difficult, because samples tend to be large, individuals are recognized to be highly complex, and many of their interrelated characteristics are of interest. Multivariate methods are important for organization, analysis, and interpretation of such voluminous and complex data.

In addition to the inability to manipulate and control relevant sources of variance in nonexperimental research, several special types of problems require attention. An understanding of the relationships among numerous variables is important and requires a more comprehensive model or theory than that afforded by simple correlations. Techniques for data reduction are important to minimize the number of variables that must be dealt with, and a meaningful model for such data reduction can substantially increase information gain in future research. Methods for examining similarities and differences among several recognized clinical populations or treatment groups in terms of multiple measurements are necessary, while empirical methods for developing meaningful classification concepts or typologies are of increasing concern in an area where the biological bases for disorders remain unknown. Given classification concepts or known clinical populations, empirical methods for assigning individuals to the most appropriate class according to clearly specified operations are of importance. Finally, methods for evaluating treatment effects and testing hypotheses in situations where experimental control over numerous relevant factors cannot be exercised are essential for adequate clinical research. These are the problems with which this text is concerned.

Following the format of several long-range programmatic research endeavors, the organization of this book deals with problems in the sequence in which they are likely to be encountered. With the exception of the first two chapters, which provide a necessary background in terms of mathematical tools and description of data sets used in examples throughout the remainder of the book, the sequence involves a logical progression from data reduction and the development of a conceptual framework for understanding relationships within the measurement domain, through problems of discrimination among groups and the study of relationships among predefined groups, to the development of classification groupings and the assignment of individuals. The final section is concerned with general linear-model techniques for testing hypotheses in the absence of conventional balanced experimental designs.

Chapter 1 provides a background for understanding the social and psychiatric variables included in the clinical data dealt with throughout the remainder of the book. Instead of adopting examples and illustrations from a wide variety of fields, we have chosen to use data from a single coordinated program of clinical research in which we have been engaged for a decade. The course of this programmatic research has followed the general sequence outlined in this text and thus provides a series of realistic examples for each section of the book. While the primary purpose is to provide a text of instruction in quantitative research methodology, a secondary purpose is to provide substantive results of importance for further use in clinical research. The Brief Psychiatric Rating Scale and the Standard History Form described in Chap. 1 are currently in wide use in clinical research in this country and abroad and have proved valuable as bases for classification methods and operational procedures for assignment of individuals among classification groups. The final section of Chap. 1 provides a description of the several large sets of clinical data that are used for illustrations, examples, and exercises throughout the remainder of the book. Understanding this book and its somewhat complex multivariate methods requires no mathematical background beyond college algebra. The essential matrix algebra necessary to understand the exposition of methods is covered in Chap. 2. The student or researcher who has not had previous exposure to matrix operations is forewarned that the assimilation of these methods requires time and study.

Many extremely competent students in the behavioral sciences are not aware of the different approaches that are necessary in the study of quantitative methods. Rather than expecting to master 30 or 40 pages of text per day, a capable student should plan to spend a week or more on a single chapter. For those who have no background in matrix algebra, the authors expect that 2 or even 3 weeks could profitably be spent in a detailed study of the material in Chap. 2. After that, frequent reference back to relevant sections may be necessary to understand the operations implied by the matrix formulas in subsequent chapters. For the student already familiar with matrix methods, a superficial reading of Chap. 2 will acquaint him with the notation and conventions used throughout the remainder of the book.

Chapter 3 provides an exposition of principal-components analysis, which we consider primarily as a method for data reduction and for orthogonal transformation of correlated variables. Chapters 4 to 7 are concerned with factor-analysis methodology. Factor analysis is viewed both as an elegant model for understanding relationships among multiple measurements and as a method for psychologically meaningful data reduction. The general method for linear factor analysis described in Chap. 4 encompasses all orthogonal-factor methods as special cases. Successive chapters describe in detail special cases including principal-axes and orthogonal factor rotation, direct cluster-oriented factor analyses, and direct oblique-factor solutions. From a clinical-content point of view, these chapters provide the empirical basis for conceiving of major psychiatric pathology in terms of four syndrome factors of thinking disturbance, withdrawal-retardation, hostile-suspiciousness, and anxious depression.

Chapter 8 is concerned with methods for deriving homogeneous groupings of individuals based upon similarities of multiple measurement profiles. This is an increasingly important and interesting type of research in psychopathology, where current diagnostic practices are based on little more than frequently occurring symptom profile patterns. Empirical methods for identifying most frequently occurring multivariate profile patterns, or groups of patients having maximum multivariate profile similarity, are generally referred to as *profile cluster-analysis techniques*. The objective of profile-clustering research is to identify naturally occurring groups or modal types that can serve as the basis for classification of future patients. Empirical research of this type has already had substantial payoff in the evaluation of treatment modalities. The use of homogeneous clustering techniques for defining patient classes has an advantage for assignment of individuals because most individuals tend to be highly similar to one of the class prototypes. The content-area results presented in this chapter define prototype profiles for empirically derived classes currently being used in clinical psychopharmacological research.

The next section of the book is concerned with discrimination between predefined groups in terms of multivariate measurement profiles. The simple discriminant function provides a basis for evaluating the difference between two predefined groups. Tests of significance for group differences in terms of multiple quantitative variables are available. Graphic and mathematical methods for defining optimal decision procedures for classification of individuals between the two groups are described. The extension of these techniques to problems involving multiple groups is discussed in Chap. 10, where the emphasis is primarily on examining the configuration of similarities and differences between several predefined groups in terms of multiple measurement profiles. In a later chapter it is shown how the multiple discriminant functions can be used for assignment of individuals among groups. Chapter 11 surveys a number of tests of multivariate hypotheses.

From the problem of discrimination among groups, attention is directed to the assignment of an individual among one of several known groups. The general theoretical model is discussed in Chap. 12, with the most elaborate of the multivariate assignment procedures based on the normal-probability-density model being presented in detail in Chap. 13. The next two chapters describe simplified decision models which can be related to the multivariate normal model under certain assumptions, and Chap. 16 presents a method for assignment using qualitative or categorical data.

The final two chapters are concerned with analysis of the regression of a single quantitative dependent variable on multiple independent variables. Multiple-regression and multiple-correlation methods useful in examining the relationship of a single quantitative dependent variable to several independent variables and the conception of analysis of variance and covariance as special cases of multiple-regression analysis are discussed. Of importance in clinical research is the evaluation of differences in multivariate prediction equations under different treatment conditions. The general linear-model analysis provides the basis for testing significance of treatment differences in arbitrary analysis-of-variance designs where there is no attempt to control cell frequencies or to balance artifically a variety of naturally occurring phenomena.

Computer programs were included in the present text only after prolonged deliberation. The objective of the book is to acquaint the student and researcher with the nature and meaning of certain advanced quantitative methods, but calculations involved in such complex methods are seldom, if ever, done by hand in this era of computers. Therefore, a valuable part of training for any student who may someday find himself in an academic or research environment is the experience of having been able to push his own data analyses through available computer facilities. From a learning point of view, however, the use of specialized computer programs that have been developed to provide only the final answers is of little help in promoting an understanding of what the methods actually involve.

We have attempted to meet the dual needs of instruction and research experience in two different ways with computer programs appended to various chapters. At the end of Chap. 2 is a general matrix-analysis program intended primarily for instructional use. It does not solve specific problems but provides the basic intermediate calculations necessary to summarize raw data in a form easily amenable to subsequent analysis. For example, means, matrices of cross products, covariances, correlations, within-groups dispersion matrices, and the inverses, determinants, and characteristic roots and vectors are basic summary statistics required to solve various problems in multivariate analysis. The computations of these various basic matrices and statistics are described in Chap. 2. In practice, we would not like to think of the modern student sitting at a desk calculator and laboriously summing, squaring, and multiplying for hours simply to summarize his raw data in a form that he can then begin to analyze. The basic matrix-analysis program provided at the end of Chap. 2 provides the intermediate summary matrices and statistics required for solutions dealt with in subsequent chapters. From a teaching and learning point of view, we recommend that this one program be used for calculations required of a student in developing an understanding of the material presented in later chapters. The specific multivariate-analysis programs appended to other chapters provide the practical solutions in a form required by the researcher, but they do not reveal in as much detail how the results are obtained. They are included for use in solving clinical research problems correctly and efficiently, not primarily as teaching aids. The one important instructional use that can be made of specialized multivariate-analysis programs is in formal or self-instruction designed to increase understanding of *what* can be accomplished in complex multivariate analysis of clinical data rather than *how* the analyses are accomplished. It is conceivable that the entire emphasis of the course could be on the evaluation and interpretation of research results. In this case, the programs for complete computer analysis might prove most useful.

Many people have contributed substantially to the development of this text. The data used as examples throughout the book were available thanks to the efforts of literally scores of senior investigators. Special acknowledgement should go to Drs. Leo E. Hollister, Alex D. Pokorny, Jonathan O. Cole, Istham Kimbell, Jr., Joe P. Tupin, B. W. Henry, and Burton J. Goldstein. Investigators in other countries who have provided psychiatric diagnostic prototype data include Drs. Pierre Pichot (Paris), Hanns Hippius (Berlin), Frank Engelsmann (Prague), and Romolo Rossi (Genoa). Drs. Donald R. Gorham and Douglas K. Spiegel contributed substantially through prior collaboration with us on methodological problems that are a primary concern of this book.

In the writing and revising of the text, the very considerable assistance of friends, consultants, and reviewers is acknowledged. Dr. Lyle V. Jones provided not only the initial encouragement to undertake the task but also continuing consultation, comment, and critical review. Others who have provided substantial technical consultation that has greatly influenced the final product are Drs. Nancy Wiggins, Elliot Cramer, Robert G. Demaree, Douglas M. McNair, Spencer M. Free, Edward F. Gocka and members of his staff at the Veterans Administration Western Research Support Center.

A most essential role was filled by Dorothy N. Snyder, who wrote many of the computer programs and actually computed most of the examples included.

Her energy, perserverence, and continued enthusiasm for the project in the face of numerous frustrations are greatly appreciated. Many other support personnel also provided material assistance.

The time and resources necessary to complete the book were available largely due to the support of the Psychopharmacology Research Branch of the National Institute of Mental Health. The computer facilities used in developing and testing programs, as well as personnel who worked on the project, were supported in part by grant DHEW 5 RO 1 MH 14675.

We are indebted to the literary executor of the late Sir Ronald A. Fisher, F.R.S. and to Oliver & Boyd, Edinburgh, for their permission to reprint Table A.3 from "Statistical Methods for Research Workers" and to Professors G. W. Snedecor and W. G. Cochran and the Iowa State University Press for permission to reproduce Table A.2 from their book "Statistical Methods."

JOHN E. OVERALL
C. JAMES KLETT

PART ONE

PRELIMINARY CONSIDERATIONS

1

Patient Evaluation and Description

1.1 Introduction

This book is concerned with a variety of methods for the analysis of multivariate data and the results obtained from their use. The methods could be presented by using artificially constructed examples or by selecting illustrations from a somewhat limited published literature. Except for certain assumptions about the statistical-distribution characteristics of the data, it is not essential that the examples used to illustrate the computational procedures have any intrinsic scientific merit. However, equal in importance to understanding the computations is an appreciation of how the methods are applied, what questions they help to answer, when one method is more appropriate than another, and what inferences can be drawn from the results. For these purposes it was considered didactically advantageous to discuss multivariate statistical methods in the context of a continuing program of clinical research and to select all examples from a single general content area so that the reader can become familiar with the data and the kind of research questions that can be asked of them. The reader will then be able to make the transition to research problems in his own field of interest.

In an effort to communicate not only computational methods but also an understanding of the nature of appropriate applications, we have chosen to illustrate each method with numerous examples drawn from extensive collections of available data. As a result, the book deliberately emphasizes content-area results and is descriptive of a programmatic research effort aimed at the clarification of problems in the description, classification, and evaluation of change in psychopathology. To some readers, the data themselves will be of inherent interest, while for others they will have only heuristic value. For the reader who is primarily interested in the data and content-area results, this book will serve as a blueprint of a systematic program of research that has gone through a rather characteristic progression from description, to evaluation, to classification, model building, and validation. The methods used in the various stages of this research are described in detail in subsequent chapters. The remaining sections of this chapter describe the nature of the data and the patient samples that will be discussed throughout the book.

1.2 The Brief Psychiatric Rating Scale

1.2.1 RATING SCALES

At the foundation of clinical research is the collection of standard, reasonably objective, and relevant data. Quantification of clinical observations is an important requirement if powerful descriptive and analytic methods are to be applied to the data. In response to these needs, a variety of psychiatric rating scales have been developed over the past 15 years (Lyerly and Abbott, 1966). Although clinicians vary in their degree of enthusiasm for the use of rating scales, there is no question of the impact they have had on psychiatric research in the last decade.

Rating scales differ in several important respects. Some scales are concerned with the frequency of disturbed behavior, while others are intended for evaluation of the severity of disturbance. Some ratings are based on completely structured interviews, while others depend upon more nondirective interviews. Several different levels of abstraction in the conception and evaluation of psychopathology are apparent in different attempts to produce adequate clinical ratings. In a general way two schools of thought currently exist among those who use rating scales for description of pathology. One school tends to doubt the reliability and consequent validity of abstract clinical judgments and attempts to reduce psychopathology to more molecular, concrete bits of behavior. As argued by the classical associationists, higher-level abstractions such as anxiety are built from the sum of specific signs and symptoms. The other school has considered that abstract concepts of psychopathology can best be evaluated directly and that the clinical abstraction may be more than the sum of specific parts.

The Brief Psychiatric Rating Scale (BPRS) is an abstract symptom-construct rating scale intended for use by qualified professionals who are used to thinking in such clinical terms. It was initially developed to provide a rapid assessment technique particularly suited for the evaluation of patient change in psychiatric drug research (Overall and Gorham, 1962). The 16 symptom-rating constructs of the original scale were derived from factor analyses of several larger sets of symptom-descriptive items, principally the Multidimensional Scale for Rating Psychiatric Patients (Lorr, 1953) and the newer Inpatient Multidimensional Psychiatric Scale (Lorr, Klett, McNair, and Lasky, 1962). The intent was to have each primary symptom factor represented by a single 7-point rating scale of severity. Some of the preliminary work that led to identification of the primary symptom constructs has been published (Gorham and Overall, 1960, 1961; Overall, Gorham, and Shawver, 1961) and will not be reviewed here, but selected references representing subsequent work with the BPRS are included at the end of this chapter. References are provided for the use of the BPRS in assessing patient change (Overall, Hollister, Shelton, Johnson, and Kimbell, 1966), classification of patients in clinical research (Overall and Hollister, 1964; Hollister, Overall, Bennett, Kimbell, and Shelton, 1965), investigations of factors affecting the nature of manifest psychopathology (Overall, 1969; Pokorny and Overall, 1970), and investigations of factors affecting clinical outcome (Overall, Hollister, Kimbell, and Shelton, 1969; Overall and Tupin, 1969). Currently, the BPRS is widely used for description, measurement, and classification of manifest psychopathology. It is employed extensively in this country for assessment of psychiatric treatments and for objective classification of patients attendant to such research. Careful translations into French, German, Czechoslovakian, Italian, and Spanish have fostered its use as a tool in cross-cultural research.

1.2.2 THE INTERVIEW

The BPRS ratings are based upon a brief unstructured interview. To increase reliability, it has generally been recommended that patients be interviewed jointly by two observers and that ratings then be completed independently. Alternative approaches involve multiple interviews conducted independently by two or more observers or consensus ratings made after discussion among observers. Each of these approaches has its merits and disadvantages. Although reliability of ratings is a problem of no small concern, a modest increase in sample size in research designs involving single raters will generally provide more power than two or more raters and smaller samples (Overall, Hollister, and Dalal, 1967; Derogatis, Bonato, and Yang, 1968). The effective use of any rating scale requires experienced and qualified raters. This is particularly true in the use of abstract construct scales, such as the BPRS, where a clear appreciation of the meaning of the constructs is demanded. It has been recommended that any rating team which is to function as a unit in a research project should first standardize procedures and achieve a consensual understanding of the rating constructs. Several training interviews are useful in providing new raters with a sharper and more literal understanding of the rating constructs and the rules governing their use. In the training sessions, independent ratings based upon common interviews should be discussed item by item to identify bases for discrepancies. In the course of such a discussion, differences in the interpretations of rating constructs will become apparent, as will obvious deviations from rating instructions. Insofar as possible, differences in interpreting the rating constructs should be eliminated during training. Remaining lack of agreement will be due to differences in judgment concerning the degree or severity of symptomatology in each rating area. These are the kinds of errors one hopes to average out in combining ratings by different observers.

The success of the interview in eliciting a satisfactory behavior sample for rating purposes also depends upon the experience and skill of the observer-interviewer. One solution to the problem of variation in ratings introduced by different levels of skill of the interviewer has been to structure the interview completely by stipulating what questions are to be asked, in what sequence, and with what allowable deviation. However, patients, as the second party in the interview, are not standard, and not all respond well to this approach. Conceiving of the interview as an interaction between patient and interviewer implies a mutual response between the two. Furthermore, denial of flexibility in the conduct of the interview can penalize the experienced and sensitive interviewer. What has been suggested when using the BPRS is an initial brief period to establish rapport, a nondirective interaction of perhaps 10 min, followed by 5 min of direct questioning in those symptom areas requiring specific responses or to fill in gaps in information appearing spontaneously.

1.2.3 THE RATING CONSTRUCTS

A facsimile of the current version of the BPRS is presented as Fig. 1.1. The two final rating constructs representing "excitement" and "disorientation" are recent additions to the BPRS and do not appear in any of the subsequent examples. The observer rates each construct on the 7-point severity continuum. The reference group used in making judgments of severity of each symptom construct is the population of patients who have the symptom in question, i.e., "As compared with patients who have this symptom, what is the degree of severity of the symptom in this patient?" Each construct in the scale is accompanied by a capsule definition as an aid in the rating. More complete definitions have been provided and are particularly helpful in training. They are included here for definitional purposes. The first six are based upon observation of the patient, rather than what he says.

Tension This construct is restricted in the BPRS to physical and motor signs commonly associated with anxiety. Tension does not involve the subjective experience or mental state of the patient. Although research psychologists, in an effort to attain a high degree of objectivity, frequently define anxiety in terms of physical signs, in the BPRS observable physical signs of tension and subjective experiences of anxiety are rated separately. Although anxiety and tension tend to vary together, developmental research with the BPRS has indicated that the degree of pathology in the two areas may be quite different

DIRECTIONS: Place an X in the appropriate box to represent level of severity of each symptom.

PATIENT ______________________

RATER ______________________

NO. ______________________

DATE ______________________

	Not Present	Very Mild	Mild	Moderate	Mod. Severe	Severe	Extremely Severe
1. SOMATIC CONCERN - preoccupation with physical health, fear of physical illness, hypochondriasis.	☐	☐	☐	☐	☐	☐	☐
2. ANXIETY - worry, fear, over-concern for present or future.	☐	☐	☐	☐	☐	☐	☐
3. EMOTIONAL WITHDRAWAL - lack of spontaneous interaction, isolation, deficiency in relating to others.	☐	☐	☐	☐	☐	☐	☐
4. CONCEPTUAL DISORGANIZATION - thought processes confused, disconnected, disorganized, disrupted.	☐	☐	☐	☐	☐	☐	☐
5. GUILT FEELINGS - self-blame, shame, remorse for past behavior.	☐	☐	☐	☐	☐	☐	☐
6. TENSION - physical and motor manifestations or nervousness, over-activation, tension.	☐	☐	☐	☐	☐	☐	☐
7. MANNERISMS AND POSTURING - peculiar, bizarre, unnatural motor behavior (not including tic).	☐	☐	☐	☐	☐	☐	☐
8. GRANDIOSITY - exaggerated self-opinion, arrogance, conviction of unusual power or abilities.	☐	☐	☐	☐	☐	☐	☐
9. DEPRESSIVE MOOD - sorrow, sadness, despondency, pessimism.	☐	☐	☐	☐	☐	☐	☐
10. HOSTILITY - animosity, contempt, belligerence, disdain for others.	☐	☐	☐	☐	☐	☐	☐
11. SUSPICIOUSNESS - mistrust, belief others harbor malicious or discriminatory intent.	☐	☐	☐	☐	☐	☐	☐
12. HALLUCINATORY BEHAVIOR - perceptions without normal external stimulus correspondence.	☐	☐	☐	☐	☐	☐	☐
13. MOTOR RETARDATION - slowed weakened movements or speech, reduced body tone.	☐	☐	☐	☐	☐	☐	☐
14. UNCOOPERATIVENESS - resistance, guardedness, rejection of authority.	☐	☐	☐	☐	☐	☐	☐
15. UNUSUAL THOUGHT CONTENT - unusual, odd, strange, bizarre thought content.	☐	☐	☐	☐	☐	☐	☐
16. BLUNTED AFFECT - reduced emotional tone, reduction in normal intensity of feelings, flatness.	☐	☐	☐	☐	☐	☐	☐
17. EXCITEMENT - heightened emotional tone, agitation, increased reactivity.	☐	☐	☐	☐	☐	☐	☐
18. DISORIENTATION - confusion or lack of proper association for person, place, or time.	☐	☐	☐	☐	☐	☐	☐

Figure 1.1 Brief Psychiatric Rating Scale. (*Overall and Gorham*, 1962).

in specific patients. A patient, especially when under the influence of a drug, may report extreme apprehension but give no external evidence of tension whatsoever, or vice versa. In rating the degree of tension, the rater should attend to the number and nature of signs of abnormally heightened activation level such as nervousness, fidgeting, tremors, twitches, sweating, frequent changing of posture, hypertonicity of movements, and heightened muscle tone.

Emotional Withdrawal This construct is defined solely in terms of the ability of the patient to relate in the interpersonal interview situation. Thus, an attempt is made to distinguish between motor aspects of general retardation, which are rated as "motor retardation," and the more mental-emotional aspects of withdrawal, even though ratings in the two areas may be expected to covary to some extent. In the factor analyses of change in psychiatric ratings, a "general retardation" factor has emerged in several different analyses, and it has included emotional, affective, and motor retardation items. It is difficult to identify the basis for rating of "ability to relate"; however, initial work has indicated that raters achieve reasonably high agreement in rating this quality. Emotional withdrawal is represented by the feeling on the part of the rater that an invisible barrier exists between the patient and other persons in the interview situation. It is suspected that eyes, facial expression, voice quality and lack of variability, and expressive movements all enter into the evaluation of this important but nebulous quality of psychiatric patients.

Mannerisms and Posturing This symptom area includes the unusual and bizarre motor behavior by which a mentally ill person can often be identified in a crowd of normal people. The severity of manneristic behavior depends both upon the nature and number of unusual motor responses. However, it is the unusualness, and not simply the amount of movement, which is to be rated. Odd, indirect, repetitive movements or movements lacking normal coordination and integration are rated on this scale. Strained, distorted, abnormal postures which are maintained for extended periods are rated. Grimaces and unusual movements of lips, tongue, or eyes are considered here also. Tics and twitches which are rated as signs of tension are not rated as manneristic behavior.

Motor Retardation "Motor retardation" involves the general slowing down and weakening of voluntary motor responses. Symptomatology in this area is represented by behavior which might be attributed to the loss of energy and vigor necessary to perform voluntary acts in a normal manner. Voluntary acts which are especially affected by reduced energy level include those related to speech as well as gross muscular behavior. With increased motor retardation, speech is slowed, weakened in volume, and reduced in amount. Voluntary movements are slowed, weakened, and less frequent.

Uncooperativeness This is the term adopted to represent signs of hostility and resistance to the interviewer and interview situation. It should be noted that "uncooperativeness" is judged on the basis of response of the patient to the interview situation while "hostility" is rated on the basis of verbal reports of hostile feelings or behavior toward others *outside* the interview situation. It was found necessary to separate the two areas because of an occasional patient who refrains from any reference to hostile feelings and who even denies them while evidencing strong animosity toward the interviewer.

Excitement Excitement refers to the emotional, mental, and psychological aspects of increased activation and heightened reactivity. The excited patient tends to be active, agitated, quick, loud, and emotionally responsive. Whereas tension is a construct concerned with physical or motor manifestations of activation, excitement has reference primarily to the mental and emotional areas. Tension usually implies a binding of the physical activation potential, while excitement is the underlying activation potential. The degree of excitement depends on the strength of arousal and heightened affect.

The remaining constructs are defined and rated primarily in terms of what the patient says in the interview and the intensity of the reported experience.

Conceptual Disorganization Conceptual disorganization involves the disruption of normal thought processes and is evidenced in confusion, irrelevance, inconsistency, disconnectedness, disjointedness, blocking, confabulation, autism, and unusual chain of associating. Ratings should be based upon the patient's spontaneous verbal products, especially those longer, spontaneous response sequences which are likely to be elicited during the initial, nondirective portion of the interview. Attention to the facial expression of the patient during the verbal response may be helpful in evaluating the degree of confusion or blocking.

Unusual Thought Content This symptom area is concerned solely with the *content* of the patient's verbalization; the extent to which it is unusual, odd, strange, or bizarre. Notice that a delusional or paranoid patient may present bizarre or unbelievable ideas in a perfectly straightforward, clear, and organized fashion. Only the unusualness of content should be rated for this item, not the degree of organization or disorganization.

Anxiety Anxiety is a term restricted to the subjective experience of worry, overconcern, apprehension, or fear. Rating of degree of anxiety should be based upon verbal responses reporting such subjective experiences on the part of the patient. Care should be taken to exclude from consideration in rating anxiety the physical signs which are included in the concept of tension, as defined in the BPRS. The sincerity of the report and the strength of the experience as indicated by the involvement of the patient may be important in evaluating degree of anxiety.

Guilt Feelings The strength of guilt feelings should be judged from the frequency and intensity of reported experiences of remorse for past behavior. The strength of the guilt feelings must be judged in part from the degree of involvement evidenced by the patient in reporting such experiences. Care should be exercised not to infer guilt feelings from signs of depression or generalized anxiety. Guilt feelings relate to specific past behavior which the patient now believes to have been wrong and the memory of which is a source of conscious concern.

Grandiosity Grandiosity involves the reported feeling of unusual ability, power, wealth, importance, or superiority. The degree of pathology should be rated relative to the discrepancy between self-appraisal and reality. The verbal report of the patient and not his demeanor in the interview situation should provide the primary basis for evaluation of grandiosity. Care should be taken not to infer grandiosity from suspicions of persecution or from other unfounded beliefs where no explicit reference to personal superiority as the basis for persecution has been elicited. Ratings should be based upon opinions currently held by the patient, even though the unfounded superiority may be claimed to have existed in the past.

Depressive Mood Depressive mood includes only the affective component of depression. It should be rated on the basis of expressions of discouragement, pessimism, sadness, hopelessness, helplessness, and gloomy thema. Facial expression, weeping, moaning, and other modes of communicating mood should be considered, but motor retardation, guilt, and somatic complaints, which are commonly associated with the psychiatric syndrome of depression, should not be considered in rating depressive mood.

Hostility Hostility is a term reserved for reported feelings of animosity, belligerence, contempt, or hatred toward other people outside the interview situation. The rater may attend to the sincerity and affect present in reporting of such experiences when he attempts to evaluate the severity of pathology in this symptom area. It should be noted that evidences of hostility toward the interviewer in the interview situation should be rated on the uncooperativeness scale and should not be considered in rating hostility as defined here.

Somatic Concern The severity of physical complaints should be rated solely on the number and nature of complaints or fears of bodily illness or malfunction, or suspiciousness of them, alleged during the interview period. The evaluation is of the degree to which the patient perceives or suspects physical ailments to play an important part in his total lack of well-being. Worry and concern over physical health is the basis for rating somatic concern. No consideration of the probability of true organic basis for the complaints is required. Only the frequency and severity of complaints are rated.

Hallucinatory Behavior The evaluation of hallucinatory experiences frequently requires judgment on the part of the rater whether the reported experience represents hallucination or merely vivid mental imagery. In general, unless the rater is quite convinced that the experiences represent true deviations from normal perceptual and imagery processes, hallucinatory behavior should be rated as *not present*.

Suspiciousness Suspiciousness is a term used to designate a wide range of mental experience in which the patient believes himself to have been wronged by another person or believes that another person has, or has had, intent to wrong. Since no information is usually available as a basis for evaluating the objectivity of the more plausible suspicions, the term "accusations" might be a more appropriate characterization of this area. The rating should reflect the degree to which the patient tends to project blame and to accuse other people or forces of malicious or discriminatory intent. The pathology in this symptom area may range from mild suspiciousness through delusions of persecution and ideas of reference.

Blunted Affect This symptom area is recognized by reduced emotional tone and apparent lack of normal intensity of feeling or involvement. Emotional expressions are apt to be absent or of marked indifference and apathy. Attempted expressions of feeling may appear to be mimetic and without sincerity.

Disorientation This rating construct has been included to provide a place for recording the particular kind of confusion that is evidenced by lack of memory or proper association for persons, places, or times. The disoriented individual may not know where he is, how to relate where he is to other points in the environment, or how to get from one place to another. The identities of persons that should be familiar may be confused. Location in time and place and even personal identity may be confused or unavailable for recall. Distortions in identity such as those that occur in delusional systems should not be rated under disorientation. Disorientation represents the type of confusion that frequently occurs in organic conditions.

1.2.4 SCORING

The BPRS is scored in different ways for different purposes. Since the symptom constructs were conceived to be an intercorrelated but relatively independent set of variables descriptive of the basic dimensions of psychopathology, they are usually scored independently. The individual 7-point scales are scored 0, 1, 2, . . . , 6 corresponding to categories ranging from *not present* to *extremely severe*. The first category is scored zero because not present appears to be a rational origin representing a true zero point on a scale of symptom severity. Also, for certain methods of profile analysis and classification, the availability of a standard rational origin is important.

It should be noted that in scoring successive categories of severity with integer values 0, 1, . . . , 6, an equal degree of separation between categories is implied. In many statistical analyses the scores are treated as if they represented true equal intervals. The fact is that equal-appearing interval scales have frequently been demonstrated to be linearly related to Thurstone type equal-interval scales except in the extreme categories. In addition, Spiegel and Overall (1970) undertook a detailed analysis in the attempt to provide more discriminating scale values for the BPRS scoring categories and found that the resulting scale values were quite consistent with the equal-category differences except at the upper extreme. Use of the "optimal" scale values actually results in poorer correlation with external criteria and poorer discrimination between different groups of patients than the simple 0, 1, 2, . . . , 6 integer values presented here. For these reasons, the simple integer score values for categories of severity are recommended.

Four higher-order syndrome factor scores can be scored, each as the sum of three separate rating items. These four factor scores can be used to represent approximately 75 percent of the reliable variance present in the full set of items. The empirical basis for definition of these factor scores will become apparent in later chapters of this book, particularly Chaps. 5 and 6. At present, it is sufficient to state that factors of thinking disturbance, withdrawal-retardation, hostile-suspiciousness, and anxious depression can be scored as the unweighted sum of item scores as follows:

Thinking disturbance	*Hostile-suspiciousness*
Conceptual disorganization	Hostility
Hallucinatory behavior	Suspiciousness
Unusual thought content	Uncooperativeness
Withdrawal-retardation	*Anxious depression*
Emotional withdrawal	Anxiety
Motor retardation	Guilt feelings
Blunted affect	Depressive mood

A total pathology score is computed as the sum of all ratings on the BPRS symptom constructs. It represents the total deviation from normality in terms of whatever kinds of symptoms the patient manifests. Especially where profile differences have been employed as a basis for classifying patients into subgroups, the total pathology score is considered to be the most adequate measure for comparing treatment responses of different types of patients. In evaluating patients of a particular type, say clinical depressions, one of the higher-order factor scores may prove a more logical basis for comparing response to treatments.

Another scoring procedure will be defined in Chap. 3 as the *schizophrenic-depressive contrast function* (S-D) and the *coping-resignation contrast function* (C-R). It will be shown that these scores are useful in delineating diagnostic regions for classification purposes, and it seems likely that they will prove appropriate for other purposes as well. In practice, these two scores are most conveniently obtained as unweighted combinations of the rating variables.

S-D = (conceptual disorganization + hallucinatory behavior + unusual thought content) − (anxiety + guilt feelings + depressive mood)

C-R = (hostility + suspiciousness + unusual thought content) − (emotional withdrawal + motor retardation + blunted affect)

1.2.5 RELIABILITY AND VALIDITY

Of interest and practical concern is the question of reliability of clinical ratings. While there is an extensive literature on the theory and measurement of reliability, use of the product-moment correlations between ratings made independently by two observers is undoubtedly the most frequent method of estimating reliability of clinical ratings. This is the so-called *interrater reliability*. Interrater reliability estimates for individual BPRS variables and the four higher-order factor scores are presented in Table 1.1 for a single rater and for scores computed as the average of two independent raters. The data used to obtain these reliability estimates consist of paired independent ratings by two experienced clinicians observing the same patient in a joint interview. Patients were male schizophrenic and depressive patients who were subjects in a series of clinical drug studies conducted in several Veterans Administration hospitals.

Most of the examples in the remainder of this book are concerned with problems related to the validity of BPRS symptom ratings. The chapters on factor analysis present results bearing on the internal validity, structural validity, or construct validity of the measurements. Chapters concerned with discriminant function and other multivariate statistical methods have relevance for the concurrent validity, correlational validity, or predictive validity of the BPRS ratings. Chapters concerned with profile classification and diagnostic decision procedures reflect validity with regard to other types of problems. In many respects, the remainder of this book provides a basis for evaluating the validity of these types of clinical ratings.

Table 1.1 Interrater Reliability Coefficients for BPRS Symptom Variables and Four Higher-order Factor Scores

	One rater	*Two raters*
Symptom rating constructs:		
Somatic concern	.79	.88
Anxiety	.80	.89
Emotional withdrawal	.75	.86
Conceptual disorganization	.81	.90
Guilt feelings	.83	.91
Tension	.70	.82
Mannerisms and posturing	.89	.94
Grandiosity	.88	.94
Depressive mood	.89	.94
Hostility	.88	.94
Suspiciousness	.84	.91
Hallucinatory behavior	.90	.95
Motor retardation	.74	.85
Uncooperativeness	.76	.86
Unusual thought content	.85	.92
Blunted affect	.58	.73
Excitement	.78	.88
Disorientation	.84	.91
Composite factor scores:		
Thinking disturbance	.91	.95
Withdrawal-retardation	.77	.87
Hostile-suspiciousness	.88	.94
Anxious depression	.88	.94

1.3 Standard Psychiatric History

Data descriptive of patient characteristics other than current symptomatology are important in clinical research, just as they are in clinical practice. Patients with different history and background characteristics tend to evidence different manifestations of psychiatric disorder, and the probabilities of occurrence of different clinical syndromes vary as a function of the social and cultural characteristics of the patients. Clinicians have long considered this to be true, and numerous isolated relationships of such variables with various aspects of psychopathology have been described repeatedly in the literature. The problem of how to integrate all the available information and how to isolate specific independent effects from the tangle of interdependent variables has not been adequately solved. It is hoped that the methods and results presented in later chapters will contribute in some measure to this issue.

Certain pieces of background information have repeatedly been found to relate to current clinical status, probable treatment outcome, and even longer-term prognosis. The most frequently referenced variables include age, race, marital status, and social class. Detailed studies undertaken to investigate variables related to differences in the nature of BPRS symptom profile patterns have revealed a substantially larger collection of variables that are potentially important, although the strongest dependencies relate to primary demographic and history variables such as age, race, marital status, educational and work achievement, and course of illness. Figure 1.2 contains a collection of history and background characteristics that have been sifted through at least one or more previous studies and found to be significantly related to differences in manifest psychopathology. This pool of items is currently being used in several large clinical studies and is being evaluated for cross-cultural differences in several countries. These are the history and background data that will be examined intensively in subsequent chapters of this book.

Most of the items in the history form are self-explanatory. They have been used successfully without a detailed glossary or more specific definitions of terms. The person recording the history is instructed to use his own judgment concerning whether a job should be considered skilled or unskilled or whether alcohol should be considered to be a problem or not with respect to the particular case.

It may be useful to note that the items on the history form are of three general types. The first group characterizes the patient and his past. The second concerns behaviors and attitudes that can be considered to be coincident with the current psychopathology and which may be, at least in part, a function of it. The final group concerns the parental home, family history, and parental social class.

Although the relationship of these variables and the BPRS rating constructs will be investigated in a variety of ways later in the book, the results of a study conducted in a psychiatric outpatient clinic may help to indicate the potential usefulness of these history variables and provide additional insight into the nature of the BPRS constructs. In this study 729 patients seen for the first time in the outpatient clinic were interviewed separately by three different observers, a senior medical student, a staff psychiatrist, and a psychiatric resident. The history was taken by the senior medical student, and the BPRS ratings made independently by the other two observers were averaged to yield a single symptom description having greater reliability. Simple one-way analyses of variance were used to test the significance of differences in symptom-rating scores for patients falling into different categories of each background variable.

Instead of attempting to describe the nature and direction of each significant relationship, only the relevance, or lack of relevance, for each of the background variables will be indicated. These results help provide an empirical basis for a standard background-data form including only the most relevant types of information. The background data included in the history form used in this study can be grouped into (1) personal data, (2) parental family-history data, (3) problems associated with the psychopathology in the social context, and (4) precipitating factors implicated by the patient.

In Table 1.2 are listed the individual background items considered and beside each item are indicated the *number* of statistically significant symptom relationships in each of the four major factor domains: thinking disturbance, withdrawal-retardation, hostile-suspiciousness, and anxious depression. In the right-hand column, the total number of all BPRS symptom ratings that varied significantly as a function of status on the background variable is indicated. The number of symptom variables that are significantly dependent on a background item is taken as an index of the relevance of the item. The number of significant symptom relationships within each major domain is taken as an index of relevance to *that* more specific domain.

The results are further condensed in Table 1.3, where the background variables most relevant for each major symptom area are listed. Relevance was judged in terms of the number of significant symptom relationships of the background variable to symptoms within the major domain. Finally, in Table 1.4 are listed the specific background variables that were found to have greatest relevance to all BPRS items.

The results reveal that certain types of information included in the history form are more relevant than others. Each of the demographic and sociocultural items pertaining directly to personal characteristics of the patient himself was found to be highly relevant with respect to differences in the nature of symptom characteristics. Certain attitudes and behaviors of the patient were also found to be related to the nature of the psychopathology.

Next in importance with regard to distinguishing between various types of psychopathology were the items in the history form which represented types of problems or difficulties the patient had been having in the social context outside the psychiatric treatment setting. Items of major importance included work, marital problems, socialization, and physical problems.

Figure 1.2 Brief Psychiatric History Form

(University of Texas Medical Branch)

Patient______________ UH#______________ Date______________

Patient's Present Age
______Below 20
______20–29
______30–39
______40–49
______50–59
______60 or Above

Ethnicity
______Not Obvious
______Oriental
______Negro
______Other Minority

Sex
______Male
______Female

Age Initial Onset
______Below 20
______20–29
______30–44
______45–60
______60 or Above

Duration This Episode
______Less Than 1 Month
______1–6 Months
______6–12 Months
______1–2 Years
______More Than 2 Years

Course of Illness
______Slow Decline
______Recurrent Acute Episodes
______First Episode

Previous Psychiatric Hospitalization
______No Previous Admission
______Single Previous Admission
______Multiple Previous Admissions

Education (Level)
______No Formal Education
______1–5 Years
______6–12 Years
______13–16 Years
______More than 16 Years

Education (Termination Point)
______High School Graduation
______College Graduation
______Graduate or Professional Degree
______Terminate at Any Other Point (Including College or Professional Dropout)

Work Level (When Working)
______Never Worked
______Housewife
______Unskilled
______Skilled
______Clerical
______Self-employed
______Professional or Managerial
______Student

Current Marital Status
______Single
______Married
______Separated
______Divorced
______Widowed

Marital History
______No Divorces
______One Divorce
______Two Divorces
______Multiple Divorces
______Widowed Remarried

Children of Patient
______No Children
______One Child
______Two to Four Children
______More Than Four

Alcohol
______Total Abstinence
______Moderate
______Problem Drinker

Religious Attitude
______Atheist
______Agnostic (Doubting)
______Unconcerned
______Moderate
______Strong Positive or Fanatic

Patient's Attitude toward Work He or She Is Prepared to Do
______Negative
______Indifferent
______Normal Interest
______Excess Involvement
______Retired

Precipitating Factors as *Identified by Patient*
______Marital Conflict
______Separation or Divorce
______Death of Loved One
______Social Relations
______Work Problems
______Alcohol Problem
______Drug Use or Abuse

Patient____________ UH#____________ Date____________

_____Other

Major Problem Areas Associated with Psychopathology as either Cause or Effect (Chief Complaints)

_____Marital Problems
_____Socialization
_____Behavior Problems
_____Work Problems
_____Alcohol Problems
_____Drug Abuse
_____Mental Problems
_____Mood Problems
_____Physical Problems (Real or Imagined)
_____Sexual Problems
_____Other

Social Class Parental Home

_____Low
_____Low Middle
_____Middle
_____Upper

Father's Education

_____Less Than High School
_____High School Graduate
_____Some College

Father's Occupation

_____Never Worked
_____Unskilled
_____Skilled
_____Clerical
_____Self-employed
_____Professional or Managerial

Mother's Education

_____Less Than High School
_____High School Graduate
_____Some College

Mother Work While Patient under 8 Years

_____Never
_____Part of Time
_____Most of Time

Problems in Parental Home while Patient Child Less Than 12 Years. (Check All That Are Appropriate.)

_____Severe Discord
_____Prolonged or Repeated Separation
_____Divorce
_____Death of Father
_____Death of Mother
_____Death of Noninfant Sibling
_____Financial Insecurity
_____Alcohol or Drug
_____Other

Family History of Psychiatric Illness

_____Father
_____Mother
_____Siblings
_____Other 1st- or 2d-degree Relatives

Of considerably less relevance with respect to the nature of presenting symptomatology were items concerned with (parental) family history and social class. While the social achievement of the patient is quite important, little or no additional information is contained in items concerned with parental social class and even with family history of psychopathology. In earlier versions of the history form, we went more deeply into parental family relationships, discipline, religion, alcohol, separation, divorce, death, and a variety of variables. These variables were discarded in the construction of the current history form because they had not previously proved useful. There appears to be very little of psychiatric significance in what the patient can report concerning parental family, its social class, attitudes, and problems. The psychopathology appears to reside in the patient and not the early family environment, contrary to much mental health literature.

The nature of life events which the patient perceives as having precipitated his current psychiatric problems appears to be of relatively minor significance with regard to differences in general levels of psychopathology. It should be recognized that this section of the history form deals with the patient's own

Table 1.2 Number of Symptom Variables in Each Factor Domain That Related Significantly to Personal, Social, and Psychiatric History Items†

	T	*W-R*	*H-S*	*AD*	*TP*
Personal characteristics of patient:					
Age		1	3	3	10
Ethnicity			1	2	3
Sex	1	1	1	3	11
Age at onset	1	2	3	3	12
Duration episode				1	1
Previous course		2		1	3
Previous hospitalization	2	2		1	7
Education			2		3
Work achievement		3	1	1	6
Marital status	2	1	2	2	10
Divorce history	2	3		1	7
Children	2	2	2	3	12
Alcohol	1	3	1	1	8
Religious involvement	3		1		4
Work attitude	3	3	2	1	15
Parental social class and psychiatric history:					
Parental social class	1	1			4
Father's education		1		1	3
Father's occupation		1			2
Mother's education		1	2	1	5
Mother's work				1	1
Father's psychiatric illness		1		1	3
Mother's psychiatric illness					1
Sibling's psychiatric illness		1			2
Social and behavioral problems:					
Marital problems		1	1	3	6
Socialization problems	3	2	3	1	13
Behavior problems	3	2	3	2	14
Work problems		1			1
Alcohol problems					
Drug abuse	1		1	1	4
Sexual problems	2	1		1	4
Physical problems	3	1	2	1	9
Precipitating factors:					
Marital conflict	1	1		3	5
Separation or divorce	1	1			2
Death of loved one		1	1	1	4
Social relations	1		2	1	5
Work problems		2			3
Alcohol abuse			1	1	3

† T = Thinking disturbance, W-R = withdrawal-retardation, H-S = hostile-suspiciousness, AD = anxious depression, and TP = total BPRS.

Table 1.3 Variables Most Relevant for Each Major Symptom Area†

Thinking disturbance	*Hostile-suspiciousness*
Work attitude	Age
Religious involvement	Age at onset
Socialization problems	Socialization problems
Behavior problems	Behavior problems
Physical problems	Marital status
Previous hospitalizations	Children
Marital status	Education
Divorce history	Work attitude
Children	Mother's education
Sexual problems	Physical problems
	Social relations as precipitating factor
Withdrawal-retardation	*Anxious depression*
Work attitude	Age
Work achievement	Sex
Alcohol	Age at onset
Divorce history	Children
Age at onset	Marital problems
Previous hospitalizations	Marital conflict as precipitating factor
Previous course of illness	Ethnicity
Children	Marital status
Socialization problems	Behavior problems
Behavior problems	

† Note that some items listed as being relevant with respect to a major symptom area may have negative relationship to the levels of symptomatology, while other items have positive relationship.

perceptions and/or rationalizations concerning precipitating events. In spite of the lack of apparent general relevance of these types of items, the attribution by the patient of causal significance to marital conflict, difficult social relations, or work problems appeared to have desirable specific relevance for depression, hostile-suspiciousness, and withdrawal-retardation.

1.4 Sources of Data

1.4.1 INTRODUCTION

The data analyzed and reported in various chapters of this book came from accumulated data files derived from several clinical research projects. Since four major sets of data will be used repeatedly to illustrate various types of analysis, it is useful to describe them here so that the data sets can then simply be identified by name or sample size in later chapters. In this way, needless redundancy in description of the samples can be avoided.

Table 1.4 Background Variables Having Greatest General Relevance for Largest Number of BPRS Symptom Ratings[†]

Age
Race
Sex
Age at onset
Previous psychiatric hospitalizations
Marital status
Children
Divorce history
Work achievement
Alcohol behavior
Work attitude (involvement)
Behavior problems
Socialization problems
Physical problems
Marital problems
Marital conflict (precipitating)
Social relations (precipitating)
Work problems (precipitating)

† Race is included on the basis of previous repeated indication of relevance for psychopathology and treatment response.

1.4.2 COMBINED SAMPLE WITH BPRS AND HISTORY DATA ($n = 3{,}498$)

One large sample of data was accumulated from a variety of sources for the primary purpose of examining the relevance of certain selected background characteristics with regard to differences in manifest psychopathology. Preliminary studies revealed that the relevance of certain variables, particularly sex, depends upon the type of treatment setting being considered. For example, in an expensive private hospital setting, females appear *on the average* less seriously ill than males. The females often have rather minor neurotic or depressive disturbances, while the males tend to evidence major psychiatric disorders. The picture tends to reverse when one examines sex-related differences in psychopathology of a state hospital population. Results such as these made it apparent that a very distorted picture of the relevance of certain nonpsychiatric factors might be obtained if the sampling were limited to a particular selected population. Social, economic, and cultural factors determine in part what type of patient goes to which type of treatment setting.

In an effort to overcome the distortions imposed by limited sampling from special psychiatric populations, a composite sample of patients drawn from several major types of treatment settings was constructed. The data sources included (1) Veterans Administration collaborative studies of chemotherapy in depression and schizophrenia, (2) an admissions office survey conducted in the Veterans Administration hospital in Houston, (3) a survey of consecutive admissions entering the Texas state hospital system through a diagnostic clinic located in Houston, (4) the administrative survey of the Texas state hospital system, (5) inpatient treatment-evaluation projects at the University of Texas Medical School (Galveston), (6) outpatient treatment-evaluation projects at the same facility, and (7) a survey of the University of Miami chronic outpatient treatment population. BPRS symptom-rating profiles plus age, race, sex, education, work achievement, and marital status were recorded for each patient. To obtain a sample with reasonably appropriate representation for each type of treatment setting, only 450 patients having less than 18 months of hospitalization for current illness were included from the Texas state hospital survey. This left a sample of 3,498 patients divided in roughly equal proportions between state hospital, Veterans Administration hospital, and university hospital (private) inpatient services and university hospital outpatient clinics. While this composite sample may not be truly representative of the total psychiatric population of this country, its breadth ensures a degree of representation that has not been achieved in most other sets of data used for similar purposes.

1.4.3 COMPOSITE SAMPLE OF 6,400 BPRS RATING PROFILES

One very large sample of data was constructed for examining relationships among BPRS symptom ratings across a heterogeneous psychiatric population. Because it was recognized that selective sampling can significantly influence the apparent relationships among variables, a composite sample was constructed to include all pretreatment or survey BPRS symptom-rating profiles available from all sources. Veterans Administration hospitals, state hospitals, university hospitals, and outpatient clinics were sources from which patients were derived. This large sample of ratings is drawn upon to illustrate various methods of correlational analysis and to provide the basis for definitive results concerning syndromes of manifest psychopathology. It is used to study sampling invariance of factor-analysis results and other methodological problems.

1.4.4 UNIVERSITY HOSPITAL INPATIENT SURVEY ($n = 670$)

A body of data which includes both pre- and posttreatment BPRS symptom ratings plus treatment record, clinical diagnosis, and the complete standard psychiatric history was obtained for a random sample of 670 new admissions to the University of Texas Medical Branch hospitals. Four clinical psychologists completed patient interviews, did the BPRS symptom ratings, and collected standard psychiatric history forms on each patient shortly after admission. After 4 weeks of treatment, the same psychologist completed a follow-up interview and a second BPRS rating profile. At the time of discharge, the psychiatry resident most closely associated with the case provided global improvement ratings and a final diagnostic impression. A record of treatment procedures was abstracted from the clinical chart for each patient. These data are used to illustrate various methods of analysis and methods for classification of individuals. Variables influencing the selection of appropriate treatment procedures are examined in analyses of these data.

1.4.5 PSYCHIATRIC DIAGNOSTIC STEREOTYPE DATA

A set of data which is considered interesting from several points of view, and which will be used to illustrate methods for classification of individuals among diagnostic groups, was derived from expert judges in several different countries. The BPRS rating profiles in the case of these data are not ratings of real patients but are rating-scale descriptions of *typical* (hypothetical) patients belonging to various diagnostic classes. Experienced clinicians were instructed to think of a typical classical paranoid schizophrenic or psychotic depressive and rate him on the BPRS.

The first study was done in America, with both psychiatrists and psychologists serving as expert professionals. Thirteen diagnostic concepts representing functional psychotic disorders described in the APA Standard Nomenclature were the subject of this investigation. It was found that expert conceptions of the different disorders could be adequately described in terms of BPRS symptom rating profiles (Overall and Gorham, 1963). Subsequently, similar studies were conducted in France, Germany, Czechoslovakia, and Italy using careful translations of the BPRS to evaluate relative similarities and differences in diagnostic concepts used in each country. Of particular interest is the Czechoslovakian study because in it the World Health Organization International Psychiatric Classification was the subject of investigation. Since the WHO system is supposedly an international classification, the results of that study should provide useful insight into concepts of classification that might at times be required for cross-cultural research. The numbers of expert professionals contributing rating profile descriptions of diagnostic classification concepts in the various countries were as follows: America, 38; France, 123; Germany, 108; Czechoslovakia, 92; and Italy, 45.

1.5 References

Derogatis, L. R., R. R. Bonato, and K. C. Yang: The Power of IMPS in Psychiatric Drug Research, *Arch. Gen. Psychiat.*, **19**:689–699 (1968).

Gorham, D. R., and J. E. Overall: Drug-action Profiles Based on an Abbreviated Psychiatric Rating Scale, *J. Nerv. Ment. Dis.*, **131**:528–535 (1960).

———, and ———: Dimensions of Change in Psychiatric Symptomatology, *Dis. Nerv. Syst.*, **22**:576–580 (1961).

Hollister, L. E., J. E. Overall, J. L. Bennett, I. Kimbell, Jr., and J. Shelton: Triperidol in Newly Admitted Schizophrenics, *Amer. J. Psychiat.*, **122**:96–98 (1965).

Lorr, M.: Multidimensional Scale for Rating Psychiatric Patients, *Vet. Admin. Tech. Bull.* TB 10–507, Nov. 16, 1953.

———, C. J. Klett, D. M. McNair, and J. J. Lasky: "Inpatient Multidimensional Psychiatric Scale, Manual," Consulting Psychologists Press, Palo Alto, Calif., 1962.

Lyerly, S. B., and P. S. Abbott: "Handbook of Psychiatric Rating Scales (1950–1964)," Government Printing Office, Washington, 1966.

Overall, J. E.: Historical and Sociocultural Factors Related to the Phenomenology of Schizophrenia, in D. V. Siva Sankar (ed.), "Schizophrenia: Current Concepts and Research," PJD Publications, Ltd., Hicksville, N.Y., 1969.

———, and D. R. Gorham: The Brief Psychiatric Rating Scale, *Psychol. Repts.*, **10**:799–812 (1962).

———, and ———: A Pattern Probability Model for the Classification of Psychiatric Patients, *Behav. Sci.*, **8**:108–116 (1963).

———, ———, and J. R. Shawver: Basic Dimensions of Change in the Symptomatology of Chronic Schizophrenics, *J. Abnor. Soc. Psychol.*, **63**:597–602 (1961).

———, and L. E. Hollister: Computer Procedures for Psychiatric Classification, *J. Amer. Med. Assoc.*, **187**:583–588 (1964).

———, L. E. Hollister, and S. N. Dalal: Psychiatric Drug Research: Sample Size Requirements for One vs. Two Raters, *Arch. Gen. Psychiat.*, **16**:152–161 (1967).

———, ———, I. Kimbell, Jr., and J. Shelton: Extrinsic Factors Influencing Responses to Psychotherapeutic Drugs, *Arch. Gen. Psychiat.*, **21**:89–94 (1969).

———, ———, J. Shelton, M. Johnson, and I. Kimbell, Jr.: Tranylcypromine Compared with Dextroamphetamine in Hospitalized Depressed Patients, *Dis. Nerv. Syst.*, **27**:653–659 (1966).

———, and J. P. Tupin: Investigation of Clinical Outcome in a Doctor's Choice Treatment Setting, *Dis. Nerv. Syst.*, **30**:305–313 (1969).

Pokorny, A. D., and J. E. Overall: Dependence of Manifest Psychopathology on Age, Sex, Ethnicity, Marital Status and Education in a State Hospital Population, *J. Psychiat. Res.*, **7**:143–152 (1970).

Spiegel, D. K., and J. E. Overall: Investigation of Optimal Scaling Applied to Psychiatric Ratings, unpublished manuscript, 1970.

2

Essential Matrix Concepts and Calculations

2.1 Introduction

Comprehension of the multivariate statistical methods discussed in the remainder of this book requires an understanding of basic matrix algebra. The present chapter provides a brief survey of all matrix operations and methods required in subsequent chapters. Clearly, a comprehensive coverage of the mathematical theory of matrices is beyond the scope or requirements of this book. The student who has already been exposed to matrix methods should read through this chapter merely to refresh his memory and to become acquainted with the notation to be used. The student who has not previously been exposed to matrix algebra should be prepared to spend what may seem like a disproportionate amount of time on this one preliminary chapter before attempting to move on to consideration of multivariate methods described in subsequent chapters. Hand calculation of examples is almost essential to fix matrix concepts and operations in the mind.

If one is willing to proceed slowly and to work through the present chapter several times, all the essential tools can be acquired. An excellent and more extensive treatment by Horst (1963) can be used to broaden understanding of matrices and matrix operations. Relevant sections in Anderson (1958), Dwyer (1951), and Rao (1952) can be used to extend knowledge of the mathematical bases for multivariate methods to be discussed. Most importantly, the student who has not had previous exposure to matrix algebra should not be impatient, realizing that gaining command of the matrix concepts presented in this chapter is in itself a major accomplishment. Such a student might profitably spend almost as much time and energy in acquiring the tools surveyed in this chapter as in studying the multivariate methods discussed in the remainder of the book.

2.2 Definitions and Notation

A *matrix* consists of numbers placed in a rectangular array. The numbers are called the *elements* of the matrix. *Rows* are horizontal lines of elements in a matrix and are numbered starting from the top. *Columns* are vertical lines of elements in a matrix and are numbered starting from the left. Matrix A of *order* $m \times n$ has m rows and n columns. By convention, the row designation is always given first when specifying order. In symbolic notation, a matrix will be represented by a capital letter and when necessary the order of the matrix will appear below the matrix symbol or in parentheses beside the designation of the matrix. The elements of a matrix are identified by double subscript according to row and column locations. For example, in the matrix A of order $m \times n$, the element a_{ij} is the ith element in the jth column, i.e., the ith element down from the top and the jth element from the left side of the matrix.

$$\underset{m \times n}{A} = \begin{bmatrix} a_{11} & a_{12} & \cdots & a_{1j} & \cdots & a_{1n} \\ a_{21} & a_{22} & \cdots & a_{2j} & \cdots & a_{2n} \\ \cdots & \cdots & \cdots & \cdots & \cdots & \cdots \\ a_{i1} & a_{i2} & \cdots & a_{ij} & \cdots & a_{in} \\ a_{m1} & a_{m2} & \cdots & a_{mj} & \cdots & a_{mn} \end{bmatrix}$$

In the matrix below, a_{13} is the third element in the first row, or 5, and a_{33} is 7.

$$A = \begin{bmatrix} 3 & 4 & 5 & 4 \\ 1 & 2 & 1 & 3 \\ 2 & 2 & 7 & 4 \\ 1 & 3 & 3 & 2 \end{bmatrix}$$

The array of elements is usually enclosed by brackets to indicate that it is a matrix.

A *vector* is a special kind of matrix, consisting of a single row or a single column. Vectors will be represented by lowercase letters in boldface type. Thus a *column vector*, or vertical array of elements, will be represented as **a** or **b**. The horizontal array of elements, or *row vector*, will be distinguished from the column vector by a prime sign; thus **a**′ is a row vector

$$\mathbf{a} = \begin{bmatrix} 3 \\ 2 \\ 5 \\ 6 \\ 4 \end{bmatrix} \qquad \mathbf{a}' = [3 \quad 2 \quad 5 \quad 6 \quad 4]$$

In later sections of the book, particular columns of a matrix will sometimes be identified by a superscript. For example, the matrix A including n columns can be represented by

$$A = [\mathbf{a}^{(1)} \quad \mathbf{a}^{(2)} \quad \cdots \quad \mathbf{a}^{(n)}]$$

The elements of a matrix or vector are usually *scalar* numbers. A scalar is a single quantity that may be an integer, a decimal number or fraction, or a complex number. Scalar quantities are designated by lowercase italic letters.

Square matrices are of special importance in statistical computations. A square matrix is one which contains an equal number of rows and columns. A *symmetric* square matrix is one in which the element in the ith row and jth column is equal to the element in the jth row and ith column

$$A = \begin{bmatrix} 5 & 2 & 1 & 4 \\ 2 & 8 & 3 & 2 \\ 1 & 3 & 6 & 1 \\ 4 & 2 & 1 & 7 \end{bmatrix}$$

Elements of a square matrix having equal row and column numbers, that is, $a_{11}, a_{22}, \ldots, a_{nn}$, make up the *principal diagonal* and are called *diagonal elements*. In the matrix above, the diagonal elements are 5, 8, 6, 7. Other elements are sometimes referred to as off-diagonal elements. The sum of the diagonal elements is called the *trace*.

A *diagonal matrix* is one in which all elements are zero except those in the principal diagonal

$$D = \begin{bmatrix} 5 & 0 & 0 & 0 \\ 0 & 2 & 0 & 0 \\ 0 & 0 & 3 & 0 \\ 0 & 0 & 0 & 1 \end{bmatrix}$$

The *identity matrix* is a special diagonal matrix in which all diagonal elements are equal to unity. The identity matrix corresponds to the number 1 in scalar algebra.

$$I = \begin{bmatrix} 1 & 0 & 0 & 0 \\ 0 & 1 & 0 & 0 \\ 0 & 0 & 1 & 0 \\ 0 & 0 & 0 & 1 \end{bmatrix}$$

A *triangular matrix* has all elements above or below the principal diagonal equal to zero. An *upper triangular matrix* is one in which all elements below the principal diagonal are equal to zero

$$T_u = \begin{bmatrix} 5 & 4 & 2 & 1 \\ 0 & 7 & 6 & 3 \\ 0 & 0 & 8 & 2 \\ 0 & 0 & 0 & 4 \end{bmatrix}$$

A *lower triangular matrix* is one in which all elements above the principal diagonal are zero

$$T_l = \begin{bmatrix} 5 & 0 & 0 & 0 \\ 4 & 8 & 0 & 0 \\ 3 & 5 & 9 & 0 \\ 2 & 4 & 6 & 2 \end{bmatrix}$$

The *transpose* of a matrix is formed by interchanging rows for columns and is designated by a prime. Thus, A' is the transpose of A. For example, the first column of A would be written as the first row of A'. The second column of A would be the second row of A' and so on. In general, the transpose of a matrix A of order $m \times n$ is matrix A' of order $n \times m$, with the element of the ith row and jth column of A appearing in the jth row and ith column of A'.

$$\underset{m \times n}{A} = \begin{bmatrix} 5 & 2 & 4 \\ 3 & 4 & 2 \\ 1 & 1 & 6 \\ 8 & 3 & 2 \end{bmatrix} \qquad \underset{n \times m}{A'} = \begin{bmatrix} 5 & 3 & 1 & 8 \\ 2 & 4 & 1 & 3 \\ 4 & 2 & 6 & 2 \end{bmatrix}$$

2.3 Addition and Subtraction of Matrices

To add or subtract two matrices, they must be of the same order. Addition of matrices consists of summing the corresponding elements.

$$\begin{bmatrix} a_{11} & a_{12} \\ a_{21} & a_{22} \\ a_{31} & a_{32} \end{bmatrix} + \begin{bmatrix} b_{11} & b_{12} \\ b_{21} & b_{22} \\ b_{31} & b_{32} \end{bmatrix} = \begin{bmatrix} a_{11} + b_{11} & a_{12} + b_{12} \\ a_{21} + b_{21} & a_{22} + b_{22} \\ a_{31} + b_{31} & a_{32} + b_{32} \end{bmatrix}$$

$$\begin{bmatrix} 3 & 7 \\ 4 & 2 \\ 1 & 5 \end{bmatrix} + \begin{bmatrix} 4 & 1 \\ 3 & 4 \\ 2 & 1 \end{bmatrix} = \begin{bmatrix} 7 & 8 \\ 7 & 6 \\ 3 & 6 \end{bmatrix}$$

Addition of vectors follows the same rules

$$\begin{bmatrix} 4 \\ 2 \\ 8 \\ 1 \end{bmatrix} + \begin{bmatrix} 3 \\ 5 \\ 7 \\ 5 \end{bmatrix} = \begin{bmatrix} 7 \\ 7 \\ 15 \\ 6 \end{bmatrix}$$

Subtraction of matrices and vectors is performed similarly; i.e., each element in the second matrix (or vector) is subtracted from the corresponding element in the first

$$\begin{bmatrix} 3 & 7 \\ 4 & 2 \\ 1 & 5 \end{bmatrix} - \begin{bmatrix} 4 & 1 \\ 3 & 4 \\ 2 & 1 \end{bmatrix} = \begin{bmatrix} -1 & 6 \\ 1 & -2 \\ -1 & 4 \end{bmatrix}$$

As in scalar algebra, two simple laws of addition apply. The *associative law of addition* states that the sequence in which adjacent matrices are added or subtracted does not matter

$$A + B - C = (A + B) - C = A + (B - C) \tag{2.1}$$

The *commutative law of addition* states that the arrangement of the matrices in the equation can be altered without affecting the result

$$A + B - C = B + A - C = B - C + A \tag{2.2}$$

Two matrices are said to be equal if they have the same number of rows and columns and if the corresponding elements of each are identical. If two matrices are equal and one is subtracted from the other, the result is a matrix containing all zero elements. This is called the *null matrix* and corresponds to the zero in scalar algebra.

2.4 Matrix Multiplication

2.4.1 THE ROW-BY-COLUMN RULE

In defining equality, addition, and subtraction, it was stated as a necessary condition that the matrices be of the same order. For multiplication they must have a *common order*. In the matrix product

$$\underset{m \times n}{A} \; \underset{n \times p}{B} = \underset{m \times p}{C}$$

the order of the *prefactor* A is $m \times n$, the order of the *postfactor* B is $n \times p$. and the common order, defined as the two adjacent interior dimensions, is n, This is equivalent to saying that the necessary condition for multiplication is that the prefactor have the same number of columns as there are rows in the postfactor. The matrices are then said to be *conformable* for multiplication. Note that the order of the product matrix C is $m \times p$, where m is the number of rows of the prefactor and p is the number of columns of the postfactor.

Multiplication follows a row-by-column rule. Corresponding elements of the first row of the prefactor and the first column of the postfactor are multiplied and summed to form the element of the first row and first column of the product matrix. Repeating this operation with the first row of the prefactor and the second column of the postfactor yields the element of the first row and second column of the product matrix. In general, the ijth element of the product matrix is obtained by multiplying elements in the jth column of the postfactor by elements in the ith row of the prefactor and summing the products

$$c_{ij} = a_{i1}b_{1j} + a_{i2}b_{2j} + \cdots + a_{in}b_{nj}$$

Thus

$$\begin{bmatrix} a_{11} & a_{12} \\ a_{21} & a_{22} \\ a_{31} & a_{32} \end{bmatrix} \begin{bmatrix} b_{11} & b_{12} \\ b_{21} & b_{22} \end{bmatrix} = \begin{bmatrix} a_{11}b_{11} + a_{12}b_{21} & a_{11}b_{12} + a_{12}b_{22} \\ a_{21}b_{11} + a_{22}b_{21} & a_{21}b_{12} + a_{22}b_{22} \\ a_{31}b_{11} + a_{32}b_{21} & a_{31}b_{12} + a_{32}b_{22} \end{bmatrix}$$

or

$$\begin{bmatrix} 3 & 0 \\ 4 & 2 \\ 1 & 1 \end{bmatrix} \begin{bmatrix} 4 & 1 \\ 3 & 4 \end{bmatrix} = \begin{bmatrix} 12 & 3 \\ 22 & 12 \\ 7 & 5 \end{bmatrix}$$

Just as any number of matrices can be added or subtracted, a product matrix can be formed by multiplying any number of matrices that are *conformable*

$$\underset{m \times m}{C} = \underset{m \times n}{A} \, \underset{n \times p}{B} \, \underset{p \times q}{X} \, \underset{q \times m}{Y}$$

As in Eq. (2.1) the *associative law of multiplication* holds

$$C = (AB)X = A(BX) \tag{2.3}$$

However, the *commutative law of multiplication* corresponding to Eq. (2.2) does not in general hold. This means that multiplication operations must usually be performed in the order specified

$$C = ABX \neq BXA \neq XAB \tag{2.4}$$

In many cases this is obvious because there would be a violation of the requirement that matrices be conformable, i.e., that they have a common interior order. The product $\underset{m \times n}{A} \, \underset{n \times p}{B}$ is defined because the two matrices have a common interior order n, but the product $\underset{n \times p}{B} \, \underset{m \times n}{A}$ is not defined. However, even in the multiplication of square matrices of the same order where conformability would be satisfied, AB does not in general equal BA.

$$AB = \begin{bmatrix} 3 & 1 \\ 0 & 2 \end{bmatrix} \begin{bmatrix} 1 & 3 \\ 2 & 1 \end{bmatrix} = \begin{bmatrix} 5 & 10 \\ 4 & 2 \end{bmatrix}$$

$$BA = \begin{bmatrix} 1 & 3 \\ 2 & 1 \end{bmatrix} \begin{bmatrix} 3 & 1 \\ 0 & 2 \end{bmatrix} = \begin{bmatrix} 3 & 7 \\ 6 & 4 \end{bmatrix}$$

The transpose of a product of two matrices is equal to the product of the transpose of the prefactor and the transpose of the postfactor *in reverse order*

$$(AB)' = B'A' \tag{2.5}$$

$$AB = \begin{bmatrix} 3 & 4 & 2 \\ 5 & 1 & 3 \\ 1 & 3 & 1 \\ 2 & 1 & 1 \end{bmatrix} \begin{bmatrix} 1 & 2 \\ 2 & 3 \\ 1 & 1 \end{bmatrix} = \begin{bmatrix} 13 & 20 \\ 10 & 16 \\ 8 & 12 \\ 5 & 8 \end{bmatrix}$$

$$(AB)' = \begin{bmatrix} 13 & 10 & 8 & 5 \\ 20 & 16 & 12 & 8 \end{bmatrix}$$

$$B'A' = \begin{bmatrix} 1 & 2 & 1 \\ 2 & 3 & 1 \end{bmatrix} \begin{bmatrix} 3 & 5 & 1 & 2 \\ 4 & 1 & 3 & 1 \\ 2 & 3 & 1 & 1 \end{bmatrix} = \begin{bmatrix} 13 & 10 & 8 & 5 \\ 20 & 16 & 12 & 8 \end{bmatrix}$$

This relationship can be generalized to any number of multipliers. The transpose of the product of any number of matrices is equal to the product of their transposes *in reverse order.*

A final relationship is known as the *distributive law of multiplication*

$$AB + AC = A(B + C) \tag{2.6}$$

The two products AB and AC can be formed and then summed, or the sum $B + C$ can be *premultiplied* by A. The distributive law is also illustrated by $BA + CA = (B + C)A$, where the sum is *postmultiplied* by A.

Multiplication of vectors follows the same rules. When a column vector is premultiplied by a row vector, the result is called a *minor product*, an *inner product*, or a *dot product* and is a scalar. The inner product of two vectors is equal to the sum of cross products of corresponding elements. Thus,

$$\mathbf{a}'\mathbf{b} = c \tag{2.7}$$

$$\begin{bmatrix} 3 & 2 & 1 \end{bmatrix} \begin{bmatrix} 1 \\ 2 \\ 1 \end{bmatrix} = 8$$

When a column vector is postmultiplied by a row vector, the result, called a *major product*, an *outer product*, or *matrix product*, is a matrix. The outer product of two vectors is a matrix in which each element is the simple product of one element from each vector

$$\mathbf{ab}' = C \tag{2.8}$$

$$\begin{bmatrix} 3 \\ 2 \\ 1 \end{bmatrix} \begin{bmatrix} 1 & 2 & 2 & 1 \end{bmatrix} = \begin{bmatrix} 3 & 6 & 6 & 3 \\ 2 & 4 & 4 & 2 \\ 1 & 2 & 2 & 1 \end{bmatrix}$$

The transpose of a major or minor product is the product of the transposes of the two vectors in reverse order, as stated in Eq. (2.5), but because the minor product is a scalar quantity, the following relationship also holds:

$$(\mathbf{a}'\mathbf{b})' = \mathbf{b}'\mathbf{a} = \mathbf{a}'\mathbf{b} = c \tag{2.9}$$

$$\begin{bmatrix} 3 & 1 & 2 \end{bmatrix} \begin{bmatrix} 1 \\ 2 \\ 1 \end{bmatrix} = \begin{bmatrix} 1 & 2 & 1 \end{bmatrix} \begin{bmatrix} 3 \\ 1 \\ 2 \end{bmatrix} = 7$$

2.4.2 SPECIAL MATRIX-VECTOR PRODUCTS

It follows from the above rules that the result of multiplying any matrix by its transpose is a square symmetric matrix. A symmetric matrix formed in this manner is referred to as a *product-moment matrix*. Such matrices will be of special interest in later sections of the book. If the matrix A $(m \times n)$, where $m > n$, is premultiplied by its transpose, the order of the resulting product-moment matrix is $n \times n$. If A $(m \times n)$ is postmultiplied by its transpose, the order of the product-moment matrix is $m \times m$. Thus,

$$A'A = \begin{bmatrix} 3 & 4 & 1 \\ 0 & 2 & 1 \end{bmatrix} \begin{bmatrix} 3 & 0 \\ 4 & 2 \\ 1 & 1 \end{bmatrix} = \begin{bmatrix} 26 & 9 \\ 9 & 5 \end{bmatrix}$$

and

$$AA' = \begin{bmatrix} 3 & 0 \\ 4 & 2 \\ 1 & 1 \end{bmatrix} \begin{bmatrix} 3 & 4 & 1 \\ 0 & 2 & 1 \end{bmatrix} = \begin{bmatrix} 9 & 12 & 3 \\ 12 & 20 & 6 \\ 3 & 6 & 2 \end{bmatrix}$$

When the order of the product-moment matrix is determined by the smaller dimension of the original matrix, as it is in $A'A$, the matrix is called a *minor* product-moment matrix or sometimes a *Gramian matrix*. If the order is determined by the larger dimension, as in AA', the matrix is called a *major* product-moment matrix. Note that the traces of both product-moment matrices are equal.

Premultiplication of any matrix by a diagonal matrix is equivalent to multiplying each element of the ith row of the matrix by the element in the corresponding row of the diagonal matrix. Postmultiplication by a diagonal matrix is equivalent to multiplying the elements of each column of the matrix by the element in the corresponding column of the diagonal matrix

$$DA = \begin{bmatrix} 1 & 0 & 0 & 0 \\ 0 & 2 & 0 & 0 \\ 0 & 0 & 3 & 0 \\ 0 & 0 & 0 & 1 \end{bmatrix} \begin{bmatrix} 3 & 1 & 2 & 2 \\ 2 & 2 & 1 & 3 \\ 1 & 1 & 1 & 2 \\ 1 & 3 & 1 & 3 \end{bmatrix} = \begin{bmatrix} 3 & 1 & 2 & 2 \\ 4 & 4 & 2 & 6 \\ 3 & 3 & 3 & 6 \\ 1 & 3 & 1 & 3 \end{bmatrix}$$

$$AD = \begin{bmatrix} 3 & 1 & 2 & 2 \\ 2 & 2 & 1 & 3 \\ 1 & 1 & 1 & 2 \\ 1 & 3 & 1 & 3 \end{bmatrix} \begin{bmatrix} 1 & 0 & 0 & 0 \\ 0 & 2 & 0 & 0 \\ 0 & 0 & 3 & 0 \\ 0 & 0 & 0 & 1 \end{bmatrix} = \begin{bmatrix} 3 & 2 & 6 & 2 \\ 2 & 4 & 3 & 3 \\ 1 & 2 & 3 & 2 \\ 1 & 6 & 3 & 3 \end{bmatrix}$$

As a special case of this relationship

$$IA = AI = A \tag{2.10}$$

The identity matrix is like the number 1 in scalar algebra.

A matrix can be premultiplied by a row vector or postmultiplied by a column vector. In the first case, the product is a row vector, and in the second it is a column vector. A matrix cannot be premultiplied by a column vector or postmultiplied by a row vector because of the requirements of conformability. If a matrix is pre- and postmultiplied by the same or different vectors, the result is a scalar quantity. Thus

$$\mathbf{a}'C\mathbf{b} = c \tag{2.11}$$

$$\mathbf{a}'C\mathbf{b} = [1 \quad 2 \quad 1] \begin{bmatrix} 1 & 2 & 1 \\ 2 & 3 & 3 \\ 1 & 1 & 1 \end{bmatrix} \begin{bmatrix} 2 \\ 1 \\ 2 \end{bmatrix} = [6 \quad 9 \quad 8] \begin{bmatrix} 2 \\ 1 \\ 2 \end{bmatrix} = 37$$

or

$$[1 \quad 2 \quad 1]\begin{bmatrix} 6 \\ 13 \\ 5 \end{bmatrix} = 37$$

When a matrix is pre- and postmultiplied by the same vector, the function is referred to as the *quadratic form*

$$\mathbf{a}'C\mathbf{a} = c \tag{2.12}$$

A matrix or vector can be multiplied by a scalar quantity by multiplying each element in the matrix or vector by that quantity. Scalar quantities, and any matrix-vector products that result in a scalar quantity, are an exception to the rule given in Eq. (2.4). Thus

$$AkBC = kABC = ABkC = ABCk \tag{2.13}$$

and

$$A\mathbf{a}(\mathbf{b}'B\mathbf{b})\mathbf{c}'C = (\mathbf{b}'B\mathbf{b})A\mathbf{a}\mathbf{c}'C = A\mathbf{a}\mathbf{c}'C(\mathbf{b}'B\mathbf{b})$$

where $\mathbf{b}'B\mathbf{b} = k$.

Matrix-vector products of special interest are those formed using the *unit vector* **1**, all the elements of which are unity. Premultiplication of a matrix by a unit vector $\mathbf{1}'A$ yields a row vector with elements which are the column sums of A. Postmultiplication of a matrix by a unit vector $A\mathbf{1}$ yields a column vector of row sums. Pre- and postmultiplication of a matrix by unit vectors $\mathbf{1}'A\mathbf{1}$ yields a scalar quantity which is the sum of all of the elements in the matrix.

If each of the elements of **a** is divided by the square root of the scalar quantity $\mathbf{a}'\mathbf{a}$, the vector is said to be *normalized* and the minor product of the resulting normalized vector **b** is equal to unity

$$\mathbf{a}'\mathbf{a} = [4 \quad 3 \quad 2]\begin{bmatrix} 4 \\ 3 \\ 2 \end{bmatrix} = 29 \qquad \sqrt{\mathbf{a}'\mathbf{a}} = \sqrt{29}$$

Let $\mathbf{b} = \mathbf{a}/\sqrt{\mathbf{a}'\mathbf{a}}$; then

$$\mathbf{b}'\mathbf{b} = \begin{bmatrix} \frac{4}{\sqrt{29}} & \frac{3}{\sqrt{29}} & \frac{2}{\sqrt{29}} \end{bmatrix}\begin{bmatrix} \frac{4}{\sqrt{29}} \\ \frac{3}{\sqrt{29}} \\ \frac{2}{\sqrt{29}} \end{bmatrix} = \frac{16}{29} + \frac{9}{29} + \frac{4}{29} = \frac{29}{29} = 1$$

The definition of a normalized vector is given by

$$\mathbf{b}'\mathbf{b} = 1 \tag{2.14}$$

Two vectors **a** and **b** are said to be *orthogonal* if

$$\mathbf{a}'\mathbf{b} = \mathbf{b}'\mathbf{a} = 0 \tag{2.15}$$

Matrices consisting of normalized, or orthogonal, vectors will be referred to occasionally throughout the book. If the columns of a matrix A are all mutually orthogonal, $\mathbf{a}_i'\mathbf{a}_j = 0$, then the minor product moment $A'A$ will be a diagonal matrix and A is referred to as an *orthogonal matrix*. If in addition, for all columns $\mathbf{a}_i'\mathbf{a}_i = 1$, the matrix is referred to as an *orthonormal matrix*. In the case of square orthonormal matrices

$$AA' = A'A = I \tag{2.16}$$

2.5 The Rank of a Matrix

2.5.1 DEFINITIONS

Any matrix C can be expressed as the product of two other matrices

$$\underset{n \times m}{C} = \underset{n \times p}{A} \; \underset{p \times m}{B} \tag{2.17}$$

in which p is less than or equal to the smaller of the two dimensions n and m. The rank of C is the *minimum* value of p required to reproduce the product matrix C. Suppose that $p = 1$ satisfies Eq. (2.17). Then the matrix C $(n \times m)$ can be reproduced as the outer product of two vectors. This means that the matrix C is of rank 1. Now suppose that there is no **a** $(n \times 1)$ and $\mathbf{b}'$ $(1 \times m)$ that will yield the product matrix C. Then various matrices A $(n \times 2)$ and B $(2 \times m)$ can be tried. If the matrix C can be reproduced by any product

$$\underset{n \times 2}{A} \; \underset{2 \times m}{B} = \underset{n \times m}{C}$$

it will be recognized to be of rank 2. In general, the rank of matrix C is the minimum value of p that will satisfy Eq. (2.17). It can be shown that p will always be less than or equal to the smaller of the two dimensions n and m.

A square matrix is said to be *singular* if its rank is less than its order. If any row or column can be accounted for as a weighted linear combination of the other rows or columns, the matrix is singular. If the rank of a square matrix is equal to its order, it is *nonsingular*. In a nonsingular matrix, no row or column can be expressed as a simple linear (weighted) function of the other rows or columns.

2.5.2 DETERMINANTS

Singular and nonsingular matrices can also be defined using determinants, and since determinants are used occasionally throughout the book, their calculation will be reviewed here briefly. A determinant is a specified function of the elements of a square matrix and is represented symbolically by enclosing the array in single vertical lines, Det $A = |A|$. The determinant of order p is defined as the sum of $p!$ terms representing all possible products obtained by taking one element from one row and column, a second element from a different row and column, and so on until each term has p elements. The appropriate algebraic sign of each product term must also be determined. When the square matrix is of the order 2 or 3, calculation of the determinant is not too difficult. When $p = 2$,

$$\begin{vmatrix} a_{11} & a_{12} \\ a_{21} & a_{22} \end{vmatrix} = a_{11}a_{22} - a_{21}a_{12}$$

when $p = 3$,

$$\begin{vmatrix} a_{11} & a_{12} & a_{13} \\ a_{21} & a_{22} & a_{23} \\ a_{31} & a_{32} & a_{33} \end{vmatrix} = a_{11}a_{22}a_{33} + a_{21}a_{32}a_{13} + a_{31}a_{12}a_{23} - a_{31}a_{22}a_{13} - a_{21}a_{12}a_{33} - a_{11}a_{32}a_{23}$$

One rule for determining the correct algebraic sign for each term in this expansion is to arrange the p elements in each term in order of the second subscript and calculate the number of interchanges required to order them by the first subscript. If the number of required interchanges is even, the algebraic sign of that term is positive. For example, the term $a_{21}a_{32}a_{13}$, ordered by the second subscript, would require two interchanges to achieve order by the first subscript, and its sign therefore is positive; the term $a_{21}a_{12}a_{33}$ would require only one interchange, and its sign therefore is negative.

It is clear that calculating determinants in this manner is tedious if p is much larger than 3, but more efficient computational methods are available for matrices of any order. In general, these methods involve transformation of the original matrix to upper triangular form. The square-root method for triangular reduction of a symmetric matrix will be described in Sec. 2.6.2. Although this transformation does not change the value of the determinant, evaluation becomes much simpler since all product terms in the above type of expansion except the one containing only diagonal elements will include at least one zero element and will therefore vanish. The value of the determinant is then equal to the product of the diagonal elements. It is apparent that if any one of the diagonal elements in the upper triangular matrix is zero, the determinant of the matrix will also be zero.

If the determinant of a matrix is zero, the matrix is singular. If the determinant has any value other than zero, the matrix is nonsingular.

2.5.3 IMPORTANCE OF MATRIX RANK

In practice it is relatively uncommon to encounter square matrices of empirical data that are not of full rank. The minor product moment of rectangular data matrices is nearly always nonsingular although it is possible to achieve rank reduction inadvertently by defining a variable as a linear combination of other variables, by including a variable that has no variance, or by putting a constraint on the data matrix such as equalizing the row sums. Thus, the square matrices derived from data matrices for statistical analyses, e.g., the correlation matrix, are nearly always nonsingular although again there are occasional exceptions. For example, a correlation matrix of p variables based upon fewer than p observation vectors or subjects is singular, and, in general, all *major* product moments of vectors or matrices are singular.

Not only are singular matrices relatively infrequent in practice, but it is generally not necessary to calculate the rank of a matrix. However, for multivariate statistical analyses, the question of matrix rank is especially important. Unique solutions for sets of linear equations cannot be defined when the matrix of the equation is singular. Symmetric matrices, such as the covariance or correlation matrices that play such an important role in multivariate analysis, cannot be handled by the usual methods unless they are nonsingular. The existence of a nonzero determinant is essential for various multivariate tests of significance. When dealing with covariance or correlation matrices, the determination that any row or column can be expressed as a linear function of other rows or columns is equivalent to the determination that one of the original variables can be expressed as a linear function of the others. Conversely, the inclusion of a composite variable, which is a simple weighted combination of other variables also included in an analysis, will result in a singular covariance or correlation matrix.

Another reason for the importance of rank is that many multivariate methods have as one of their purposes the reduction in rank of the original set of p measurements or, more precisely stated, the approximation of the original matrix by matrices of lower rank. For example it will be shown that a matrix A $(p \times p)$ can be reproduced with reasonable accuracy by the major product moment of another matrix F $(p \times r)$, where $p > r$. Thus $\hat{A} = FF'$. The rank of the product of two or more matrices cannot be greater than the rank of the matrix of lowest rank. Therefore the rank of $\hat{A}$, the approximation of A, is equal to or less than r.

2.6 The Inverse of a Matrix

2.6.1 DEFINITION

In general, one matrix cannot be divided by another. Instead, this operation is performed in a manner analogous to multiplication by the reciprocal in scalar algebra, that is, $a/b = (1/b)a$. The reciprocal quantity $1/b$ is equal to 1 when multiplied by b. By analogy, the inverse of a matrix A is the matrix A^{-1} that satisfies

$$A^{-1}A = AA^{-1} = I \qquad (2.18)$$

All square nonsingular matrices have an inverse of this form. The inverse of a matrix is a matrix which when multiplied by the original matrix yields an identity matrix as a product. Thus, multiplication of a matrix by an inverse matrix is analogous to dividing the first matrix by a second matrix, although the operation of division does not exist per se in matrix algebra.

The operation corresponding to division of matrices is illustrated by solving the following equation for B:

$$ABC = D$$

"Divide" by premultiplying both sides by A^{-1}

$$A^{-1}ABC = A^{-1}D$$

$$IBC = A^{-1}D$$

$$BC = A^{-1}D$$

and postmultiplying by C^{-1}

$$BCC^{-1} = A^{-1}DC^{-1}$$

$$BI = A^{-1}DC^{-1}$$

$$B = A^{-1}DC^{-1}$$

Similarly, solving for C

$$ABC = D$$

$$BC = A^{-1}D$$

$$C = B^{-1}A^{-1}D$$

or, solving for A

$$ABC = D$$

$$AB = DC^{-1}$$

$$A = DC^{-1}B^{-1}$$

The inverse of a product is equal to the product of the inverses in reverse order

$$(AB)^{-1} = B^{-1}A^{-1} \tag{2.19}$$

For example,

$$(AB)^{-1}AB = (B^{-1}A^{-1})AB = B^{-1}(A^{-1}A)B = B^{-1}IB = B^{-1}B = I$$

and the inverse of an inverse equals the original matrix

$$AB = (B^{-1}A^{-1})^{-1} \tag{2.20}$$

There are a number of ways to solve for the inverse of a matrix, and all require considerable calculation, particularly if the order of the matrix is large. In practice matrix inverses are rarely computed by hand but are easily obtained by electronic computers. The square-root method of obtaining the inverse of a symmetric matrix is described in the next section to illustrate the procedure. This method, rather than a more general solution for the inverse of square nonsymmetric matrices, was chosen because nearly all statistical procedures involve symmetric matrices and because the square-root method also yields an upper triangular transformation of the original matrix which is useful in solving for the value of determinants.

It may be helpful first to identify some special cases of inverses that do not require extensive solution. The inverse of a scalar has already been defined as its reciprocal. If the elements of a diagonal matrix are expressed as reciprocals, this new matrix is the inverse of D

$$D^{-1}D = DD^{-1} = \begin{bmatrix} 2 & 0 & 0 & 0 \\ 0 & 3 & 0 & 0 \\ 0 & 0 & 4 & 0 \\ 0 & 0 & 0 & 5 \end{bmatrix} \begin{bmatrix} \frac{1}{2} & 0 & 0 & 0 \\ 0 & \frac{1}{3} & 0 & 0 \\ 0 & 0 & \frac{1}{4} & 0 \\ 0 & 0 & 0 & \frac{1}{5} \end{bmatrix} = \begin{bmatrix} 1 & 0 & 0 & 0 \\ 0 & 1 & 0 & 0 \\ 0 & 0 & 1 & 0 \\ 0 & 0 & 0 & 1 \end{bmatrix}$$

Square orthonormal matrices were defined in Eq. (2.16) as matrices that satisfy $AA' = A'A = I$. Therefore pre- or postmultiplying a square orthonormal matrix by its transpose is equivalent to multiplying by the inverse.

2.6.2 THE SQUARE-ROOT METHOD OF CALCULATING THE INVERSE

In the partitioned matrix in Table 2.1, the a_{ij}'s are the elements of a symmetric matrix A. The corresponding elements below the diagonal of A are not required for the solution and are therefore not entered. To the right of this matrix is an identity matrix of the same order as A. The matrix below A is the upper triangular factor of A, and the matrix to its right is the inverse of T_u in transposed form. These bottom two matrices are unknown but can be obtained using the rules given in Table 2.1. In general, each diagonal element in the upper triangular matrix S which appears directly beneath the original matrix A in Table 2.1 can be obtained using the following general rule:

$$s_{ii} = \sqrt{a_{ii} - s_{1i}s_{1i} - s_{2i}s_{2i} - \cdots - s_{(i-1)i}s_{(i-1)i}}$$

Each off-diagonal element appearing to the right of the diagonal elements in the rows of S can be calculated as follows:

$$s_{ij} = (a_{ij} - s_{1i}s_{1j} - s_{2i}s_{2j} - \cdots - s_{(i-1)i}s_{(i-1)j})\frac{1}{s_{ii}}$$

where s_{ii} is the diagonal entry in the same row that was previously calculated as shown above.

A numerical example appears in Table 2.2. The lower left-hand matrix in this table is the upper triangular factor of A, and it can be verified that $A = T_u'T_u = T_uT_u'$. The lower right-hand matrix in Table 2.2 is the inverse of T_u'. A check on the accuracy of the calculation of this matrix can be performed

Table 2.1 The Square-root Method of Matrix Inversion

1	a_{11}	a_{12}	a_{13}	a_{14}	1	0	0	0
2		a_{22}	a_{23}	a_{24}	0	1	0	0
3			a_{33}	a_{34}	0	0	1	0
4				a_{44}	0	0	0	1
1*a*	s_{11}	s_{12}	s_{13}	s_{14}	s_{15}	0	0	0
2*a*		s_{22}	s_{23}	s_{24}	s_{25}	s_{26}	0	0
3*a*			s_{33}	s_{34}	s_{35}	s_{36}	s_{37}	0
4*a*				s_{44}	s_{45}	s_{46}	s_{46}	s_{48}

Each element in row 1*a* is equal to the corresponding element of row 1 multiplied by the reciprocal of the square root of the lead element a_{11}.

The elements in row 2*a* are obtained as follows:

$$s_{22} = \sqrt{a_{22} - s_{12}s_{12}}$$

$$s_{23} = (a_{23} - s_{12}s_{13})\frac{1}{s_{22}}$$

$$s_{24} = (a_{24} - s_{12}s_{14})\frac{1}{s_{22}}$$

$$s_{25} = (.0 - s_{12}s_{15})\frac{1}{s_{22}}$$

$$s_{26} = [1.0 - s_{12}(.0)]\frac{1}{s_{22}}$$

The elements in row 3*a* are obtained as follows:

$$s_{33} = \sqrt{a_{33} - s_{13}s_{13} - s_{23}s_{23}}$$

$$s_{34} = (a_{34} - s_{13}s_{14} - s_{23}s_{24})\frac{1}{s_{33}}$$

$$s_{35} = (.0 - s_{13}s_{15} - s_{23}s_{25})\frac{1}{s_{33}}$$

$$s_{36} = [.0 - s_{13}(.0) - s_{23}s_{26}]\frac{1}{s_{33}}$$

$$s_{37} = [1.0 - s_{13}(.0) - s_{23}(.0)]\frac{1}{s_{33}}$$

The elements in row 4*a* are obtained as follows:

$$s_{44} = \sqrt{a_{44} - s_{14}s_{14} - s_{24}s_{24} - s_{34}s_{34}}$$

$$s_{45} = .0 - s_{14}s_{15} - s_{24}s_{25} - s_{34}s_{35}$$

$$s_{46} = .0 - s_{14}(.0) - s_{24}s_{26} - s_{34}s_{36}$$

$$s_{47} = .0 - s_{14}(.0) - s_{24}(.0) - s_{34}s_{37}$$

$$s_{48} = 1.0 - s_{14}(.0) - s_{24}(.0) - s_{34}(.0)$$

Table 2.2 An Example of the Square-root Method

1	4	2	3	2	1.0000	.0000	.0000	.0000
2		5	3	1	.0000	1.0000	.0000	.0000
3			6	2	.0000	.0000	1.0000	.0000
4				3	.0000	.0000	.0000	1.0000
1*a*	2.0000	1.0000	1.5000	1.0000	.5000	.0000	.0000	.0000
2*a*		2.0000	.7500	.0000	−.2500	.5000	.0000	.0000
3*a*			1.7854	.2800	−.3151	−.2100	.5601	.0000
4*a*				1.3862	−.2971	.0424	−.1131	.7214

A	I
T_u	$(T_u^{-1})'$

$s_{11} = \sqrt{4} = 2$

$s_{22} = \sqrt{5 - 1(1)} = 2.000$

$s_{23} = [3 - 1(1.5)].5 = .7500$

$s_{24} = [1 - 1(1)].5 = .0000$

$s_{25} = [0 - 1(.5)].5 = -.2500$

$s_{26} = [1 - 1(0)].5 = .5000$

$s_{33} = \sqrt{6 - 1.5(1.5) - .75(.75)} = 1.7854$

$s_{34} = [2 - 1.5(1) - .75(0)].5601 = .2800$

$s_{35} = [0 - 1.5(.5) - .75(-.25)].5601 = -.3151$

$s_{36} = [0 - 1.5(0) - .75(.5)].5601 = -.2100$

$s_{37} = [1 - 1.5(0) - .75(0)].5601 = .5601$

$s_{44} = \sqrt{3 - 1(1) - 0(0) - .28(.28)} = 1.3862$

$s_{45} = [0 - 1(.5) - 0(-.25) - .28(-.3151)].7214 = -.2971$

$s_{46} = [0 - 1(0) - 0(.5) - .28(-.21)].7214 = .0424$

$s_{47} = [0 - 1(0) - 0(0) - .28(.5601)].7214 = -.1131$

$s_{48} = [1 - 1(0) - 0(0) - .28(0)].7214 = .7214$

Table 2.3 The Inverse of Matrix A

$$A^{-1} = T_u^{-1}(T_u^{-1})'$$

$$\begin{bmatrix} .4999 & -.0714 & -.1428 & -.2142 \\ -.0714 & .2959 & -.1224 & .0306 \\ -.1428 & -.1224 & .3264 & -.0816 \\ -.2142 & .0306 & -.0816 & .5203 \end{bmatrix}$$

from any of the following relationships: $I = T_u T_u^{-1} = T_u^{-1} T_u = T'_u (T'_u)^{-1} = (T'_u)^{-1} T'_u$. The inverse T_u^{-1} can be thought of as the square root of the inverse of A since $A^{-1} = T_u^{-1}(T'_u)^{-1}$. To obtain the inverse of A, premultiply the lower right-hand matrix by its transpose. A check on the computational accuracy of A^{-1} is obtained from $A^{-1}A = AA^{-1} = I$. The inverse of A is shown in Table 2.3. It can be seen that the inverse of a symmetric matrix is a symmetric matrix.

2.6.3 DELETION OF VARIABLES FROM A PREVIOUSLY COMPUTED MATRIX INVERSE

In the analysis of empirical data it is not uncommon to encounter applications that require the inverse of a matrix A of order p and subsequently require the inverse of A after the deletion of one or more of the p variables. Familiar examples are the stepwise regression techniques that begin with a large set of predictor variables and successively discard those which contribute little to the prediction of the criterion or regression techniques that evaluate several subsets of predictor variables. Rather than computing a new inverse at each stage in these analyses, it is often more practical to modify the matrix inverse already available.

Suppose that the inverse of the matrix of covariances among p variables has been obtained and that it is desirable to obtain the inverse corresponding to the covariance matrix after the deletion of the kth variable. Let c_{ij} be the element in the ith row and jth column of the original inverse and $\hat{c}_{ij}$ be the corresponding element in the new reduced matrix inverse

$$\hat{c}_{ij} = c_{ij} - \frac{c_{ik}c_{jk}}{c_{kk}} \tag{2.21}$$

When $j = k$ or $i = k$, the value of the adjusted element $\hat{c}_{ik}$ or $\hat{c}_{jk}$ will be zero, and when all elements of the matrix have been corrected, the kth row and column will contain all zero elements. Deletion of this row and column leaves an adjusted inverse for the matrix of covariances among the remaining $p - 1$ variables.

2.6.4 THE SQUARE-ROOT METHOD FOR EVALUATING DETERMINANTS

In Sec. 2.5.2 the determinant of a matrix was defined. Although determinants can be evaluated in the nonsymmetric case, all the applications with which we shall be concerned in this text involve symmetric covariance or cross-product matrices. The square-root method, described in Sec. 2.6.2, provides an easy and readily available method for calculating the value of the determinant of any square symmetric matrix.

In Table 2.1 the calculation of the inverse of a symmetric matrix was illustrated schematically. In the course of obtaining the matrix inverse, the terms necessary to compute the value of the determinant are obtained. The triangular matrix appearing in the lower left-hand partition can be conceived as a triangular square root of the original matrix. Since it is triangular in form, only one of the possible products involved in the determinant (Sec. 2.5.2) has a nonzero value. This is the product of the diagonal entries. The value of the determinant associated with the original symmetric matrix is the *square* of the product of diagonal entries in the square-root transformation

$$|A| = (s_{11}s_{22}s_{33} \cdots s_{pp})^2$$

where s_{11}, s_{22}, s_{33}, . . . , s_{pp} are diagonal elements in the square-root factor matrix as shown in Table 2.1.

For the square symmetric matrix shown in upper triangular form in Table 2.2, the value of the determinant calculated as the sum of $p! = 4(3)(2)(1)$ terms representing all possible products obtained by taking one element from one row and column is

$$|A| = 4(5)(6)(3) + 2(2)(2)(2) - 2(3)(2)(2) - 2(2)(6)(3) + 2(3)(3)(3) \\ -2(1)(3)(2) + \cdots + 4(1)(3)(2) = 98$$

This same value can be obtained as the square of the product of diagonal entries in the lower left partition of Table 2.2

$$|A| = [2.0000(2.0000)(1.7854)(1.3862)]^2 = 98.00$$

Whereas the direct evaluation of determinants of rank 3 or 4 is practical, the utility of the square-root method in evaluating determinants of larger rank will be obvious.

Literally scores of other methods exist for evaluating determinants of large rank. The intent here has been to present only one efficient method that can be used to solve a practical problem. The interested reader is directed to Dwyer (1951) for a survey of other methods. The square-root method is recommended here primarily because a single computational sequence can be used to evaluate the determinant, triangularize a symmetric matrix, and calculate the square-root inverse and the complete inverse of a symmetric matrix. Each of these results is important for one purpose or another in the statistical analyses to be discussed.

2.7 Some Familiar Statistical Forms

2.7.1 SUMS, SQUARES, AND PRODUCTS

Some of the simple statistical quantities are defined directly from the discussion of Sec. 2.4.2. Consider **a** and **b** as two column vectors of observations on n individuals, representing two clinical variables. Premultiplication of either vector by the unit vector yields the sum of that column. Division by n (or multiplication by n^{-1}) yields the mean

$$\sum a = \mathbf{1}'\mathbf{a} \qquad \bar{a} = \mathbf{1}'\mathbf{a}n^{-1}$$
$$\sum b = \mathbf{1}'\mathbf{b} \qquad \bar{b} = \mathbf{1}'\mathbf{b}n^{-1} \tag{2.22}$$

The minor product moment of either vector yields the familiar sum of squares

$$\sum a^2 = \mathbf{a}'\mathbf{a} \qquad \sum b^2 = \mathbf{b}'\mathbf{b} \tag{2.23}$$

and the minor product of the two vectors is the usual cross-product term

$$\sum ab = \mathbf{a}'\mathbf{b} = \mathbf{b}'\mathbf{a} \tag{2.24}$$

Extending these definitions from single vectors to matrices, it is easy to verify that if X $(n \times p)$ is a data matrix of n individuals and p measurements, the minor product moment $X'X$ is a square symmetric matrix of order p, the diagonal elements of which are sums of squares and the off-diagonal elements are sums of cross products.

2.7.2 THE SP MATRIX OR CORRECTED SUMS-OF-PRODUCTS MATRIX

Expressing sums of squares or cross products as deviations from the mean requires the subtraction of a correction term. The usual formula for the sum of squared deviations is $\sum a^2 - (\sum a)^2/n$, and the usual formula for the sum of cross products in deviation form is

$$\sum ab - \frac{\sum a \sum b}{n}$$

Premultiplication of the matrix X by a unit vector $\mathbf{1}'X$ yields a row vector of column sums of the p measurements. The major product moment of this vector divided by the scalar quantity n is a square symmetric matrix of correction terms for the corresponding elements of the raw products matrix previously described, that is, $n^{-1}(X'\mathbf{1})(\mathbf{1}'X)$ is a matrix of order p, the diagonal elements are of the form $(\sum a)^2/n$, and the off-diagonal elements are of the form $\sum a \sum b/n$.

The SP matrix is a matrix of sums of squares and cross products corrected in this manner. In matrix notation Y is an SP matrix if

$$Y = X'X - n^{-1}(X'\mathbf{1})(\mathbf{1}'X) \tag{2.25}$$

2.7.3 THE COVARIANCE MATRIX

Corrected sums of squares or cross products when divided by the degrees of freedom, $n - 1$, equal variance or covariances. Similarly if the SP matrix is multiplied by the reciprocal of $n - 1$, it is converted to a covariance matrix

$$C = \frac{1}{n-1} Y \tag{2.26}$$

2.7.4 THE CORRELATION MATRIX

A familiar and computationally handy formula for the product-moment correlation coefficient is $r_{xx} = \sum x_i x_j / \sqrt{\sum x_i^2 \sum x_j^2}$ where the x's are in deviation form. If the diagonal elements of Y in Eq. (2.25) are arranged in a diagonal matrix D_Y, the elements, which are corrected sums of squares of the p variates, can be manipulated as scalars. The operation $D_Y^{\frac{1}{2}}$ denotes that each diagonal element of Y has been expressed as a square root. $D_Y^{-\frac{1}{2}}$ denotes that each of the elements in $D_Y^{\frac{1}{2}}$ has been expressed as a reciprocal. Pre- and postmultiplication of the Y matrix by $D_Y^{-\frac{1}{2}}$ yields the correlation matrix R

$$R = D_Y^{-\frac{1}{2}} Y D_Y^{-\frac{1}{2}}$$

$$R = \begin{bmatrix} \frac{1}{\sqrt{\sum y_1^2}} & 0 & 0 \\ 0 & \frac{1}{\sqrt{\sum y_2^2}} & 0 \\ 0 & 0 & \frac{1}{\sqrt{\sum y_3^2}} \end{bmatrix} \begin{bmatrix} \sum y_1^2 & \sum y_1 y_2 & \sum y_1 y_3 \\ \sum y_2 y_1 & \sum y_2^2 & \sum y_2 y_3 \\ \sum y_3 y_1 & \sum y_3 y_2 & \sum y_3^2 \end{bmatrix}$$

$$\times \begin{bmatrix} \frac{1}{\sqrt{\sum y_1^2}} & 0 & 0 \\ 0 & \frac{1}{\sqrt{\sum y_2^2}} & 0 \\ 0 & 0 & \frac{1}{\sqrt{\sum y_3^2}} \end{bmatrix}$$

$$= \begin{bmatrix} \frac{\sum y_1^2}{\sqrt{\sum y_1^2 \sum y_1^2}} & \frac{\sum y_1 y_2}{\sqrt{\sum y_1^2 \sum y_2^2}} & \frac{\sum y_1 y_3}{\sqrt{\sum y_1^2 \sum y_3^2}} \\ \frac{\sum y_2 y_1}{\sqrt{\sum y_2^2 \sum y_1^2}} & \frac{\sum y_2^2}{\sqrt{\sum y_2^2 \sum y_2^2}} & \frac{\sum y_2 y_3}{\sqrt{\sum y_2^2 \sum y_3^2}} \\ \frac{\sum y_3 y_1}{\sqrt{\sum y_3^2 \sum y_1^2}} & \frac{\sum y_3 y_2}{\sqrt{\sum y_3^2 \sum y_2^2}} & \frac{\sum y_3^2}{\sqrt{\sum y_3^2 \sum y_3^2}} \end{bmatrix} \quad (2.27)$$

2.7.5 MULTIPLE AND PARTIAL CORRELATIONS

When analyzing the relationships among a number of variables, it is often desirable to know the extent to which each variable is unique or, conversely, the extent to which it can be predicted from other variables in the matrix. The uniqueness of a variable is often defined by the coefficient of nondetermination $1 - R^2$, where R is a zero-order or multiple-correlation coefficient. A predictor variable that correlates reasonably well with a criterion variable but is relatively unique from the other predictor variables (a high coefficient of nondetermination) will generally be a useful variable to include in a predictive battery. In Chap. 4, the concept of communality, which has an important place in factor analysis, will be introduced, and the squared multiple-correlation coefficient (coefficient of determination) will be discussed as one way of estimating the communality. For these and other applications it is helpful to have an efficient way of simultaneously computing all multiple correlations among a group of p variables, i.e., the multiple correlation of each variable and the remaining $p - 1$ variables. If the inverse of the correlation matrix is available, this is an easy matter.

Let D be a diagonal matrix the elements of which are the diagonal elements of the inverse of the correlation matrix. Then the matrix $I - D^{-1}$ will be a diagonal matrix of squared multiple-correlation coefficients. The reciprocal of a diagonal element of the inverse of a correlation matrix subtracted from 1 is equal to the squared multiple correlation of the variable identified by that row and column and the remaining variables in the matrix.

The regression coefficients associated with these multiple correlations are not so commonly required, but they can easily be obtained by premultiplying D^{-1} by the inverse of the correlation matrix and changing all the signs of the resulting matrix. The columns of this matrix are the standard regression coefficients associated with each multiple correlation. A more common computing formula for obtaining standard regression coefficients is $R^{-1}\mathbf{y}$, where R is the correlation matrix of predictor variables and $\mathbf{y}$ is a vector of correlations between the criterion and the predictors. If several criteria were to be predicted from the same set of predictors, there would be a corresponding number of $\mathbf{y}$ vectors and the formula would be written $\beta = R^{-1}Y$.

Let $D^{-\frac{1}{2}}$ be a diagonal matrix the elements of which are the reciprocals of the square roots of the diagonal elements of the inverse of the correlation matrix. If the inverse of the correlation matrix is pre- and postmultiplied by $D^{-\frac{1}{2}}$, the off-diagonal elements of the resulting matrix will be correlations of two variables with the remaining $p - 2$ variables partialed out. The signs of these partial correlations will be reversed, but subtracting the matrix from an identity matrix will correct the signs and give zero entries in the principal diagonal.

These are all very useful computing formulas. They have been introduced here to illustrate the efficiency of matrix computations and the compactness of matrix notation. Multiple correlation and regression are discussed at greater length in Chap. 17.

2.7.6 WITHIN-GROUPS MATRICES

In simple analysis of variance the total sum of squares can be partitioned into between-groups and within-groups components such that total = between + within. This can be accomplished by calculating the mean corrected sum of squares separately for each of k groups, summing across the k groups to obtain the within-groups sum of squares, and subtracting this quantity from the total corrected sum of squares to get the between-groups sum of squares

$$\text{Total} = \sum x_t^2 - \frac{(\sum x_t)^2}{n_t}$$

$$\text{Within} = \left[\sum x_1^2 - \frac{(\sum x_1)^2}{n_1}\right] + \left[\sum x_2^2 - \frac{(\sum x_2)^2}{n_2}\right] \cdots \left[\sum x_k^2 - \frac{(\sum x_k)^2}{n_k}\right]$$

$$\text{Between} = \text{total} - \text{within}$$

In multivariate analysis there is also a need to partition the sums of squares in this manner, and the corresponding total, between-, and within-groups matrices are formed in exactly the same way, either element by element or by manipulating matrices. The SP matrix given by Eq. (2.25) is a total SP matrix if the sums, squares, and cross products are obtained over the total number of individuals. Another way of expressing this is that the number of rows n in X, the number of elements in the unit vector, and the value of the scalar quantity n are equal to the total number of individuals included in all k groups. The within-groups SP matrix is formed by computing an SP matrix for each of the k groups separately using Eq. (2.25), where the number of rows n in X, the number of elements in the unit vector, and the value of the scalar quantity n are equal to the number of individuals in a particular group. Summing the square matrices of order p over k groups equals the within-groups SP matrix

$$\mathrm{SP}(W) = \mathrm{SP}_{(1)} + \mathrm{SP}_{(2)} \cdots \mathrm{SP}_{(k)} \tag{2.28}$$

Although the between-groups SP matrix could also be calculated directly, it is usually more convenient to obtain it by subtraction

$$\mathrm{SP}(B) = \mathrm{SP}(T) - \mathrm{SP}(W) \tag{2.29}$$

Between- and within-groups covariance matrices are defined by extension of these equations

$$C(W) = (n_1 + n_2 \cdots + n_k - k)^{-1}\mathrm{SP}(W) \tag{2.30}$$

$$C(B) = (k - 1)^{-1}\mathrm{SP}(B) \tag{2.31}$$

The within-groups covariance matrix will often be referred to as the *dispersion* matrix.

Finally if a dispersion matrix is used for Y in Eq. (2.27), the result is called a *within-groups correlation matrix*.

2.7.7 THE TRACE OF A MATRIX

The sum of the diagonal elements of a square matrix is called the *trace* of the matrix. The trace of product-moment matrices is frequently of interest. The trace of the major product moment of A $(m \times n)$ is equal to the trace of the minor product moment even though the two product matrices are of different order. The trace in this instance is equal to the sum of every element in A squared.

$$\mathrm{tr}\, A'A = \mathrm{tr}\, AA' = \sum^{mn} a_{ij}^2 \tag{2.32}$$

The traces of other matrices are also meaningful; for example, tr SP is the sum of squares of the p variates each corrected for its own mean, and the trace of the covariance matrix is the sum of the variances.

2.7.8 QUADRATIC FORMS

Recognizing that $\mathbf{a}'C\mathbf{a}$ is a scalar quantity is useful in the reduction of a matrix equation, but this form also has important implications in multivariate statistics. For example, if y is a linear combination of p correlated measurements obtained by multiplying the original measurements $\mathbf{x}$ (in deviation form) by a vector of weighting coefficients $\mathbf{a}$, the variance of y over n individuals can be calculated directly from the covariance matrix of p measurements by pre- and postmultiplying by the vector of weighting coefficients. Let

$$y = \mathbf{a}'\mathbf{x}$$

then

$$\frac{1}{n}\sum y^2 = \frac{1}{n}\sum \mathbf{a}'\mathbf{x}\mathbf{x}'\mathbf{a}$$

$$= \mathbf{a}'(n^{-1}\sum \mathbf{x}\mathbf{x}')\mathbf{a}$$

The expression enclosed in parentheses will be recognized as the covariance matrix where the x's are in deviation form. Similarly, if $y = \mathbf{a}'\mathbf{z}$, where the original measurements are in standard-score form,

$$\frac{1}{n}\sum y^2 = \frac{1}{n}\sum \mathbf{a}'\mathbf{z}\mathbf{z}'\mathbf{a}$$

$$= \mathbf{a}'(n^{-1}\sum \mathbf{z}\mathbf{z}')\mathbf{a}$$

$$= \mathbf{a}'R\mathbf{a}$$

where R is the correlation matrix.

2.8 Characteristic Roots and Vectors

This section introduces some additional matrix concepts that have had great significance in empirical science and important applications in the social, behavioral, and biological sciences. They will be further elaborated in the succeeding chapter.

First, the *characteristic roots* (latent roots, eigenvalues) of a matrix are defined as the scalar values that satisfy the determinantal equation $|A - \lambda I| = 0$. If A is any square matrix of order n, then the matrix $A - \lambda I_{(n)}$ is called the *characteristic matrix* of A

$$\begin{bmatrix} a_{11} & a_{12} & a_{13} \\ a_{21} & a_{22} & a_{23} \\ a_{31} & a_{32} & a_{33} \end{bmatrix} - \begin{bmatrix} \lambda & 0 & 0 \\ 0 & \lambda & 0 \\ 0 & 0 & \lambda \end{bmatrix} = \begin{bmatrix} a_{11} - \lambda & a_{12} & a_{13} \\ a_{21} & a_{22} - \lambda & a_{23} \\ a_{31} & a_{32} & a_{33} - \lambda \end{bmatrix}$$

For any square matrix that is not null there is at least one value of λ that will satisfy the equation, and there will be more depending upon the order and rank of the matrix. Consider the following matrix of order 2:

$$|A_2 - \lambda I_{(2)}| = \begin{vmatrix} a_{11} - \lambda & a_{12} \\ a_{21} & a_{22} - \lambda \end{vmatrix} = (a_{11} - \lambda)(a_{22} - \lambda) - a_{12}a_{21}$$

$$= \lambda^2 - (a_{11} + a_{22})\lambda + (a_{11}a_{22} - a_{12}a_{21})$$

Note that the expanded equation has three terms: λ^2, $(a_{11} + a_{22})\lambda$, where the coefficient is the trace of the matrix A, and $a_{11}a_{22} - a_{12}a_{21}$ which is the determinant of A. For example,

$$\begin{vmatrix} 3-\lambda & 2 \\ 1 & 4-\lambda \end{vmatrix} = (3-\lambda)(4-\lambda) - 2$$

$$= \lambda^2 - 7\lambda + 10$$

$$= (\lambda - 5)(\lambda - 2)$$

The characteristic roots of A are 5 and 2. The expansion of $|A_3 - \lambda I_3|$ is

$$|A_3 - \lambda I_3| = (a_{11} - \lambda)(a_{22} - \lambda)(a_{33} - \lambda) + a_{21}a_{32}a_{13} + a_{31}a_{12}a_{23} - a_{31}(a_{22} - \lambda)a_{13} - a_{21}a_{12}(a_{33} - \lambda) - (a_{11} - \lambda)a_{32}a_{23}$$

Expanding further and combining terms yields

$$|A_3 - \lambda I_{(3)}| = -\lambda^3 + (a_{11} + a_{22} + a_{33})\lambda^2 - (a_{11}a_{33} - a_{31}a_{13} + a_{22}a_{33} - a_{32}a_{23} + a_{11}a_{22} - a_{21}a_{12})\lambda + (a_{11}a_{22}a_{33} + a_{21}a_{32}a_{13} + a_{31}a_{12}a_{23} - a_{31}a_{22}a_{13} - a_{21}a_{12}a_{33} - a_{11}a_{32}a_{23})$$

This expansion has four terms: the coefficient of the first term is -1, as the order of the matrix is odd; the coefficient of the second term is again the trace of A_3; the coefficient of the third term is the sum of the three minor determinants formed by deleting the ith row and column of A; and the last term is the determinant of A. The reader can intuitively see what the expansion of the determinant $|A_{(n)} - \lambda I_{(n)}|$ would be for higher orders of the matrix. In general the *characteristic function* of A is

$$f(\lambda) = (-1)^n[\lambda^n - p_1\lambda^{n-1} + p_2\lambda^{n-2} + \cdots + (-1)^n p_n]$$

where $p_1, p_2, \ldots, p_n$ are the values described above. Factoring this expression yields the characteristic roots. It is obvious, however, that as the order of the matrix increases, it becomes less feasible to solve directly for the characteristic roots by expanding the determinantal equation.

Some additional relationships concerning characteristic roots are worth noting in passing. The sum of the characteristic roots of A is equal to the trace of A. The continued product, $\lambda_1(\lambda_2) \cdots (\lambda_n)$, of the roots is equal to the determinant of A. If the matrix is nonsingular, the number of characteristic roots will equal the rank (and also the order of the matrix). If the matrix is a nonsingular covariance matrix, the roots will all be positive.

The *characteristic vectors* (latent vectors, eigenvectors) of a matrix are defined as any vectors that satisfy the equation $A\mathbf{x} = \lambda\mathbf{x}$ or $(A - \lambda I)\mathbf{x} = 0$. Solutions can be found where λ is a characteristic root by solving simultaneous equations of the form

$$\begin{aligned}
(a_{11} - \lambda)x_1 + a_{12}x_2 \quad &+ \cdots + a_{1n}x_n &= 0 \\
a_{21}x_1 \quad + (a_{22} - \lambda)x_2 &+ \cdots + \ a_{2n}x_n &= 0 \\
a_{n1}x_1 \quad + \ a_{n2}x_2 \quad &+ \cdots + (a_{nn} - \lambda)x_n &= 0
\end{aligned}$$

These equations are obtained by postmultiplication of $A - \lambda I$ by $\mathbf{x}$. As the solution of a set of such equations is required to obtain each characteristic vector associated with each characteristic root, it is clear that direct calculation of characteristic roots and vectors even using computers is a formidable undertaking. In the next chapter, an iterative procedure will be described which will yield the characteristic roots and vectors of a covariance or correlation matrix. It will also be made clear how these rather abstract matrix relationships are of value in empirical science.

2.9 Computer Program for Basic Matrix Analyses

Rather than attempting to present a separate specific computer program for each type of complex multivariate analysis, it is more practical and useful to present a single general matrix-analysis program that will provide the student or researcher with the intermediate results, or building blocks, from which more complex analyses can be developed. With reasonable hand calculations, any analysis described in this book can be completed starting with appropriate dispersion matrices, determinant values, or other intermediate results.

The preceding approach has several advantages: (1) It is not always easy to adapt a computer program written for one computer to a different type of computer, and there is an advantage in having only one such conversion to make. (2) We feel that although the complete "canned" programs now available are very useful, the student or researcher is not likely to achieve much increased understanding of the methodology by their use. By providing only the basic calculations of matrices, inverses, and determinant values, the approach employed here leaves the student to solve the problem for himself. Although the student should understand how basic matrices are computed, there seems to be little value in requiring that these preliminary calculations be done manually.

For the reader not familiar with computer programming, the remainder of the chapter may be somewhat obscure. Teaching programming and other arcane aspects of modern computer technology is beyond the scope of this book, but several references have been included for the serious student.

The program MATRX was written in Fortran IV for the IBM 1130 disk-oriented computer. It should be reasonably easy to adapt for use on any computer having Fortran IV compiler and disk capabilities. MATRX consists of a mainline program and five subroutines. It was written to accomplish basic matrix calculations for problems involving multiple measurements on individuals in several different groups. The program computes within-groups, between-groups, and total SP matrices, covariance matrices, and correlation matrices. It computes the triangular square-root inverse and the full inverse of each (nonsingular) matrix and evaluates the determinant. The characteristic roots and vectors (principal components) are also computed for each matrix. Not all these results will be required for any particular problem, so that the user must be prepared to select from the output only portions that are relevant.

Preparation of data for processing by MATRX is simple. The multiple measurements on each individual are punched into consecutive fields of a single card. (Actually, any type of data organization can be handled by modification of the format in statement 1 of the SP subprogram.) The data cards are sorted into groups and the number in each group counted. For processing, the data cards must be preceded by two control cards: (1) A card specifying the number of variables NVAR and the number of groups NGPS. These two numerical parameters must be right-justified in the first two four-column fields. (2) A card specifying the number of individuals in each group in the order in which data cards are to be read. These group frequencies should be right-justified in consecutive four-column fields of the second card. For some types of problems only a single group will be specified in card 1, in which case the total N should be entered in the first four-column field of card 2. The data should follow immediately behind the two control cards and will read according to format 1. (A *format* instruction specifies for the computer the organization of data fields in the punched card.)

The printout resulting from a single computer run is organized according to the following outline.

1 Group means for the NVAR original measurements
2 Analyses of SP matrices
 a Within-groups SP matrices
  (1) Within-groups matrix
  (2) Determinant
  (3) Square-root inverse
  (4) Complete inverse
  (5) Principal components and principal axes
 b Between-groups SP matrices
  (1) Between-groups matrix
  (2) Principal components and principal axes
 c Total SP matrices
  (1) Total matrix
  (2) Determinant
  (3) Square-root inverse
  (4) Complete inverse
  (5) Principal components and principal axes
3 Analyses of covariance matrices
 a Within-groups covariance matrices
  (1) Within-groups matrix
  (2) Determinant
  (3) Square-root inverse
  (4) Complete inverse

- (5) Principal components and principal axes
- **b** Between-groups covariance matrices
  - (1) Between-groups matrix
  - (2) Principal components and principal axes
- **c** Total covariance matrices
  - (1) Total matrix
  - (2) Determinant
  - (3) Square-root inverse
  - (4) Complete inverse
  - (5) Principal components and principal axes

4 Analyses of correlation matrices
- **a** Within-groups correlation matrices
  - (1) Within-groups matrix
  - (2) Determinant
  - (3) Square-root inverse
  - (4) Complete inverse
  - (5) Principal components and principal axes
- **b** Between-groups correlation matrices
  - (1) Between-groups matrix
  - (2) Principal components and principal axes
- **c** Total correlation matrices
  - (1) Total matrix
  - (2) Determinant
  - (3) Square-root inverse
  - (4) Complete inverse
  - (5) Principal components and principal axes

The complete computer program is presented below. The mainline appears first, followed by five subroutines. Anyone wanting to use this program should consult a computer specialist familiar with Fortran coding for instructions on general format requirements. Advice may also be needed with regard to modification of Format 1 in order to read the data properly from punched cards. For SP matrices involving large elements, the format for printout of inverses may need to be modified to preserve enough decimal places. Otherwise, all that should be required is proper setup of job control cards, which depends upon the particular computer system being used.

Once again it should be stressed that our main purpose in providing this general matrix-analysis program is for instructional purposes. Given this program, the student should be capable of preparing multivariate data for almost all the analyses described in later sections of this book. The program MATRX does not complete the multivariate analysis; it simply reduces the data and provides basic intermediate results. The student can gain insight by taking the problem from there.

Mainline Computer Program MATRX

```
      DIMENSION W(18),B(18),T(18)
     1,NG(15)
      DEFINE FILE 1(40,40,U,L1),2(40,40,U,L2),3(40,40,U,L3),4(40,40,U,L4
     1), 5(40,40,U,L5), 6(40,40,U,L6), 7(40,40, U,L7)
      COMMON L1,L2,L3,L4,L5,L6,L7
       NTIM=1
      READ(2,1) NVAR,NGPS
      READ(2,1)(NG(I),I=1,NGPS)
    1 FORMAT(20I4)
      CALL SP(NVAR,NGPS,NG,B,T,W)
      L1=1
      WRITE(3,51)
   51 FORMAT(//,1X,'ANALYSES OF SUMS OF PRODUCTS MATRICES')
      WRITE(3,19)
   19 FORMAT(1X,'*************************************')
      DO 500 I=1,3
      WRITE(3,22)
   22 FORMAT(//,10X,'WITHIN GROUPS MATRICES')
      DO 100 II=1,3
      DO 20 K=1,NVAR
      READ(II'K)(W(J),J=1,NVAR)
   20 WRITE(3,4)(W(J),J=1,NVAR)
    4 FORMAT(21X,10F9.3)
      IF(II-2)21,29,21
   21 CALL INVER(II,NVAR)
      WRITE(3,202)
  202 FORMAT(/,15X,'TRIANGULAR SQUARE ROOT INVERSE')
      DO 40 M=1,NVAR
      READ(6'M)(B(M2),M2=1,NVAR)
   40 WRITE(3,4)(B(M2),M2=1,NVAR)
      WRITE(3,201)
  201 FORMAT(/,15X,'COMPLETE INVERSE')
      DO 90 M=1,NVAR
      READ(5'M)(B(M2),M2=1,NVAR)
   90 WRITE(3,4)(B(M2),M2=1,NVAR)
   29 CALL AXES(NVAR,II,MM)
      DO 60 M=1,MM
      READ(7'M)(B(M2),M2=1,NVAR)
   60 WRITE(3,4)(B(M2),M2=1,NVAR)
      GO TO (61,65,100),II
   61 WRITE(3,62)
   62 FORMAT(//,10X,'BETWEEN GROUPS MATRICES')
      GO TO 100
   65 WRITE(3,66)
   66 FORMAT(//,10X,'TOTAL GROUPS MATRICES')
  100 CONTINUE
      GO TO (101,103,500),I
  101 CALL COVS(NG,NVAR,NGPS)
      WRITE(3,102)
  102 FORMAT(//,1X,'ANALYSES OF COVARIANCE MATRICES')
      WRITE(3,18)
   18 FORMAT(1X,'*******************************')
      GO TO 500
  103 CALL CORR(NVAR)
      WRITE(3,104)
  104 FORMAT(//,1X,'ANALYSES OF CORRELATION MATRICES')
      WRITE(3,17)
   17 FORMAT(1X,'********************************')
  500 CONTINUE
      CALL EXIT
      END
```

Subroutine to Calculate Within-groups, Between-groups, and Total Matrices

```
C      COMPUTE WITHIN, BETWEEN AND TOTAL SP MATRICES
C      STORE IN FILES 1, 2, AND 3, RESPECTIVELY.
       SUBROUTINE SP (NVAR,NGPS,NG,B,T,W)
       DIMENSION
      1A(18,36),T(18),W(18),B(18),BSUM(18),XDAT(18),TSUM(18),XBAR(18),NG(
      215)
       COMMON L1,L2,L3,L4,L5,L6,L7
     1 FORMAT(16F2.0)
     2 FORMAT(18F6.2)
     3 FORMAT(/22H MEAN VECTOR FOR GROUP,I5,10X,3H N=,I5)
C      INITIALIZE FILE RECORD NUMBER INDICES
       L1=1
       L2=1
       L3=1
       L4=1
C      ZERO ELEMENTS FOR SUMMATION
       TOT=0.0
       M4=2*NVAR
       DO 5 M=1,NVAR
       TSUM(M)=0.0
       B(M)=0.0
       T(M)=0.0
       DO 8 M2=1,M4
       IF(M2-NVAR) 9,9,8
     9 W(M2)=0.0
     8 A(M,M2)=0.0
     5 WRITE(1'M)(W(M2),M2=1,NVAR)
       NOBS=0
C      COMPUTE MEANS AND RAW CROSS PRODUCTS FOR K GROUPS SEPARATELY
       DO 15 K=1,NGPS
       DO 6 JJ=1,NVAR
     6 BSUM(JJ)=0.0
       N=NG(K)
       NOBS=NOBS+N
C      READ N DATA VECTOR IN EACH GROUP AND COMPUTE SUMS AND PRODUCTS
       DO 10 J=1,N
       READ(2,1)(XDAT(M),M=1,NVAR)
       DO 7 M=1,NVAR
       BSUM(M)=BSUM(M)+XDAT(M)
       TSUM(M)=TSUM(M)+XDAT(M)
       DO 7 M2=M,NVAR
       M3=M2+NVAR
       A(M,M2)=A(M,M2)+XDAT(M)*XDAT(M2)
     7 A(M,M3)=A(M,M3)+XDAT(M)*XDAT(M2)
    10 CONTINUE
       FN=N
       TOT=TOT+FN
       DO 13 M=1,NVAR
       XBAR(M)=BSUM(M)/FN
       READ(1'M) (W(M2) ,M2=M,NVAR)
       DO 12 M2=M,NVAR
       W(M2)=W(M2)+A(M,M2)-BSUM(M)*BSUM(M2)/FN
    12 A(M,M2)=0.0
    13 WRITE(1'M) (W(M2) ,M2=M,NVAR)
C      PRINT VECTOR OF GROUP MEANS AND SAMPLE SIZE
C      STORE GROUP MEANS IN FILE 4
       WRITE(3,3) K,N
       WRITE(3,2)(XBAR(M),M=1,NVAR)
    15 WRITE(4'L4)(XBAR(M),M=1,NVAR),FN
C      COMPUTE TOTAL CORRECTED SP MATRIX AND THEN
C      COMPUTE BETWEEN SP MATRIX BY SUBTRACTION
       IF(NGPS-1) 20,20,16
    16 DO 37 M=1,NVAR
       M4=M+NVAR
       DO 37 M2=M,NVAR
       M3=M2+NVAR
       M5=M3-NVAR
    37 A(M5,M4)=A(M,M3)
       DO 25 M=1,NVAR
       DO 17 M2=1,NVAR
       M3=M2+NVAR
    17 T(M2)=A(M,M3)-TSUM(M)*TSUM(M2)/TOT
```

```
25 WRITE(3'L3)(T(M2),M2=1,NVAR)
   L1=1
   L2=1
   L3=1
   DO 27 M=1,NVAR
   READ(1'M)(A(M,M2),M2=M,NVAR)
   DO 27 M2=M,NVAR
27 A(M2,M)=A(M,M2)
   DO 32 M=1,NVAR
   M3=NVAR+1
   READ(3'M)(A(M,M5),M5=M3,M4)
32 WRITE(1'M)(A(M,M2),M2=1,NVAR)
   DO 36 M=1,NVAR
   DO 35 M2=1,NVAR
   M3=M2+NVAR
35 B(M2)=A(M,M3)-A(M,M2)
36 WRITE(2'M)(B(M2),M2=1,NVAR)
20 CONTINUE
   RETURN
   END
```

Subroutine to Compute Matrix Inverse and to Evaluate Determinant

```
      SUBROUTINE INVER (NUM,NVAR)
      DIMENSION A(18,36),B(18)
      COMMON L1,L2,L3,L4,L5,L6,L7
      N2=2*NVAR
      N1=NVAR+1
      DO 10 I=1,NVAR
   10 READ(NUM'I)(A(I,M),M=1,NVAR)
C     FORM IDENTITY MATRIX IN RIGHT HALF OF A
      DO 11 J=N1,N2
      DO 11 I=1,NVAR
   11 A(I,J)=0.0
      DO 12 J=N1,N2
      I=J-NVAR
   12 A(I,J)=1.0
C     DIVIDE THRU BY SQ ROOT OF FIRST PIVOTAL ELEMENT
      X=1.0/SQRT(A(1,1))
      DO 13 J=1,N2
   13 A(1,J)=A(1,J)*X
C     SUCCESSIVELY SWEEP OUT OTHER PIVOTAL ELEMENTS
      DO 15 K=2,NVAR
      NSTP=K-1
      DO 14 M=1,NSTP
   14 A(K,K)=A(K,K)-A(M,K)*A(M,K)
      A(K,K)=SQRT(A(K,K))
      NSTR=K+1
      DO 16 I=NSTR,N2
      DO 17 M=1,NSTP
   17 A(K,I)=A(K,I)-A(M,K)*A(M,I)
   16 A(K,I)=A(K,I)/A(K,K)
   15 CONTINUE
      MULTIPLY TRIANGULAR SQ ROOT MATRIX BY TRANSPOSE
      DET=1.0
      DO 20 M=1,NVAR
   20 DET=DET*A(M,M)*A(M,M)
      DO 101 M=1,NVAR
      M3=M+NVAR
      WRITE(6'M)(A(M,M5),M5=N1,N2)
      DO 105 M2=1,NVAR
  105 A(M,M2)=A(M2,M3)
      B(M)=0.0
  101 CONTINUE
      WRITE(3,1)
    1 FORMAT(//,15X,'DETERMINANT')
      WRITE(3,2)   DET
    2 FORMAT(/,15X,F10.5)
      DO 104 M=1,NVAR
      DO 102 M2=1,NVAR
      B(M2)=0.0
      M4=NVAR+M2
      DO 102 M3=1,NVAR
  102 B(M2)=B(M2)+A(M,M3)*A(M3,M4)
  104 WRITE(5'M)(B(M2),M2=1,NVAR)
      RETURN
      END
```

Subroutine to Compute Principal Axes

```
      SUBROUTINE AXES(NVAR,NUM,MM)
      DIMENSION A(18,36),B(18),C(18)
      COMMON L1,L2,L3,L4,L5,L6,L7
      L7=1
      WRITE(3,2)
    2 FORMAT(/,15X,'PRINCIPAL AXES')
      TOLSQ=.000004
      FNVAR=NVAR
      SVAL=0.0
      SUM2=0.0
      DO 10 M=1,NVAR
   10 READ(NUM' M)(A(M,M2),M2=1,NVAR)
      MM=1
      DO 111 J=1,NVAR
  111 SUM2=SUM2+A(J,J)
      TEST=SUM2*.95
    8 DO 20 J=1,NVAR
   20 B(J)=1.0
    6 SCL=0.0
      IT=0
      DO 22 J=1,NVAR
      C(J)=0.0
      DO 21 K=1,NVAR
   21 C(J)=C(J)+A(J,K)*B(K)
   22 SCL=SCL+C(J)*C(J)
      SCALE=SQRT(SCL)
      SCL=SQRT(1.0/SCL)
      DO 23 J=1,NVAR
      C(J)=C(J)*SCL
      DIF=(C(J)-B(J))**2
      B(J)=C(J)
      IF(DIF-TOLSQ)23,23,24
   24 IT=1
   23 CONTINUE
      IF(IT)99,99,6
C     COMPUTE EIGENVALUE
   99 VAL=0.0
      DO 30 J=1,NVAR
      DO 30 K=1,NVAR
   30 VAL=VAL+A(J,K)*B(J)*B(K)
      WRITE(7'L7)(B(J),J=1,NVAR)
C     COMPUTE RESIDUAL MATRIX
      SVAL=SVAL+VAL
      IF(SVAL-TEST)97,98,98
   97 DO 95 J=1,NVAR
      DO 95 K=1,NVAR
   95 A(J,K)=A(J,K)-C(J)*C(K)*VAL
      MM=MM+1
      GO TO 8
   98 CONTINUE
      RETURN
      END
```

Subroutine to Convert SP Matrix into Covariance Matrix

```
      SUBROUTINE COVS(NG,NVAR,NGPS)
      DIMENSION W(18),NG(15)
      COMMON L1,L2,L3,L4,L5,L6,L7
      L1=1
      NOBS=0
      DO 1 I=1,NGPS
    1 NOBS=NOBS+NG(I)
      DF=NOBS-NGPS
      DO 2 I=1,NVAR
      READ(1'I)(W(J),J=1,NVAR)
      DO 3 J=1,NVAR
    3 W(J)=W(J)/ DF
    2 WRITE (1'I)(W(J),J=1,NVAR)
      DF=NGPS-1
      L2=1
      DO 5 I=1,NVAR
      READ(2'I)(W(J),J=1,NVAR)
      DO 4 J=1,NVAR
    4 W(J)=W(J)/ DF
    5 WRITE (2'I)(W(J),J=1,NVAR)
      DF=NOBS-1
      L3=1
      DO 7 I=1,NVAR
      READ(3'I)(W(J),J=1,NVAR)
      DO 6 J=1,NVAR
    6 W(J)=W(J)/ DF
    7 WRITE (3'I)(W(J),J=1,NVAR)
      RETURN
      END
```

Subroutine to Convert Covariance Matrix into Correlation Matrix

```
      SUBROUTINE CORR (NVAR)
      DIMENSION A(18,18),D(18)
      COMMON L1,L2,L3,L4,L5,L6,L7
      NUM=1
      DO 1 I=1,NVAR
    1 READ (1'I)(A(I,J),J=1,NVAR)
      DO 2 J=1,NVAR
    2 D(J)=1/SQRT(A(J,J))
      DO 5 I=1,NVAR
      DO 4 J=1,NVAR
    4 A(I,J)=A(I,J)*D(I)*D(J)
    5 WRITE (1'I)(A(I,J),J=1,NVAR)
      DO 7 I=1,NVAR
    7 READ (2'I)(A(I,J),J=1,NVAR)
      DO 8 J=1,NVAR
    8 D(J)=1/SQRT(A(J,J))
      DO 10 I=1,NVAR
      DO 9 J=1,NVAR
    9 A(I,J)=A(I,J)*D(I)*D(J)
   10 WRITE (2'I)(A(I,J) ,J=1,NVAR)
      DO 12 I=1,NVAR
   12 READ (3'I)(A(I,J),J=1,NVAR)
      DO 13 J=1,NVAR
   13 D(J)=1/SQRT(A(J,J))
      DO 15 I=1,NVAR
      DO 14 J=1,NVAR
   14 A(I,J)=A(I,J)*D(I)*D(J)
   15 WRITE (3'I)(A(I,J),J=1,NVAR)
      RETURN
      END
```

2.10 Exercises

1 Construct a data matrix containing four scores for each of 10 subjects such that the fourth score for each subject is a weighted combination of the first three scores. Then compute the matrix of covariances among the four scores. Is the covariance matrix singular? What is its rank? Can the fourth column be expressed as a weighted combination of the other three columns? How do the weights compare to the weights used in defining the fourth variable as a function of the first three variables? What implications do these results have for correlational analyses in which some variables are weighted combinations of other variables?

2 In Eq. (2.25) the corrected SP matrix is defined as the minor product moment of a data matrix and its transpose minus the outer product of the vector of means and the vector of sums. Assuming that Z $(n \times p)$ is a data matrix containing p scores for each of n subjects and that the column mean has been subtracted from all elements in each of the p columns, what is the nature of the matrix $C = [1/(n-1)]Z'Z$? If c_{ij} is the element in the ith row and jth column of C, how can it be defined in scalar notation as a function of the elements in Z? What if the p columns of Z $(n \times p)$ have been normalized as well as mean-corrected? Given such a data matrix containing normalized and mean-corrected columns, represent in matrix notation the calculation of the complete $p \times p$ matrix of intercorrelations.

3 Given that the matrix of intercorrelations among p variables can be represented as the product of a normalized data matrix and its transpose, show that the correlation matrix will be singular if the number of subjects upon which it is based is less than p. Support your conclusion by relevant equations from this chapter.

4 Given the matrix equation $C\mathbf{b} = \mathbf{c}$, solve for the vector $\mathbf{b}$ where

$$C = \begin{bmatrix} 1.00 & .50 & .25 \\ .50 & 1.00 & .55 \\ .25 & .55 & 1.00 \end{bmatrix} \qquad \text{and} \qquad \mathbf{c}' = [.40 \quad .35 \quad .25]$$

It will be recognized in reading Chaps. 9, 16, and 17 that discriminant-function and multiple-regression problems require precisely these calculations.

5 Write the matrix equation for computing the between-groups SP matrix directly rather than using Eq. (2.29).

2.11 References

Anderson, T. W.: "Introduction to Multivariate Statistical Analysis," Wiley, New York, 1958.

Dwyer, P. S.: "Linear Computations," Wiley, New York, 1951.

Hohn, F. E.: "Elementary Matrix Algebra," Macmillan, New York, 1958.

Horst, P.: "Matrix Algebra for Social Scientists," Holt, New York, 1963.

Lehman, R. S., and D. E. Bailey: "Digital Computing: Fortran IV and Its Applications in Behavioral Science," Wiley, New York, 1968.

Louden, R. K.: "Programming the IBM 1130 and 1800," Prentice-Hall, Englewood Cliffs, N.J., 1967.

McCracken, D. D.: "A Guide to Fortran IV Programming," Wiley, New York, 1965.

Rao, C. R.: "Advanced Statistical Methods in Biometric Research," Wiley, New York, 1952.

Searle, S. R.: "Matrix Algebra for the Biological Sciences," Wiley, New York, 1966.

3

Principal-components Analysis

3.1 Introduction

Principal-components analysis is a method for reducing p correlated measurement variables to a smaller set of statistically independent linear combinations having certain unique properties with regard to characterizing individual differences. The first principal component is that weighted combination of the several original variables which accounts for a maximum amount of the total variation, or individual differences, represented in the complete set of original variables. The second principal component is that weighted combination of the several original variables which of all possible weighted combinations uncorrelated with the first principal component accounts for a maximum amount of the remaining variation or individual differences. The rth principal component is that weighted combination which of all possible weighted combinations independent of the first $r - 1$ accounts for a maximum amount of the remaining variation among individuals in terms of their original score values. The properties of statistical orthogonality (independence) and maximization of variance uniquely define principal components.

A preliminary word may be in order concerning the use of principal components, and the related factor-analysis techniques, for the representation of individual differences. Measurement permits one to characterize differences between individuals on any particular trait or variable. The variance computed from scores on such a measurement variable represents individual differences on that variable; and indeed, if there were no individual differences, there would be no variance. Principal components of a matrix of covariances or intercorrelations among p variables represent new composite variables which account for or describe a maximum amount of the total variability among individuals on all the original measurements.

Principal-components analysis should be used where the aim is to describe differences between individuals in a heterogeneous sample in terms of a relatively few composite variables and should not be used where the aim is to characterize similarities among individuals within a specially selected homogeneous sample, e.g., similarities among psychiatric depressed patients. Of course, even psychiatric depressed patients are heterogeneous in some respects, and principal-components analysis might be used appropriately to define composite variables representing maximally the differences that do exist.

Psychologists, and behavioral scientists in general, have probably not exploited the method of principal-components analysis as fully as might be appropriate. Part of the reason lies in the obvious similarity of components analysis to the methodology of factor analysis, which because of its development by psychologists has been more familiar to behavioral scientists. In spite of computational similarities (most striking in the principal-axes method of factor analysis, which is formally identical to principal-components analysis when both are applied to correlation matrices with unity in the diagonal), the theoretical basis and emphasis of principal-components analysis are quite different. The major concern in factor analysis involves understanding the structure and relationships among the multiple measurements. Factor scores, defined as some function of the original measurements, have usually been given only secondary consideration by most investigators. In principal-components analysis, the primary emphasis is on definition of composite or factor-score variates that have certain desirable statistical properties. Formally, principal-components analysis and principal-axes factor analysis differ only in terms of the scaling of the vectors.

Although principal components are fundamentally abstract mathematical or statistical constructs and the derivations and proofs can become quite technical, the computational procedures required for application of the method are relatively straightforward. In the remaining sections of this chapter four aims are distinguished. The first is to present technical definitions that specify precisely what principal components are and make explicit some important properties of principal components. Although no attempt will be made initially to prove the validity of any of the relationships, the student or researcher planning to use principal-components analysis should be familiar with these basic definitions and relationships, whether he pursues their mathematical basis or not. The second section provides description of an iterative method for computing principal components and associated statistics. The applied researcher should understand the computational method, even though in actual research computations are likely to be accomplished by an electronic computer using standard computer programs. Following the presentation of computational procedures, the mathematical basis of principal-components analysis will be provided in greater detail for the student of applied statistics who wishes to understand how the computational procedure results in solution vectors which satisfy previously stated definitions of principal components. Readers not interested in these necessarily more technical mathematical proofs can skip this section with full confidence that the proofs will not be required for effective use of principal-components analysis in applied research. Finally, an application to clinical data will be presented in detail to illustrate the use of the method.

3.2 Definitions

In the preceding chapter it was shown that the variance of a linear combination of p correlated measurements can be expressed as a quadratic form in terms of the variances and covariances of the original variables

$$V(y) = \sum \sum c_{ij} a_i a_j = \mathbf{a}'C\mathbf{a} \tag{3.1}$$

The first principal component is defined as the vector of weighting coefficients a_i which maximizes variance $V(y)$ subject to the restriction that the sum of squares for the a_i equals a constant, which conventionally has been taken as $\sum a_i^2 = 1.0$. This restriction is obviously necessary, as the value of $V(y)$ could otherwise be increased without limit simply by multiplying the coefficients by a larger and larger constant. The function to be maximized is thus

$$f(a) = \frac{\mathbf{a}'C\mathbf{a}}{\mathbf{a}'\mathbf{a}} \tag{3.2}$$

Maximizing this function with regard to the a_i yields the following matrix equation, in which λ is a Lagrange multiplier:

$$(C - \lambda I)\mathbf{a} = 0 \quad \text{and} \quad \mathbf{a}'\mathbf{a} = 1 \tag{3.3}$$

A solution to this matrix equation implies that λ must satisfy $|C - \lambda I| = 0$. In Eq. (3.3), λ is a single (scalar) number and I is the identity matrix. This means that $\lambda I = D$ is a diagonal matrix in which all diagonal elements equal λ and all other elements are zero. In principal-components analysis, the matrix C is usually the matrix of covariances between the original variables, although it may be a correlation matrix. The vector **a** is the desired vector of weighting coefficients which will maximize the criterion function $f(a)$. If a correlation matrix is analyzed by the method of principal components, the elements of the vector **a** are weighting coefficients to be applied to original variables in standard-score form. If a covariance matrix is used, the weighting coefficients in **a** are applied to variables in raw-score form. In deference to the historical development of principal-components analysis, the following discussion of the method will assume the use of the covariance matrix. It should be recognized, however, that the method is applicable to either matrix. A brief discussion of the application of the method to correlation matrices rather than covariance matrices appears in Sec. 3.3.

It follows from Eq. (3.3) that the principal-component vector satisfies the relationship $C\mathbf{a} = \lambda\mathbf{a}$, where λ is a single scalar multiplier,

$$\begin{aligned}(C - \lambda I)\mathbf{a} &= C\mathbf{a} - \lambda\mathbf{a} \\ C\mathbf{a} &= \lambda\mathbf{a}\end{aligned} \tag{3.4}$$

This important relationship specifies that the result of multiplying the covariance matrix C by a principal-component vector is another vector the elements of which are precisely proportional to the elements in the principal-component vector. Whenever this important relationship holds, the vector $\mathbf{a}$ is a principal-component vector. The iterative solution for principal components to be presented in the next section is based upon this fundamental property of principal components. The reader should be aware that numerous methods exist for obtaining principal components of a covariance or correlation matrix. The direct mathematical solution for characteristic roots and vectors (Sec. 2.8) provides the principal components and their associated variances, but such a solution is not practical for matrices of large rank. The Jacobi method and the tridiagonalization method are feasible alternatives that may actually have some advantages over the iterative method discussed here (Greenstadt, 1960). Just as in the case of matrix inversion, where only the square-root method was described in detail, we have chosen to discuss the iterative solution for principal components and principal factors because it is straightforward and readily understood. Large matrices can be analyzed by this method, and there is no requirement that the matrix be of full rank. The number of iterations required in order for the method to stabilize on a principal component is greater when two components have similar variances. A problem with the method is that it may result in no solution if two principal components having precisely equal variances happen to be present.

The variance associated with a principal component is precisely the multiplier λ, given the restriction that $\mathbf{a}'\mathbf{a} = 1$

$$V(y) = \mathbf{a}'C\mathbf{a} = \mathbf{a}'\lambda\mathbf{a} = \lambda\mathbf{a}'\mathbf{a} = \lambda \tag{3.5}$$

This is demonstrated by premultiplying Eq. (3.4) by $\mathbf{a}'$. When a vector $\mathbf{a}$ has been obtained which satisfies the proportionality required in Eq. (3.4), the constant of proportionality λ is the variance of the principal component. From a computational point of view, let the vector obtained by postmultiplying C by the principal component vector $\mathbf{a}$ be represented by $\mathbf{b}$. Then the variance of the principal component is $V(y) = \mathbf{a}'\mathbf{b}$.

The final relationships to be considered are those of geometric and statistical orthogonality. Geometric orthogonality is present when the inner product of two vectors, say $\mathbf{a}^{(1)}$ and $\mathbf{a}^{(2)}$, is equal to zero, that is, $\mathbf{a}^{(1)\prime}\mathbf{a}^{(2)} = 0$. Statistical orthogonality refers to zero correlation or covariance between two linear combinations, $\mathbf{a}^{(1)\prime}C\mathbf{a}^{(2)} = 0$. A unique property of principal components is that successive principal-component vectors possess *both* geometric and statistical orthogonality

$$\mathbf{a}^{(i)\prime}C\mathbf{a}^{(j)} = \mathbf{a}^{(i)\prime}\mathbf{a}^{(j)} = 0 \tag{3.6}$$

To this point, relationships implicit in principal-components analysis have been defined for single vectors. These same relationships can be defined for the full set of principal components using matrix notation. Let A be a matrix containing $r \leq p$ columns which are the r principal-component vectors $\mathbf{a}^{(1)}$, $\mathbf{a}^{(2)}, \ldots, \mathbf{a}^{(r)}$. (Although in most cases a maximum of p principal components can be defined, the first r usually account for almost all variance.) Let Λ be a diagonal matrix containing, in order, the principal-component variances $\lambda_1, \lambda_2, \ldots, \lambda_r$

$$\Lambda = \begin{bmatrix} \lambda_1 & 0 & \cdots & 0 \\ 0 & \lambda_2 & \cdots & 0 \\ \cdots & \cdots & \cdots & \cdots \\ 0 & 0 & \cdots & \lambda_r \end{bmatrix}$$

It follows that

$$CA = A\Lambda \qquad \text{where } A'A = I \tag{3.7}$$

Thus,

$$A'CA = A'A\Lambda = \Lambda \tag{3.8}$$

Since the off-diagonal elements of Λ are zero, Eq. (3.8) defines statistical independence among the r principal components. The condition $A'A = I$ defines geometric orthogonality, as well as unit sums of squares, for all principal-component vectors. Only in principal-components analysis, and in related principal-axes factor analysis, is simultaneous geometric and statistical orthogonality present

$$A'A = I \qquad \text{and} \qquad A'CA = \Lambda$$

3.3 Computing Principal Components and Associated Variances

The iterative method recommended for computing principal-component vectors and associated variances involves only very simple matrix-vector multiplication. One starts with an arbitrary trial vector, multiplies the covariance matrix by it, and then normalizes the result to obtain the next trial vector. The process is repeated until two successive trial vectors are obtained which are identical to any desired number of decimal places, at which point the stabilized vector is accepted as the principal-component vector. Successive principal components are computed by partialing out variance associated with previously computed principal components and operating in this iterative manner on a residual matrix.

Let $\mathbf{t}_0 = [1 \quad 1 \quad 1 \quad 1 \quad \cdots \quad 1 \quad 1]$ be any arbitrary trial vector, not orthogonal to the first principal-component vector. It is convenient to start with $\mathbf{t}_0$ containing all unit elements or $\mathbf{t}_0$ equal to a column of the covariance matrix

$$C\mathbf{t}_0 = \mathbf{x}_0 \qquad \text{and} \qquad \mathbf{t}_1 = \frac{1}{\sqrt{\mathbf{x}_0'\mathbf{x}_0}}\mathbf{x}_0 = \mathbf{x}_0 k_0$$

$$C\mathbf{t}_1 = \mathbf{x}_1 \qquad \text{and} \qquad \mathbf{t}_2 = \frac{1}{\sqrt{\mathbf{x}_1'\mathbf{x}_1}}\mathbf{x}_1 = \mathbf{x}_1 k_1$$

$$C\mathbf{t}_2 = \mathbf{x}_2 \qquad \text{and} \qquad \mathbf{t}_3 = \frac{1}{\sqrt{\mathbf{x}_2'\mathbf{x}_2}}\mathbf{x}_2 = \mathbf{x}_2 k_2$$

....

$$C\mathbf{t}_n = \mathbf{x}_n \qquad \text{and} \qquad \mathbf{t}_{n+1} = \frac{1}{\sqrt{\mathbf{x}_n'\mathbf{x}_n}}\mathbf{x}_n = \mathbf{x}_n k_n$$

Iteration is terminated when $\mathbf{t}_n = \mathbf{t}_{n+1}$ to a prespecified number of decimal places. The final trial vector $\mathbf{t}_n$ or $\mathbf{t}_{n+1}$ is then accepted as the principal-component vector having the largest associated variance. When $\mathbf{t}_n = \mathbf{t}_{n+1} = \mathbf{a}$, we have

$$\mathbf{t}_{n+1} = \mathbf{a} = \frac{1}{\sqrt{\mathbf{x}_n'\mathbf{x}_n}}\mathbf{x}_n$$

and thus

$$\mathbf{x}_n = \sqrt{\mathbf{x}_n'\mathbf{x}_n}\,\mathbf{a} = \lambda\mathbf{a}$$

The final iteration yielding $C\mathbf{t}_n = \mathbf{x}_n$ thus clearly satisfies the definition represented in Eq. (3.4)

$$C\mathbf{a} = \lambda\mathbf{a} \qquad \text{where } \mathbf{a} = \mathbf{t}_n = \mathbf{t}_{n+1} \text{ and } \lambda = \sqrt{\mathbf{x}_n'\mathbf{x}_n}$$

In order to compute a second principal-component vector orthogonal to the first, the variance associated with the first vector is partialed out of the covariance matrix C

$$C^{(1)} = C - \lambda_1\mathbf{a}^{(1)}\mathbf{a}^{(1)\prime}$$

In scalar notation, each element in the residual matrix $C^{(1)}$ is equal to the corresponding element in C minus λ_1 times the product of the two elements in the principal-component vector.

$c_{ij}^{(1)} = c_{ij} - \lambda_1 a_i^{(1)} a_j^{(1)}$, where c_{ij} is the element in the ith row and jth column of C and $a_i^{(1)}$ and $a_j^{(1)}$ are the ith and jth elements, respectively, of the first principal-component vector.

The second principal-component vector and associated variance λ_2 are computed exactly as described for the first principal component, except that trial vectors are applied to $C^{(1)}$ instead of C. Successive other principal components are obtained in the same manner from successive residual matrices in which all variance associated with previously obtained vectors has been partialed out

$$C^{(1)} = C - \lambda_1 \mathbf{a}^{(1)}\mathbf{a}^{(1)\prime}$$
$$C^{(2)} = C^{(1)} - \lambda_2 \mathbf{a}^{(2)}\mathbf{a}^{(2)\prime}$$
$$C^{(3)} = C^{(2)} - \lambda_3 \mathbf{a}^{(3)}\mathbf{a}^{(3)\prime}$$
$$\cdots\cdots\cdots\cdots\cdots\cdots$$
$$C^{(r)} = C^{(r-1)} - \lambda_r \mathbf{a}^{(r)}\mathbf{a}^{(r)\prime}$$

It will be noted that, at the rth stage, the residual matrix is

$$C^{(r)} = C - \lambda_1 \mathbf{a}^{(1)}\mathbf{a}^{(1)\prime} - \lambda_2 \mathbf{a}^{(2)}\mathbf{a}^{(2)\prime} \cdots - \lambda_r \mathbf{a}^{(r)}\mathbf{a}^{(r)\prime}$$

Without offering proof, it should be noted that the maximum number of possible principal components is equal to the number of variables represented in the covariance matrix. Thus, when p principal components have been defined, all elements in the final residual matrix will be equal to zero. This means that all information in the covariance matrix can be represented in terms of the p principal components and their associated variances

$$C = \lambda_1 \mathbf{a}^{(1)}\mathbf{a}^{(1)\prime} + \lambda_2 \mathbf{a}^{(2)}\mathbf{a}^{(2)\prime} + \cdots + \lambda_p \mathbf{a}^{(p)}\mathbf{a}^{(p)\prime}$$

In matrix notation, this result can be written

$$C = A\Lambda A' \tag{3.9}$$

Thus, the complete covariance matrix can be reconstructed as the product of the matrix containing principal-component vectors and the Λ matrix.

In actuality, the statement that the maximum number of principal components is equal to the number of variables represented in the covariance matrix is true only if the covariance matrix is of full rank, which it usually is in the case of real data. A more accurate statement is that the total number of principal components is equal to the rank of the covariance matrix. Reference to Sec. 2.5 will reveal that Eq. (3.9) is equivalent to Eq. (2.17) if one defines $B = \Lambda A'$. The remainder of Sec. 2.5 concerned with definition of the rank of a matrix follows. From this it can be concluded that the method of principal components is a convenient method for determining the rank of a covariance or correlation matrix. The rank of the matrix is equal to the number of principal components whose associated variances λ_i exceed zero. The *effective rank* of a covariance or correlation matrix is equal to the number of principal components having associated λ_i greater than some predefined small value, say $\lambda_i \geq 1.0$. As will be seen, determination of the effective rank of a matrix becomes important in factor analysis, where a primary problem centers about the number of factors necessary to account for correlations among p variables.

In the preceding discussion, the method of principal components was described with reference to analysis of a covariance matrix. It should be emphasized that the method can be applied to a matrix of intercorrelations among p variables and that all the relationships defined in Sec. 3.3 hold when R

is substituted for C in the various equations. The question of when to apply principal-components analysis to a correlation matrix and when to a covariance matrix deserves consideration.

Historically, principal-components analysis was developed by statisticians (Hotelling, 1933, 1936) who discussed its application to covariance matrices. However, users have frequently been cautioned that application to problems in which the several variables have grossly different variances may not be meaningful (Anderson, 1958). The method of principal-components analysis is specifically concerned with accounting for a maximum portion of the *variances* represented in the several measurements. If one or two variables have disproportionately large variances, the principal-components solution will be "pulled" in the direction of accounting for those variances.

In many practical applications, the interest is not one of accounting for variance per se but is instead a concern for the individual differences represented by the multiple measurements. The statistical variance of a variable is dependent on the unit of measurement. For example, height can be recorded in inches or millimeters. Given careful measurement, the same individual differences can be represented in either unit; however, the *variance* of millimeter measurements will be much larger. Where interest is in accounting for individual differences, it is frequently desirable to convert all measurements to units which are standard in terms of the individual differences. Calculation of product-moment correlation coefficients is precisely equivalent to calculation of covariances among variables that have first been standardized, or put in z-score form. Where variances are very disparate among several measurements, and where the aim is to define principal components which maximally account for individual differences, not just statistical variance, the application of principal-components analysis to the matrix of intercorrelations is recommended.

If the matrix which is subjected to principal-components analysis is a correlation matrix, it should be noted that the normalized principal-component vectors (the $\mathbf{a}$ vectors of Sec. 3.3) contain weighting coefficients to be applied to *standard scores* in defining the new composite functions. To obtain sets of weighting coefficients that will define the same composite variables in terms of the original *raw scores*, one must divide each element by the standard deviation of the variable to which it is applied.

3.4 Example of Principal-components Calculations

A small matrix of order 4×4 has been chosen to illustrate the actual calculations involved in the iterative solution for principal components. Scores for symptoms of thinking disturbance, withdrawal-retardation, hostile-suspiciousness, and anxious depression were computed from BPRS symptom ratings for each of 6,000 patients. Covariances among the four scores are shown in Table 3.1.

The iterative solution involved an initial trial vector $\mathbf{t}_0' = [1 \quad 1 \quad 1 \quad 1]$, shown as the first row in Table 3.1. Multiplication of the matrix C by the first

Table 3.1 Computation of First Principal Component†

TRIAL VECTORS

$\mathbf{t}_0' = [1.000 \quad 1.000 \quad 1.000 \quad 1.000]$
$\mathbf{t}_1' = [\ .541 \quad .458 \quad .542 \quad .449]$
$\mathbf{t}_2' = [\ .564 \quad .441 \quad .557 \quad .419]$
. . . .
$\mathbf{t}_8' = [\ .593 \quad .432 \quad .563 \quad .379]$
$\mathbf{t}_9' = [\ .594 \quad .432 \quad .563 \quad .378]$

ORIGINAL COVARIANCE MATRIX C

$$\begin{bmatrix} 20.7504 & 5.9007 & 7.9047 & 0.6875 \\ 5.9007 & 17.3257 & 3.9803 & 2.6663 \\ 7.9047 & 3.9803 & 17.6308 & 5.8351 \\ 0.6875 & 2.6663 & 5.8351 & 20.0666 \end{bmatrix}$$

PRODUCT VECTORS AND SCALING COEFFICIENTS

$\mathbf{x}_0' = [35.2433 \quad 29.8730 \quad 35.3509 \quad 29.2555] \qquad \mathbf{x}_0'\mathbf{x}_0 = 4{,}240.0567$
$\mathbf{x}_1' = [18.5383 \quad 14.5009 \quad 18.2976 \quad 13.7787] \qquad \mathbf{x}_1'\mathbf{x}_1 = 1{,}078.6061$
$\mathbf{x}_2' = [19.0107 \quad 14.3168 \quad 18.4902 \quad 13.2351] \qquad \mathbf{x}_2'\mathbf{x}_2 = 1{,}083.4421$
. . . .
$\mathbf{x}_8' = [19.5886 \quad 14.2443 \quad 18.5527 \quad 12.4390] \qquad \mathbf{x}_8'\mathbf{x}_8 = 1{,}085.5524$
$\mathbf{a}^{(1)\prime} = [.594 \quad .432 \quad .563 \quad .378]$ and $\lambda_1 = \sqrt{\mathbf{x}_8'\mathbf{x}_8} = 32.947$

† To duplicate these results exactly by hand calculations, it will be necessary to carry more decimal places at each stage than are shown in the example. Calculations for this example were actually done by computer with numbers truncated at three and four decimal places in tabling to conserve space.

trial vector resulted in the product vector labeled $\mathbf{x}_0'$, which has been entered below the original covariance matrix. According to the row-by-column rule for vector-matrix multiplication, each element in $\mathbf{x}_0'$ is the sum of products of elements in $\mathbf{t}_0$ with corresponding elements in the *column* of C appearing directly above. For example,

$$35.2433 = 1.000(20.7594) + 1.000(5.9007) + 1.000(7.9047) + 1.000(.6875)$$

The next step was to compute the scaling coefficient $1/\sqrt{\mathbf{x}_0'\mathbf{x}_0} = 1/\sqrt{\sum x^2}$. The value for $\mathbf{x}_0'\mathbf{x}_0 = \sum x^2 = 4{,}240.0567$ appears beside row $\mathbf{x}_0'$. The second trial vector was obtained by dividing each element in $\mathbf{x}_0'$ by $\sqrt{\mathbf{x}_0'\mathbf{x}_0}$, which is equivalent, of course, to multiplying each element in $\mathbf{x}_0'$ by the scalar constant $k_0 = 1/\sqrt{\mathbf{x}_0'\mathbf{x}_0}$. The second trial vector $\mathbf{t}_1'$ is entered below the first trial vector $\mathbf{t}_0'$.

Multiplication of matrix C by the vector $\mathbf{t}_1'$ yielded the vector labeled $\mathbf{x}_1'$ that is entered below the covariance matrix. Again, each element in $\mathbf{x}_1'$ is equal to the sum of elements in one *column* of C, each multiplied by the corresponding element from $\mathbf{t}_1$. The sum of squares of elements in $\mathbf{x}_1'$ is 1,078.6061. Division of each element in $\mathbf{x}_1'$ by the square root of this value resulted in the third trial vector labeled $\mathbf{t}_2'$.

The process of multiplying the original covariance matrix C by successive trial vectors and scaling the product vector by $1/\sqrt{\sum x^2}$ to obtain the next trial vector was continued until two consecutive trial vectors $\mathbf{t}_8'$ and $\mathbf{t}_9'$ agreed to three decimal places. At this point, the final trial vector $\mathbf{t}_9'$ was accepted as the first principal-component vector, and the value $\sqrt{\mathbf{x}_8'\mathbf{x}_8} = \sqrt{\sum x^2} = 32.947$ was accepted as the principal-component variance $\lambda_1 = 32.947$.

Next, the first residual covariance matrix was computed by subtracting the product of the ith and jth elements in the first principal-component vector multiplied by the variance λ_1

$$c_{ij}^{(1)} = c_{ij} - \lambda_1 a_i^{(1)} a_j^{(1)}$$

For example, the element in the second row and third column of the first residual matrix is $c_{23}^{(1)} = -4.0406$. This value was obtained from the corresponding value in the original covariance matrix as follows:

$$c_{23}^{(1)} = 3.9803 - 32.947(.432)(.563) = -4.041$$

Other elements in $C^{(1)}$ were obtained from the corresponding elements in C by a similar operation. Actual calculations were carried out with greater precision than indicated by the three decimal places provided in the example.

To compute the second principal-component vector and associated variance, iteration was commenced with trial vector $\mathbf{t}_0' = [1 \quad 1 \quad 1 \quad 1]$, and the sequence of operations was exactly as before. The vector $\mathbf{x}_0'$ was computed from the

Table 3.2 Computation of Second Principal Component

TRIAL VECTORS

$$\begin{aligned}
\mathbf{t}_0' &= [\ 1.000 \quad 1.000 \quad 1.000 \quad 1.000] \\
\mathbf{t}_1' &= [-.531 \quad .297 \quad -.185 \quad .770] \\
\mathbf{t}_3' &= [-.554 \quad .182 \quad -.096 \quad .806] \\
&\cdots \\
\mathbf{t}_{13}' &= [-.516 \quad -.149 \quad .096 \quad .837] \\
\mathbf{t}_{14}' &= [-.515 \quad -.151 \quad .097 \quad .837]
\end{aligned}$$

FIRST RESIDUAL COVARIANCE MATRIX $C^{(1)}$

$$\begin{bmatrix} 9.1042 & -2.5681 & -3.1256 & -6.7079 \\ -2.5681 & 11.1673 & -4.0406 & -2.7114 \\ -3.1256 & -4.0406 & 7.1837 & -1.1692 \\ -6.7079 & -2.7114 & -1.1692 & 15.3704 \end{bmatrix}$$

PRODUCT VECTORS AND SCALING COEFFICIENTS

$$\begin{aligned}
\mathbf{x}_0' &= [\ -3.2975 \quad 1.8470 \quad -1.1518 \quad 4.7816] & \mathbf{x}_0'\mathbf{x}_0 &= 38.4763 \\
\mathbf{x}_1' &= [-10.1951 \quad 3.3507 \quad -1.7769 \quad 14.8243] & \mathbf{x}_1'\mathbf{x}_1 &= 338.0856 \\
&\cdots \\
\mathbf{x}_{13}' &= [-10.2375 \quad -3.0004 \quad 1.9271 \quad 16.6332] & \mathbf{x}_{13}'\mathbf{x}_{13} &= 394.1939 \\
\mathbf{a}^{(2)\prime} &= [-.515 \quad -.151 \quad .097 \quad .837] \quad \text{and} & \lambda_2 = \sqrt{\mathbf{x}_{13}'\mathbf{x}_{13}} &= 19.854
\end{aligned}$$

equation $\mathbf{t}_0'C^{(1)} = \mathbf{x}_0'$. The sum of squares $\sum x^2 = \mathbf{x}_0'\mathbf{x}_0 = 38.4763$ was computed from elements in $\mathbf{x}_0'$. The second trial vector $\mathbf{t}_1'$ was obtained by dividing elements in $\mathbf{x}_0'$ by the square root of $\sum x^2 = 38.4763$. Iteration was continued until elements in trial vectors $\mathbf{t}_{13}'$ and $\mathbf{t}_{14}'$ were found to agree closely in the third decimal place. At this point, the final trial vector $\mathbf{t}_{14}'$ was accepted as the second principal-component vector, and the square root of the final sum of squares $\mathbf{x}_{13}'\mathbf{x}_{13} = 394.1882$ was taken as the associated variance, $\lambda_2 = 19.854$.

Table 3.3 Computation of Third Principal Component

TRIAL VECTORS

$\mathbf{t}_0' = [\ 1.000 \quad 1.000 \quad 1.000 \quad 1.000]$
$\mathbf{t}_1' = [-.175 \quad .828 \quad -.521 \quad .103]$
. . . .
$\mathbf{t}_9' = [-.261 \quad .867 \quad -.420 \quad .045]$
$\mathbf{t}_{10}' = [-.263 \quad .867 \quad -.418 \quad .044]$

SECOND RESIDUAL COVARIANCE MATRIX $C^{(2)}$

$$\begin{bmatrix} 3.8395 & -4.1365 & -2.1220 & 1.8561 \\ -4.1365 & 10.7001 & -3.7417 & -.1601 \\ -2.1321 & -3.7487 & 6.9987 & -2.7841 \\ 1.8561 & -.1601 & -2.8018 & 1.4392 \end{bmatrix}$$

PRODUCT VECTORS AND SCALING COEFFICIENTS

$\mathbf{x}_0' = [-.5532 \quad 2.6492 \quad -1.6662 \quad 0.3211] \qquad \mathbf{x}_0'\mathbf{x}_0 = 10.3122$
. . . .
$\mathbf{x}_9' = [-3.6171 \quad 11.9315 \quad -5.7578 \quad .6186] \qquad \mathbf{x}_9'\mathbf{x}_9 = 188.9825$
$\mathbf{a}^{(3)\prime} = [-.263 \quad .867 \quad -.418 \quad .044]$ and $\lambda_3 = \sqrt{\mathbf{x}_9'\mathbf{x}_9} = 13.747$

Table 3.4 Computation of Fourth Principal Component

TRIAL VECTORS

$\mathbf{t}_0' = [1.000 \quad 1.000 \quad 1.000 \quad 1.000]$
$\mathbf{t}_1' = [\ .558 \quad -.190 \quad -.706 \quad .391]$
$\mathbf{t}_2' = [\ .558 \quad -.190 \quad -.706 \quad .391]$

THIRD RESIDUAL COVARIANCE MATRIX $C^{(3)}$

$$\begin{bmatrix} 2.8780 & -.9794 & -3.6386 & 2.0163 \\ -.9794 & .3333 & 1.2382 & -.6861 \\ -3.6386 & 1.2382 & 4.6001 & -2.5491 \\ 2.0163 & -.6861 & -2.5491 & 1.4125 \end{bmatrix}$$

PRODUCT VECTORS AND SCALING COEFFICIENTS

$\mathbf{x}_0' = [\ .2762 \quad -.0940 \quad -.3493 \quad .1935] \qquad \mathbf{x}_0'\mathbf{x}_0 = 0.2446$
$\mathbf{x}_1' = [5.1524 \quad -1.7534 \quad -6.5140 \quad 3.6096] \qquad \mathbf{x}_1'\mathbf{x}_1 = 85.1375$
$\mathbf{a}^{(4)\prime} = [.558 \quad -.190 \quad -.706 \quad .391]$ and $\lambda_4 = \sqrt{x_1 x_1} = 9.224$

It can be verified that the sum of the four principal-component variances $\lambda_1 + \lambda_2 + \lambda_3 + \lambda_4 = 75.772$ is equal to the sum of the elements in the principal diagonal of the original covariance matrix. The diagonal elements of a covariance matrix are variances of the several variables, and the sum of the diagonal elements is frequently referred to as the *total variance* represented in the matrix. The method of principal-components analysis partitions the total variance into p additive components. The proportion of total variance accounted for by each principal component can be calculated by dividing the principal-component variance λ_i by the sum of the diagonal elements from the original covariance matrix.

It is worth noting that the iterative multiplication by trial vectors is tantamount to raising the covariance matrix C to the nth power

$$C\mathbf{t}_0 = \mathbf{x}_0$$

and

$$\mathbf{t}_1 = \mathbf{x}_0 k_0 \qquad \text{where } k_0 = \frac{1}{\sqrt{\mathbf{x}_0'\mathbf{x}_0}}$$

$$C\mathbf{t}_1 = C\mathbf{x}_0 k_0 = CC\mathbf{t}_0 k_0 = \mathbf{x}_1$$

and

$$\mathbf{t}_2 = \mathbf{x}_1 k_1 \qquad \text{where } k_1 = \frac{1}{\sqrt{\mathbf{x}_1'\mathbf{x}_1}}$$

$$C\mathbf{t}_2 = C\mathbf{x}_1 k_1 = CCC\mathbf{t}_0 k_0 k_1 = \mathbf{x}_2$$

and

$$\mathbf{t}_3 = \mathbf{x}_2 k_2 \qquad \text{where } k_2 = \frac{1}{\sqrt{\mathbf{x}_2'\mathbf{x}_2}}$$

$$C\mathbf{t}_3 = C\mathbf{x}_2 k_2 = CCCC\mathbf{t}_0 k_0 k_1 k_2 = \mathbf{x}_3$$

and

$$\mathbf{t}_4 = \mathbf{x}_3 k_3 \qquad \text{where } k_3 = \frac{1}{\sqrt{\mathbf{x}_3'\mathbf{x}_3}}$$

In general,

$$C\mathbf{t}_n = CC^n\mathbf{t}_0 k_0 k_1 \cdots k_n = CC^n\mathbf{t}_0 \left(\prod_j^n k_j\right) \tag{3.10}$$

where $\prod$ denotes the continued product, in this instance, $k_0 k_i \cdots k_n$, so that $\mathbf{t}_n = C^n\mathbf{t}_0 q$, where $q = k_0 k_1 k_2 \cdots k_n = \prod_j^n k_j$.

The fact that iteration with an arbitrary trial vector is equivalent to powering the covariance matrix is the basis for understanding how the iterative procedure actually converges on the largest principal component of either the original or the residual matrix. This fact also provides the basis for a computational shortcut, frequently recommended for calculation of principal components, which involves squaring or cubing the initial covariance matrix prior to iteration with trial vectors.

Let $\hat{C} = CC = C^2$. Iteration with trial vector 1, 2, . . . , n yields the result

$$\hat{C}\mathbf{t}_n = \hat{C}\hat{C}^n\mathbf{t}_0\left(\prod^n k_i\right)$$

Since $\hat{C} = C^2$, n iterations performed on $\hat{C}$ are equivalent to $2n + 1$ iterations using the original matrix C

$$\hat{C}\hat{C}^n\mathbf{t}_0\left(\prod^n k_i\right) = \hat{C}C^{2n}\mathbf{t}_0\left(\prod^n k_i\right) = \lambda^2\mathbf{a}$$

Similarly, if one first raises C to the third power, $\hat{\hat{C}} = CCC = C^3$, then n iterations performed on $\hat{\hat{C}}$ are equivalent to $3n + 2$ iterations performed on the original matrix C

$$\hat{\hat{C}}\hat{\hat{C}}^n\mathbf{t}_0\left(\prod^n k_i\right) = \hat{\hat{C}}C^{3n}\mathbf{t}_0\left(\prod^n k_i\right) = \lambda^3\mathbf{a}$$

The extent to which powering of C prior to iteration for principal components facilitates computation will depend on the characteristics of the particular matrix being analyzed, its size, the number of significant principal components, and the differences in variance accounted for by successive principal components. While powering of the covariance matrix may not be universally important, the mathematical basis for this shortcut is interesting. It should be noted that λ_i variances associated with normalized principal-component vectors obtained from matrices that have been squared or cubed prior to iterative solution are thus raised to the second or third power also.

3.5 Proof That Iteration with an Arbitrary Trial Vector Results in Identification of the Principal Component Having the Largest Variance

A very important attribute of the iterative solution for principal components is the fact that the vector accounting for the largest (remaining) variance is obtained at each stage. This is important because interest is frequently limited to defining only the largest components of variance. In this section a proof is offered that iteration converges on the largest remaining principal component. Since it is obvious that any vector which is proportional to a principal-component vector can be normalized to satisfy $\mathbf{a}'\mathbf{a} = 1$, it is necessary to prove only that iteration leads to a vector that is proportional within a scaling factor to the largest principal component.

It was previously specified in Eq. (3.9) that $C = A\Lambda A'$, where A is a matrix containing as columns the successively defined principal components of C and Λ is a diagonal matrix containing variances associated with the principal components. From the definition of principal components given in Eq. (3.7), it can be deduced that raising C to a successively higher power is reflected in corresponding powering of the variances of the principal components

$$CA = A\Lambda$$

$$CCA = CA\Lambda = A\Lambda\Lambda$$

$$CCCA = CA\Lambda\Lambda = A\Lambda\Lambda\Lambda$$

In general,

$$C^n A = A\Lambda^n \tag{3.11}$$

From Eq. (3.10), it is apparent that for the nth iteration we have

$$C\mathbf{t}_{n-1} = CC^{n-1}\mathbf{t}_0\left(\prod^{n-1} k_i\right) = C^n\mathbf{t}_0 q$$

where $q = \prod\limits^{n-1} k_i$ is a composite scaling constant. Substituting $A\Lambda^n A' = C^n$, we obtain

$$C^n\mathbf{t}_0 q = A\Lambda^n A'\mathbf{t}_0 q$$

The aim is to show that $C\mathbf{t}_{n-1}$ approaches $\lambda_1\mathbf{a}^{(1)}$ as n is increased without limit. In mathematical terms, the goal is to prove that

$$C\mathbf{t}_{n-1} = A\Lambda^n A'\mathbf{t}_0 q \rightarrow \lambda_1\mathbf{a}^{(1)}$$

where $\mathbf{a}^{(1)}$ is the principal-component vector having largest variance λ_1. That is, the aim is to prove that trial vector $\mathbf{t}_{n-1}$ approaches the principal-component vector $\mathbf{a}^{(1)}$ which has largest associated variance λ_1.

Inserting the trivial ratio $\lambda_1{}^n/\lambda_1{}^n = 1$ into the above equation, we obtain

$$C\mathbf{t}_{n-1} = \lambda_1{}^n A\left(\frac{1}{\lambda_1}\Lambda\right)^n A'\mathbf{t}_0 q$$

The limit approached by $[(1/\lambda_1)\Lambda]^n$ is a diagonal matrix containing nth powers of the ratios of successive principal-component variances to the largest variance λ_1

$$\lim_{n\to\infty}\left(\frac{1}{\lambda_1}\Lambda\right)^n = \lim_{n\to\infty}\begin{bmatrix} \left(\frac{\lambda_1}{\lambda_1}\right)^n & \cdots & \cdots \\ \cdots & \left(\frac{\lambda_2}{\lambda_1}\right)^n & \cdots \\ \cdots & \cdots & \cdots \\ \cdots & \cdots & \left(\frac{\lambda_p}{\lambda_1}\right)^n \end{bmatrix}$$

It should be noted that every ratio except the first is a fraction with value less than 1, since λ_1 is the largest principal variance. As n is increased, fractional values raised to the nth power become smaller and approach zero as a limit

$$\lim_{n\to\infty}\left(\frac{\lambda_i}{\lambda_1}\right)^n = 0 \qquad \text{for all } i \neq 1$$

$$\lim_{n\to\infty}\left(\frac{1}{\lambda_1}\Lambda\right)^n = \begin{bmatrix} 1 & 0 & 0 & \cdots & 0 \\ 0 & 0 & 0 & \cdots & 0 \\ 0 & 0 & 0 & \cdots & 0 \\ 0 & 0 & 0 & \cdots & 0 \end{bmatrix}$$

$$\lim_{n\to\infty} \lambda_1{}^n A\left(\frac{1}{\lambda_1}\Lambda\right)^n A'\mathbf{t}_0 q = \lim_{n\to\infty} \lambda_1{}^n A \begin{bmatrix} 1 & & & \\ & 0 & & \\ & & 0 & \\ & & & \cdot \\ & & & & \cdot \end{bmatrix} A'\mathbf{t}_0 q$$

and the matrix

$$A \begin{bmatrix} 1 & & & \\ & 0 & & \\ & & 0 & \\ & & & \cdot \\ & & & & \cdot \end{bmatrix} A' = [\mathbf{a}^{(1)} \quad 0 \quad 0 \quad \cdots] A' = \mathbf{a}^{(1)}\mathbf{a}^{(1)\prime}$$

Thus,

$$\lim_{n\to\infty} C\mathbf{t}_{n-1} = \lim_{n\to\infty} \lambda_1{}^n \mathbf{a}^{(1)}\mathbf{a}^{(1)\prime}\mathbf{t}_0 q = \lim_{n\to\infty} \lambda_1{}^n \mathbf{a}^{(1)} cq$$

where $c = \mathbf{a}^{(1)\prime}\mathbf{t}_0$ is the scalar product of two previously defined vectors.

This proves that the vector approached by the product $C\mathbf{t}_{n-1}$ is a vector that is proportional to the largest principal-component vector, since the product $s = \lambda_1{}^n cq$ is a scalar number. If $\mathbf{t}_{n-1}$ is a normalized trial vector with elements proportional to the elements in the principal-component vector, then from Eq. (3.4) it follows that the scaling constant approaches $1/\lambda_1^{n-1}$ as n is increased without limit

$$cq \rightarrow \frac{1}{\lambda_1^{n-1}}$$

Thus, if

$$cq = a^{(1)\prime}\mathbf{t}_0\left(\prod^{n-1} k_i\right) = \frac{1}{\lambda_1^{n-1}}$$

in the limit, then

$$\lim_{n\to\infty} C\mathbf{t}_{n-1} = \lambda_1 \mathbf{a}^{(1)}$$

3.6 An Extended Example

3.6.1 PRINCIPAL COMPONENTS OF PSYCHIATRIC SYMPTOM PROFILE PATTERNS

The purpose of this section is to discuss an application of principal-components analysis in defining the most descriptive contrast functions for representing differences in symptom profile *patterns.* The method of principal-components analysis was chosen to define the single weighted combination of symptom-rating variables which accounts for maximum variance in symptom profile patterns. Extended beyond the first dimension, the method was used to define a maximally parsimonious descriptive model. This application is somewhat different from most principal-components analyses, which are concerned with accounting for *total* multivariate profile differences, because in this investigation interest was in deriving composite functions descriptive of differences in *pattern* only.

The profile attributes of *elevation* and *pattern* have been distinguished frequently in psychological and psychiatric research (Cronbach and Gleser, 1953). *Elevation* refers to the average level of all scores, e.g., symptom ratings in the profile. The index of elevation is the mean computed across all scores within the individual profile. The concept of profile *pattern* is concerned with the configuration of deviations of individual scores about the mean value for all scores in the profile. With reference to psychiatric symptom profiles, elevation is associated with general level of severity. Pattern is associated with the particular nature or type of psychopathology. In the investigation to be described, interest was in definition of composite functions having maximum utility for description of differences in type of psychopathology *independent of average level of severity.*

The data consisted of BPRS profiles for 6,000 patients available in an accumulated data bank. The BPRS contains 16 symptom-rating constructs, each represented by a 7-point scale of severity (Chap. 1). Preliminary to the principal-components analysis, the mean value on all 16 ratings for each patient was subtracted from each score within his profile. It will be recognized that the individual profile mean is a linear combination of the several symptom ratings

$$y_m = ax_1 + ax_2 + \cdots + ax_{16} \qquad \text{where } a = \tfrac{1}{16}$$

By subtraction of the individual profile mean, a set of 16 deviation scores was obtained, which are independent of the linear function y_m. Any linear combination of the resulting *deviation* scores will be statistically orthogonal to the equally weighted function y_m. Thus, linear functions of the elevation-corrected profile scores will be descriptive of pattern, independent of elevation.

In view of the large data bank available, it was considered worthwhile to divide the total sample into four independent, randomly drawn subsamples $n = 1{,}500$ for purposes of evaluating the stability of the principal-components results. A separate principal-components analysis was computed using the *covariance* matrix computed from the 1,500 mean-corrected profiles in each subsample.

3.6.2 THE SCHIZO-DEPRESSIVE CONTRAST FUNCTION

The first principal-component vectors from each of the four analyses are presented in Table 3.5. These vectors contain the weighting coefficients which define composite variables having maximum utility for describing individual differences in profile patterns within the four independent samples. Examination of the four sets of coefficients reveals remarkable consistency in patterns. On the average, they accounted for 22.8 percent of the total variance; i.e., the variance λ_1 was equal to 22.8 percent of the sum of diagonal elements in the covariance matrix. The first principal component of BPRS profile pattern can be described as the *contrast* between symptoms of anxious depression, on the one hand, and symptoms of thinking disturbance on the other. In a heterogeneous clinical population, the single most descriptive profile characteristic involves the relative prominence of these two types of symptoms.

A good approximation to the first principal component of BPRS symptom profile patterns is the simple unweighted difference in ratings for symptoms of conceptual disorganization, hallucinatory behavior, and unusual thought content vs. ratings for anxiety, guilt feelings, and depressive mood. Because these two sets of symptoms correspond generally to recognized target symptoms of schizophrenia and depression, the function will be referred to as the schizo-depression (S-D) contrast.

$$\text{S-D} = (\text{conceptual disorganization} + \text{hallucinatory behavior} + \text{unusual thought content}) - (\text{anxiety} + \text{guilt feelings} + \text{depressive mood})$$

Table 3.5 First Principal-component Vectors from Analyses of Four Independent Samples of Elevation-corrected BPRS Profiles

	1	*2*	*3*	*4*
Somatic concern	−.32	−.30	−.31	−.32
Anxiety	−.40	−.39	−.39	−.39
Emotional withdrawal	.23	.23	.17	.24
Conceptual disorganization	.39	.38	.38	.39
Guilt feelings	−.26	−.24	−.27	−.27
Tension	−.15	−.13	−.12	−.09
Mannerisms and posturing	.14	.15	.17	.13
Grandiosity	.06	.07	.08	.06
Depressive mood	−.45	−.48	−.49	−.48
Hostility	−.12	−.14	−.07	−.10
Suspiciousness	.10	.07	.14	.09
Hallucinatory behavior	.18	.22	.22	.21
Motor retardation	−.03	−.03	−.08	−.06
Uncooperativeness	.10	.04	.09	.10
Unusual thought content	.31	.34	.30	.31
Blunted affect	.23	.20	.19	.17

To examine the validity of the simple unweighted contrast for representation of the first principal component, product-moment correlation coefficients were computed across all 6,000 cases between the simple contrast and each of the four weighted functions defined in Table 3.5. The simple unweighted S-D contrast was found to correlate above $r = .98$ with each of the four empirically derived first principal-component functions. It has long been recognized that very precise differential weighting is not important in dealing with relatively unreliable psychological measurements (Guilford, 1954). On the basis of these results, it is proposed that the simple S-D contrast, which is an index of the relative prominence of three symptoms of thinking disturbance vs. three symptoms of anxious depression, should be used to represent the major dimension of profile-pattern difference. In the general psychiatric population, possibly excluding private outpatients, the most descriptive single statement that can be made is to characterize the patient with regard to relative prominence of thinking disturbance vs. anxious depression. This suggests that the first and most important diagnostic classification decision in the area of functional psychiatric disorders should involve consideration of the relative prominence of these two types of symptoms.

3.6.3 THE COPING-RESIGNATION CONTRAST FUNCTION

Table 3.6 lists vectors representing the second principal component from each of the four independent analyses. On the average, they accounted for 15.1 percent of the total variance; i.e., the variance λ_2 was 15.1 percent of the sum of diagonal elements in the original covariance matrix. Once again, the degree of consistency among the four sets of coefficients is impressive. The second

Table 3.6 Second Principal-component Vectors from Analyses of Four Independent Samples of Elevation-corrected BPRS Profiles

	1	*2*	*3*	*4*
Somatic concern	−.03	.01	−.02	−.03
Anxiety	.13	.15	.22	.15
Emotional withdrawal	−.45	−.45	−.45	−.41
Conceptual disorganization	.09	.07	.03	.05
Guilt feelings	−.03	−.04	−.03	−.01
Tension	.05	.09	.15	.14
Mannerisms and posturing	−.12	−.08	−.10	−.14
Grandiosity	.25	.18	.12	.19
Depressive mood	−.15	−.16	−.12	−.14
Hostility	.28	.26	.28	.29
Suspiciousness	.33	.34	.36	.32
Hallucinatory behavior	.17	.16	.15	.16
Motor retardation	−.40	−.42	−.39	−.41
Uncooperativeness	−.02	.03	−.04	.03
Unusual thought content	.33	.32	.31	.30
Blunted affect	−.43	−.45	−.46	−.49

principal component of mean-corrected profile patterns in the general psychiatric population appears well characterized as a contrast between symptoms of withdrawal, retardation, and blunted affect on the one hand and hostility, suspiciousness, and paranoid-type thinking disturbance on the other. Among patients who score toward the thinking-disturbance end of the first principal dimension (S-D), the second principal component appears to characterize the difference between hostile-suspicious paranoid types and primary withdrawn blunted-affect core schizophrenics (Overall et al., 1963). Among patients scoring toward the depression end of the first principal dimension, the second principal component appears to characterize the difference between hostile depressions and withdrawn-retarded depressions, with those in the center falling into the simple anxious-depression group (Overall et al., 1966).

For practical purposes, the second principal component of BPRS profile patterns can be represented as a simple unweighted contrast between symptoms of hostility, suspiciousness, and unusual thought content on the one hand and symptoms of emotional withdrawal, motor retardation, and blunted affect on the other. This continuum can be conceived as representing the contrast between active, but maladaptive, coping behavior vs. withdrawal and resignation. This second major dimension of individual difference in BPRS profile pattern will be termed the coping-resignation (C-R) contrast.

$$\text{C-R} = (\text{hostility} + \text{suspiciousness} + \text{unusual thought content}) - (\text{emotional withdrawal} + \text{motor retardation} + \text{blunted affect})$$

A product-moment correlation coefficient between the simple unweighted C-R contrast and each of the four weighted functions defined by the second principal component of each of the four sets of mean-corrected BPRS profiles was computed. In each instance, the correlation between the simple unweighted contrast and the empirically derived second principal component exceeded .98. As a result, it is concluded that the simple C-R contrast is an adequate representation of the second major dimension of difference among BPRS profile patterns.

3.6.4 TWO-DIMENSIONAL MODEL OF PSYCHOPATHOLOGY

The first two principal components of symptom profile pattern provide the bases for a maximally powerful model useful for examination of differences in multivariate profile patterns and for classification of patients. The two linear functions define coordinate axes in a two-dimensional plane. Because these axes are the first two principal components of profile pattern, the two-dimensional space provides for maximum separation of individuals who differ in profile pattern

$$\begin{aligned} y^{(1)} = -&.31x_1 - .40x_2 + .24x_3 + .39x_4 - .26x_5 - .13x_6 + .16x_7 + .06x_8 \\ &- .47x_9 - .12x_{10} + .08x_{11} + .19x_{12} - .03x_{13} + .09x_{14} + .30x_{15} + .22x_{16} \end{aligned}$$

$$\begin{aligned} y^{(2)} = -&.03x_1 + .13x_2 - .43x_3 + .07x_4 - .04x_5 + .10x_6 - .11x_7 + .20x_8 \\ &- .17x_9 + .27x_{10} + .34x_{11} + .18x_{12} - .42x_{13} + .01x_{14} + .35x_{15} - .42x_{16} \end{aligned}$$

In these two linear functions, $x_1, x_2, \ldots, x_{16}$ are scores on the 16 symptom-rating variables of the BPRS, and the weights are the average of those found in the four independent samples. It is possible to compute the value of $y^{(1)}$ and $y^{(2)}$ for any patient by applying the indicated weights to his BPRS rating scores. Using these two derived scores as coordinate values, the patient can be located at a point in a classification diagram. In the two-dimensional model, nine tentative classification regions were delineated by partitioning the space one-half standard deviation above and below the mean for each principal-component function. This yielded eight differentiated patterns corresponding to the eight extreme regions, plus one central undifferentiated category, as shown in Fig. 3.1.

A patient whose profile falls into region A has a high value for $y^{(1)}$ and a low value for $y^{(2)}$. A high score on $y^{(1)}$ indicates relative prominence of thinking disturbance as opposed to depressive symptoms; i.e., the patient profile lies toward the schizophrenic end of the S-D contrast function. A low or negative score on $y^{(2)}$ means that withdrawal, retardation, and blunting are more prominent symptoms than hostility, suspiciousness, and unusual thought content. Patients falling into region A can be described phenomenologically as withdrawn, disorganized, thinking-disturbance types.

Patients falling into region B have strong positive scores on $y^{(1)}$ but neutral scores on $y^{(2)}$. These patients can be characterized as thinking-disturbance

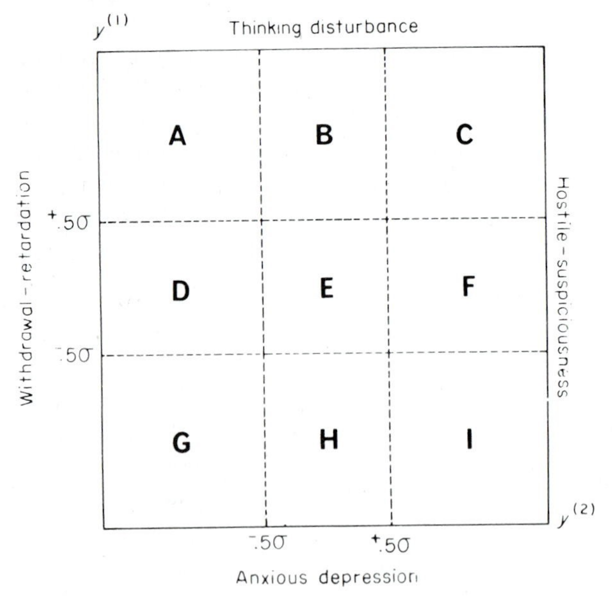

Figure 3.1 Classification regions established by principal-components model.

types with little suggestion of differentiation toward either the primary, withdrawn, blunted end or the hostile, suspicious, paranoid end of the second continuum. Patients falling into region B will be designated as simple thinking-disturbance types.

Region C is characterized by strong positive scores for both $y^{(1)}$ and $y^{(2)}$. Patients falling into this region can be characterized phenomenologically as disorganized, active, hostile, suspicious, thinking-disturbance types. Patients in region C will be designated as disorganized paranoid thinking-disturbance types.

Patients in region D can be characterized as having relative prominence of primary withdrawal, retardation, and blunting of affect without substantial imbalance between thinking disturbance and depressive symptoms. It should be noted that a neutral value on any contrast function can result from equally high levels of both positive and negative symptoms. Patients falling into region D will be characterized as primary withdrawn schizophrenic types. Patients falling into region E have relatively neutral scores on both primary contrast functions, so that they will be designated as undifferentiated types.

Patients in region F have prominent hostile, suspicious, thinking-disturbance symptoms without marked disorganization. The level of disorganized thinking disturbance, if any, is balanced by equal prominence of depressive symptoms. Patients in this region will be termed simple paranoid types.

All patients falling into regions G, H, and I evidence relative prominence of anxious depressive symptoms, as contrasted with thinking disturbance. Patients in region G will be designated withdrawn, retarded depressions because of the concomitant presence of a strong negative balance on function $y^{(2)}$. Patients in region H have anxious-depressive symptomatology without an imbalance in either the withdrawal-retardation or the hostile-suspiciousness paranoid direction. Finally, patients falling into region I are characterized by a combination of hostile, suspicious, paranoid symptomatology and depression. These will be designated as hostile depressions. It is interesting to note that the tripartite classification of depressions growing out of partitioning the principal-component space appears very similar to the three-class division earlier derived by empirical profile-clustering techniques (Overall et al., 1966).

Mean BPRS profiles for patients falling into the nine classification regions, computed using the total sample of 6,000 profiles, are shown in Table 3.7. These nine profiles can be used as classification prototypes for assigning patients to classification groups, as will be described in later chapters. Diagnostic-prototype profiles derived from standard nomenclature will also be examined, and at that time it will be relevant to consider the relationships between the principal-component-derived prototypes and prototypes representing conventional diagnostic groups.

Based directly on the principal-component functions $y^{(1)}$ and $y^{(2)}$, each defined as a weighted combination of all 16 rating scores, rules for classification of

Table 3.7 Mean BPRS Profiles for Patients in Nine Regions of the Principal-components Space

	Type								
	A	*B*	*C*	*D*	*E*	*F*	*G*	*H*	*I*
Somatic concern	.33	1.06	1.05	1.71	2.85	1.84	2.44	3.06	2.09
Anxiety	.38	1.15	1.34	1.47	2.28	3.29	2.88	3.52	3.95
Emotional withdrawal	4.24	3.02	1.36	3.20	2.30	1.87	2.40	1.06	1.45
Conceptual disorganization	3.14	4.13	3.70	1.57	2.06	2.86	1.08	.95	1.46
Guilt feelings	.10	.56	.61	.79	1.40	1.22	2.31	1.78	1.83
Tension	1.45	1.66	1.94	1.37	1.68	2.91	1.81	2.69	2.57
Mannerisms and posturing	2.11	1.69	1.27	.90	1.18	1.32	.77	.72	.86
Grandiosity	.11	1.00	3.15	.62	1.40	1.56	.96	.68	.98
Depressive mood	.49	.73	.56	2.57	1.99	1.88	4.18	3.21	3.56
Hostility	.37	.90	1.93	1.04	1.74	3.33	1.38	.91	3.84
Suspiciousness	.57	1.85	2.58	.90	1.70	3.91	.73	.70	2.54
Hallucinatory behavior	.45	2.21	3.03	.52	1.32	1.75	.65	.67	.72
Motor retardation	2.40	1.17	.39	2.49	1.41	.77	2.53	1.10	.98
Uncooperativeness	1.80	1.11	1.26	1.25	1.54	2.35	.73	.63	1.74
Unusual thought content	.72	3.74	4.15	.77	1.53	3.40	.75	.84	1.16
Blunted affect	3.41	3.62	1.39	3.58	2.31	1.41	2.75	1.00	1.16

patients among the nine categories can be specified as follows. For each patient, calculate $y^{(1)}$ and $y^{(2)}$ as defined earlier in this chapter. Then assign the patients as follows:

Class A	if $y^{(1)} > 1.44$	and $y^{(2)} < -1.19$
Class B	if $y^{(1)} > 1.44$	and $-1.19 < y^{(2)} < 1.19$
Class C	if $y^{(1)} > 1.44$	and $y^{(2)} > 1.19$
Class D	if $-1.44 < y^{(1)} < 1.44$	and $y^{(2)} < -1.19$
Class E	if $-1.44 < y^{(1)} < 1.44$	and $-1.19 < y^{(2)} < 1.19$
Class F	if $-1.44 < y^{(1)} < 1.44$	and $y^{(2)} > 1.19$
Class G	if $y^{(1)} < -1.44$	and $y^{(2)} < -1.19$
Class H	if $y^{(1)} < -1.44$	and $-1.19 < y^{(2)} < 1.19$
Class I	if $y^{(1)} < -1.44$	and $y^{(2)} > 1.19$

For example, these rules state that a patient should be placed in class B if his score on $y^{(1)}$ is greater than 1.44 and his score on $y^{(2)}$ falls between -1.19 and 1.19.

3.6.5 A TEST OF THE TWO-DIMENSIONAL MODEL

It has been demonstrated that individual patient profiles can be classified into nine categories, or classes, derived from principal-components analysis simply by computing scores on the two major principal components of profile pattern and then using these scores as coordinate values to locate each patient within the two-dimensional space shown in Fig. 3.1. According to the principal-components model, this two-dimensional space should provide for maximum display of individual differences in symptom profile pattern.

To assess the validity of this model, it is useful to examine where prototype profiles for various clinical diagnostic groups are located within the space. Does the model provide for adequate separation among groups that are recognized to have substantially different phenomenological characteristics? Which groups that should be discriminably different are not well separated in the simple two-dimensional space? Does the analysis point to any "holes" in the standard diagnostic nomenclature; e.g., are there regions within the principal-components model where no diagnostic prototypes are located?

The data used to examine these questions were drawn from parallel studies conducted in America, France, Germany, Czechoslovakia, and Italy. In each study, several professional experts were asked to describe hypothetical typical patients belonging to a variety of functional diagnostic categories frequently used in the respective countries. A mean profile was computed to represent the BPRS prototype for each diagnostic concept.

A score for each diagnostic profile was calculated for the two principal components $y^{(1)}$ and $y^{(2)}$ using the exact weighted functions previously defined. These weighted composites were then accepted as y and x coordinate values, respectively, and each diagnostic prototype was located within the two-dimensional space. Results are shown in Figs. 3.2 to 3.6. The numerals in the figures indicate the specific diagnostic group in each country.

Examination of the distribution of diagnostic prototypes within the space reveals potentially important information. The various regions in the central and upper portion of the model tend to be occupied by diagnostic prototypes from the various countries. In fact, within the set of profiles from each country, prototypes exist which should describe reasonably well patients within the various regions associated with relative elevation of thinking-disturbance symptoms. The same is not true in the lower region of the space, where diagnostic-prototype profiles tend to bunch much more tightly in the central or left central region. No diagnostic-prototype profile from any country was found to fall into the lower right-hand region of the model. This is the region that has been described as being occupied by the hostile-depression profile type. Although there are sizable numbers of patients who have the hostile-depression profile pattern ($n = 551$ in the total sample of 6,000), there is no standard diagnostic concept corresponding to the phenomenological description suggested by the two-dimensional model.

In Fig. 3.7 are plotted all the diagnostic prototypes from all five countries to emphasize the general configuration of diagnostic concepts across all countries and to point out more clearly certain types of patients that do not appear to be represented in any of the diagnostic typologies. Specifically, regions D and I (center left and lower right) are not represented by the diagnostic stereotypes,

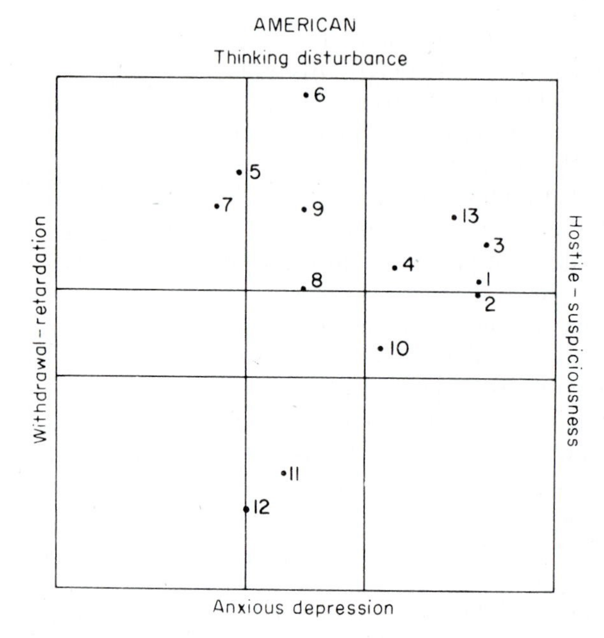

Figure 3.2 Location of American diagnostic stereotypes in principal-components model: (1) paranoia, (2) paranoid state, (3) paranoid schizophrenic, (4) acute undifferentiated schizophrenic, (5) catatonic schizophrenic, (6) hebephrenic schizophrenic, (7) simple schizophrenic, (8) chronic undifferentiated schizophrenic, (9) residual schizophrenic, (10) schizo-affective, (11) psychotic depressive, (12) manic-depressive, depressive, and (13) manic-depressive, manic.

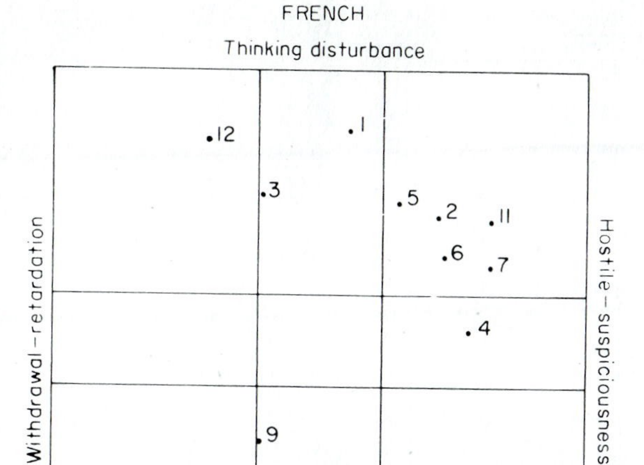

Figure 3.3 Location of French diagnostic stereotypes in principal-components model: (1) schizophrénie paranoïde, (2) manie, (3) état confusionnel, (4) psychose paranoïaque, (5) excitation atypique, (6) bouffée délirante, (7) psychose hallucinatoire chronique, (8) dépression réactionnelle, (9) dépression atypique, (10) mélancolie, (11) paraphrénie, and (12) hébéphrénie.

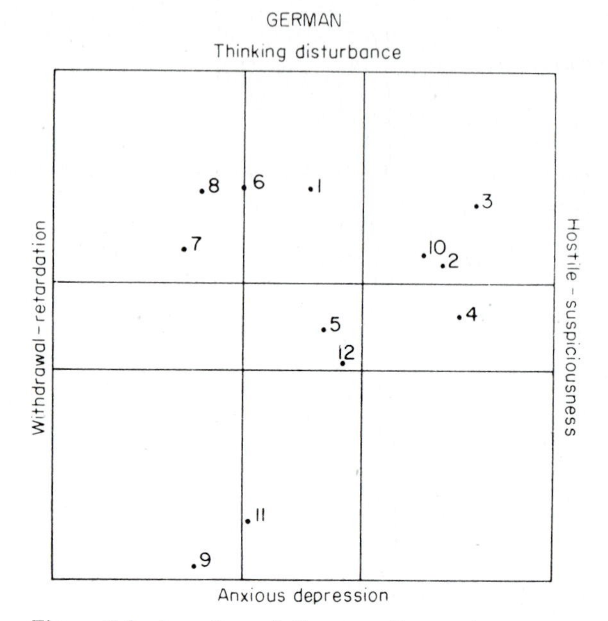

Figure 3.4 Location of German diagnostic stereotypes in principal-components model: (1) schubförmige katatone Schizophrenie, (2) schubförmige paranoide Schizophrenie, (3) schubförmige paranoide-halluzinatorische Schizophrenie, (4) Paranoia, (5) caenästhetische Schizophrenie, (6) hebephrene Schizophrenie, (7) Schizophrenie simplex, (8) schizophrener Personlichkeitswandel, (9) endogene Depression, (10) Manie, (11) endoreaktive Dysthymie, and (12) Mischpsychose.

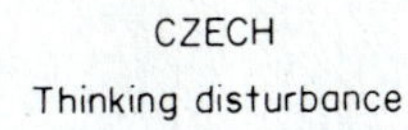

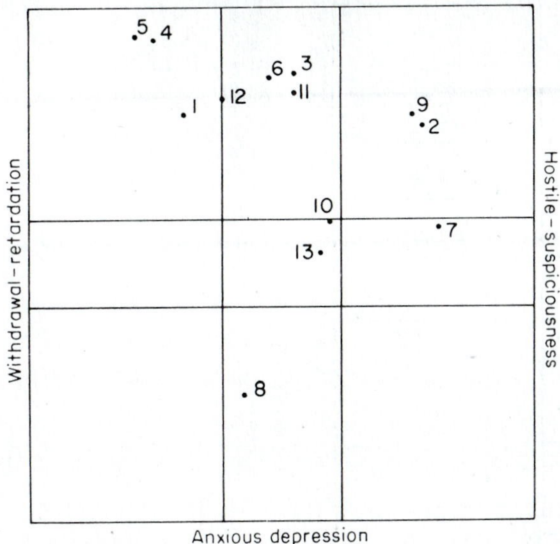

Figure 3.5 Location of Czechoslovakian diagnostic stereotypes in principal-components model: (1) schizophrenia simplex, (2) schizophrenia paranoidní, (3) schizophrenia hebefrenická, (4) schizophrenia katatonická, (5) schizophrenia katatonická stuporozní, (6) schizophrenia katatonická agitovaná, (7) paranoidní stavy, (8) deprese psychotická, (9) manie, (10) schizoafektivní psychóza, (11) chronická schizophrenia s floridními příznaky, (12) chronická schizophrenia bez výraznějšich příznaky, and (13) jiné funkční psychotické stavy.

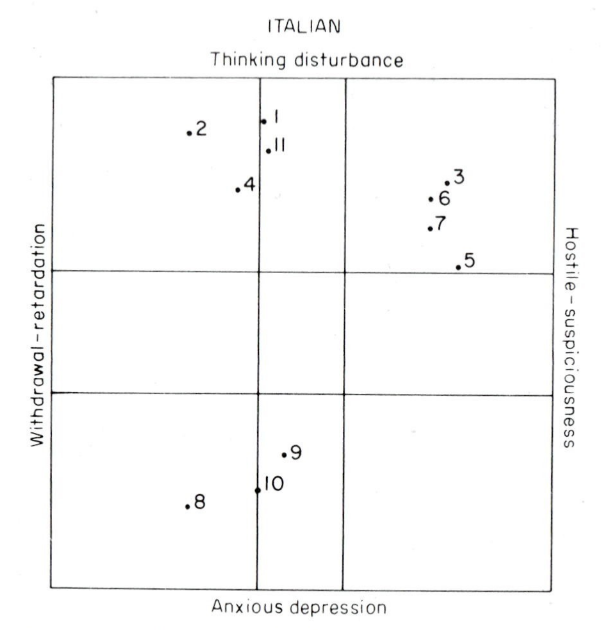

Figure 3.6 Location of Italian diagnostic stereotypes in principal-components model: (1) schizophrenia ebefrenica, (2) schizophrenia catatonica, (3) schizophrenia paranoide, (4) schizophrenia simplex, (5) paranoia, (6) paraphrenia, (7) mania, (8) melancolia, (9) depressione atipica, (10) melancolia involutiva, (11) stato confusionale o amenza.

although substantial numbers of actual patients were found to fall into each of the regions. By way of contrast, the nine mean vectors presented in Table 3.7 were located in the model according to scores on the two principal components, and the results are presented in Fig. 3.8. As would be expected, the mean profiles calculated from BPRS ratings for patients actually assigned by profile analysis to the nine categories tend to spread symmetrically across the space. From the point of view of describing individual differences in the phenomenology of psychiatric disorder, the nine types defined from partitioning of the principal-components space would appear more representative than any of the sets of clinical stereotypes. At the very least, the empirical analysis suggests deficiencies in the clinical diagnostic nomenclature which could be remedied by the addition of one or two new classes.

3.7 Summary

Principal-components analysis has been presented as a method for reducing p correlated measurement variables to a smaller set of statistically independent linear combinations having certain unique properties with regard to characterizing individual differences. The student will appreciate the relationship of this

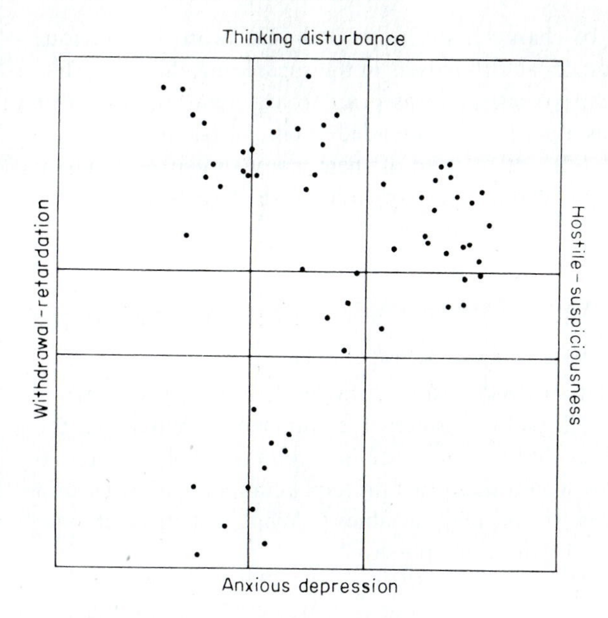

Figure 3.7 Location of diagnostic stereotypes from five countries in principal-components model.

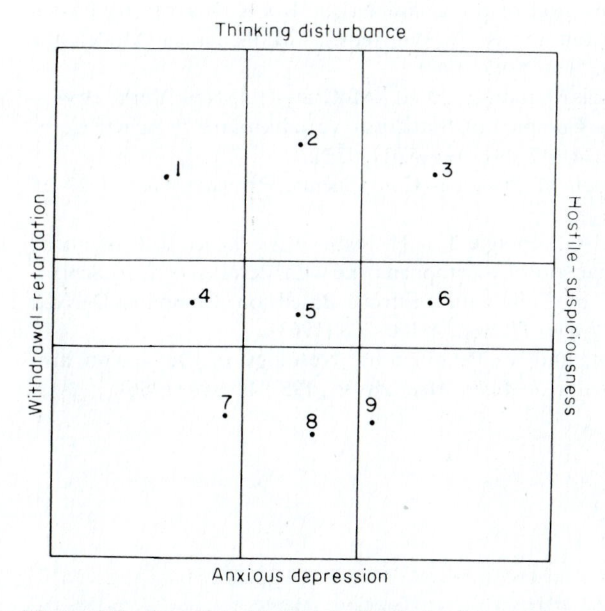

Figure 3.8 Location of empirically derived prototypes in principal-components model.

method to the description of characteristic roots and vectors in the previous chapter and to the discussion of factor analysis in the succeeding chapters. The presentation of principal-components analysis prior to a general discussion of factor analysis rather than as a part of it is intended to emphasize that the goals of the two approaches are different in spite of their many similarities. These similarities and differences will become more apparent as the key issues in factor analysis are presented in the following chapters.

3.8 Exercises

1 Show that the four principal components in Tables 3.1 to 3.4 are statistically orthogonal. Next evaluate the geometric orthogonality.

2 Construct a 5×5 correlation matrix using variables 2, 4, 5, 9, and 15 from Table 5.1 and compute the first two principal components. What are the associated λ's? How much of the total variance is accounted for by these two principal components? How would these two principal components be used in computing linear composites of the five variables? What statistical characteristics would the two score distributions possess?

3.9 References

Anderson, T. W.: "Introduction to Multivariate Statistical Analysis," Wiley, New York, 1958.

Cronbach, L. J., and G. C. Gleser: Assessing Similarity between Profiles, *Psychol. Bull.*, **50**:456–473 (1953).

Greenstadt, J.: The Determination of the Characteristic Roots of a Matrix by the Jacobi Method, in A. Ralston and R. A. Wilf (eds.), "Mathematical Models for Digital Computers," Wiley, New York, 1960.

Guilford, J. P.: "Psychometric Methods," 2d ed., McGraw-Hill, New York, 1954.

Hotelling, H.: Analysis of a Complex of Statistical Variables into Principal Components, *J. Educ. Psychol.*, **24**:417–441, 498–520 (1933).

———: Simplified Calculation of Principal Components, *Psychometrika*, **1**:27–35 (1936).

Overall, J. E., L. E. Hollister, G. Honigfeld, I. H. Kimbell, F. Meyer, J. L. Bennett, and E. M. Caffey: Comparison of Acetophenazine with Perphenazine in Schizophrenics: Demonstration of Differential Effects Based on Computer-Derived Diagnostic Models, *Clin. Pharm. Therap.*, **4**:200–208 (1963).

———, ———, M. Johnson, and V. Pennington: Nosology of Depression and Differential Response to Drugs, *J. Amer. Med. Assoc.*, **195**:946–948 (1966).

PART TWO

FACTOR ANALYSIS AND RELATED METHODOLOGY

4

General Algebraic Formulation of Factor Analysis

4.1 Introduction

Factor analysis is a powerful method of statistical analysis that has as its aim the explanation of relationships among numerous correlated variables in terms of a relatively few underlying factor variates. In practice, it is used for a variety of purposes in social and psychological research ranging from statistical data reduction to development of elaborate theories of personality. The primary result of factor analysis is a factor matrix F which contains coefficients of relationship between the original variables and the derived factor variates. The coefficients in the factor matrix are called *factor loadings*. The nature of the derived factors can be *interpreted* or inferred from examination of the nature of variables most highly related to each factor as contrasted with variables not so related. *Factor scores* are composite variables that represent the status of individuals on factor dimensions. They are measurements of the factor, and they are usually defined as weighted combinations of several original variables. Most of the differences between individuals that were originally represented in terms of numerous correlated measurements can be represented in terms of the smaller number of factor scores. The method of factor analysis involves defining new factor variates as linear transformations of the original correlated variables. Rather than deriving the desired transformation from direct analysis of the original variables, it is computationally more efficient to derive the factor matrix as a simple transformation of the matrix of intercorrelations among the original variables. The correlations of p variables with any composite variable that can be defined as a linear transformation of the original variables can be obtained directly from the correlation matrix. Factor analysis thus involves defining a new reduced-rank factor matrix as a linear transformation of the matrix of intercorrelations among the original variables.

Factors are conceived as primary dimensions of *individual difference*. Desirable properties of a good factor solution (transformation) include (1) parsimony, (2) orthogonality or at least relative independence, and (3) psychological or conceptual meaningfulness. There would be little point in the analysis if the number of factors required to explain individual differences in the measurement domain was not substantially smaller than the number of original variables. For both conceptual and statistical reasons, relative independence among factors is important. Each factor should represent a distinctly different underlying source of variance. Finally, in order to understand the nature of factors and thus to understand the nature of major differences between individuals, it is important to be able to conceptualize what each factor represents. These objectives are sought by judicious choice of the transformation matrices, the nature of which will be discussed more concretely for various specific types of analyses in later chapters. In the present chapter, we shall be concerned with a general algebraic formulation of factor analysis in terms of transformation matrices $\hat{W}$, W, and T, which can at this point be considered to be completely arbitrary.

In this chapter a very general mathematical definition of factor analysis and factor-analysis methodology will be presented. The content is thus necessarily more technical and abstract than that of succeeding chapters, which deal with specific methods of analysis. Instead of discussing factor analysis in terms of numerous specific methods, as done so frequently in other texts, it is the aim here to provide a general model that encompasses the variety of "different" methods as special cases. The student who takes the time to become familiar with the general mathematical formulation will find that he can relate the various methods or even devise new factor-analysis methods at will. The reader not really practiced in matrix algebra and matrix notation can sharpen his skills by working through the present chapter slowly with repeated reference to the material in Chap. 2.

4.2 Definitions

4.2.1 FACTOR LOADINGS AND FACTOR COSINES

The method of factor analysis, as opposed to theory or interpretation, involves simple transformation of a product-moment matrix to a matrix of lesser rank. Although the method can be applied to covariance or even cross-product matrices, most applications in clinical research involve analyses of product-moment correlation matrices. The major reason that correlations are used in factor analyses arises from the fact that units of measurement are usually different and frequently arbitrary for psychological variables. In working with correlations, the variables are in effect transformed to standard-score form. For simplicity in this discussion, consider the product-moment matrix to be a matrix of intercorrelations among p measurement variables. The correlation matrix usually has communality estimates inserted into the principal diagonal for factor analysis, but that is a problem to which we shall address ourselves in a later section.

The matrix of factor loadings is obtained by applying a nonsingular transformation matrix $\hat{W}D_1$ to the product-moment matrix R

$$\underset{p \times p}{R} \; \underset{p \times r}{\hat{W}} \; \underset{r \times r}{D_1} = \underset{p \times p}{R} \; \underset{p \times r}{W} = \underset{p \times r}{F_1} \tag{4.1}$$

where R is a product-moment matrix, $\hat{W}$ is a transformation matrix of rank r, D_1 is a diagonal scaling matrix containing the reciprocals of the square roots of the diagonal elements in $C = \hat{W}'R\hat{W}$, and F_1 is the matrix of factor loadings. It will be shown that the factor loadings are indices of relationship between the original variables and the derived factor variates. Different methods of factor analysis arise from different definitions of $W = \hat{W}D$.

Before proceeding, it may be well to examine the role of the diagonal matrix D_1. In Eq. (4.1), the only restriction placed on the transformation matrix $\hat{W}$ is that none of the r columns be linearly dependent on the other $r - 1$ columns. The elements in $\hat{W}$ can be of any arbitrary size and still satisfy the requirements of Eq. (4.1). To facilitate the interpretation of factors, it is most meaningful to conceive of factor loadings as representing *correlations* between original measurements and the factors. To permit interpretation of factor loadings as correlation coefficients, a rescaling of the arbitrary variances is required. (Product-moment correlation coefficients are z-score covariances; hence the need to standardize factor variances.) The matrix $C = \hat{W}'R\hat{W}$ contains variances and covariances among r unscaled factor variates, with the variances appearing in the principal diagonal. To scale factor variates to unit length, D_1 is defined to contain the reciprocals of the square roots of the unscaled factor variances. The student should verify that postmultiplication of $\hat{W}$ by D_1 in Eq. (4.1) is equivalent to multiplying each *column* of $\hat{W}$ by a scaling coefficient. If $W = \hat{W}D_1$, the columns in W contain coefficients defining factor variates with unit variance. For simplicity, $W = \hat{W}D_1$ will be used in developing other relationships.

The matrix of factor cosines, or correlations among the r factor variates, is specified by[1]

$$D_1\hat{W}'R\hat{W}D_1 = W'RW = \boldsymbol{\Sigma}_1 \tag{4.2}$$

[1] Throughout the chapters on factor analysis, the symbol $\boldsymbol{\Sigma}$ will be used to represent matrices of factor cosines or correlations. In the chapters on classification $\boldsymbol{\Sigma}$ is also used to represent the variance-covariance matrix. This notation should not be confused with the common use of sigma ($\boldsymbol{\Sigma}$) as a summation operator in scalar algebra.

It can be verified by considering the row-by-column rule for matrix multiplication that each element in $\mathbf{\Sigma}_1$ is the scalar product obtained from premultiplying R by one *row* of W' and then postmultiplying by one *column* of W.

It has been noted that the only formal restriction imposed on the transformation matrix $\hat{W}$ in the most general definition of factor analysis is that it be of rank r. This is, of course, an extremely broad conception of factor analysis. To satisfy one purpose for which factor analysis is used, $\hat{W}$ should be defined to achieve good representation of the relationships in the original product-moment matrix with the number of factors r considerably smaller than p. That is, *parsimony* requires that the number of factors be considerably smaller than the number of variables, and maximum parsimony is achieved by defining $\hat{W}$ so that r is as small as possible. For the r factor variates to account for most of the individual differences represented in the several original variables, the matrix $\hat{R} = F\mathbf{\Sigma}_1^{-1}F'$ should closely approximate the original R matrix. To satisfy the requirement of *independence* among factors, $\hat{W}$ must be defined to minimize the off-diagonal elements in $\mathbf{\Sigma}_1$ of Eq. (4.2). Complete independence among factors is achieved by defining $\hat{W}$ such that $\mathbf{\Sigma}_1$ is an identity matrix. This represents a so-called *orthogonal-factor solution.* To provide for *meaningful* representation of relationships among the original variables and to facilitate understanding of what the factor variates measure, variables which do or do not relate substantially to each factor should be clearly distinguishable. Meaningful interpretation is facilitated by defining $\hat{W}$ in such manner that a relatively few variables relate highly to each factor while the relationships of other variables to the factor are minimal. Meaningful interpretation is facilitated if $\hat{W}$ can be defined so that variables within homogeneous clusters relate highly to only one factor. The manner of achieving these objectives of factor analysis will be discussed in later sections. The point to be made here is that the method of factor analysis can be defined in very general terms. The real problem is to achieve a *good* factor solution through appropriate definition of the transformation matrix $\hat{W}D_1 = W$ to realize objectives of parsimony, independence, and psychological meaningfulness.

4.2.2 ROTATION OF FACTOR AXES

Rotation of factors is the process of redefining the factor axes so that they represent the original variables in a simpler and more meaningful way. Traditionally, factor analysis has been accomplished in two stages. First, factors are defined to achieve maximum parsimony and independence without regard for the question of meaningful relationship to the original variates. Subsequently, the factor axes are repositioned (rotated) to achieve a more meaningful relationship to these original variates. In its most general definition, factor rotation is the process of applying a nonsingular transformation matrix $\hat{T}$ to the matrix of factor loadings. Again, rescaling by a diagonal matrix D_2 is required to maintain unit variance for the factors

$$\underset{n\times n}{R}\ \underset{n\times r}{\hat{W}}\ \underset{r\times r}{D_1}\ \underset{r\times r}{\hat{T}}\ \underset{r\times r}{D_2} = \underset{n\times r}{F_1}\ \underset{r\times r}{\hat{T}}\ \underset{r\times r}{D_2} = \underset{n\times r}{F_2} \tag{4.3}$$

where $\hat{T}$ is the nonsingular rotational transformation matrix, D_2 is a diagonal scaling matrix containing the reciprocals of the square roots of the diagonal elements in $C_T = \hat{T}'D_1\hat{W}'R\hat{W}D_1\hat{T} = \hat{T}'\mathbf{\Sigma}_1\hat{T}$, and F_2 is the matrix of factor loadings relating the original variables to the rotated factors.

Cosines, or correlations, among the rotated factors can be obtained by pre- and postmultiplying the original product-moment matrix R by the product of the factor and rotation transformations

$$D_2\hat{T}'D_1\hat{W}'R\hat{W}D_1\hat{T}D_2 = \mathbf{\Sigma}_2 \tag{4.4}$$

Alternatively, since $D_1\hat{W}'R\hat{W}D_1 = \mathbf{\Sigma}_1$ of Eq. (4.2),

$$D_2\hat{T}'\mathbf{\Sigma}_1\hat{T}D_2 = \mathbf{\Sigma}_2 \tag{4.5}$$

It is important to recognize that the process of factor rotation is equivalent to redefining the factor-transformation matrix and refactoring R. Instead of the original factor-transformation matrix $\hat{W}$ of Eq. (4.1), we can consider the composite factor-transformation matrix $\hat{Q} = \hat{W}D_1\hat{T}$ of Eq. (4.3). Thus, the complete factor-transformation matrix in the traditional two-stage analysis consists of two parts. The matrix $\hat{W}$ is chosen to ensure objectives of parsimony and independence, and then $\hat{T}$ is defined to satisfy objectives of simplicity and meaningfulness. An alternative to the two-stage factor-rotation sequence is to define $\hat{W}D_1$ of Eq. (4.1) initially such that it approximates the $\hat{W}D_1\hat{T}D_2$ of Eq. (4.3). Such direct factor-analysis methods will be considered in later chapters.

4.2.3 ORTHOGONAL AND OBLIQUE FACTORS

The term *orthogonality*, when used with regard to factor analysis, refers to statistical independence or lack of correlation among the factor variates. For factors to be statistically orthogonal, the cosine matrix defined in Eq. (4.2) must be an identity matrix. Thus, the requirement that factors be orthogonal imposes a restriction on the definition of the factor-transformation matrix. To define *orthogonal* factors, the matrix $\hat{W}$ must be defined to satisfy

$$\underset{r\times r}{D_1}\ \underset{r\times n}{\hat{W}'}\ \underset{n\times n}{R}\ \underset{n\times r}{\hat{W}}\ \underset{r\times r}{D_1} = \underset{r\times r}{I} \tag{4.6}$$

If $\hat{W}$ is defined so that $D_1\hat{W}'R\hat{W}D_1 \neq I$, the factors are said to be *oblique*.

Even though the original factor transformation results in orthogonal factors as defined in Eq. (4.6), the subsequent rotational transformation may be either orthogonal or oblique. For the rotated factors to be orthogonal, the matrix $\mathbf{\Sigma}_2$ of Eq. (4.5) must be an identity matrix. Since orthogonal rotation is applied only to factors which were originally defined as orthogonal, an orthogonal rotational transformation implies

$$D_2\hat{T}'\mathbf{\Sigma}_1\hat{T}D_2 = D_2\hat{T}'\hat{T}D_2 = I \tag{4.7}$$

where $\mathbf{\Sigma}_1 = I$ from the original orthogonal factor transformation. Thus, for orthogonal rotation of originally orthogonal factors, the transformation matrix must be defined such that each pair of columns has zero sum of cross products.

4.3 Correlations and Factor Variates

It is convenient to think of factors as artificial mathematical variables. When the matrix which is analyzed is a product-moment correlation matrix with unities in the principal diagonal, the factor loadings can be conceived as correlations between the original test variates and the derived factor variates. Each factor variate can be defined as a weighted combination of z-score values for the original test variates

$$y = w_1 z_1 + w_2 z_2 + \cdots + w_p z_p = \mathbf{w}'\mathbf{z} \tag{4.8}$$

where the elements $w_1, w_2, \ldots, w_p$ are weighting coefficients and $z_1, z_2, \ldots, z_p$ are standardized scores on the original test variables. The product-moment correlation coefficient is conventionally defined as the mean of z-score cross products

$$r_{ij} = \frac{1}{n} \sum^{n} z_i z_j \tag{4.9}$$

This same definition holds where one of the variables is a linear combination, such as $y = \mathbf{w}'\mathbf{z}$

$$r_{yj} = \frac{1}{n} \sum^{n} \mathbf{w}'\mathbf{z} z_j = \mathbf{w}' \frac{1}{n} \sum^{n} \mathbf{z}' z_j = \mathbf{w}'\mathbf{r} \tag{4.10}$$

where $y = \mathbf{w}'\mathbf{z}$ has been scaled to unit variance (z-score form) and z_j is the jth variable in the vector $\mathbf{z}$. The vector $\mathbf{r}$ contains the product-moment correlations between each of the p variables and the jth variable

$$\mathbf{r} = \frac{1}{n} \sum^{n} \mathbf{z}' z_j = \frac{1}{n} \begin{bmatrix} \sum z_1 z_j \\ \sum z_2 z_j \\ \cdot \\ \cdot \\ \cdot \\ \sum z_p z_j \end{bmatrix} = \begin{bmatrix} r_{1j} \\ r_{2j} \\ \cdot \\ \cdot \\ \cdot \\ r_{pj} \end{bmatrix}$$

Thus, the product-moment correlation between a linear combination $y = \mathbf{w}'\mathbf{z}$ and a particular variable can be obtained as the inner, or dot, product of the vector of (properly scaled) weighting coefficients and the vector containing correlations between the particular variable z_j and each of the variables represented in the vector $\mathbf{z}$

$$r_{yj} = \mathbf{w}'\mathbf{r} = \sum_{i}^{p} w_i r_{ij} \tag{4.11}$$

In this same manner, the correlations between a linear combination of p variables, $y = \mathbf{w}'\mathbf{z}$, and each of the variables entering into the linear combination can be defined in terms of the intercorrelations among the original variables

$$\mathbf{r}_y = \frac{1}{n}\sum^{n} \mathbf{w}'\mathbf{z}\mathbf{z}' = \mathbf{w}'\frac{1}{n}\sum^{n} \mathbf{z}\mathbf{z}' = \mathbf{w}'R \tag{4.12}$$

where the vector $\mathbf{w}'$ has been scaled so that variance $V(y) = \mathbf{w}'R\mathbf{w} = 1$. The outer or major product $\mathbf{z}\mathbf{z}'$ is a matrix containing cross products between all possible pairs of elements in $\mathbf{z}$. Summed over individuals and divided by n, the matrix $\frac{1}{n}\sum^{n} \mathbf{z}\mathbf{z}' = R$ contains mean z-score cross products, or product-moment correlations, among the p tests. Thus, a vector containing as elements the product-moment correlation coefficients relating p original variables to a linear combination of the p variables can be obtained by multiplying the matrix of intercorrelations among the p variables by the vector of properly scaled weighting coefficients. Proper scaling implies $\mathbf{w}'R\mathbf{w} = 1$.

The product-moment correlation between two linear combinations of the same p variables can be defined in terms of the matrix of intercorrelations among the original variables

$$r_{y_1y_2} = \frac{1}{n}\sum^{n} \mathbf{w}_1'\mathbf{z}\mathbf{z}'\mathbf{w}_2 = \mathbf{w}_1'\left(\frac{1}{n}\sum^{n} \mathbf{z}\mathbf{z}'\right)\mathbf{w}_2 = \mathbf{w}_1'R\mathbf{w}_2 \tag{4.13}$$

where $y_1 = \mathbf{w}_1'\mathbf{z}$ and $y_2 = \mathbf{z}'\mathbf{w}_2$ have both been scaled to unit variance; that is, $V(y_1) = \mathbf{w}_1'R\mathbf{w}_1 = 1$, and $V(y_2) = \mathbf{w}_2'R\mathbf{w}_2 = 1$.

In this section it has been consistently required for the sake of simplicity that the linear combinations be scaled to unit variance. It may be well to remind the reader that such scaling can be achieved for any arbitrary vector, say $\hat{\mathbf{w}}$, by computing the variance of the unscaled vector and then dividing each element in $\hat{\mathbf{w}}$ by the square root of $V(\hat{y}) = \hat{\mathbf{w}}'R\hat{\mathbf{w}}$

$$\mathbf{w} = \frac{1}{\sqrt{\hat{\mathbf{w}}'R\hat{\mathbf{w}}}}\,\hat{\mathbf{w}} \tag{4.14}$$

In factor analysis, this scaling is accomplished by the diagonal matrices D_1 and D_2 of Eqs. (4.1) and (4.3). The elements in diagonal matrix D_1 of Eq. (4.1) are the reciprocals of the square roots of the variances of the unscaled factors, and similarly for the elements in D_2 of Eq. (4.3).

From the relationships developed in this section, it can be seen that factor loadings defined by Eq. (4.1) can properly be conceived as product-moment correlation coefficients relating p original test variates to r derived factor variates

$$R\hat{W}D_1 = RW = F_1$$

where the columns of W are scaled such that

$$\mathbf{w}_i = \frac{1}{\sqrt{\hat{\mathbf{w}}_i' R \hat{\mathbf{w}}_i}} \hat{\mathbf{w}}_i$$

Each column of F_1 contains correlations between the p original test variates and a derived factor variate, as defined in Eq. (4.12). The matrix R is postmultiplied by W to yield factor loadings in columns rather than rows, because that is the conventional manner of presentation

$$R(\mathbf{w}_1, \mathbf{w}_2, \ldots, \mathbf{w}_r) = (\mathbf{f}_1, \mathbf{f}_2, \ldots, \mathbf{f}_r)$$

where $\mathbf{w}_1' R \mathbf{w}_1 = \mathbf{w}_2' R \mathbf{w}_2 = \cdots = \mathbf{w}_r' R \mathbf{w}_r = 1$.

The factor cosine matrix $\mathbf{\Sigma}_1$ defined in Eq. (4.2) contains the product-moment correlations between various pairs of factor variates. Consideration of the row-by-column rule for matrix multiplication reveals that each element in $\mathbf{\Sigma}_1$ is a product of the type $\mathbf{w}_i' R \mathbf{w}_j$ defined in Eq. (4.13), where $\mathbf{w}_i' R \mathbf{w}_i = \mathbf{w}_j' R \mathbf{w}_j = 1$ are the diagonal elements of $\mathbf{\Sigma}_1$.

Since factor loadings following rotational transformation can be thought of as resulting from a refactoring of the original correlation matrix using $\hat{Q} = \hat{W} D_1 \hat{T}$, the rotated factor loadings can be considered as representing correlation coefficients relating original test variates to the derived factors and the cosines among rotated factors can be interpreted as product-moment correlations between the factor variates. Let $Q = \hat{W} D_1 \hat{T} D_2$ of Eq. (4.3). The rotated factor loadings are defined by $RQ = F_2$, which is analogous to $RW = F_1$ of Eq. (4.1)

$$R(\mathbf{q}_1, \mathbf{q}_2, \ldots, \mathbf{q}_r) = (\mathbf{f}_1, \mathbf{f}_2, \ldots, \mathbf{f}_r)$$

where $\mathbf{q}_1' R \mathbf{q}_1 = \mathbf{q}_2' R \mathbf{q}_2 = \cdots = \mathbf{q}_r' R \mathbf{q}_r = 1$. Each column vector $\mathbf{f}_i$ is a matrix-vector product of the type defined in Eq. (4.12). Cosines among the rotated factors are defined by $Q'RQ = \mathbf{\Sigma}_2$

$$(\mathbf{q}_1', \mathbf{q}_2', \ldots, \mathbf{q}_r') R (\mathbf{q}_1, \mathbf{q}_2, \ldots, \mathbf{q}_r) = \mathbf{\Sigma}_2$$

in which each element of $\mathbf{\Sigma}_2$ is of the form $\sigma_{ij} = \mathbf{q}_i' R \mathbf{q}_j$ given by Eq. (4.13) and $\mathbf{q}_1' R \mathbf{q}_1 = \mathbf{q}_2' R \mathbf{q}_2 = \cdots = \mathbf{q}_r' R \mathbf{q}_r = 1$.

4.4 General Method for Orthogonal-factor Analysis

The two-stage factor-rotation approach to factor analysis has traditionally involved a preliminary orthogonal-factor analysis followed by rotation to achieve more meaningful positioning of factor axes. In Sec. 4.3 an orthogonal-factor solution was said to require that the transformation matrix $\hat{W}$ be chosen such that $\mathbf{\Sigma}_1 = D_1 \hat{W}' R \hat{W} D_1 = I$. The reader may well have felt some concern how such a transformation matrix can be defined in practice. It is the purpose of this section to present a general computational algorithm. Numerous specific orthogonal-factor-analysis methods can be conceived as special cases of the general approach. In fact, any student or researcher who becomes acquainted with the general method can formulate new specific methods. The problem, of course, is to develop good new methods which accomplish what the factor analysis is intended to accomplish.

In actual practice, one does not define the complete matrix $\hat{W}$, subject to the restriction that $\mathbf{\Sigma}_1 = I$ before calculating the factor loadings. Instead, factors are defined successively in such manner that each factor is statistically independent of all preceding factors. Given the complete matrix of factor loadings, one can by simple matrix multiplication solve for the transformation matrix W that will yield F_1.

Loadings of the p test variates on the first factor axis are computed by applying a *hypothesis vector* $\mathbf{h}_1$ to the original matrix of intercorrelations among the test variates R_1

$$R_1\mathbf{h}_1s_1 = \mathbf{f}_1 \tag{4.15}$$

where R_1 contains intercorrelations among the p test variates, $\mathbf{h}_1$ is a vector chosen to satisfy objectives of the analysis, s_1 is a scaling constant equal to the reciprocal of the square root of $v_1 = \mathbf{h}_1'R_1\mathbf{h}_1$, and $\mathbf{f}_1$ contains the correlations of the p tests with the first factor. The hypothesis vector $\mathbf{h}_1$ can be chosen to maximize variance accounted for, to establish the factor axis in close relationship to a homogeneous cluster of test vectors, or to accomplish some other objective.

The first residual correlation matrix R_2 is computed by partialing out all variance associated with the first factor

$$R_2 = R_1 - \mathbf{f}_1\mathbf{f}_1' \tag{4.16}$$

where $\mathbf{f}_1\mathbf{f}_1'$ is a matrix containing the cross products of all possible pairs of elements in the vector $\mathbf{f}_1$. This adjustment is tantamount to correcting each of the original standard-score variables for all linear relationships to the first factor variate and then computing a *covariance* matrix between the residual scores. Since the residual variates are completely independent of the first factor, any second factor defined in terms of the residual variates will be orthogonal to the first factor. Proof of orthogonality will be considered in Sec. 4.5.

Loadings of the several test vectors on a second factor, orthogonal to the first, can be obtained by applying a second hypothesis vector to R_2

$$R_2\mathbf{h}_2s_2 = \mathbf{f}_2 \tag{4.17}$$

where $\mathbf{h}_2$ is a column vector not linearly dependent on $\mathbf{h}_1$ and s_2 is a scaling coefficient equal to the reciprocal of the square root of $v_2 = \mathbf{h}_2'R_2\mathbf{h}_2$. The vector $\mathbf{f}_2$ contains correlations of the original test variates with the second factor variate.

Factor loadings for successive additional factors, each orthogonal to those previously defined, can be computed as follows:

$$R_3 = R_2 - \mathbf{f}_2\mathbf{f}_2'$$
$$R_3\mathbf{h}_3s_3 = \mathbf{f}_3$$
$$R_4 = R_3 - \mathbf{f}_3\mathbf{f}_3'$$
$$R_4\mathbf{h}_4s_4 = \mathbf{f}_4$$
$$R_5 = R_4 - \mathbf{f}_4\mathbf{f}_4'$$
. . . .
$$R_r = R_{r-1} - \mathbf{f}_{r-1}\mathbf{f}_{r-1}'$$

and

$$R_r\mathbf{h}_rs_r = \mathbf{f}_r \tag{4.18}$$

The problem of when to stop defining additional factors will be discussed later. In general, the aim is to account for most of the variance in R with a relatively small number of factors.

4.5 Orthogonality of Factors

In this section, a proof is presented that factors computed from successively residualized matrices as just described are statistically orthogonal. From Eq. (4.1),

$$R_1\hat{W}D_1 = R_1W = F_1$$

so that

$$W = R_1^{-1}F_1 \tag{4.19}$$

and the columns of W are

$$\mathbf{w}_1 = R_1^{-1}\mathbf{f}_1, \quad \mathbf{w}_2 = R_1^{-1}\mathbf{f}_2, \ldots, \quad \mathbf{w}_r = R_1^{-1}\mathbf{f}_r \tag{4.20}$$

The condition of orthogonality defined in Sec. 4.3 requires that $W'R_1W = \mathbf{\Sigma}_1 = I$. This is equivalent to requiring that $\mathbf{w}_i'R_1\mathbf{w}_j = 0$ for all $i \neq j$.

The first two factors are orthogonal if $\mathbf{w}_1'R_1\mathbf{w}_2 = 0$, where $\mathbf{w}_1 = R_1^{-1}\mathbf{f}_1$ and $\mathbf{w}_2 = R_1^{-1}\mathbf{f}_2$. Recalling that $R_1\mathbf{h}_1s_1 = \mathbf{f}_1$ and $R_2\mathbf{h}_2s_2 = \mathbf{f}_2$, where $R_2 = R_1 - \mathbf{f}_1\mathbf{f}_1' = R_1 - R_1\mathbf{w}_1\mathbf{w}_1'R_1$, leads to

$$\begin{aligned}
\mathbf{w}_1'R_1\mathbf{w}_2 &= \mathbf{w}_1'R_1R_1^{-1}(R_1 - R_1\mathbf{w}_1\mathbf{w}_1'R_1)\mathbf{h}_2s_2 \\
&= \mathbf{w}_1'(R_1 - R_1\mathbf{w}_1\mathbf{w}_1'R_1)\mathbf{h}_2s_2 \\
&= \mathbf{w}_1'R_1\mathbf{h}_2s_2 - \mathbf{w}_1'R_1\mathbf{w}_1\mathbf{w}_1'R_1\mathbf{h}_2s_2 \\
&= \mathbf{w}_1'R_1\mathbf{h}_2s_2 - \mathbf{w}_1'R_1\mathbf{h}_2s_2 = 0
\end{aligned}$$

since

$$\mathbf{w}_1'R_1\mathbf{w}_1 = \mathbf{f}_1'R_1^{-1}R_1R_1^{-1}\mathbf{f}_1 = s_1\mathbf{h}_1'R_1\mathbf{h}_1s_1 = 1$$

where

$$s_1 = \frac{1}{\sqrt{\mathbf{h}_1' R_1 \mathbf{h}_1}}$$

Thus, the first two factors satisfy the condition of orthogonality. A similar algebraic proof of absence of correlation between other factors is possible, although the task becomes more cumbersome as remote residual matrices are considered. In general, by induction, if it can be proved that the first $r - 1$ factors are uncorrelated, it can be shown that the rth factor is independent of each of the preceding $r - 1$. Since orthogonality of the first two factors can be proved, this result can be used in proving the orthogonality of the first three, and so on. In general, the rth factor will be orthogonal to the jth factor if it can be shown that $\mathbf{w}_j' R_1 \mathbf{w}_r = 0$. Recall that $\mathbf{w}_r = R_1^{-1}\mathbf{f}_r$ and that $\mathbf{f}_r = R_r \mathbf{h}_r s_r$, where the residual matrix $R_r = (R_1 - R_1\mathbf{w}_1\mathbf{w}_1'R_1 - \cdots - R_1\mathbf{w}_{r-1}\mathbf{w}_{r-1}'R_1)$. Thus,

$$\begin{aligned}\mathbf{w}_j' R_1 \mathbf{w}_r &= \mathbf{w}_j' R_1 R_1^{-1}(R_1 - R_1\mathbf{w}_1\mathbf{w}_1'R_1 - \cdots - R_1\mathbf{w}_{r-1}\mathbf{w}_{r-1}'R_1)\mathbf{h}_r s_r \\ &= \mathbf{w}_j'(R_1 - R_1\mathbf{w}_1\mathbf{w}_1'R_1 - \cdots - R_1\mathbf{w}_{r-1}\mathbf{w}_{r-1}'R_1)\mathbf{h}_r s_r \\ &= \mathbf{w}_j' R_1 \mathbf{h}_r s_r - \mathbf{w}_j' R_1 \mathbf{h}_r s_r = 0 \end{aligned} \tag{4.21}$$

because

$$\mathbf{w}_j' R_1 \mathbf{w}_1 = \mathbf{w}_j' R_1 \mathbf{w}_2 = \cdots = \mathbf{w}_j' R_1 \mathbf{w}_{r-1} = 0$$

and

$$\mathbf{w}_j' R_1 \mathbf{w}_j = 1$$

4.6 Calculation of Factor-transformation Matrix and Factor-score Equations Given Factor Loadings

Factor analysis has been defined in Eqs. (4.1) and (4.3) as a simple linear transformation of the original correlation matrix. In actual practice, however, one can arrive at the matrix of factor loadings without first defining the factor-transformation matrix. For example, in the general method for orthogonal-factor analysis as described in Sec. 4.4, the factor loadings are computed by applying hypothesis vectors $\mathbf{h}_1, \mathbf{h}_2, \ldots, \mathbf{h}_r$ to successive residual matrices $R_1, R_2, \ldots, R_r$. Except in principal-axes factor analysis, the matrix $H = (\mathbf{h}_1, \mathbf{h}_2, \ldots, \mathbf{h}_r)$ will not in general transform the original R_1 directly. That is, $R_1 H \neq F_1$ because the columns of F_1 were defined by applying columns of H to successive *residual* matrices.

If F is a $p \times r$ matrix of factor loadings, then from Eqs. (4.1) and (4.3) we know that there exists a simple transformation matrix W

$$R_1 W = F$$

or

$$R_1 H T = F \tag{4.22}$$

where H contains the successive hypothesis vectors $\mathbf{h}_1, \mathbf{h}_2, \ldots, \mathbf{h}_r$ and $W = HT$. There exists a transformation matrix T such that $R_1 HT = F$, where R_1 is the original correlation matrix, H contains the vectors $\mathbf{h}_1$, $\mathbf{h}_2, \ldots, \mathbf{h}_r$ used in computing F, and F contains the vectors of factor loadings as defined in either Eq. (4.1) or (4.3). The matrices R_1, H, and F are available as the result of any factor analysis computed using the general method of Sec. 4.4. The problem is to define the matrix T.

Multiplying Eq. (4.22) on the left by H' yields

$$H'R_1HT = H'F$$

The matrix $H'R_1H$ is of rank r and contains the covariances among linear functions defined by the columns of H. Premultiplying by the inverse of $H'R_1H$, the transformation matrix T is defined

$$T = (H'R_1H)^{-1}H'F \tag{4.23}$$

and

$$HT = H(H'R_1H)^{-1}H'F \tag{4.24}$$

The matrix HT is the factor-transformation matrix required to define the matrix of factor loadings as a function of R_1 in Eq. (4.22). A computationally simpler definition of T than that given in Eq. (4.23) can be derived from the relationship

$$H'R_1H = H'FF'H = H'F(H'F)'$$

The matrix $H'F$ is of rank r. Let $H'F = A$. Then

$$(H'R_1H)^{-1} = (AA')^{-1} = A'^{-1}A^{-1}$$

and Eq. (4.23) becomes

$$T = (AA')^{-1}A = A'^{-1}A^{-1}A = A'^{-1} = (H'F)'^{-1} \tag{4.25}$$

That $H'R_1H = H'FF'H$ can be shown as follows: $R_1 = R_{r+1} + FF'$. Pre- and postmultiplication of both sides by H' and H, respectively, yields $H'R_1H = H'R_{r+1}H + H'FF'H$. From equations given in Sec. 4.5, $H'R_{r+1}H = 0$. Therefore, $H'R_1H = H'FF'H$.

Since the columns of F have been shown to contain correlation coefficients relating the p original variables to the r factors, the columns of HT contain weighting coefficients defining factor variates having these same correlations with the original variables. If one wants to compute for each individual a set of derived factor scores such that the derived scores will have correlations with the original variables as specified in F, then Eq. (4.24) should be used. (Alternatively, one could compute $HT = W = R_1^{-1}F$, but this solution is computationally burdensome with large matrices and may be laden with error when R_1 is nearly singular.) The columns of the $p \times r$ matrix HT contain weighting coefficients to be applied to z scores on the original measures in computing factor scores. For application to raw scores, the coefficient associated with each variable should be divided by the standard deviation of that variable.

The notions of orthogonality of factor solutions and statistical independence among associated factor variates may need some additional clarification at this point. The "orthogonality" of an orthogonal-factor solution as defined by Eq. (4.6) refers precisely to the statistical independence of appropriately computed factor scores. In this section, methods for obtaining factor-transformation matrices $W = HT$ satisfying Eq. (4.1) were described. If the original factor analysis, as well as any subsequent factor rotations, were "orthogonal," then $HT = W$ also satisfies

$$T'H'RHT = I \tag{4.26}$$

If factor scores are obtained by applying the factor transformation to the original (z-score) data, the transformed factor variates can be represented as follows in the matrix Y:

$$Z'HT = Y$$

where Z' contains standardized scores on the p original variables and HT is the factor-transformation matrix.

Noting that the original variables in the matrix Z' are in z-score form, the intercorrelations among them can be represented

$$R = \frac{1}{n} Z'Z$$

where each off-diagonal element in R is a mean z-score cross product. Upon substituting $(1/n)Z'Z = R$ into Eq. (4.26), the following relationships are apparent:

$$T'H'RHT = \frac{1}{n} THZ'ZHT = \frac{1}{n} Y'Y = I$$

This shows that the mean factor-variate cross products are zero; hence, the factor-score intercorrelations are zero.

4.7 Communality Estimates in Factor Analysis

For most factor theorists, the method of factor analysis is concerned only with accounting for the *relationships* (correlations) among several variables in terms of a factor matrix of lesser rank. Factor analysis is undertaken to explain relationships among many variables in terms of a relatively few factors. The general linear computation method described in Sec. 4.4 involves a successive partialing out of variance associated with the successive factors. Intercorrelations among the multiple variables are said to be accounted for when the off-diagonal elements in the last residual correlation matrix are reduced to near-zero value. The residual correlation matrix, after r factors have been partialed out, can be represented as the difference between the original correlation matrix and a series of *rank-1* matrices that are outer products of the vectors of factor loadings

$$\begin{aligned} R_{r+1} &= R - \mathbf{f}_1\mathbf{f}_1' - \mathbf{f}_2\mathbf{f}_2' - \cdots - \mathbf{f}_r\mathbf{f}_r' \\ &= R - (\mathbf{f}_1\mathbf{f}_1' + \mathbf{f}_2\mathbf{f}_2' + \cdots + \mathbf{f}_r\mathbf{f}_r') \end{aligned} \tag{4.27}$$

Assuming the objective of accounting for *relationships* among variables, the factors should be defined so that the off-diagonal elements in R_r are reduced essentially to zero. The sum of the rank-1 product matrices can be represented in matrix notation as the product of the factor matrix F and its transpose

$$FF' = \mathbf{f}_1\mathbf{f}_1' + \mathbf{f}_2\mathbf{f}_2' + \cdots + \mathbf{f}_r\mathbf{f}_r' \tag{4.28}$$

where each element in FF' is recognized as the sum of r simple scalar products, according to the row-by-column rule for multiplication of matrices. Thus, the final residual correlation matrix can be defined as

$$R_{r+1} = R - FF' \tag{4.29}$$

It should be emphasized that the relationships among variables are represented entirely in the *off-diagonal* elements of the correlation matrix R. The objective of accounting for relationships among variables does not require that the diagonal elements of R_{r+1} be reduced to zero. In one sense, this objective does not involve the diagonal elements at all. Disregard of the diagonal elements in the correlation matrix leads to a fundamental equation in factor analysis. Let $\hat{R}$ be a matrix in which the off-diagonal elements are the correlations among the original variables and the elements in the principal diagonal are called *communalities*. The communalities are elements defined to satisfy the requirement that the correlations (off-diagonal) be accounted for by a parsimonious number of factors. The fundamental factor-analysis equation becomes

$$\hat{R} = FF' \tag{4.30}$$

The objective of factor analysis, in certain circumstances and for some investigators, is to define a factor transformation [Eq. (4.1)] such that the factor matrix F satisfies Eq. (4.30), where $\hat{R}$ is the original correlation matrix with communality estimates in the principal diagonal. The communalities are the diagonal elements in the matrix FF'. It can be seen from Eq. (4.30) that the communality for each variable is the sum of squares of factor loadings for that variable on the $r < p$ "common factors." The purpose of the factor analysis is to define a factor transformation that will account completely for correlations among the several variables with a minimum number of factors, and *communalities* are the diagonal elements of $\hat{R}$ required to satisfy Eq. (4.30) with a minimum number of factors. The general linear computation method described in Sec. (4.4) can be employed to analyze a "reduced" correlation matrix with communality estimates in the principal diagonal in the same manner that it can be used to analyze a matrix of intercorrelations with unities in the principal diagonal. Admittedly, estimating communality values in advance of the analysis poses a problem.

One frequently used procedure is to begin by factoring the original correlation matrix R to determine the number of factors for which the sum of squares of loadings for all variables on each factor exceeds 1.0. The communality for each variable is then estimated as the sum of squares of loadings on the r factors, and these values are subsequently inserted into the principal diagonal to define $\hat{R}$. The complete factor analysis is then repeated using $\hat{R}$ instead of R. In fact, some investigators repeat this process of factor analysis, estimating communalities from the resulting factor loadings and subsequently by re-analyzing $\hat{R}$ several times to obtain revised estimates at each stage.

Another frequently used procedure, which has only modest mathematical and statistical justification, involves insertion into the principal diagonal of the original correlation matrix R the largest (absolute) off-diagonal element in each row or column. The primary justification for this procedure is historical precedent and the fact that it works reasonably well. The correlation between two variables which measure the same thing, like the correlation of a variable with itself, can be conceived as an estimate of the reliability of the measurements. In a well-designed factor analysis enough redundancy is present so that communalities should tend to approach the reliabilities of the measurements. In this sense, use of the largest absolute value has some theoretical justification. The matrix $\hat{R}$ is thus defined as a matrix that contains as off-diagonal elements the original correlations among variables and in the principal diagonal the largest absolute value found among off-diagonal elements in the row or column. The factor analysis, using methods described in Sec. (4.4), can then be undertaken on $\hat{R}$ in an attempt to satisfy Eq. (4.30).

A procedure for estimation of communality values that has stronger statistical justification involves the use of the multiple R^2 (square of the multiple-correlation coefficient) relating a given variable to all others in the set. As is well known, the multiple R^2 can be interpreted as the proportion of variance in one variable that can be accounted for by regression on the other variables. Thus, the multiple R^2 represents the proportion of variance for each variable that is shared in common by other variables. From a computational point of view, the multiple R^2 of each variable with all others can be obtained readily from the diagonal elements of the inverse of the matrix of intercorrelations among all p variables (Sec. 2.7.5). Let D be a diagonal matrix containing the diagonal elements from the inverse of the complete correlation matrix. Then D^{-1} is a diagonal matrix containing the reciprocals of the diagonal elements in the inverse of the complete correlation matrix. A diagonal matrix containing the multiple R^2 of each variable with all others can be obtained by subtracting D^{-1} from an identity matrix

$$D_r = I - D^{-1} \tag{4.31}$$

where D_r is a diagonal matrix containing the multiple R^2 relating each variable to all others and D^{-1} is a diagonal matrix containing the *reciprocals* of the diagonal elements in the inverse of the complete correlation matrix.

While the use of multiple R^2 values as communality estimates has statistical appeal, several problems should be mentioned: (1) It is not practical to calculate the inverse of a very large correlation matrix. (2) As is well known, multiple R^2 values computed from *sample* data tend to approach 1.0 as the number of variables approaches the sample size. Thus, the R^2 values calculated from sample data will tend to yield exaggerated estimates of communalities unless the sample n is substantially larger than the number of variables in the matrix. (3) Variables are usually chosen for inclusion in a factor analysis because of recognized high correlations with other variables. Particularly in a large matrix, there is always the danger that at least one of the R^2 values will approach unity, which means that the matrix will be essentially singular and the inverse will be unstable (Sec. 2.5).

It should be recognized that the use of communality estimates in the principal diagonal changes the statistical properties of factor analysis. One is, in effect, analyzing covariances among theoretical common-part scores that do not exist in nature, a concept developed in greater detail in the next section. If one performs a factor analysis of a matrix $\hat{R}$ which has communality estimates in the principal diagonal, the analysis satisfies $\hat{R}\hat{W}D_1 = F_1$ [Eq. (4.1)] and $D_1\hat{W}'\hat{R}\hat{W}D_1 = \Sigma_1$ [Eq. (4.2)]. Factors which are orthogonal as defined by $D_1\hat{W}'\hat{R}\hat{W}D_1 = I$ of Eq. (4.2) do not, in general, satisfy $D_1\hat{W}'R\hat{W}D_1 = I$ when the original correlation matrix with unities in the principal diagonal is used to define correlations among factors. In geometric terms, factors which are statistically orthogonal in the theoretical "common" test space are not, in general, orthogonal in the original measurement space. These problems do not arise when unity or the total variance is used in the principal diagonal of the matrix analyzed (Overall, 1962). Since factor variates are composite variables defined to represent the factor dimensions, the correlations among precisely defined factor variates (Sec. 4.6) should be the same as the correlations among the factors in the original measurement space. Thus, if the purpose of the factor analysis is to define factor variates that have precisely known statistical relationships, the use of communality estimates should perhaps be avoided. Factor analysis can be used for the purpose of achieving a meaningful orthogonal transformation preliminary to other multivariate analyses (Chaps. 8, 14, and 15), and for this purpose strict statistical orthogonality of factors may be desired. On the other hand, if the purpose of the factor analysis is a more theoretical one of explaining the correlations among variables in terms of a minimum number of factors and the investigator is not concerned with definition of factor variates having precisely defined statistical independence, the use of communality estimates is recommended.

4.8 The Reduced Correlation Matrix as a Matrix of Covariances among Common-part Scores

In the preceding section, the communality of a variable was defined as the value of the diagonal element in $\hat{R}$ necessary to satisfy Eq. (4.30) with a minimum number of factors. We shall now attempt to present a more theoretical, if not more practical, definition of communalities and the associated reduced correlation matrix.

Each variable can be conceived as composed of several, say r, common factor components plus a residual unique component that is not represented in any other variable. Let the standardized z score for the ith variable be represented as

$$z_i = c_{1i} + c_{2i} + \cdots + c_{ri} + u_i \tag{4.32}$$

where the $c_{1i}, c_{2i}, \ldots, c_{ri}$ are common factor components representing the status of the individual on each of r common factors. We shall refer to these as common-part components and to the sum of all r common-part components for a variable as the *common-part score.*

The purpose of factor analysis, at least in the more traditional applications, is to determine the factorial composition of each variable. Let us begin by assuming a model in which the score on the ith variable, excluding the unique component, is specified in terms of common-part components only

$$\hat{z}_i = c_{1i} + c_{2i} + \cdots + c_{ri} \tag{4.33}$$

where $\hat{z}_i$ is the common-part score for the ith variable. The common-part components are assumed to be mutually independent and to sum to zero across all individuals

$$\frac{1}{n}\sum c_{1i}c_{2i} = \frac{1}{n}\sum c_{1i}c_{ri} = \cdots = \frac{1}{n}\sum c_{2i}c_{ri} = 0 \tag{4.34}$$

The *variance* of any score is defined as the mean squared deviation. Since the common-part components sum to zero across all individuals, the common-part score $\hat{z}_i$ has zero mean. The variance of $\hat{z}_i$ is thus

$$\begin{aligned} c_{ii} &= \frac{1}{n}\sum \hat{z}_i\hat{z}_i \\ &= \frac{1}{n}\sum (c_{1i} + c_{2i} + \cdots + c_{ri})(c_{1i} + c_{2i} + \cdots + c_{ri}) \\ &= \frac{1}{n}\sum c_{1i}c_{1i} + \frac{1}{n}\sum c_{1i}c_{2i} + \cdots + \frac{1}{n}\sum c_{ri}c_{ri} \end{aligned}$$

Since all mean product terms involving *different* factors drop out under the assumption of independence of factors [Eq. (4.34)], the variance of the common-part score for the ith variable is simply the sum of the variances of the r common-part components

$$c_{ii} = \frac{1}{n}\sum c_{1i}c_{1j} + \frac{1}{n}\sum c_{2i}c_{2j} + \cdots + \frac{1}{n}\sum c_{ri}c_{rj}$$

or

$$c_{ii} = \sigma_{1i}^2 + \sigma_{2i}^2 + \cdots + \sigma_{ri}^2 \tag{4.35}$$

The variance of the common-part scores is less than the variance of the total score because the unique component is not included. The variance of the observed z score of Eq. (4.32) is the mean squared deviation

$$\begin{aligned}\sigma_{ii}^2 &= \frac{1}{n}\sum z_i z_i \\ &= \frac{1}{n}\sum (c_{1i} + c_{2i} + \cdots + c_{ri} + u_i)(c_{1i} + c_{2i} + \cdots + c_{ri} + u_i) \\ &= \frac{1}{n}\sum c_{1i}c_{1i} + \frac{1}{n}\sum c_{1i}c_{2i} + \cdots + \frac{1}{n}\sum c_{ri}u_i + \frac{1}{n}\sum u_i u_i\end{aligned}$$

After deleting the terms that involve mean products between different factors (which are equal to zero), the total variance of z_i is seen to be equal to the sum of the variances of the separate independent components *including* the unique component

$$\sigma_{ii}^2 = \frac{1}{n}\sum z_i z_i = \sigma_{1i}^2 + \sigma_{2i}^2 + \cdots + \sigma_{ri}^2 + \sigma_{ui}^2 \tag{4.36}$$

The *communality* of a variable is equal to the variance of the common-part scores. It is obvious from comparison of Eqs. (4.35) and (4.36) that the variance of the common-part scores is smaller than the total variance of z_i by the magnitude of the variance of the unique component.

The covariance between common-part scores for two variables, say the ith and jth, is equal to the mean cross product of the common-part components. Since the unique component of each variable measures something that is not measured by any other variable, unique components do not contribute to the correlation between variables. Thus, the *covariance* between common-part scores is equal to the *correlation* between the total scores with unique components included. The covariance between common-part scores for the ith and jth variables is defined as the mean cross product

$$c_{ij} = \frac{1}{n} \sum \hat{z}_i \hat{z}_j$$

$$= \frac{1}{n} \sum (c_{1i} + c_{2i} + \cdots + c_{ri})(c_{1j} + c_{2j} + \cdots + c_{rj})$$

$$= \frac{1}{n} \sum c_{1i}c_{1j} + \frac{1}{n} \sum c_{2i}c_{2j} + \cdots + \frac{1}{n} \sum c_{ri}c_{rj}$$

Once again, all terms involving products of components representing different factors drop out because of the postulated independence of factors. Thus,

$$c_{ij} = \frac{1}{n} \sum c_{1i}c_{1j} + \frac{1}{n} \sum c_{2i}c_{2j} + \cdots + \frac{1}{n} \sum c_{ri}c_{rj} \tag{4.37}$$

The product-moment correlation between two variables is the mean z-score cross product. Since

$$\frac{1}{n} \sum c_{1i}u_j = \frac{1}{n} \sum c_{2i}u_j = \frac{1}{n} \sum u_i c_{1j} = \cdots = \frac{1}{n} \sum u_i u_j = 0$$

assuming independence of the unique components, the mean z-score cross product

$$r_{ij} = \frac{1}{n} \sum z_i z_j$$

$$= \frac{1}{n} \sum (c_{1i} + c_{2i} + \cdots + c_{ri} + u_i)(c_{1j} + c_{2j} + \cdots + c_{rj} + u_j)$$

can easily be shown to be precisely the same as the *covariance* of the common-part scores. Thus, the product-moment correlations among pairs of variables are estimates of the *covariances* among common-part scores, and the communalities are estimates of the *variances* of the common-part scores. The reduced correlation matrix $\hat{R}$, which has communality estimates in the principal diagonal and correlations as the off-diagonal elements, can be conceived as a matrix of variances and covariances among common-part scores.

Since common-part components cannot be measured directly, there is no way to obtain by direct measurement a set of scores that will yield such a variance-covariance matrix. In that sense, factor analysis of a reduced correlation matrix $\hat{R}$ results in transformation of hypothetical variables that do not exist in nature. Factor transformations that result in orthogonal factors when applied to the hypothetical common-part scores do not usually represent truly orthogonal transformations when applied to real measurement data. It is for this reason that the distinction between theoretical and more practical statistical applications of factor-analysis methods was made in the preceding section. The use of communality estimates is consistent with an elegant theoretical model that has the purpose of defining common-part components for each variable in a hypothetical common-factor domain from which all unique variance has been excluded.

4.9 The Number of Factors

The problem of determining the number of factors that should be extracted is a very difficult one and is obviously highly related to the communality problem. The purpose of inserting communality estimates into the principal diagonal is to reduce the rank of the correlation matrix so that the off-diagonal elements can be accounted for by a relatively small number of factors. The question considered here is: How many factors adequately account for $\hat{R}$?

Guttman (1954) has offered a proof that the minimum number of factors required to explain completely the intercorrelations among p variables is equal to the number of principal components of the matrix R (unities in diagonal) having associated variances greater than $\lambda_j = 1.0$. The rule of thumb that the variance accounted for by each factor (sum of the squared loadings on each factor) should exceed 1.0 has been used most frequently as an arbitrary basis for deciding how many factors to define. As far as the present authors are aware, there is absolutely no mathematical or statistical justification for this arbitrary cutting point when a matrix $\hat{R}$ with communality estimates in the principal diagonal is being analyzed. In spite of this, the most frequently employed rule for deciding how many factors to consider has been that the sum of squares of factor loadings should exceed 1.0 for each factor separately.

Consideration of Eqs. (4.27) to (4.30) suggests that the basis for deciding how many factors to define should involve consideration of the extent to which $\hat{R}$ can be approximated by the product FF'. That is, factors should be defined until the elements in the residual matrix $R_{r+1} = \hat{R} - FF'$ approach zero within desired tolerance. One can define arbitrarily a maximum acceptable value for the mean of absolute values of all off-diagonal elements in the final residual matrix. Better, one can require that the average of the three largest (absolute) residual correlations be less than a predefined value.

In chapters which follow, direct cluster-oriented factor solutions are discussed as special cases of the general method. Where such direct solutions, which require no rotation, are used, attention to the number of variables having "significant" loadings on each factor is a meaningful basis for the decision of when to stop defining additional factors. If one wants each factor to represent a primary underlying source of variance that is measured in common by several variables, and if factor loadings are conceived as indices of relationship of variables to the factors, then it is desirable to have at least three variables with "significant" loadings on each factor. The problem is, of course, that the definition of "significant" has not been developed statistically for the case where communality estimates are used in the principal diagonal of the matrix being analyzed. In lieu of a statistical definition of significance, it is reasonably easy after experience in a particular research area to define what should be considered a minimum adequate loading for a single variable on a factor. In our own experience, factors defined by three or more variables having loadings exceeding .35 have been found to be stable and replicable.

Where factor analysis is being employed for statistical data reduction and transformation, the question of number of factors is less critical. Even where factor analysis is being used to develop a conceptual model or a set of theoretical constructs, the problem is frequently exaggerated. In many areas of investigation, there will be found to be good three-factor solutions, good four-factor solutions, and good five-factor solutions. All can be good and appropriate! Obviously, the five-factor solution will account in more detail for total individual differences represented in the larger set of variables than the three-factor solution will. On the other hand, the three-factor solution will be more abstract and parsimonious. From a data-reduction point of view, the goal is to account for a large percentage of the total variance with a relatively small number of factors. This view of the problem is useful when the matrix analyzed contains unities in the principal diagonal. Statistical data reduction is usually considered to be adequate and effective when the number of factors is approximately one-fourth the number of original variables and the variance accounted for is 50 to 75 percent of the total variance. One can predict with a twinge of wryness that the results obtained from most factor analyses in the psychological and psychiatric domain will fall within these broad limits.

4.10 Summary

The purpose of this chapter has been to introduce the method of factor analysis in terms of a very general model that has many variations, depending upon the definition of the transformation matrices. Several concepts which are important in factor analysis and which lead to further variation in the application of the method have also been introduced. For some investigators, further transformational rotation is an essential part of the method. Similarly, for some theorists, the use of communality estimates is part of the unique definition of factor analysis. The place of both these concepts in the general model has been discussed, but a rigid position on whether either is required has been avoided.

Principal-components analysis was presented prior to introducing factor analysis instead of being discussed as one of several variations of the general model because the goal of components analysis is somewhat different. In the next chapter the principal-axes method of factor analysis, which is formally equivalent to principal components under certain circumstances, will be presented as a special case of the general model, and in the next few chapters other variations will be described. This treatment of factor analysis, while not exhaustive, should help consolidate the central idea of this chapter that factor analysis is capable of many variations under the same general model.

In considering the relationships and differences between principal-components analysis, as discussed in Chap. 3, and factor analysis, as defined in the present chapter, the difference in scaling should be noted. Factors are scaled so that the associated factor variates all have unit variance. Principal-component vectors are normalized in such manner that the linear composites have different variances $\lambda_1, \lambda_2, \ldots, \lambda_r$. As we shall see in the discussion of principal-axes factor analysis, under certain conditions one can go directly from a principal-components solution to a principal-factor solution simply by rescaling. In the view of the present authors, the two fundamental distinctions between principal-components analysis and factor analysis are represented in the scaling and in the emphasis with regard to interpretation. The emphasis in factor analysis is on understanding the nature and structure of complex measurements through examination of their relationships to a relatively few underlying factors. Only secondary emphasis is usually accorded the definition of factor scores. By way of contrast, the emphasis in principal-components analysis is on the definition of composite variables which have certain optimal statistical properties. With this different emphasis, the use of communality estimates in a principal-components analysis is not considered meaningful, although the factor analysis of a matrix with unities in the principal diagonal is desirable at times.

4.11 References

Guttman, L.: Some Necessary Conditions for Common-factor Analyses, *Psychometrika*, **19**:149–161 (1954).

Overall, J. E.: Orthogonal Factors and Uncorrelated Factor Scores, *Psychol. Repts.*, **10**:651–662 (1962).

5
Principal-axes Factor Analysis

5.1 Introduction

Principal-axes factor analysis is computationally so similar to principal-components analysis that little discussion of the method of computation will be necessary. Conceptually, however, there is a difference in emphasis. Whereas principal-components analysis is primarily concerned with the definition of linear functions of composite scores having certain optimal statistical properties, factor analysis is concerned with relationships among test variates and relationships of the test variates to derived factors.

Formally, the difference between principal components and principal axes is only one of scaling, although principal-axes factor analysis is frequently accomplished on a correlation matrix with communality estimates in the principal diagonal while the logic of principal-components analysis requires total variances (or unity) in the principal diagonal of the matrix analyzed. As described in Chap. 3, principal-component vectors are normalized so that the sum of squares for elements in each vector is equal to 1.0. The magnitudes of individual elements in a principal-component vector thus depend on the number of elements in the vector. The variance accounted for by a principal component is represented in the associated λ_i [Eq. (3.5)]. For principal-axes factor analysis, the vectors are rescaled so that the sum of squares of the factor loadings is equal to λ_i. Principal-axes factor loadings can be obtained by multiplying each element of the corresponding (normalized) principal-component vector by the square root of λ_i. From a formal point of view, the scaling is the only difference between vectors of principal components and vectors of principal-axes factor loadings. After this type of scaling, the vectors of factor loadings can be considered to contain correlations between the original tests and the derived factor variates. Principal-axes factor analysis is usually applied to a matrix of intercorrelations, rather than to a matrix of covariances. Psychological measurements tend to have arbitrary scale units; hence factor analyses are usually undertaken using correlation matrices to remove the influence of differing scales of measurement among the variables.

5.2 Definitions

Principal-axes factor analysis, like all methods of linear factor analysis, results in a transformation of the original correlation matrix as defined in Eq. (4.1)

$$R\hat{W}D_1 = F_1$$

In principal-axes analysis, the matrix $\hat{W}$ is defined to satisfy the conditions

$$\hat{W}'R\hat{W} = \Lambda \tag{5.1}$$

where Λ is a diagonal matrix containing the principal variances

$$D_1\hat{W}'R\hat{W}D_1 = I \tag{5.2}$$

where $D_1 = \Lambda^{-\frac{1}{2}}$ is a diagonal matrix containing the reciprocals of the square roots of the principal variances $\lambda_1, \lambda_2, \ldots, \lambda_r$, and

$$\hat{W}'\hat{W} = I \tag{5.3}$$

It will be recognized that the factor-transformation matrix $\hat{W}$ contains as columns the principal-component vectors satisfying $(R - \lambda I)W = 0$ of Sec. 3.2. Thus, principal-axes analysis is a special case of the general linear factor-analysis model in which hypothesis vectors are chosen to maximize the variance accounted for. Principal-axes factors are defined in such a manner that the variance accounted for by each successive orthogonal factor is maximum

$$\mathbf{w}_i'R\mathbf{w}_i = \lambda_i \tag{5.4}$$

This means that there exists no other set of r factors that will account for more of the total variance in a correlation matrix than the first r principal-axes factors do.

5.3 Computing Principal-axes Factor Loadings

The computation of principal-axes factors follows the general method for orthogonal-factor analysis described in Sec. (4.4)

$$R_1\mathbf{h}_1 s_1 = \mathbf{f}_1$$
$$R_2 = R_1 - \mathbf{f}_1\mathbf{f}_1'$$
$$R_2\mathbf{h}_2 s_2 = \mathbf{f}_2$$
$$R_3 = R_2 - \mathbf{f}_2\mathbf{f}_2'$$
$$\cdots\cdots\cdots\cdots$$
$$R_r\mathbf{h}_r s_r = \mathbf{f}_r$$

At each stage in the analysis, the vector $\mathbf{h}_i$ is obtained by the iterative method described in Sec. 3.3.

$$R_i\mathbf{t}_0 = \mathbf{x}_0 \qquad \mathbf{t}_1 = \frac{1}{\sqrt{\mathbf{x}_0'\mathbf{x}_0}}\mathbf{x}_0$$

$$R_i\mathbf{t}_1 = \mathbf{x}_1 \qquad \mathbf{t}_2 = \frac{1}{\sqrt{\mathbf{x}_1'\mathbf{x}_1}}\mathbf{x}_1$$

$$R_i t_{n-1} = \mathbf{x}_{n-1} \qquad \mathbf{t}_n = \frac{1}{\sqrt{\mathbf{x}_{n-1}'\mathbf{x}_{n-1}}}\mathbf{x}_{n-1}$$

The process is continued until $\mathbf{t}_{n-1} = \mathbf{t}_n$ to a prespecified number of decimal places. When the iterative process has resulted in a stable vector, that vector is taken as $\mathbf{h}_i = \mathbf{t}_n$ and the factor loadings on the ith factor are computed

$$R_i\mathbf{h}_i s_i = \mathbf{f}_i \tag{5.5}$$

where R_i is the residual correlation matrix at the ith stage, $\mathbf{h}_i$ is the ith principal-component vector scaled so that $\mathbf{h}_i'\mathbf{h}_i = 1$, s_i is a scaling coefficient equal to $1/\sqrt{\mathbf{h}_i'R_i\mathbf{h}_i}$, and f_i contains the projections of the p tests on the ith principal factor. Since $\mathbf{h}_i'R_i\mathbf{h}_i = \lambda_i$ in principal-components analysis [Eq. (3.5)], the scaling coefficient $s_i = \lambda_i^{-\frac{1}{2}} = 1/\sqrt{\lambda_i}$. Similarly, since $R_i\mathbf{a}_i = \lambda_i\mathbf{a}_i$ in principal-components analysis [Eq. (3.4)] and $s_i = \lambda_i^{-\frac{1}{2}}$, the vector of factor loadings is $\mathbf{f}_i = \lambda_i^{\frac{1}{2}}\mathbf{h}_i = \sqrt{\lambda_i}\mathbf{h}_i$, where $\mathbf{h}_i$ is the ith principal-component vector.

In practice, since principal-axes factor loadings are precisely proportional to the principal-components vectors computed from the same correlation matrix, the computational procedure described in detail in Sec. 3.3 can be recommended. As a final step, the elements in each principal-component vector should be multiplied by the square root of the associated λ_i.

5.4 A Computational Example

There would seem to be no need for a trivial computational example since the iterative procedure for computing principal components has been described in detail in Sec. 3.4. The major function of the following example is to illustrate in a practical way what principal-axes analysis does and does not achieve.

Because factor analysis is concerned with the analysis of relationships among variables, it is necessary that a sufficient number of variables be included to provide enough correlations to work with. The 16 symptom ratings of the BPRS represent such a set of variables.

Intercorrelations among the 16 BPRS symptom rating variables computed using the sample of 6,000 patient profiles are presented in Table 5.1. The diagonal elements in the table are communality estimates, in this case the highest correlation in each column. The first step in the analysis was to compute the principal-component vectors and associated variances. This was accomplished by the iterative method described in Chap. 3. All principal components for which the variance λ_i exceeded 5 percent of the sum of the diagonal communality estimates were computed. The first four principal-component vectors, stable to four decimal places, are presented in Table 5.2. It will be noted, incidentally, that the first two principal-component vectors differ somewhat from those of the previous components analysis of the 16 symptom-rating variables shown in Tables 3.5 and 3.6. There are several reasons for this difference. The previous analysis was undertaken on elevation-corrected profiles. In addition, principal components computed from correlation matrices will not in general be the same as those computed from covariance matrices even though both are derived from the same data. Finally, communality estimates were used in the principal diagonal for the factor analysis.

Principal-axes factor loadings were computed from the principal components by multiplying each element in the normalized principal-component vector by the square root of the associated λ_i. The principal-component vectors (Table 5.2) were scaled so that the sum of squares for elements in each vector equaled unity. Multiplication by $\lambda_i^{\frac{1}{2}}$ results in vectors of factor loadings for which the sum of squares is λ_i. The resulting principal-axes factor loadings are presented in Table 5.3. The coefficients in each column can be conceived as representing correlation coefficients between the 16 symptom-rating variables and one of the principal factors. The sum of squares of factor loadings in each column represents the variance accounted for by the factor and is equal to the λ_i value for the corresponding principal component. The column labeled h^2 is a reestimate of the communality derived from the analysis by squaring the loadings for each variable and summing across all factors. As pointed out in Chap. 4, it is a common practice to repeat the entire analysis using these new communality estimates as diagonal entries.

An important insight into the method of principal-axes factor analysis can be gained from examination of Table 5.3. Although principal-axes factors are statistically orthogonal and account for maximum possible variance, they tend to be complex and difficult to interpret. Most variables relate substantially to the first principal-axes factor. Other factors are not as complex as the first, but interpretation is still difficult. In the usual course of factor analysis, a principal-axes factor analysis is only the first step. Rotational transformation is required to bring the factors into more meaningful relationship to the original test variates.

Table 5.1 Intercorrelations among 16 Symptom-rating Variables

$n = 6{,}000$

	1	*2*	*3*	*4*	*5*	*6*	*7*	*8*	*9*	*10*	*11*	*12*	*13*	*14*	*15*	*16*
1	.31	.28	−.06	−.08	.17	.07	−.06	.14	.31	.14	−.01	−.02	.05	−.03	−.03	−.02
2	.28	.49	−.09	−.11	.36	.46	−.05	−.10	.49	.22	.25	.06	.00	.02	.03	−.16
3	−.06	−.09	.57	.36	.00	.13	.42	−.00	−.00	.03	.14	.20	.42	.38	.16	.57
4	−.08	−.11	.36	.53	−.06	.13	.34	.24	−.19	.09	.29	.35	.09	.23	.53	.33
5	.17	.36	.00	−.06	.44	.22	−.01	.05	.44	.12	.05	.03	.07	−.05	.01	.01
6	.07	.46	.13	.13	.22	.46	.25	−.02	.17	.16	.29	.16	−.00	.18	.17	−.02
7	−.06	−.05	.42	.34	−.01	.25	.42	.07	−.10	.05	.18	.27	.17	.32	.25	.29
8	.14	−.10	−.00	.24	.05	−.02	.07	.30	−.11	.27	.16	.19	−.13	.10	.30	.03
9	.31	.49	−.00	−.19	.44	.17	−.10	−.11	.49	.18	.00	−.09	.22	−.03	−.16	−.00
10	.14	.22	.03	.09	.12	.16	.05	.27	.18	.47	.47	.07	−.12	.35	.16	−.03
11	−.01	.25	.14	.29	.05	.29	.18	.16	.00	.47	.47	.37	−.05	.34	.47	.09
12	−.02	.06	.20	.35	.03	.16	.27	.19	−.09	.07	.37	.52	.03	.14	.52	.19
13	.05	.00	.42	.09	.07	−.00	.17	−.13	.22	−.12	−.05	.03	.42	.12	−.03	.39
14	−.03	.02	.38	.23	−.05	.18	.32	.10	−.03	.35	.34	.14	.12	.38	.16	.16
15	−.03	.03	.16	.53	.01	.17	.25	.30	−.16	.16	.47	.52	−.03	.16	.53	.18
16	−.02	−.16	.57	.33	.01	−.02	.29	.03	−.00	−.03	.09	.19	.39	.16	.18	.57

Table 5.2 Principal-component Vectors

	I	*II*	*III*	*IV*
Somatic concern	−.0117	.2684	.0715	.0959
Anxiety	.0399	.5251	.0488	.1056
Emotional withdrawal	.3412	−.1466	.4255	−.1723
Conceptual disorganization	.3784	−.1567	−.0828	.1872
Guilt feelings	.0300	.3707	.1517	.2315
Tension	.1971	.3331	.0097	.0062
Mannerisms and posturing	.3163	−.0863	.1247	−.0687
Grandiosity	.1580	.0013	−.2498	.0349
Depressive mood	−.0383	.4476	.3420	.0898
Hostility	.1889	.2699	−.1951	−.4798
Suspiciousness	.3357	.2029	−.2672	−.2008
Hallucinatory behavior	.3305	−.0055	−.1581	.3599
Motor retardation	.1244	−.0384	.4655	.0578
Uncooperativeness	.2856	.0237	.0319	−.5407
Unusual thought content	.3714	−.0199	−.2967	.3765
Blunted affect	.2819	−.1822	.3821	.0967
Eigenvalue λ_i	2.8872	1.9360	1.4322	.6970

Table 5.3 Principal-axes Factor Loadings

	I_0	II_0	III_0	IV_0	h^2
Somatic concern	−.0200	.3735	.0856	.0801	.1536
Anxiety	.0678	.7307	.0584	.0881	.5497
Emotional withdrawal	.5798	−.2039	.5093	−.1439	.6578
Conceptual disorganization	.6429	−.2181	−.0991	.1563	.4951
Guilt feelings	.0510	.5158	.1816	.1933	.3390
Tension	.3349	.4635	.0116	.0052	.3272
Mannerisms and posturing	.5374	−.1201	.1493	−.0574	.3288
Grandiosity	.2685	.0019	−.2990	.0291	.1623
Depressive mood	−.0651	.6228	.4093	.0750	.5653
Hostility	.3211	.3756	−.2335	−.4006	.4592
Suspiciousness	.5705	.2823	−.3198	−.1677	.5356
Hallucinatory behavior	.5617	−.0077	−.1893	.3005	.4417
Motor retardation	.2114	−.0534	.5571	.0483	.3602
Uncooperativeness	.4854	.0331	.0382	−.4514	.4419
Unusual thought content	.6312	−.0277	−.3550	.3143	.6240
Blunted affect	.4791	−.2535	.4573	.0807	.5094
Variance	2.8867	1.9356	1.4319	.6968	

5.5 Approximate Principal Axes

For most practical purposes, there is little concern that orthogonal factors represent *precisely* the principal axes or associated principal components. Even in the modern computer era, the iterative procedure can require excessive computation when large matrices are to be analyzed. It is possible to arrive at a good orthogonal solution that is generally only very slightly less powerful in accounting for total variation than the exact principal-axes solution. This can be accomplished using only a fraction of the computation required for a complete principal-axes analysis.

If one wants to choose an original measurement variable to approximate the first principal-component or first principal-factor variate, which variable should one choose? Factor loadings can be conceived as correlations between a factor variate and the several original variables. The total variance accounted for by a factor is equal to the sum of squares of the factor loadings, and the first principal-axes factor accounts for maximum total variance. Thus, the measurement variable having the greatest sum of squared correlations with other variates is closest to the first principal axis.

The correlation coefficients relating each original variable to the one for which the sum of squares is greatest can be taken as the hypothesis vector $\mathbf{h}_1$ to define a factor that is close to the first principal axis. Variance associated with the first factor can be partialed out by computing a residual matrix in the usual way. The process can then be repeated to define a second approximate principal-axes factor. The whole procedure can be summarized in terms of the general orthogonal factor-analysis method of Sec. 4.4

$$R_1\mathbf{h}_1 s_1 = \mathbf{f}_1$$

where $\mathbf{h}_1$ is a vector containing as elements the correlation coefficients from that column of R_1 which has largest sum of squares and

$$s_1 = \frac{1}{\sqrt{\mathbf{h}_1' R_1 \mathbf{h}_1}}$$

$$R_2 = R_1 - \mathbf{f}_1\mathbf{f}_1'$$

$$R_2\mathbf{h}_2 s_2 = \mathbf{f}_2$$

where $\mathbf{h}_2$ is a vector with elements equal to the residual correlation coefficients in that column of R_2 for which the sum of squares is greatest and

$$s_2 = \frac{1}{\sqrt{\mathbf{h}_2' R_2 \mathbf{h}_2}}$$

$$R_3 = R_2 - \mathbf{f}_2\mathbf{f}_2'$$

. . . .

$$R_r\mathbf{h}_r s_r = \mathbf{f}_r$$

where $\mathbf{h}_r$ is that column of R_r in which the sum of squared elements is greatest and

$$s_r = \frac{1}{\sqrt{\mathbf{h}_r' R_r \mathbf{h}_r}}$$

5.6 Orthogonal Rotation of Principal-axes Factors

5.6.1 DEFINITIONS

Principal-axes factors are calculated to maximize the variance accounted for by a minimum number of factors. The first factor accounts for a maximum amount of the original variation, and each successive factor accounts for a maximum of the residual variation. Thus for a specified number of factors, no other solution will account for more of the total variance, and conversely no other solution will account for a specified amount of the total variance with fewer factors. In achieving this objective no attention is paid the psychological meaningfulness of the initial, unrotated orthogonal principal-axes factors. In point of fact, principal-axes factors tend to be maximally complex with regard to the configuration of variables because of the emphasis on maximizing the variance accounted for by each successive factor. Orthogonal rotation is the process of redefining factors for more meaningful reference to the several original measurement variables in such manner that the total variance accounted for by a fixed number of factors remains constant and statistical independence is maintained. The objective of the rotation is to increase the simplicity and meaningfulness of relationships of factors to the original measurement variables.

Simple structure is a concept first advanced by Thurstone (1947) to represent the goal of factor rotation. A factor structure is simple and meaningful to the extent that most variables relate highly to only one factor and each factor can be identified as representing that which is measured in common by a relatively small number of variables. Simple structure is achieved when, for each factor, the factor loadings (factor-variate correlations) for most variables are near zero and the remaining factor loadings are relatively large. The factor can then be conceived as representing the variance shared in common by the subset of variables that relate highly to it and not to represent that which is measured by other variables. In somewhat different terms, the objective of factor rotation is to display more clearly the configuration of relationships among the original measurement variables.

The process of factor rotation has been formally defined in Chap. 4 as the application of a transformation matrix $\hat{T}D$ to the matrix of unrotated factor loadings. From Eq. (4.1)

$$R\hat{W}D_1 = F_1$$

and from Eq. (4.3)

$$R\hat{W}D_1\hat{T}D_2 = F_1\hat{T}D_2 = F_1T = F_2 \tag{5.6}$$

For *orthogonal* rotation, the matrix $T = \hat{T}D_2$ must be chosen such that

$$D_2\hat{T}'\hat{T}D_2 = T'T = I \tag{5.7}$$

That is, the sum of squares within each column of T must equal 1.0, and the sum of cross products between each pair of columns must equal zero.

The process of factor rotation can be most easily understood as the process of defining new factors as simple linear combinations of the original factors. The columns of $\hat{T}$, and consequently the columns of $T = \hat{T}D_2$, consist of weighting coefficients which define new factors as weighted combinations of the r original factors. Factor loadings for the p original variables on the *first* new rotated factor are obtained by applying the weights in the first column of $T = \hat{T}D_2$ to the p rows of the original factor matrix. Factor loadings for the p original variables on the *second* rotated factor are obtained by applying the weights in the second column of $T = \hat{T}D_2$, and so on. This can be verified by considering the row-by-column matrix multiplication $F_1\hat{T}D_2 = F_1T = F_2$. The problem for orthogonal factor rotation is one of defining coefficients in $\hat{T}$ to achieve a rotated factor structure which is simple and meaningful under the restriction that $T'T = I$.

In actual practice, orthogonal rotation is accomplished by successive pairwise transformations of one pair of factors at a time until the desired simple structure is achieved. Although the final rotated-factor solution can be represented as a single composite transformation of the type defined in Eq. (5.6), the usual computations involve defining the rotated loadings for one factor as a weighted combination of the loadings on two previously defined factors.

At any stage in the process of rotation, the new rotated loadings on factor j are defined as a weighted combination of the preceding loadings on factors j and k

$$\hat{f}_{ij} = \frac{1}{\sqrt{1 + \phi^2}} f_{ij} + \frac{\phi}{\sqrt{1 + \phi^2}} f_{ik} \tag{5.8}$$

The new rotated loadings on factor k are also defined as a weighted combination of the preceding loadings on factors j and k

$$\hat{f}_{ik} = \frac{-\phi}{\sqrt{1 + \phi^2}} f_{ij} + \frac{1}{\sqrt{1 + \phi^2}} f_{ik} \tag{5.9}$$

It should be stressed that Eqs. (5.8) and (5.9) will result in an orthogonal rotation in which the total variance accounted for remains constant for *any* arbitrary choice of ϕ. In practice, the proper choice of ϕ is important in order for the rotation to achieve the desired goal of simple structure. In the next few sections, several alternative ways of determining ϕ will be described, and then it will be shown how pairwise rotation can be expressed in terms of a complete transformation matrix.

5.6.2 GRAPHIC METHOD FOR DEFINING COEFFICIENT OF ROTATION

Although the concept of rotation has been developed as a series of matrix operations which are conveniently defined and summarized by Eq. (4.3), a geometric representation of pairwise rotation is helpful in illustrating the results of orthogonal rotation as well as the desired characteristics of the coefficient ϕ. Historically, geometric concepts have played an important role in the development of factor analysis.

Factor loadings can be conceived as *projections* on orthogonal reference axes. For each pair of factors, the factor loadings can be taken as cartesian coordinate values to locate each variable in the plane defined by the two axes, as illustrated in Fig. 5.1 for the four principal-axes factors of Table 5.3. *Rotation* of the factor axes then involves relocating them by clockwise or counterclockwise movement around the origin. Both axes must be shifted in the same direction and to the same degree in order to maintain the geometric orthogonal relationship.

The coefficient ϕ of Eqs. (5.8) and (5.9) defines the left or right displacement of the vertical axis. A *negative* value of coefficient ϕ results in a counterclockwise rotation, and a *positive* value of coefficient ϕ results in a clockwise rotation. The object of the rotation is to increase the magnitude of loadings for certain variables, while at the same time decreasing their cross-factor loadings. In general, variables which load highly on one factor should have minimum loadings on the other, insofar as possible.

Consider the plot of principal-axes factors II_0 vs. III_0 in Fig. 5.1. The desirable first rotation is indicated by the dotted lines. The value of ϕ is the distance from the end of axis II_0 to the end of axis II_1 as measured at the perimeter of the unit rectangle. (By unit rectangle it is specified that the original vertical and horizontal axes are of unit length and thus that the rotated axes project to length greater than 1.0 in the rectangle.) Another way to define ϕ is to specify that it is equal to the ratio of the horizontal distance to the vertical distance for any point on the rotated axis. This ratio is unnecessary if the horizontal distance is evaluated at a vertical distance of 1.0.

Given the value of ϕ derived from consideration of the plot of loadings on factors II_0 vs. III_0, the rotational transformation Eqs. (5.8) and (5.9) are defined as follows:

$$(\hat{f}_i)_{\mathrm{II}_1} = \frac{1}{\sqrt{1 + .36^2}} (f_i)_{\mathrm{II}_0} + \frac{.36}{\sqrt{1 + .36^2}} (f_i)_{\mathrm{III}_0}$$

$$(\hat{f}_i)_{\mathrm{III}_1} = \frac{-.36}{\sqrt{1 + .36^2}} (f_i)_{\mathrm{II}_0} + \frac{1}{\sqrt{1 + .36^2}} (f_i)_{\mathrm{III}_0}$$

Calculation of rotated loadings for factors II_1 and III_1 resulted in the entries shown in the first two columns of the left-hand section of Table 5.4. The reader should verify how these transformed loadings were obtained from the original factor loadings shown in Table 5.3. The plots of the principal-axes factors after this first rotation are shown in Fig. 5.2.

Inspection of the plots in Fig. 5.2 and the projections on factors I_0 and III_1 of Table 5.4 suggests a counterclockwise rotation of the magnitude indicated by

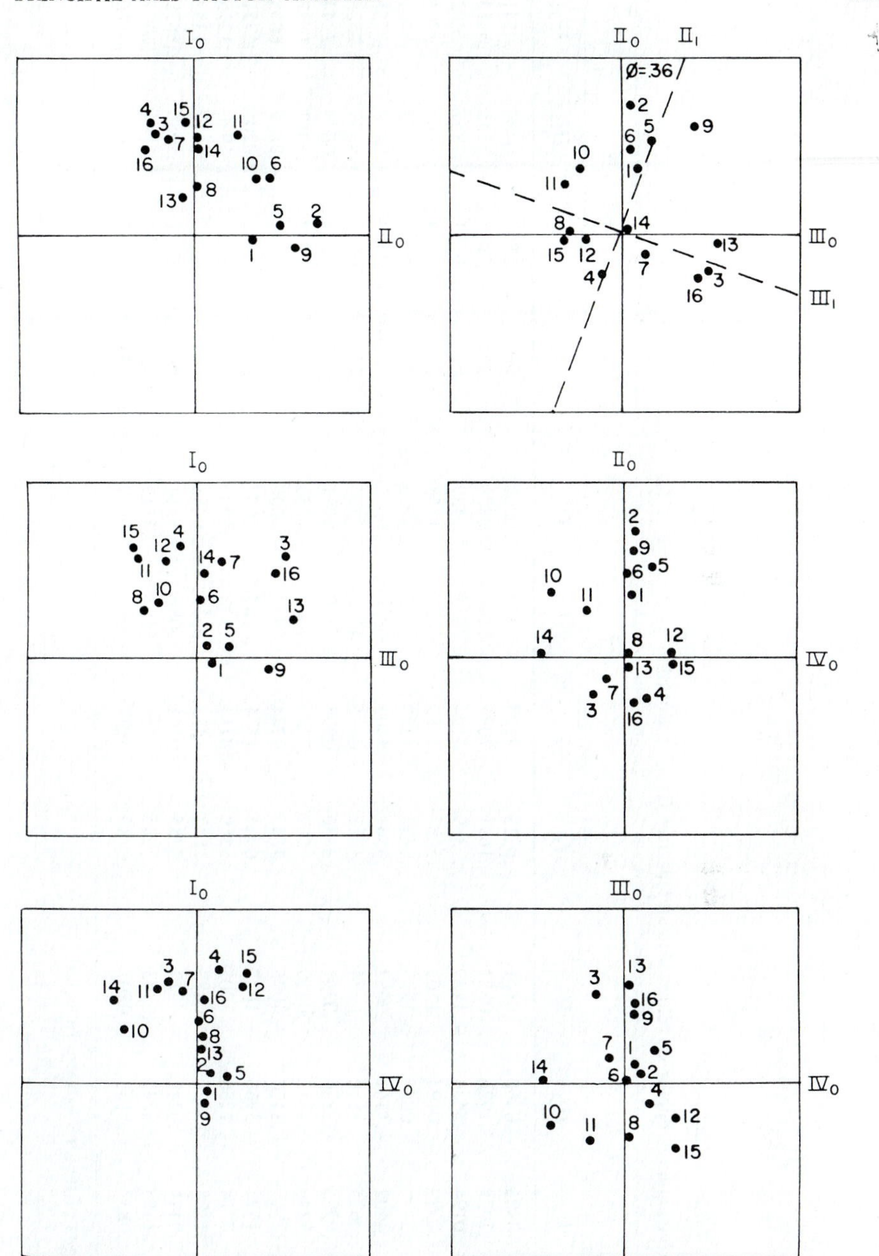

Figure 5.1 Plots of unrotated principal-axes factors with indication of one pairwise rotation using $\phi = .36$.

Table 5.4 Principal-axes Factor Loadings Calculated after Graphic Rotation

$\phi = +.36$				$\phi = -.74$				$\phi = +.84$			
I_0	II_1	III_1	IV_0	I_1	II_1	III_2	IV_0	I_2	II_1	III_2	IV_1†
−.0200	.3805	−.0460	.0801	.0113	.3805	−.0489	.0801	.0602	.3805	−.0489	−.0541
.0678	.7075	−.1926	.0881	.1691	.7075	−.1145	.0881	.1861	.7075	−.1145	.0413
.5798	−.0194	.5484	−.1439	.1399	−.0194	.7857	−.1439	.0146	−.0194	.7857	.2002
.6429	−.2389	−.0194	.1563	.5283	−.2389	.3668	.1563	.5051	−.2389	.3668	.2201
.0510	.5470	−.0038	.1933	.0433	.5470	.0273	.1933	.1575	.5470	.0273	−.1201
.3349	.4402	−.1461	.0052	.3561	.4402	.0818	.0052	.2760	.4402	.0818	.2251
.5374	−.0625	.1812	−.0574	.3242	−.0625	.4653	−.0574	.2113	−.0625	.4653	.2525
.2685	−.0995	−.2820	.0291	.3836	−.0995	−.0670	.0291	.3124	−.0995	−.0670	.2244
−.0651	.7249	.1742	.0750	−.1559	.7249	.1013	.0750	−.0711	.7249	.1013	−.1577
.3211	.2744	−.3470	−.4006	.4645	.2744	−.0879	−.4006	.0980	.2744	−.0879	.6055
.5705	.1574	−.3966	−.1677	.6945	.1574	.0205	−.1677	.4239	.1574	.0205	.5751
.5617	−.0714	−.1756	.3005	.5559	−.0714	.1930	.3005	.6189	−.0714	.1930	.1275
.2114	.1385	.5424	.0483	−.1527	.1385	.5617	.0483	−.0859	.1385	.5617	−.1352
.4854	.0441	.0247	−.4514	.3755	.0441	.3086	−.4514	−.0028	.0441	.3086	.5872
.6312	−.1463	−.3247	.3143	.7005	−.1463	.1144	.3143	.7385	−.1463	.1144	.2099
.4791	−.0837	.5163	.0807	.0780	−.0837	.7000	.0807	.1116	−.0837	.7000	−.0116

† Algebraic signs of factor IV_1 were reflected.

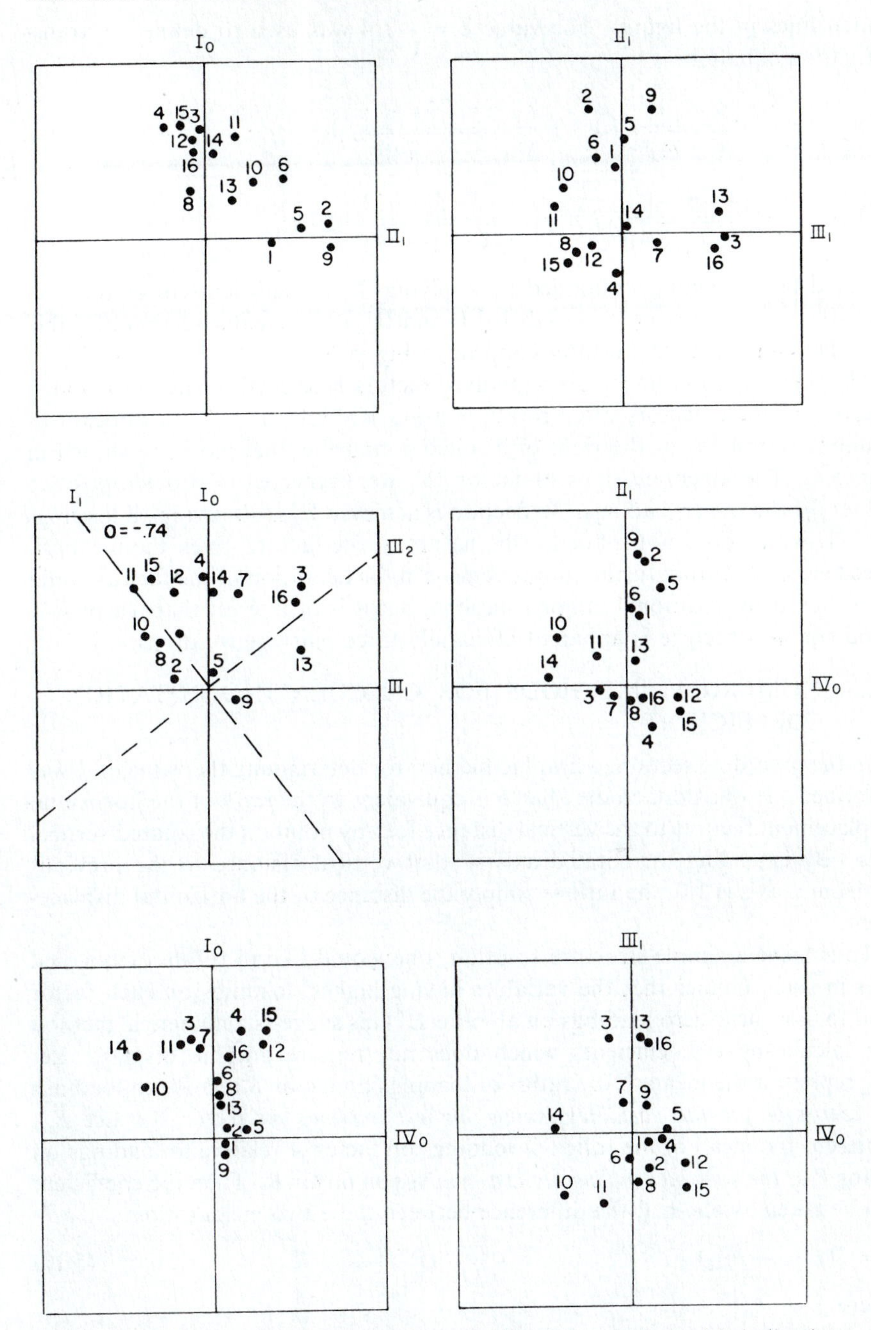

Figure 5.2 Plots of principal-axes factors after first rotation with indication of a second pairwise rotation using $\phi = -.74$.

dotted lines in the figure. The value $\phi = -.74$ was used to define the transformation equations.

$$(\hat{f}_i)_{\text{I}_1} = \frac{1}{\sqrt{1 + (-.74)^2}} (f_i)_{\text{I}_0} + \frac{-.74}{\sqrt{1 + (-.74)^2}} (f_i)_{\text{III}_1}$$

$$(\hat{f}_i)_{\text{III}_2} = \frac{.74}{\sqrt{1 + (-.74)^2}} (f_i)_{\text{I}_0} + \frac{1}{\sqrt{1 + (-.74)^2}} (f_i)_{\text{III}_1}$$

Rotated factor loadings obtained by applying these coefficients to columns I^0 and III_1 are shown in columns labeled I_1 and III_2 of the center section of Table 5.4. The plots after this rotation appear in Fig. 5.3.

The third rotation undertaken involved factors I_1 and IV_0. The recalculated loading of these factors after rotation using a value of $\phi = .84$ appears as columns I_2 and IV_1 at the right of Table 5.4, and the final plots are shown in Fig. 5.4. The algebraic signs of factor IV_1 were reflected to provide positive values for the larger loadings. Reflection is achieved by multiplying all loadings by -1, which does not influence the nature of the factor. Even the inexperienced eye can see that further improvement in the definition of the factors could be achieved by additional minor rotations, but it is of interest that reasonably good simple structure is apparent after only three pairwise rotations.

5.6.3 NUMERICAL METHODS FOR CALCULATING ROTATION COEFFICIENT

In the preceding section, a graphic method for determining the value of ϕ was described. It was pointed out that ϕ is equivalent to the *ratio* of the horizontal displacement relative to the vertical distance for any point on the rotated vertical axis. By extending the rotated axis so that vertical distance to the previous horizontal axis is 1.0, the ratio is simply the distance of the horizontal displacement.

To achieve a simple-structure solution, one would like to rotate orthogonal axes in such manner that the variables having highest loadings on each factor tend to have near-zero loadings on all others. This suggests a numerical method for calculating ϕ coefficients which does not require graphic display. Let $\bar{F}_{b/a}$ represent the mean of the ratios of loadings on factor B relative to loadings on factor A *for the variables having highest loadings on factor A*. Let $\bar{F}_{a/b}$ represent the mean of the ratios of loadings on factor A relative to loadings on factor B *of the variables having highest loadings on factor B*. Then the coefficient can be taken as one-half the difference between these two mean ratios

$$\phi = \tfrac{1}{2}(\bar{F}_{b/a} - \bar{F}_{a/b}) \tag{5.10}$$

where

$$\bar{F}_{b/a} = \frac{1}{n} \sum^{n} \frac{f_{ib}}{f_{ia}}$$

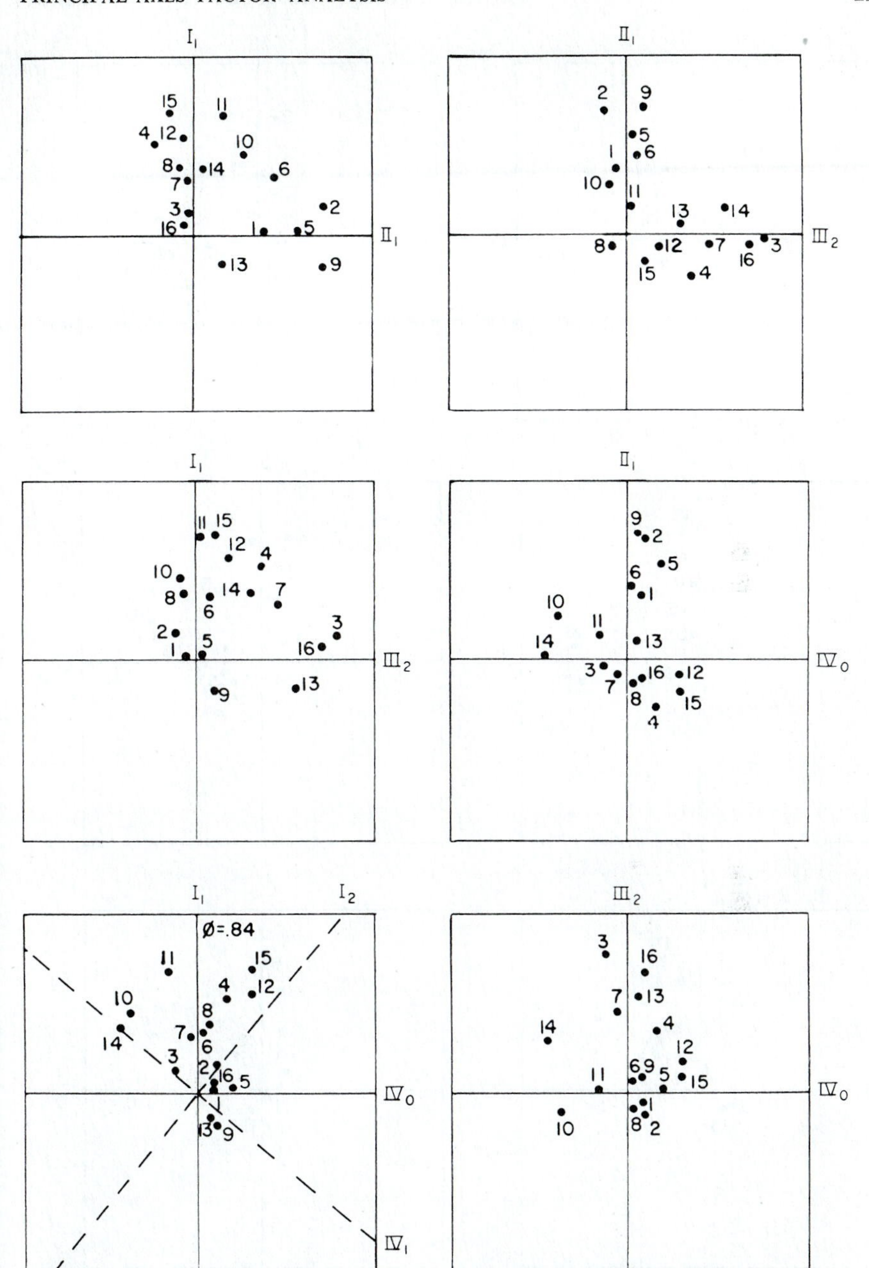

Figure 5.3 Principal-axes factors after second rotation with indication of third pairwise rotation using $\phi = .84$.

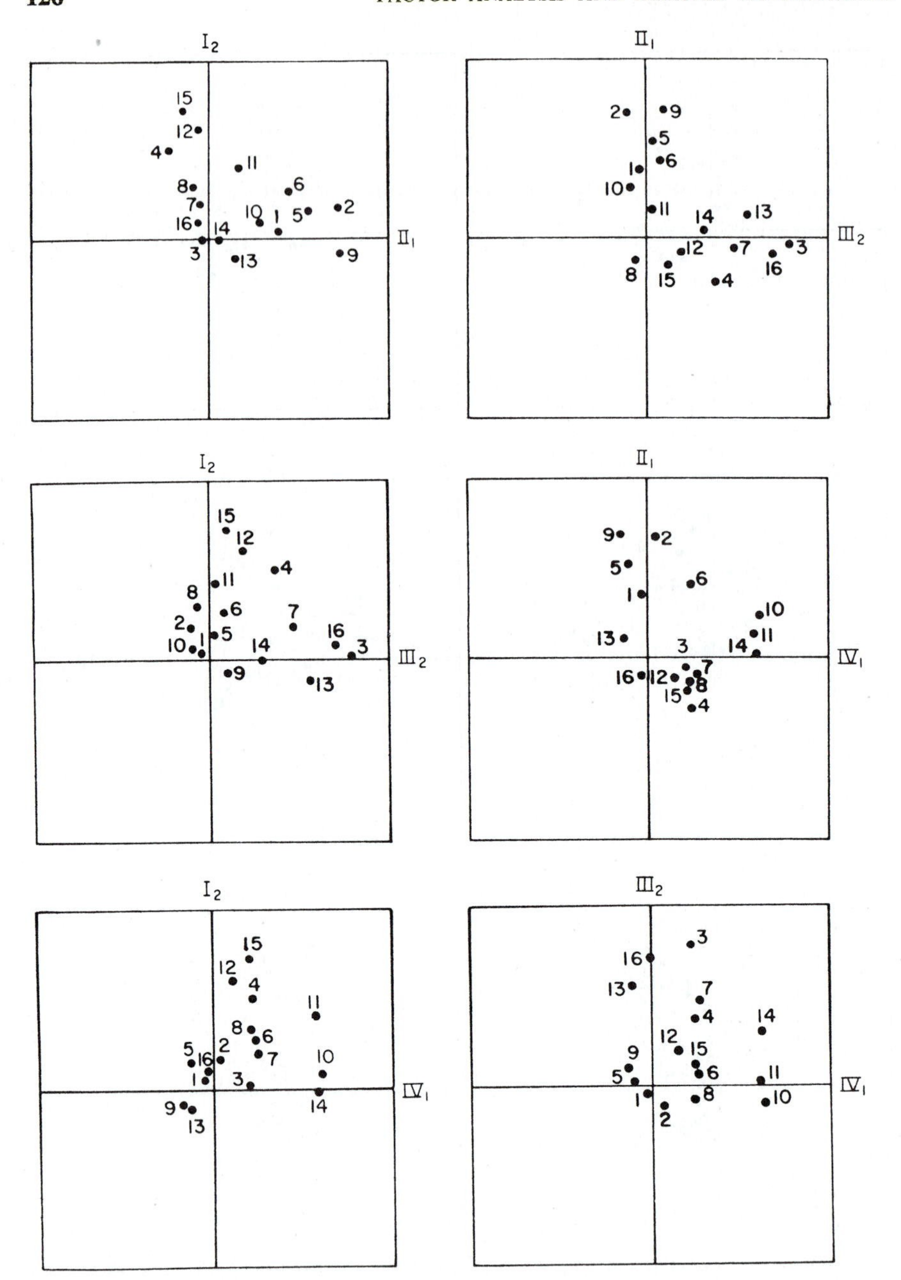

Figure 5.4 Principal-axes factors after third rotation.

and

$$\bar{F}_{a/b} = \frac{1}{n}\sum^{n}\frac{f_{ia}}{f_{ib}}$$

for the variables having highest loadings on factors A and B, respectively.

Equation (5.10) is tantamount to defining an *average* of the angles of rotation which will pass an axis collinear with each of the several tests having highest loading on each factor. This method of calculating ϕ is not recommended for practical problems because it emphasizes only a few of the total number of loadings and disregards the others entirely; but as a heuristic device, it can be used effectively to illustrate a simple numerical calculation of a ϕ coefficient.

Consider the loadings of the 16 BPRS variables on the first two principal-axes factors from Table 5.3.

I_0	II_0
−.02	.37
.07	.73
.58	−.20
.64	−.22
.05	.52
.33	.46
.54	−.12
.27	.00
−.07	.62
.32	.38
.57	.28
.56	−.01
.21	−.05
.49	.03
.63	−.03
.48	−.25
1.00	.00
.00	1.00

$$\bar{F}_{b/a} = \frac{1}{3}\left(\frac{-.20}{.58} + \frac{-.22}{.64} + \frac{-.03}{.63}\right) = -.2454$$

$$\bar{F}_{a/b} = \frac{1}{3}\left(\frac{.07}{.73} + \frac{.05}{.52} + \frac{-.07}{.62}\right) = .0264$$

$$\phi = \tfrac{1}{2}(-.2454 - .0264) = -.1359$$

The effect of this counterclockwise rotation can be visualized by referring back to the plot of I_0 and II_0 in Fig. 5.1.

5.6.4 WEIGHTED CROSS-FACTOR DEFINITION OF ROTATION COEFFICIENT

The definition of simple structure in terms of same-factor vs. cross-factor loadings *for variables having the highest loadings on each factor* is intuitively simple. However, in practice, there is a disadvantage in arbitrarily considering only the three variables having the greatest relationship to each factor in determining the direction and extent of rotation, as was the case in the previous example. Instead, consider a slight variation on this approach which is less arbitrary and also has advantages from a computer programming point of view.

Let f_{ia}^3 be the cube of the loading for the ith variable on factor A and f_{ib}^3 be the cube of the loading for the ith variable on factor B. Raising factor loadings to the third power preserves the sign but greatly increases the difference between larger and smaller values. Since factor loadings derived from analysis of a correlation matrix are fractional values, cubing the smaller loadings reduces them to essentially zero. Larger values are less reduced.

The coefficient ϕ required to calculate the rotational transformation as specified in Eqs. (5.8) and (5.9) can be defined as follows:

$$\phi = \frac{1}{2}\left(\frac{\sum f_{ia}^3 f_{ib}}{\sum f_{ia}^3} - \frac{\sum f_{ib}^3 f_{ia}}{\sum f_{ib}^3}\right) \tag{5.11}$$

where the first term is the sum of factor loadings on factor B each weighted by the cube of the corresponding loading on factor A relative to the sum of third powers of loadings of all variables on factor A, and where the second term is defined similarly for the other factor. The similarity of the rationale

Table 5.5 Rotated Principal-axes Factor Loadings

	I	*II*	*III*	*IV*	h^2
Somatic concern	.0328	.3854	−.0599	−.0207	.1536
Anxiety	.2003	.6999	−.1353	.0386	.5497
Emotional withdrawal	.1536	−.0293	.7893	.1018	.6578
Conceptual disorganization	.1448	−.1645	.3266	.5835	.4951
Guilt feelings	.0011	.5803	.0015	.0468	.3390
Tension	.3116	.4295	.0617	.2044	.3272
Mannerisms and posturing	.2103	−.0556	.4542	.2742	.3288
Grandiosity	.2001	−.0852	−.0859	.3281	.1623
Depressive mood	−.0000	.7172	.0948	−.2048	.5653
Hostility	.6543	.1493	−.0688	.0640	.4592
Suspiciousness	.5938	.1106	.0056	.4132	.5356
Hallucinatory behavior	.0993	.0310	.1366	.6420	.4417
Motor retardation	−.1281	.1655	.5590	−.0632	.3602
Uncooperativeness	.5673	−.0720	.3359	.0460	.4419
Unusual thought content	.1669	−.0381	.0513	.7694	.6240
Blunted affect	−.0628	−.0302	.6843	.1907	.5094
Variance	1.4277	1.7827	1.8977	1.8429	

to that previously described for minimizing cross-factor loadings should be obvious. Given that raising factor loadings to the third power tends to approach the ideal of a zero-one vector, each term inside the brackets is essentially a weighted mean of cross-factor loadings for variables having highest loadings on factors A and B, respectively.

Results from application of this modified rotation procedure to the matrix of principal-axes factor loadings given in Table 5.3 are presented in Table 5.5. The nature of these factors will be discussed in Sec. 5.7. The purpose of this section is to demonstrate that rational alternative definitions of ϕ can easily be conceived and that they can result in good orthogonal simple-structure solutions. As an exercise in understanding rotation, the student is encouraged to consider the ideal of simple structure and to devise new methods of his own. The normalized varimax criterion described next is a good objective definition of simple structure, but it is just one of many.

5.6.5 NORMALIZED VARIMAX ROTATION

As has been shown, projections of variables on pairs of factor axes can be increased or decreased by defining new (rotated) factors as weighted combinations of the original unrotated factors. The choice of appropriate transformation weights can be determined to achieve a clear simple structure. Many numerical criteria of good simple structure have been proposed. For example, one can appreciate that the *variance* of squared factor loadings will be substantial when some loadings are large and the others are near zero. The so-called *varimax* criterion is essentially such an index. Rotation is continued as long as the value of the *objective criterion* is increased. When the value of the objective-criterion function stabilizes and cannot be further increased, the rotation is complete.

Although numerous computer rotation methods are available, only the *normalized varimax* method will be described because of its current popularity. The objective-criterion function is the variance of squared (normalized) factor loadings

$$V = n \sum_{p}^{r} \sum_{i}^{n} \left(\frac{f_{ip}}{h_i} \right)^4 - \sum_{p}^{r} \left(\sum_{i}^{n} \frac{f_{ip}^{\,2}}{h_i^{\,2}} \right)^2 \tag{5.12}$$

where f_{ip} is the factor loading for the ith variable on the pth factor and h_i is the square root of the sum of squares of loadings for the ith variable on all factors. The role of h_i in this function is to normalize factor loadings so that the sum of the squared loadings of each variable is equal to unity.

The normalized varimax rotation can be accomplished by first normalizing factor loadings for each variable, completing all rotations, and then at the completion of the rotational procedure rescaling back to original sums of squares. Let b_{ip} represent the *normalized* loading of the ith variable on the pth factor.

The normalized varimax-criterion function then becomes

$$V = n \sum^{r} \sum^{n} b_{ip}^4 - \sum^{r} \left(\sum^{n} b_{ip}^2 \right)^2 \tag{5.13}$$

It would be possible to maximize the objective-criterion function by simple trial-and-error rotation of pairs of factors. For example, factor A could be rotated slightly toward factor B and the criterion function evaluated. If the criterion function improved, the new rotation would be accepted. Obviously, this would be a laborious way of arriving at the maximum value of the objective criterion. Kaiser (1959) has shown that in trigonometric terms the desired degree of rotation for any pair of factors is the tangent of 4 times the proper angle of rotation

$$\tan 4\Omega = \frac{D - 2AB/n}{C - [(A^2 - B^2)/n]} \tag{5.14}$$

where

$$A = \sum^{n} (b_{ip}^2 - b_{iq}^2)$$

$$B = \sum^{n} 2b_{ip}b_{iq}$$

$$C = \sum^{n} [(b_{ip}^2 - b_{ip}^2)^2 - (2b_{ip}b_{iq})^2]$$

$$D = 2 \sum^{n} (b_{ip}^2 - b_{iq}^2) 2b_{ip}b_{iq}$$

Having defined the proper angle of rotation Ω, the coefficient ϕ is the sine of that angle

$$\phi = \sin \Omega$$

and the new rotated (normalized) factor loadings are

$$[\hat{b}_{ip} \quad \hat{b}_{iq}] = [b_{ip} \quad b_{iq}] \begin{bmatrix} \cos \Omega & -\sin \Omega \\ \sin \Omega & \cos \Omega \end{bmatrix}$$

Having computed $\phi = \sin \Omega$, the general orthogonal rotation procedure is the same as that previously defined in Eqs. (5.10) and (5.11).

Following completion of each pairwise rotation of two factors, the value of the objective criterion [Eq. (5.13)] can be calculated to determine whether it has been increased. The procedure is continued across all possible pairs of factors until the variance of the squared (normalized) factor loadings no longer increases. As a final step, each factor loading is multiplied by the original normalizing coefficient to rescale back to the original sum of squares for each variable

$$f_{ip} = \hat{b}_{ip} h_i \tag{5.15}$$

Computer programs to accomplish orthogonal normalized varimax rotation are widely available and will not be included here. It should be obvious that ϕ as defined here could be substituted for ϕ as defined in Sec. 5.6.4 and that the general orthogonal rotational program described at the end of this chapter would then become a normalized varimax rotation program.

5.6.6 CHECKING ACCURACY OF ORTHOGONAL ROTATION

It is quite easy to make an error in calculations when undertaking rotational transformation by any of the methods just described. The fact that the total variance accounted for by rotated orthogonal factors should be the same as that accounted for by the unrotated orthogonal factors provides the basis for a convenient check on calculations. Following orthogonal rotation, the sum of squares of factor loadings for any given test variate across all factors should be identical to the sum of squares of loadings for the same test variate across the original unrotated principal-axes factors. Inspection of the right-hand columns in Tables 5.3 and 5.5 reveals consistency to the third decimal place. Such a check should always be run following hand calculation of rotated factor loadings. The variance of the first four unrotated principal-axes factors was 6.9510. After rotation the factors in Table 5.5 still account for that amount of variance, but it will be noticed that the percent of variance accounted for by each factor has been redistributed.

5.6.7 CALCULATION OF THE COMPLETE ROTATIONAL TRANSFORMATION MATRIX

Since the actual calculation of rotated factor loadings is usually accomplished through a succession of simple pairwise factor transformations of the type defined in Eqs. (5.8) and (5.9), it is of some interest to examine how the final rotational transformation matrix $T = \hat{T}D$ of Eq. (5.6) can be computed. A convenient mechanism is to append an identity matrix to the bottom of the original factor matrix. Consider each row of the identity matrix to represent factor loadings on the r orthogonal factors of a variable that is collinear with one of the factors. Since the original factors are orthogonal, a variable which correlates perfectly with one of them will have zero correlations with the others. In the course of successive pairwise rotations, the same operations can be performed on the elements of the appended identity matrix as are performed on the other factor loadings. At the conclusion of the series of pairwise rotational transformations, the appended matrix will contain the cosines (correlations) between the final rotated factors and the original unrotated factors. The resulting matrix is the desired transformation matrix $T = \hat{T}D$. If one multiplies the original factor matrix by the transformation matrix, the matrix of factor loadings on the rotated factors will result.

Alternatively, if a final rotated factor matrix is available and one wants to define the transformation matrix that will bring the original unrotated-factor matrix to that form, the following matrix equation can be used:

$$F_1T = F_2$$
$$F_1'F_1T = F_1'F_2$$
$$T = (F_1'F_1)^{-1}F_1'F_2 \qquad (5.16)$$

where F_1 is the original unrotated-factor matrix and F_2 is the final rotated-orthogonal-factor matrix. In the case of principal-axes factor analysis, the matrix $F_1'F_1$ is a diagonal matrix containing the factor variances (sums of squares of loadings on each factor). Consequently, the matrix $(F_1'F_1)^{-1}$ is also a diagonal matrix containing the reciprocals of the factor variances. Thus, the computation of the rotational transformation matrix from the original and rotated factor loadings is even simpler than it might at first appear.

As an exercise to verify these relationships, the reader is encouraged to append an identity matrix to the bottom of the principal-axes factor matrix in Table 5.3 and to accomplish a series of pairwise rotations using any arbitrary value of ϕ in Eqs. (5.8) and (5.9). Next, use Eq. (5.16) to compute the rotational transformation matrix and compare the results.

5.6.8 CONSTRUCTING THE TRANSFORMATION MATRIX

The application of the simple formulas for pairwise rotation described in earlier sections can also be conceived as a series of matrix operations. The principal problem in achieving a satisfactory rotational transformation is one of choosing elements of $\hat{T}D$ such that the objective of simple structure is approached while the orthogonality of factors is retained. Orthogonality of factors will be retained as long as the columns of $\hat{T}D$ are defined in such manner that they are mutually orthonormal, i.e., cross products are zero and sums of squares are unity.

Orthogonality can be maintained by transforming one pair of factors at a time. The simple transformation matrix for one pairwise rotation of factors I and II is of the type

$$\hat{T}D = \begin{bmatrix} \dfrac{1.0}{\sqrt{1+\phi^2}} & \dfrac{-\phi}{\sqrt{1+\phi^2}} & .0 & \cdots & .0 \\ \dfrac{\phi}{\sqrt{1+\phi^2}} & \dfrac{1.0}{\sqrt{1+\phi^2}} & .0 & \cdots & .0 \\ .0 & .0 & 1.0 & \cdots & .0 \\ \cdots & \cdots & \cdots & \cdots & \cdots \\ .0 & .0 & .0 & \cdots & 1.0 \end{bmatrix} \tag{5.17}$$

The simple transformation matrix for rotation of factors I and III is of the type

$$\hat{T}D = \begin{bmatrix} \dfrac{1.0}{\sqrt{1+\phi^2}} & .0 & \dfrac{-\phi}{\sqrt{1+\phi^2}} & .0 & \cdots & .0 \\ .0 & 1.0 & .0 & .0 & \cdots & .0 \\ \dfrac{\phi}{\sqrt{1+\phi^2}} & .0 & \dfrac{1.0}{\sqrt{1+\phi^2}} & .0 & \cdots & .0 \\ .0 & .0 & .0 & 1.0 & \cdots & .0 \\ \cdots & \cdots & \cdots & \cdots & \cdots & \cdots \\ .0 & .0 & .0 & .0 & \cdots & 1.0 \end{bmatrix} \tag{5.18}$$

In general, each pairwise rotation is accomplished by defining a simple transformation matrix which is an identity matrix *except for four elements*, two in each of the columns corresponding to factors being rotated at that stage. The two complementary nonzero off-diagonal elements are of equal magnitude but opposite sign. For *any* such matrix, one can quickly verify that $D\hat{T}'\hat{T}D = I$. Thus, as has been stated earlier, the only real problem is one of choosing the value of ϕ appropriately.

A complete orthogonal factor rotation is actually undertaken as a series of simple pairwise rotations. Thus, the complete rotational transformation matrix $T = \hat{T}D_2$ of Eq. (4.3) or (5.6) is the product of a series of simple transformation matrices of the type illustrated in Eqs. (5.17) and (5.18)

$$T = (\hat{T}^{(1)}D^{(1)})(\hat{T}^{(2)}D^{(2)}) \cdots (\hat{T}^{(n)}D^{(n)}) \qquad (5.19)$$

It can easily be verified that

$$T'T = I \qquad \text{if } (D^{(1)}\hat{T}^{(1)})'(\hat{T}^{(1)}D^{(1)}) = I$$

$$\text{and } (D^{(2)}\hat{T}^{(2)})'(\hat{T}^{(2)}D^{(2)}) = I$$

and in general

$$(D^{(i)}\hat{T}^{(i)})'(\hat{T}^{(i)}D^{(i)}) = I$$

For example, after three simple two-factor rotations

$$T'T = (D^{(3)}\hat{T}^{(3)})'(D^{(2)}\hat{T}^{(2)})'(D^{(1)}\hat{T}^{(1)})'(\hat{T}^{(1)}D^{(1)})(\hat{T}^{(2)}D^{(2)})(\hat{T}^{(3)}D^{(3)})$$

$$= (D^{(3)}\hat{T}^{(3)})'(D^{(2)}\hat{T}^{(2)})'(I)(\hat{T}^{(2)}D^{(2)})(\hat{T}^{(3)}D^{(3)})$$

and

$$T'T = (D^{(3)}\hat{T}^{(3)})'(I)(\hat{T}^{(3)}D^{(3)}) = I$$

The actual process of orthogonal rotation involves applying one simple transformation matrix to obtain a first rotated-factor matrix $F^{(1)}$, then a second simple transformation matrix is applied to the first rotated-factor matrix to obtain the second rotated-factor matrix $F^{(2)}$, and so on

$$F_1\hat{T}D^{(1)} = F^{(1)}$$

$$F_1\hat{T}^{(1)}D^{(1)}\hat{T}^{(2)}D^{(2)} = F^{(1)}\hat{T}^{(2)}D^{(2)} = F^{(2)}$$

$$F_1\hat{T}^{(1)}D^{(1)}\hat{T}^{(2)}D^{(2)}\hat{T}^{(3)}D^{(3)} = F^{(1)}\hat{T}^{(2)}D^{(2)}\hat{T}^{(3)}D^{(3)}$$

$$= F^{(2)}\hat{T}^{(3)}D^{(3)} = F^{(3)}$$

where after n simple rotations

$$F_1(\hat{T}^{(1)}D^{(1)}\hat{T}^{(2)}D^{(2)} \cdots \hat{T}^{(n)}D^{(n)}) = F_1\hat{T}D_2 = F_2$$

of Eq. (4.3) or (5.6).

What needs to be emphasized is that the complete rotational transformation matrix $T = \hat{T}D_2$ of Eq. (4.3) or (5.6) is actually obtained through a series of simple rotations of pairs of factors and that the final complete rotational transformation matrix is the product of several simple transformation matrices of the type illustrated in Eqs. (5.17) and (5.18). Since orthogonality can be maintained by assigning opposite signs to the two complementary nonzero off-diagonal elements of each simple transformation matrix [Eq. (5.17) or (5.18)], the only real problem is how to specify values for ϕ which will lead to a desirable rotated solution.

5.7 Interpretation of Rotated Principal-axes Factors

As pointed out earlier, unrotated principal-axes factors tend to be maximally complex, and the variables tend to have high loadings, particularly on the first factor. It is instructive to compare Tables 5.3 and 5.5 and see how rotation of the factor axes has facilitated the interpretation of the factors. It will be noted initially that the variance associated with the unrotated factors has been distributed by the rotation.

Factor I has substantial loadings on three symptom variables: hostility, suspiciousness, and uncooperativeness. There is a modest loading on tension, but all the other symptoms are essentially uncorrelated with this factor. This factor identified in analyses of many other sets of data has been referred to as *hostile-suspiciousness.*

Factor II is correlated most highly with depressive mood, anxiety, and guilt feelings. Of the other variables, only somatic concern and tension show any tendency to be related to this factor, and even these variables are consistent with the interpretation of this factor as *anxious depression.*

Factor III is defined by the high loadings on emotional withdrawal, blunted affect, and motor retardation. This factor has been named *withdrawal-retardation*, and it can be inferred from the smaller loadings on conceptual disorganization, mannerisms and posturing, and uncooperativeness that this factor is more likely to be associated with symptoms of schizophrenia than those of depressive reactions.

Factor IV has been named *thinking disturbance* as it correlated most prominently with unusual thought content, hallucinatory behavior, and conceptual disorganization. There are smaller correlations with grandiosity and suspiciousness which are consistent with this interpretation. This factor clearly represents a central part of the diagnostic conception of schizophrenia.

It would seem from the above discussion that the rotation has produced a meaningful simple structure. Each factor is defined by a few high loadings, and the remaining variables are essentially uncorrelated with the factor. Looking at the rows, we see that most variables have high loadings on only a single factor. The few exceptions are not inconsistent with the interpretation of the factors. The few variables that do not contribute substantially to the definition of a single factor are mannerisms and posturing, tension, somatic concern, and grandiosity. All these, and particularly the last two, have generally lower communalities (h^2) than the other variables. To the extent that they are related to other variables in the set, however, the relationships are meaningful.

Computer Program for Orthogonal Factor Rotation

```
      DIMENSION A(40,10),X(40)
    1 FORMAT(4I4)
    2 FORMAT(16F5.3)
      IT=1
      READ(2,1) NVAR,NFAC,NSTP
      SUMT2=((NFAC*(NFAC-1))/2)*.0025
      NSTAR= NVAR+1
      TEST=NFAC
      DO 10 J=1,NFAC
   10 READ(2,2)(A(K,J),K=1,NVAR)
      NVAR2=NVAR+NFAC
C     AUGMENT FACTOR MATRIX WITH IDENTITY
      DO 300 J=NSTAR,NVAR2
      DO 299 K=1,NFAC
  299 A(J,K)=0.0
      II=J-NVAR
  300 A(J,II)=1.0
      DO 305  J=1,NVAR2
  305 WRITE(3,306)(A(J,K),K=1,NFAC)
  306 FORMAT(10F12.4)
C     COMPUTE PHI AS WEIGHTED CROSS-PRODUCT
      NFAC2=NFAC-1
      ITER=0
  400 SUMT=0.0
      DO 201 K=1,NFAC2
      K3=K+1
      DO 200 K2=K3,NFAC
      R= 0.0
      OR=0.0
      AD=0.0
      AD2=0.0
      DO 25 J=1,NVAR
      PR=A(J,K)**3
      PR2=A(J,K2)**3
      AD=AD+PR
      AD2=AD2+PR2
      R=R+PR*A(J,K2)
   25 OR=OR+PR2*A(J,K)
      R=R/AD
      OR=OR/AD2
      PHI1=0.5*(R-OR)
C     COMPUTE TRANSFORMED FACTOR LOADINGS
  230 SCL=1.0 + PHI1*PHI1
      T1=1.0/ SQRT(SCL)
      T2=PHI1/SQRT(SCL)
      SUMT=SUMT+T2*T2
      DO 235 J=1,NVAR2
      X1=A(J,K)
      X2=A(J,K2)
      A(J,K)=X1*T1+X2*T2
  235 A(J,K2)=X2*T1-X1*T2
C     COMPUTE COMMUNALITY
      DO 236 J=1,NVAR2
      X(J)=0.0
      DO 237 K4=1,NFAC
  237 X(J)=X(J)+A(J,K4)*A(J,K4)
  236 CONTINUE
  200 CONTINUE
  201 CONTINUE
      ITER=ITER +1
      IF(SUMT-SUMT2) 99,99,401
  401 IF(NSTP-ITER) 99,99,400
    7 FORMAT(1X,10F10.5)
   99 CONTINUE
C     PRINT ROTATED FACTOR LOADINGS
      WRITE(3,5) K,K2
    5 FORMAT(24H ROTATED FACTOR LOADINGS ,2I5)
      DO 240 J=1,NVAR2
  240 WRITE(3,7)(A(J,K4),K4=1,NFAC),X(J)
      CALL EXIT
      END
```

5.8 Computer Program for Orthogonal Factor Rotation

In this section a computer program that accomplishes orthogonal factor rotation is presented. Although the program is written to calculate the coefficient ϕ in a particular way (Sec. 5.6.4), it can easily be modified by substituting some other method of calculating ϕ into the section of the program between statements 400 and 230. This program is not recommended for general use because other good orthogonal-rotation programs, such as normalized varimax, are available in almost all computer centers. It is introduced here to provide the basis for experimentation in factor rotation which can lead to greater understanding. The student can consider the objectives that he wishes to achieve by factor rotation and devise a method of calculating the ϕ coefficient that will accomplish his purposes. The definition of ϕ presented in Sec. 5.6.4 is one alternative. The normalized varimax definition of ϕ described in Sec. 5.6.5 could be substituted into the program. Any arbitrary definition of ϕ can be inserted between statements 400 and 230.

Use of the orthogonal-rotation program requires one control card specifying the number of factors (NFAC), the number of variables (NVAR), and the outside number of iterations to be permitted in case the solution has not stabilized (NSTP). Satisfactory results are usually obtained with NSTP = 10, thus preventing the use of excessive computer time. These three parameters are punched in successive four-column fields of the single control card. The vectors of factor loadings are entered next, by factor.

5.9 References

Kaiser, H. F.: Computer Program for Varimax Rotation in Factor Analysis, *Educ. Psychol. Meas.*, **19**:413–420 (1959).

Thurstone, L. L.: "Multiple Factor Analysis," University of Chicago Press, Chicago, 1947.

6 Orthogonal Powered-vector Factor Analysis

6.1 Introduction

Whereas rotation is almost always required to establish a simple and meaningful relationship between factors and original variables following principal-axes analysis, the orthogonal powered-vector method tends to position factors meaningfully without rotation. This is not to say that rotational transformations cannot be applied to results from the powered-vector analysis; however, rotation is usually not necessary to establish meaningful factor representation of the original variables. Parsimony, orthogonality, and psychological meaningfulness are all objectives of the initial direct orthogonal solution. In comparison with the principal-axes method, powered-vector analysis places less emphasis on parsimony (maximizing variance accounted for by a limited number of factors) and more emphasis on psychologically meaningful definition of factors. Both methods lead to factors that are statistically orthogonal. The powered-vector method requires very much less computation than is required for principal-axes analysis and subsequent rotation, a consideration that becomes important in analyses of matrices of large order.

The powered-vector method of factor analysis yields an orthogonal cluster-oriented solution in which each factor tends to represent a distinct homogeneous subset of variables. The first factor is defined in such a manner that it correlates highly with variables forming a distinct correlation cluster, the cluster being a subset of variables that correlate relatively highly and are relatively independent of other such subsets. The second powered-vector factor is defined in such manner that it correlates highly with variables in a second distinct cluster and is orthogonal to the first factor. The third powered-vector factor is orthogonal to the first two and tends to represent a third homogeneous subset (if such is present). Since considerable confusion exists in the minds of some users of factor analysis concerning orthogonality of factors vs. independence among clusters of variables, it should be emphasized that factors can be orthogonal, i.e., factor variates can be defined so that they are statistically independent, even though the clusters of variables that they represent are not completely independent of one another.

A parsimonious set of orthogonal factor variates is desired such that each tends to represent a distinct subset or cluster of variables. That is, each factor variate should correlate highly with variables belonging to one and only one distinct cluster, insofar as this is possible under the restriction of orthogonality. The solution really involves answers to two separate questions: How can homogeneous clusters of variables be identified quickly and easily? Given that a cluster of variables has been identified, how can a factor be defined which represents the cluster and yet is independent of all other factors?

6.2 Group-centroid Method

Neglecting for a moment the question of how to do it, assume that homogeneous clusters have been identified. A composite variable which is a simple sum of the scores for variables within a cluster will correlate highly with all variables in the cluster. Recalling the general method for computation of orthogonal factors and factor loadings (Sec. 4.4), successive hypothesis vectors $\mathbf{h}_1, \mathbf{h}_2, \ldots, \mathbf{h}_r$ can be defined to include nonzero coefficients for variables within previously identified clusters and zeros elsewhere. The nonzero elements will ordinarily be 1.0; however, in the case of a variable that correlates negatively with the majority of variables in the group or cluster, a coefficient of -1.0 should be used. For example, suppose that variable 4 has large negative correlation with all other variables entering into group 2. The hypothesis vector $\mathbf{h}_2$ should appear as shown below:

$$R_1\mathbf{h}_1 s_1 = \mathbf{f}_1 \qquad \text{where } \mathbf{h}_1 = [0 \quad 1 \quad 0 \quad 0 \quad \cdots \quad 1 \quad 0] \text{ and } s_1 = \frac{1}{\sqrt{\mathbf{h}_1' R_1 \mathbf{h}_1}}$$

$$R_2 = R_1 - \mathbf{f}_1\mathbf{f}_1'$$

$$R_2\mathbf{h}_2 s_2 = \mathbf{f}_2 \qquad \text{where } \mathbf{h}_2 = [1 \quad 0 \quad 0 \quad -1 \quad \cdots \quad 0 \quad 1] \text{ and } s_2 = \frac{1}{\sqrt{\mathbf{h}_2' R_2 \mathbf{h}_2}}$$

$$R_3 = R_2 - \mathbf{f}_2\mathbf{f}_2'$$

$$\cdots\cdots\cdots\cdots\cdots\cdots\cdots\cdots\cdots\cdots$$

$$R_r\mathbf{h}_r s_r = \mathbf{f}_r \qquad \text{where } \mathbf{h}_r = [0 \quad 0 \quad 1 \quad 0 \quad \cdots \quad 0 \quad 0] \text{ and } s_r = \frac{1}{\sqrt{\mathbf{h}_r' R_r \mathbf{h}_r}}$$

In each hypothesis vector, only those variables belonging to a distinct homogeneous subset or cluster are given nonzero weight. This is tantamount to defining successive factor variates as linear composites of z-score values for variables within a distinct cluster after adjustment has been made for the relationship to previously defined factors. This method of analysis will be called the *orthogonal group-centroid method.* It presupposes a meaningful basis for grouping variables. The orthogonal powered-vector method of factor analysis is a close variant of the group-centroid approach. Although the successive hypothesis vectors are defined as functions of columns in the (residual) correlation matrix, it will be appreciated that they are closely similar to the group-centroid zero-one vectors.

6.3 Homogeneous Subsets

There are numerous ways in which homogeneous subsets of variables can be identified within a correlation matrix. In the powered-vector method of analysis such cluster identification is accomplished mechanically. It is assumed that if distinct clusters exist, variables within each cluster will be more highly correlated with other variables in the cluster than with variables outside the cluster. Thus, in the presence of a distinct cluster configuration, most columns of the correlation matrix should contain some relatively large coefficients (relating the particular variable to other variables in the same cluster) and numerous relatively small coefficients. A variable is a good "cluster marker" if the difference between cluster and noncluster correlations is great. It may be helpful for the reader to examine the 16 × 16 matrix of intercorrelations presented in Table 5.1 with an eye for the pattern of large and small coefficients in each column. For example, variable 15 correlates highly with variables 4 and 12. These three variables constitute one of the symptom clusters identified in the principal-axes analysis. Variable 9 tends to correlate most highly with variables 2 and 5. These three variables form the nucleus for another cluster.

6.4 Definition of Method

In the powered-vector method, a simple numerical index computed as the sum of fourth powers of coefficients in each column of the (residual) correlation matrix is used to identify the column having the greatest potential for defining a homogeneous cluster. (For hand calculations, the sum of squares can be used effectively to identify the column that will be used in defining the hypothesis vector at each stage.)

$$v_j = \sum_{i=1}^{p} r_{ij}^4 \tag{6.1}$$

By raising elements in the jth column to the fourth power, the distinction between positive and negative correlations is eliminated and the smaller values approach zero. Thus, the coefficient v_j is essentially an index of the number of relatively large coefficients within the jth column of the (residual) correlation matrix at any stage in the analysis. The column yielding the largest v_j index is accepted as the cluster marker. It is important to appreciate, in the absence of more complicated notation, that the r_{ij} elements in Eq. (6.1), and similarly in Eq. (6.2), are the elements from a particular column of the residual correlation matrix being operated upon to define a particular factor. For the first factor, the r_{ij} are elements from the original correlation matrix. For the kth factor, the r_{ij} are elements of the $k - 1$ residual correlation matrix.

Having identified a column of the correlation matrix in which several relatively large coefficients identify variables in a cluster, the hypothesis vector used in computing factor loadings is defined. *The hypothesis vector required to compute factor loadings at any stage in the powered-vector analysis consists of the cubes of coefficients in that column of the (residual) correlation matrix having largest sum of fourth powers*

$$\mathbf{h}_1' = [r_{1j}{}^3 \quad r_{2j}{}^3 \quad \cdots \quad r_{pj}{}^3] \tag{6.2}$$

where j is the column in R_i for which the sum of fourth powers of elements is largest. The factor loadings representing relationships of the p original variables to the ith orthogonal powered-vector factor are obtained as follows:

$$R_i\mathbf{h}_i s_i = \mathbf{f}_i \tag{6.3}$$

where $\mathbf{h}_i$ is as defined above and $s_i = 1/\sqrt{\mathbf{h}_i' R_i \mathbf{h}_i}$.

The complete orthogonal powered-vector factor analysis is simply a special case of the general method described in Sec. 4.4. The successive hypothesis vectors are chosen to represent the cluster configuration

$$R_1\mathbf{h}_1 s_1 = \mathbf{f}_1$$

where $\mathbf{h}_1$ contains the cubes of elements in that column of R_1 for which the sum of fourth powers is greatest

$$R_2 = R_1 - \mathbf{f}_1\mathbf{f}_1'$$

$$R_2\mathbf{h}_2 s_2 = \mathbf{f}_2$$

where $\mathbf{h}_2$ contains the cubes of the elements in that column of R_2 for which the sum of fourth powers is greatest

$$R_3 = R_2 - \mathbf{f}_2\mathbf{f}_2'$$

. . . .

$$R_r\mathbf{h}_r s_r = \mathbf{f}_r$$

where $\mathbf{h}_r$ contains the cubes of the elements in that column of R_r for which the sum of fourth powers is greatest.

6.5 Use of Communality Estimates

One further point requires discussion before proceeding to a computational example. In group-centroid or cluster-type factor solutions, it is important that the effect of "specific" or "unique" variance associated with individual variables be minimized. The communality of a variable was defined in Chap. 4 as the proportion of the total variance for the variable that could be accounted for by regression on the several factor variates. Conversely, the unique variance is the proportion of total variability for each original variable that is independent of the factors.

It is of considerable utility in factor-analysis computations to reduce or eliminate the unique component for each variable by substituting communality estimates into the principal diagonal of the correlation matrix. This is accomplished by replacing the unit elements in the principal diagonal of the correlation matrix with coefficients which represent an estimate of the proportion of variance that will be shared among the factors, as discussed in Sec. 4.7.

Although the use of the squared multiple R for initial communality estimates is often recommended, a simple and convenient method for reducing the unique variance is to substitute for the diagonal elements of the correlation matrix the largest *absolute* value appearing in each row or column. This procedure should be repeated after each residual matrix is computed. While this procedure is often employed by the authors, certain limitations must be pointed out. Use of these crude estimates of communality can produce a matrix which is not a true product-moment, or Gramian, matrix. That is, if any of the estimates inserted in the principal diagonal are too small, then the *reduced* matrix may not be subject to representation as the product of a matrix and its transpose. The reestimation of communality values in successive residual matrices enhances the danger that the successive factors will not be precisely orthogonal as defined by $D_1W'\tilde{R}WD_1 = \mathbf{\Sigma}_1$. In general, however, for the uses to which factor analysis is put in practical research, minor departures from orthogonality are of no consequence.

6.6 A Computational Example

Even for purposes of illustrating the computational method for orthogonal powered-vector factor analysis, it is necessary to choose a problem large enough to ensure that multiple common factors do exist. Rather than using the same data that were used in the principal-components example, another significant set of data has been selected. A total of 2,900 BPRS rating profiles representing various subtypes of paranoid, schizophrenic, and depressive diagnoses were selected from the diagnostic-stereotype data provided by over 300 psychiatrists in four European countries. Manic and schizo-affective types were omitted in the selection of these data in order that only three major subtypes would be represented. It will be recalled that these data did not constitute ratings for real psychiatric patients but represent psychiatrists' conceptions of various diagnostic types. The 16×16 matrix of intercorrelations among the 16 BPRS symptom-rating constructs computed across 2,900 rating profiles for hypothetical patients in three major diagnostic categories is presented in Table 6.1.

The first step in the analysis involved identification of the column having the largest sum of fourth powers. The sums of fourth powers for the elements in each column are entered at the bottom of the table. Column 9 of the original

Table 6.1 Intercorrelations among 16 Psychiatric Symptom-rating Variables Derived from Diagnostic Profiles

	1	*2*	*3*	*4*	*5*	*6*	*7*	*8*	*9*	*10*	*11*	*12*	*13*	*14*	*15*	*16*
1	.579	.579	−.093	−.199	.525	.135	−.201	−.212	.564	−.148	−.024	−.071	.265	−.131	−.122	−.135
2	.579	.629	−.162	−.253	.629	.371	−.265	−.196	.663	−.080	.097	−.012	.193	−.137	−.141	−.340
3	−.093	−.162	.574	.454	−.124	−.013	.451	−.080	−.158	.168	.033	.069	.284	.535	.247	.547
4	−.199	−.253	.454	.601	−.248	.068	.601	.237	−.331	.273	.156	.374	−.006	.426	.523	.413
5	.525	.629	−.124	−.248	.773	.253	−.243	−.278	.773	−.278	−.136	−.193	.407	−.134	−.250	−.210
6	.135	.371	−.013	.068	.253	.371	.141	.070	.213	.165	.194	.120	−.064	.115	.035	−.223
7	−.201	−.265	.451	.601	−.243	.141	.601	.175	−.296	.261	.080	.254	.054	.486	.434	.415
8	−.212	−.196	−.080	.237	−.278	.070	.175	.473	−.367	.473	.443	.387	−.472	.109	.414	−.043
9	.564	.663	−.158	−.331	.773	.213	−.296	−.367	.773	−.297	−.162	−.240	.451	−.181	−.323	−.199
10	−.148	−.080	.168	.273	−.278	.165	.261	.473	−.297	.639	.639	.373	.313	.394	.393	.112
11	−.024	.097	.033	.156	−.136	.194	.080	.443	−.162	.639	.639	.506	−.340	.246	.403	−.055
12	−.071	−.012	.069	.374	−.193	.120	.254	.387	−.240	.373	.506	.545	−.266	.250	.545	.023
13	.265	.193	.284	−.006	.407	−.064	.054	−.472	.451	−.313	−.340	−.266	.472	.151	−.191	.269
14	−.131	−.137	.535	.426	−.134	.115	.486	.109	−.181	.394	.246	.250	.151	.535	.374	.417
15	−.122	−.141	.247	.523	−.250	.035	.434	.414	−.323	.393	.403	.545	−.191	.374	.545	.243
16	−.135	−.340	.574	.413	−.210	−.223	.415	−.043	−.199	.112	−.055	.023	.269	.417	.243	.574
$\sum r^4$	.41	.70	.40	.49	1.00	.05	.45	.27	**1.12**	.49	.48	.32	.22	.34	.41	.33

correlation matrix was identified as having the largest index value, $\sum r_{i9}^4 = 1.12$. Examination of the elements in column 9 reveals that variables 1, 2, and 5 (in addition to variable 9) are represented by relatively large values. The first hypothesis vector $\mathbf{h}_1$ was defined to contain the *cubes* of elements in column 9 of the original correlation matrix. The ninth (diagonal) element was replaced by the largest other absolute value in the vector to reduce the effect of unique variance

$$\mathbf{h}_1' = [.179 \quad .291 \quad -.004 \quad -.036 \quad .462 \quad .010 \quad -.026 \quad -.049 \quad .462 \quad -.026 \quad .004 \quad -.014 \quad .092 \quad -.006 \quad -.034 \quad .008]$$

It should be noted that all the elements in $\mathbf{h}_1$ except those associated with variables 1, 2, 5, and 9 were reduced to near zero by the cubing process. Thus, the hypothesis vector in powered-vector factor analysis *tends* to be proportional to a zero-one cluster-centroid vector.

Computation of factor loadings for the first orthogonal powered-vector factor involved postmultiplying the original correlation matrix R_1 by the hypothesis vector $\mathbf{h}_1$ and then scaling the factor to unit variance

$$R_1\mathbf{h}_1 s_1 = \mathbf{f}_1$$

The vector written in transposed row form obtained from postmultiplying R_1 by $\mathbf{h}_1$ was

$$(R_1\mathbf{h}_1)' = [.835 \quad .952 \quad -.216 \quad -.464 \quad 1.082 \quad .329 \quad -.435 \quad -.506 \quad 1.116 \quad -.424 \quad -.214 \quad -.319 \quad .581 \quad -.263 \quad -.439 \quad -.334]$$

The scaling factor s_1 is the reciprocal of the square root of the product of $\mathbf{h}_1'$ and the product vector $R_1\mathbf{h}_1$

$$s_1 = \frac{1}{\sqrt{\mathbf{h}_1' r \mathbf{h}_1}} = \frac{1}{\sqrt{1.58823}} = .7935$$

The factor loadings for the first orthogonal powered-vector factor, obtained by multiplying each element in the vector $R_1\mathbf{h}_1$ by the scalar quantity $s_1 = .7935$, are (again written in transposed row form)

$$\mathbf{f}_1' = [.663 \quad .755 \quad -.172 \quad -.368 \quad .859 \quad .261 \quad -.345 \quad -.401 \quad .855 \quad -.337 \quad -.170 \quad -.253 \quad .461 \quad -.209 \quad -.348 \quad -.265]$$

It is obvious that the powered-vector algorithm resulted in a factor which correlated highly with variables 1, 2, 5, and 9 and to a lesser extent with 13 and 8 (negative). Consideration of these variables (somatic concern, anxiety, guilt feelings, depressive mood, and motor retardation, with grandiosity negative) leads to the interpretation that the first factor represents the psychiatrists' clinical concept of depression.

The next step in the analysis involved partialing out all variance associated with the first factor to obtain a residual matrix R_2

$$R_2 = R_1 - \mathbf{f}_1\mathbf{f}_1'$$

Table 6.2 Residual Correlation Matrix Following Extraction of First Powered-vector Factor

	1	*2*	*3*	*4*	*5*	*6*	*7*	*8*	*9*	*10*	*11*	*12*	*13*	*14*	*15*	*16*
1	.109	.078	.021	.045	−.044	−.038	.028	.054	−.023	.075	.088	.097	−.040	.008	.109	.041
2	.078	.225	−.032	.025	−.019	.174	−.004	.107	−.005	.174	.225	.179	−.155	.021	.122	−.140
3	.021	−.032	.529	.391	.023	.032	.392	−.149	−.006	.110	.004	.026	.363	.499	.187	.529
4	.045	.025	.391	.474	.068	.164	.474	.089	−.005	.149	.094	.281	.164	.349	.395	.315
5	−.044	−.019	.023	.068	.068	.029	.054	.067	.013	.011	.010	.024	.011	.045	.049	.017
6	−.038	.174	.032	.164	.029	.253	.231	.175	−.018	.253	.238	.186	−.184	.170	.126	−.154
7	.028	−.004	.392	.474	.054	.231	.474	.036	.010	.145	.021	.167	.213	.414	.314	.324
8	.054	.107	−.149	.089	.067	.175	.036	.375	−.012	.338	.375	.286	−.287	.025	.274	−.149
9	−.023	−.005	−.006	−.005	.013	−.018	.010	−.012	.043	.001	−.012	−.016	.043	.004	−.015	.035
10	.075	.174	.110	.149	.011	.253	.145	.338	.001	.582	.582	.288	−.158	.324	.276	.023
11	.088	.225	.004	.094	.010	.238	.021	.375	−.012	.582	.582	.463	−.262	.211	.344	−.100
12	.097	.179	.026	.281	.024	.186	.167	.286	−.016	.288	.463	.463	−.150	.197	.457	−.044
13	−.040	−.155	.363	.164	.011	−.184	.213	−.287	.043	−.158	−.262	−.150	.391	.247	−.031	.391
14	.008	.021	.499	.349	.045	.170	.414	.025	.004	.324	.211	.197	.247	.499	.301	.362
15	.109	.122	.187	.395	.049	.126	.314	.274	−.015	.276	.344	.457	−.031	.301	.457	.151
16	.041	−.140	.529	.315	.017	−.154	.324	−.149	.035	.023	−.100	−.044	.391	.362	.151	.529
$\sum r^4$	.00	.01	.28	.18	.00	.02	.18	.07	.00	.27	**.32**	.16	.08	.21	.16	.22

Each element in the first residual matrix presented in Table 6.2 is equal to the corresponding element in Table 6.1 minus the product of two factor loadings. The element in the second row and third column is equal to the corresponding element minus the product of the second and third factor loadings. In general,

$$r_{ij}^{(2)} = r_{ij}^{(1)} - f_i^{(1)} f_j^{(1)}$$

where superscripts indicate the first and second matrices, respectively. The complete first residual correlation matrix R_2 is presented in Table 6.2.

At this point, a close examination of the residual matrix is indicated. Off-diagonal elements in the rows and columns associated with variables related most strongly to factor I were reduced to near zero. The diagonal entries in rows 1, 2, 5, and 9 were substantially reduced. Little of the common variance was partialed out from the other variables. The correlations among variables that did not relate highly to factor I were altered little. This matrix provides insight into the effect of partialing out variance associated with a single cluster-oriented common factor.

The sums of fourth powers for elements in each column of the first residual matrix are entered at the bottom of Table 6.2. The sum of fourth powers for elements in column 11 of the first residual correlation matrix was found to be largest. The second hypothesis vector $\mathbf{h}_2$ with elements that are the cubes of elements from column 11 of the first residual matrix was defined

$$\mathbf{h}_2' = [.001 \quad .011 \quad .000 \quad .001 \quad .000 \quad .014 \quad .000 \quad .053 \quad .000 \quad .197 \quad .197 \quad .099 \quad -.018 \quad -.009 \quad .041 \quad .001]$$

It is worthy of note that the variables loading most heavily on the first factor (1, 2, 5, and 9) are represented by near-zero coefficients in the vector $\mathbf{h}_2$. Only variables 10, 11, and 12 were assigned relatively heavy weight in $\mathbf{h}_2$.

Factor loadings on the second orthogonal powered-vector factor were computed by postmultiplying matrix R_2 (Table 6.2) by the vector $\mathbf{h}_2$ and scaling the result to unit variance. The matrix-vector product $R_2\mathbf{h}_2$ was computed to contain the following elements (written as a row vector):

$$(R_2\mathbf{h}_2)' = [.050 \quad .115 \quad .023 \quad .099 \quad .012 \quad .140 \quad .067 \quad .209 \quad -.006 \quad .298 \quad .322 \quad .237 \quad -.123 \quad .141 \quad .207 \quad -.029]$$

The scaling coefficient s_2 required to scale the factor to unit variance was computed from the inner product of vectors $\mathbf{h}_2'$ and $R_2\mathbf{h}_2$

$$s_2 = \frac{1}{\sqrt{\mathbf{h}_2' R_2 \mathbf{h}_2}} = \frac{1}{\sqrt{.17202}} = 2.4111$$

where $v_2 = \mathbf{h}_2' R_2 \mathbf{h}_2$ is the sum of cross products of elements in the vectors $\mathbf{h}_2$ and $R_2\mathbf{h}_2$. The second vector of factor loadings $\mathbf{f}_2$ was obtained by multiplying each element in the vector $R_2\mathbf{h}_2$ by the scalar quantity $s_2 = 2.4111$. In row form, the second vector of factor loadings computed as described above was

$$\mathbf{f}_2' = [.121 \quad .278 \quad .055 \quad .240 \quad .030 \quad .338 \quad .162 \quad .504 \quad -.015 \quad .719 \quad .776 \quad .571 \quad -.297 \quad .340 \quad .500 \quad -.070]$$

Inspection of the second vector of factor loadings reveals that variables 8, 10, 11, 12, and 15 have relatively high loadings, while the remaining variables have relatively low loadings. Examination of the nature of these variables (grandiosity, hostility, suspiciousness, hallucinatory behavior, and unusual thought content) leads to the interpretation that the second factor represents the clinical syndrome of paranoid hostile-suspiciousness. It is interesting to note that in Europe, where these rating profiles came from, paranoid states and paranoid schizophrenics tend to be conceived as involving more hallucinatory symptomatology than in our own conception. This will become evident in other analyses of the psychiatric diagnostic-stereotype data.

The second residual correlation matrix R_3 was computed by partialing out from the matrix R_2 all variance associated with the second orthogonal powered-vector factor. This is equivalent to partialing out both factors I and II from the original correlation matrix

$$R_3 = R_2 - \mathbf{f}_2\mathbf{f}_2' = R_1 - \mathbf{f}_1\mathbf{f}_1' - \mathbf{f}_2\mathbf{f}_2'$$

Each element in R_3 is equal to the corresponding element in R_2 minus the product of factor loadings associated with the row and column variables

$$r_{ij}^{(3)} = r_{ij}^{(2)} - f_i^{(2)} f_j^{(2)}$$

The second residual correlation matrix is presented in Table 6.3. The extent to which the elimination of two factors has resulted in reduction of selected correlations should be examined. Intercorrelations among variables 1, 2, 5, 6, 8, 9, 10, 11, 12, and 15 have been reduced to near zero. Substantial correlations remain only among variables 3, 4, 7, 13, 14, and 16.

Disregarding the sums of fourth powers for elements in each column, which have been entered at the bottom of Table 6.3, the reader is invited to scan the various columns to see which appears to represent the best cluster marker. In which column are the larger coefficients most distinctly different from the smaller coefficients? By inspection, which column would have the largest sum of fourth powers? Calculations entered at the bottom of the table verify that it is column 3.

The third hypothesis vector defined with elements equal to the cubes of entries in column 3 of the second residual matrix was

$$\mathbf{h}_3' = [.000 \quad .000 \quad .151 \quad .054 \quad .000 \quad .000 \quad .056 \quad -.006 \quad .000 \quad .000 \quad .000 \quad .000 \quad .055 \quad .111 \quad .004 \quad .151]$$

The product vector obtained by postmultiplying the second residual correlation matrix by the third hypothesis vector (in transpose form) was

$$(R_3\mathbf{h}_3)' = [.007 \quad -.042 \quad .278 \quad .199 \quad .017 \quad -.002 \quad .211 \quad -.073 \quad .008 \quad .034 \quad -.031 \quad .012 \quad .201 \quad .239 \quad .102 \quad .262]$$

Table 6.3 **Residual Correlation Matrix Following Extraction of Two Powered-vector Factors**

	1	*2*	*3*	*4*	*5*	*6*	*7*	*8*	*9*	*10*	*11*	*12*	*13*	*14*	*15*	*16*
1	.079	.045	.014	.016	−.048	−.079	.008	−.007	−.021	−.012	−.006	.027	−.004	−.034	.048	.049
2	.045	.121	−.048	−.041	−.028	.080	−.049	−.033	−.001	−.025	.010	.020	−.072	−.074	−.017	−.121
3	.014	−.048	.532	.378	.022	.013	.383	−.176	−.005	.071	−.039	−.006	.379	.480	.160	.532
4	.016	−.041	.378	.435	.061	.083	.435	−.032	−.002	−.023	−.092	.144	.235	.268	.275	.332
5	−.048	−.028	.022	.061	.061	.019	.049	.052	.013	−.010	−.013	.007	.020	.035	.034	.020
6	−.079	.080	.013	.083	.019	.176	.176	.004	−.013	.010	−.024	−.007	−.084	.055	−.043	−.130
7	.008	−.049	.383	.435	.049	.176	.435	−.045	.012	.028	−.104	.074	.261	.359	.233	.335
8	−.007	−.033	−.176	−.032	.052	.004	−.045	.176	−.004	−.024	−.016	−.002	−.138	−.146	.022	−.114
9	−.021	−.001	−.005	−.002	.013	−.013	.012	−.004	.039	.012	−.000	−.008	.039	.009	−.008	.034
10	−.012	−.025	.071	−.023	−.010	.010	.028	−.024	.012	.123	.024	−.123	.056	.079	−.084	.073
11	−.006	.010	−.039	−.092	−.013	−.024	−.104	−.016	−.000	.024	.104	.020	−.031	−.053	−.044	−.046
12	.027	.020	−.006	.144	.007	−.007	.074	−.002	−.008	−.123	.020	.171	.020	.003	.171	−.004
13	−.004	−.072	.379	.235	.020	−.084	.261	−.138	.039	.056	−.031	.020	.379	.348	.118	.370
14	−.034	−.074	.480	.268	.035	.055	.359	−.146	.009	.079	−.053	.003	.348	.480	.131	.385
15	.048	−.017	.160	.275	.034	−.043	.233	.022	−.008	−.084	−.044	.171	.118	.131	.275	.186
16	.049	−.121	.532	.332	.020	−.130	.335	−.114	.034	.073	−.046	−.004	.370	.385	.186	.532
$\sum r^4$	.00	.00	**.28**	.12	.00	.00	.13	.00	.00	.00	.00	.00	.08	.17	.02	.23

The variance of the unscaled factor $v_3 = \mathbf{h}_3' R_3 \mathbf{h}_3$ was obtained as the sum of products of elements in the vectors $\mathbf{h}_3'$ and $R_3\mathbf{h}_3$. The scaling coefficient s_3, which is the reciprocal of the square root of the variance of the unscaled factor, was computed to be

$$s_3 = \frac{1}{\sqrt{\mathbf{h}_3' R_3 \mathbf{h}_3}} = \frac{1}{\sqrt{.14232}} = 2.6507$$

Correlations of the 16 original variables with the third orthogonal powered-vector factor, i.e., the factor loadings, were obtained by multiplying elements in the vector $(R_3\mathbf{h}_3)'$ by the scalar quantity $s_3 = 2.6507$. The resulting vector of factor loadings, written in transposed row form, was

$$\mathbf{f}_3' = [.019 \quad -.112 \quad .737 \quad .527 \quad .045 \quad -.005 \quad .560 \quad -.193 \quad .021 \quad .089 \quad -.083 \quad .033 \quad .533 \quad .633 \quad .270 \quad .694]$$

Symptom-construct ratings with highest loadings on the third orthogonal powered-vector factor include variables 3, 4, 7, 13, 14, and 16. These variables represent emotional withdrawal, conceptual disorganization, mannerisms and posturing, motor retardation, uncooperativeness, and blunted affect. The symptom variables related to factor III correspond to the clinical conception of the schizophrenic syndrome. Thus, the orthogonal powered-vector factor analysis provided a third major syndrome factor corresponding to the final major diagnostic syndrome represented in the input data.

A third residual matrix, R_4, was computed by partialing out variance associated with the third powered-vector factor

$$\begin{aligned} R_4 &= R_3 - \mathbf{f}_3\mathbf{f}_3' \\ &= R_1 - \mathbf{f}_1\mathbf{f}_1' - \mathbf{f}_2\mathbf{f}_2' - \mathbf{f}_3\mathbf{f}_3' \end{aligned}$$

Examination of the elements of this residual matrix, presented in Table 6.4, revealed that essentially all correlations among the 16 symptom variables in the diagnostic-stereotype data had been accounted for by only three orthogonal factors. Although a fourth factor could be defined, there seemed to be little residual correlation in need of representation. Only a couple of small *diads*, or two-variable clusters, suggest themselves. (A fourth factor was actually extracted but was found to include only two nontrivial loadings, and the total variance accounted for was small.) Thus, the decision was made to consider only three orthogonal powered-vector factors as necessary and meaningful in this analysis. These factors are shown in Table 6.5. It is worthy of note that the three factors defined in this analysis of *selected* diagnostic-stereotype profiles clearly represented the depressive, schizophrenic, and paranoid syndromes that were selected for inclusion. However, factor analysis can also be used to identify distinct syndromes when one does not know the composition of the heterogeneous population from which input data were derived.

Table 6.4 Residual Correlation Matrix Following Extraction of Three Powered-vector Faetors

	1	*2*	*3*	*4*	*5*	*6*	*7*	*8*	*9*	*10*	*11*	*12*	*13*	*14*	*15*	*16*
1	.079	.478	.000	.006	−.049	−.079	−.002	−.003	−.021	−.014	−.004	.026	−.014	−.046	.043	.036
2	.047	.080	.035	.018	−.023	.080	.014	−.054	.001	−.015	.000	.024	−.012	−.002	.013	−.043
3	.000	.035	.039	−.011	−.012	.017	−.030	−.034	−.021	.005	.022	−.030	−.014	.014	−.039	.021
4	.006	.018	−.011	.140	.037	.086	.140	.070	−.013	−.070	−.049	.126	−.046	−.066	.133	−.033
5	−.049	−.023	−.012	.037	.060	.019	.023	.060	.012	−.014	−.010	.005	−.004	.006	.022	−.012
6	−.079	.080	.017	.086	.019	.179	.179	.003	−.013	.010	−.024	−.007	−.081	.058	−.042	−.127
7	−.002	.014	−.030	.140	.023	.179	.179	.063	.000	−.022	−.058	.056	−.037	.004	.081	−.054
8	−.003	−.054	−.034	.070	.060	.003	.063	.075	−.000	−.007	−.039	.004	−.035	−.024	.075	.019
9	−.021	.001	−.021	−.013	.012	−.013	.000	−.000	.027	.010	.001	−.009	.027	−.005	−.013	.020
10	−.014	−.015	.005	−.070	−.014	.010	−.022	−.007	.010	.126	.031	−.126	.008	.022	−.108	.011
11	−.004	.000	.022	−.049	−.010	−.024	−.058	−.032	.001	.031	.058	.022	.013	−.001	−.021	.012
12	.026	.024	−.030	.126	.005	−.007	.056	.004	−.009	−.126	.022	.162	.002	−.018	.162	−.027
13	−.014	−.012	−.014	−.046	−.004	−.081	−.037	−.035	.027	.008	.013	.002	.081	.011	−.026	.001
14	−.046	−.002	.014	−.066	.006	.058	.004	−.024	−.005	.022	−.001	−.018	.011	.066	−.040	−.054
15	.043	.013	−.039	.133	.022	−.042	.081	.075	−.013	−.108	−.021	.162	−.026	−.040	.162	−.002
16	.036	−.043	.021	−.033	−.012	−.127	−.054	.019	.020	.011	.012	−.027	.001	−.054	−.002	.127
$\sum r^4$	.000	.000	.000	.001	.000	.002	.002	.000	.000	.000	.000	.002	.000	.000	.002	.001

Table 6.5 Orthogonal Powered-vector Factors of the Diagnostic-stereotype Data

	I	*II*	*III*
Somatic concern	.663	.121	.019
Anxiety	.755	.278	−.112
Emotional withdrawal	−.172	.055	.737
Conceptual disorganization	−.368	.240	.527
Guilt feelings	.859	.030	.045
Tension	.261	.338	−.005
Mannerisms and posturing	−.345	.162	.560
Grandiosity	−.401	.504	−.193
Depressive mood	.885	−.015	.021
Hostility	−.337	.719	.089
Suspiciousness	−.170	.776	−.083
Hallucinatory behavior	−.253	.571	.033
Motor retardation	.461	−.297	.533
Uncooperativeness	−.209	.340	.633
Unusual thought content	−.348	.500	.270
Blunted affect	−.265	−.070	.694

6.7 Variations on the Basic Method

The reader should by now begin to feel comfortable with the general linear computation method for defining orthogonal factors (Sec. 4.4) and be able to recognize that the orthogonal powered-vector method of factor analysis is simply one of an infinite number of possible variations which involve different rules for defining elements in the hypothesis vectors $\mathbf{h}_1, \mathbf{h}_2, \ldots, \mathbf{h}_r$.

Even within the general context of the powered-vector method there are two rather arbitrary parameters that one might choose to vary. Specifically, it has been recommended that the sum of fourth powers of elements in each column of the original or residual matrix be used as the criterion for choosing that column to become the basis for definition of the hypothesis vector. The logic is to identify a column in which the variables entering into one correlation cluster stand out clearly. The sum of fourth powers which has been adopted to achieve this goal is a convenient index for computer calculation. Earlier in the development of the method, the *variance* of squared elements in each column was used as the criterion for selecting a cluster marker. It has been proposed in this chapter that even the simple sum of squares will prove quite adequate in most cases and can be recommended for hand calculations. The reader who makes the distinction between the general logic and purpose on the one hand and the convenient numerical index chosen to accomplish that purpose on the other will be in a position to vary the method and to search for even better alternatives.

Another arbitrary parameter chosen to accomplish a logical purpose is the *cubing* of elements from a column of the correlation matrix to define the hypothesis vector at each stage. The idea is that a vector is needed in which variables within a homogeneous cluster are represented by relatively large elements and other variables are represented by near-zero elements. Signs should be retained because negative correlations are just as important as positive ones. The use of cubes of correlations among variables tends not only to reduce the smaller values essentially to zero but also to provide some distinction among the larger values. Some other odd power, say the fifth power, could be specified. In this event the distinction among larger coefficients would be enhanced, and the factor vector would tend to be defined solely in terms of one or two central variables. It is possible to zero out all small values and set all large values equal to unity, in which case the factor vector would pass through the cluster centroid. Again, the reader who is aware of the distinction between the logic and objectives and the arbitrary numerical methods chosen in an attempt to achieve those objectives and who is familiar with the general linear computation method for orthogonal-factor analysis will be able to examine various alternatives for himself.

6.8 Iterative Powered-vector Factor Analysis

A variation of the general powered-vector method that has particular intuitive appeal is the iterative solution. Beginning with a trial vector which contains the cubes of elements in a selected column of the correlation matrix, factor loadings are computed. These will be the correlations of the original variables with the powered-vector factor as defined in Sec. 6.4. Instead of stopping at this point, the cubes of the factor loadings can be used to define a second trial vector, and a new vector of factor loadings can be computed. This process can be repeated any specified number of times or until two successive vectors of factor loadings are identical to any desired number of decimal places. Because only a single cluster tends to be represented to any substantial degree even in the first powered trial vector, stabilization occurs quite rapidly. Usually three or four iterations will prove quite adequate

$$R_1\mathbf{h}_1^{(1)}s_1^{(1)} = \mathbf{f}_1^{(1)}$$

where $\mathbf{h}_1^{(1)}$ contains the cubes of elements in that column of R_1 having largest sum of fourth powers

$$R_1\mathbf{h}_1^{(2)}s_1^{(2)} = \mathbf{f}_1^{(2)}$$

where $\mathbf{h}_1^{(2)}$ contains the cubes of elements in $\mathbf{f}_1^{(1)}$, and

$$R_1\mathbf{h}_1^{(t)}s_1^{(t)} = \mathbf{f}_1^{(t)}$$

where $\mathbf{h}_1^{(t)}$ contains the cubes of elements in $\mathbf{f}_1^{(t-1)}$ from the previous iteration. The first residual matrix is computed in the usual manner

$$R_2 = R_1 - \mathbf{f}_1\mathbf{f}_1'$$

where $\mathbf{f}_1 = \mathbf{f}_1^{(t)}$ from the final iteration. The second iterated powered-vector factor is defined as follows:

$$R_2\mathbf{h}_2^{(1)}s_2^{(1)} = \mathbf{f}_2^{(1)}$$

where $\mathbf{h}_2^{(1)}$ contains the cubes of elements in that column of R_2 having largest sum of fourth powers

$$R_2\mathbf{h}_2^{(2)}s_2^{(2)} = \mathbf{f}_2^{(2)}$$

where $\mathbf{h}_2^{(2)}$ contains the cubes of elements in $\mathbf{f}_2^{(1)}$, and

$$R_2\mathbf{h}_2^{(t)}s_2^{(t)} = \mathbf{f}_2^{(t)}$$

where $\mathbf{h}_2^{(t)}$ contains the cubes of elements in $\mathbf{f}_2^{(t-1)}$ from the previous iteration. A third residual matrix can be computed and the iteration carried out as shown above. In general, the method is identical to the simple orthogonal powered-vector method except that the hypothesis vector is modified somewhat by the iterative multiplication before it is used to define the factor vectors.

The advantage of the iterative powered-vector solution is that it permits adjustment of the initial hypothesis vector to provide better allocation of variance among the principal vectors within a homogeneous cluster. Because of this, the iterated powered-vector factors tend to account for slightly more variance than the simple orthogonal powered-vector factors and to approach closely the principal-axes solution in this regard.

6.9 Comparison with the Principal-axes Method

To provide a comparison of the two methods the correlation matrix that was used to illustrate the principal-axes method was also factored by the orthogonal powered-vector method. Table 6.6 presents the two sets of factor loadings derived by applying the two methods to the data of Table 5.1. The two solutions are essentially the same as far as pattern of high and low factor loadings is concerned.

Although the unrotated powered-vector solution is considered satisfactory, it was rotated by the method described in Sec. 5.6.5 to illustrate that possible improvement can be obtained by rotation. The resulting loadings shown in Table 6.7 are more similar to the results of the principal-axes factor analysis. We leave it to the reader to decide whether the rotated solution is more meaningful and whether it is a better approximation to orthogonal simple structure. A major purpose here has been to emphasize that rotation of powered-vector factors can be accomplished just as with any other orthogonal factors. Even so, orthogonal rotation may not always provide the best fit to a particular set of data. In the next chapter the concept of oblique rotation is developed and will be applied to the orthogonal powered-vector factors to determine whether they can be further improved.

6.10 Computer Program for Orthogonal Powered-vector Factor Analysis

This program is included because it is not generally available elsewhere and because hand calculations become too laborious for even realistically large exercises in factor analysis. With the aid of this program, it is hoped that the

Table 6.6 Comparison of Rotated Principal-axes Factors and Orthogonal Powered-vector Factors

	Rotated principal-axes factors†				*Orthogonal powered-vector factors*			
	III	*II*	*IV*	*I*	*I*	*II*	*III*	*IV*
Somatic concern	−.060	.385	−.021	.033	−.053	.350	−.057	.024
Anxiety	−.135	.700	.039	.200	−.129	.750	.043	−.035
Emotional withdrawal	.789	−.029	.102	.154	.778	.009	−.082	.057
Conceptual disorganization	.327	−.164	.584	.145	.503	−.076	.500	−.077
Guilt feelings	.001	.580	.047	.001	.003	.546	−.058	−.117
Tension	.062	.430	.204	.312	.144	.560	.151	−.029
Mannerisms and posturing	.454	−.056	.274	.210	.527	.040	.145	.021
Grandiosity	−.086	−.085	.328	.200	.052	−.057	.388	.201
Depressive mood	.095	.717	−.205	−.000	−.008	.614	−.284	−.004
Hostility	−.069	.149	.064	.654	.052	.334	.273	.549
Suspiciousness	.006	.111	.413	.594	.226	.299	.537	.356
Hallucinatory behavior	.137	.031	.642	.099	.312	.087	.574	−.190
Motor retardation	.559	.164	−.063	−.128	.506	.096	−.302	−.146
Uncooperativeness	.336	−.072	.046	.567	.414	.096	.123	.479
Unusual thought content	.051	−.038	.769	.167	.299	.054	.695	−.083
Blunted affect	.684	−.030	.191	−.063	.714	−.085	−.059	−.112

† Columns of principal-axes factors have been permuted to be consistent with orthogonal powered-vector factors.

Table 6.7 Powered-vector Factors after Rotation

	I	*II*	*III*	*IV*
Somatic concern	−.034	.342	−.065	.080
Anxiety	−.137	.736	.033	.144
Emotional withdrawal	.759	−.028	.132	.147
Conceptual disorganization	.315	−.131	.614	.148
Guilt feelings	.025	.561	.007	.009
Tension	.086	.527	.201	.180
Mannerisms and posturing	.449	−.003	.274	.156
Grandiosity	−.087	−.142	.283	.299
Depressive mood	.083	.626	−.236	.054
Hostility	−.066	.174	.064	.672
Suspiciousness	.021	.147	.418	.599
Hallucinatory behavior	.115	.052	.670	.079
Motor retardation	.583	.135	−.065	−.127
Uncooperativeness	.329	−.041	.055	.559
Unusual thought content	.058	−.015	.733	.204
Blunted affect	.699	−.082	.193	−.031

student will analyze multivariate data of his own choosing for a better understanding of what the method accomplishes.

The orthogonal powered-vector factor-analysis program presented here is written in Fortran IV for the IBM 1130/1800 digital computer with disk. For this particular version of the program, all computations are done in core to provide maximum speed. Thus, the size of the matrix that can be analyzed depends upon the core capacity of the particular computer. Approximately 40 variables can be analyzed with an 8-K core, and approximately 60 with a 16-K core. There is essentially no limit on the number of individuals for whom data can be analyzed.

The program consists of two parts: first the correlation matrix is computed from raw data, or a previously computed correlation matrix is read into core, and then the simple or iterated orthogonal powered-vector factor analysis is accomplished. Only a single control card is required. It must include in successive four-column fields (1) the number of variables, (2) the number of individuals if a correlation matrix is to be computed or the entry 0001 if a previously computed correlation matrix is to be analyzed, and (3) the number of iterations of the initial powered-vector solution the user wants. The third field should be left blank for the usual noniterated solution and punched 0001 or 0002 for an iterated solution. (Here iteration implies the use of previously computed factor loadings to define new hypothesis vectors, as described in Sec. 6.8.) Any number of iterations can be specified, but our experience has been that results change little after two or three iterations. In fact, we seldom employ this option because of the negligible effect it has.

The single control card is followed either by data cards read in format 3 or by previously computed correlations read in format 2. The option of reading in a previously computed correlation matrix facilitates checking the program by applying it to correlation matrices presented in this book. Analysis of the correlation matrix presented in Table 6.1 should yield the results shown in Table 6.5.

The output from the program includes (1) the complete correlation matrix listed in 16-column partitions, (2) the orthogonal powered-vector factor loadings, and (3) communality estimates calculated from the factor loadings.

Program for Orthogonal Powered-vector Factor Analysis

```
      DIMENSION A(40,40), B(40), C(40), H(40)
      DEFINE FILE 1(80,160,U,L1),2(80,160,U,L2)
      L1=1
    1 FORMAT (4I4)
C     READ CONTROL CARD FROM CARD READER
      READ(2,1)NVAR,NOBS,NIT
C     TEST TO DETERMINE WHETHER CORRELATIONS MUST BE COMPUTED
      IF(NOBS-1)10,10,20
C     READ PREVIOUSLY COMPUTED CORRELATION MATRIX
    2 FORMAT (F8.4)
   10 DO 11 J=1,NVAR
      DO 11 J2=J,NVAR
      READ(2,2)A(J,J2)
   11 A(J2,J)=A(J,J2)
      L1 =1
```

```
      DO 12 J= 1,NVAR
   12 WRITE(1'L1)(A(J,J2),J2=1,NVAR)
      GO TO 40
C     COMPUTE CORRELATION MATRIX FROM RAW DATA
   20 DO 21 J=1,NVAR
      B(J)=0.0
      DO 21 J2=1,NVAR
   21 A(J,J2)=0.0
      FNOBS=NOBS
      FNM1=FNOBS-1.0
C     READ RAW DATA AND COMPUTE SUMS AND PRODUCTS
    3 FORMAT(17F1.0,27X,17F1.0,7X,3F2.0)
      DO 24 K=1,NOBS
      READ(2,3)(C(J),J=1,NVAR)
      DO 26 J=1,NVAR
      DO 25 J2=J,NVAR
   25 A(J,J2)=A(J,J2)+C(J)*C(J2)
   26 B(J)=B(J)+C(J)
   24 CONTINUE
C     COMPUTE MEAN CORRECTED SUMS OF PRODUCTS AND CORRELATIONS
      WRITE(3,4)
    4 FORMAT(1X,30H VAR      MEAN        SD          )
      DO 28 J=1,NVAR
      DO 27 J2=J,NVAR
   27 A(J,J2)=A(J,J2)-B(J)*B(J2)/FNOBS
      C(J)=B(J)/FNOBS
      B(J)=SQRT(A(J,J)/FNM1)
   28 WRITE(3,5)J,C(J),B(J)
    5 FORMAT (I4,2F10.4)
      DO 29 J=1,NVAR
   29 B(J)=SQRT(1.0/A(J,J))
      DO 31 J=1,NVAR
      DO 30 J2=J,NVAR
      A(J,J2)=A(J,J2)*B(J)*B(J2)
   30 A(J2,J)=A(J,J2)
   31 WRITE(1'L1)(A(J,J2),J2=1,NVAR)
C     PRINT CORRELATION MATRIX IN SETS OF 16 COLUMNS
      NSTR=1
   36 NSTP=NSTR+15
      IF(NVAR-NSTP)39,39,32
   39 NSTP=NVAR
   32 DO 33 J=NSTR,NSTP
   33 C(J)=J
      WRITE(3,6)(C(J),J=NSTR,NSTP)
    6 FORMAT (5X,16F6.0)
      DO 34 J=1,NVAR
      XROW=J
   34 WRITE(3,7)XROW,(A(J,J2),J2=NSTR,NSTP)
    7 FORMAT(F4.0,3X,16F6.3)
      NSTR=NSTP+1
      IF(NSTR-NVAR)36,36,40
   40 L1=1
C     GO THROUGH COMPLETE POWERED VECTOR ANALYSIS ITER TIMES
C     WITH COMMUNALITIES REESTIMATED EACH TIME
      DO 53 J=1,NVAR
   53 H(J)=0.0
      NSTOP= NVAR
      NFAC=0
      L1= 1
      L2=1
      DO 52 J=1,NVAR
      READ(1'L1)(A(J,J2),J2=1,NVAR)
      A(J,J)= H(J)
   52 H(J)= 0.0
      DO 75 I= 1,NSTOP
      MAX= 1
C     REPLACE DIAGONALS WITH NEW COMMUNALITY ESTIMATES
      DO 43 J=1,NVAR
      AMAX=0.0
      DO 41 J2=1,NVAR
      ATEST=ABS(A(J,J2))
      IF(ATEST-AMAX)41,41,42
   42 AMAX=ATEST
   41 CONTINUE
   43 A(J,J)= AMAX
```

```
      S4=0.0
      DO 56 J=1,NVAR
      TEST=0.0
      DO 54 J2=1,NVAR
   54 TEST=TEST+A(J,J2)**4
      IF(TEST-S4)56,56,55
   55 MAX=J
      S4 = TEST
   56 CONTINUE
C     DEFINE HYPOTHESIS VECTOR CONTAINING CUBES OF ELEMENTS
C     IN SELECTED COLUMN
      DO 58 J=1,NVAR
   58 B(J)=A(J,MAX)**3
C     COMPUTE FACTOR LOADINGS AND STORE ON DISC
      DO 63 M=1,NIT
      VAR=0.0
      DO 60 J=1,NVAR
      C(J)=0.0
      DO 59 J2=1,NVAR
   59 C(J)=C(J)+A(J,J2)*B(J2)
   60 VAR=VAR+C(J)*B(J)
      SCL=SQRT(1.0/VAR)
      NLOAD=0
      DO 61 J=1,NVAR
      C(J)=C(J)*SCL
      B(J)=C(J)**3
      TEST=ABS(C(J))
      IF(TEST-.30)61,61,62
   62 NLOAD=NLOAD+1
   61 CONTINUE
   63 CONTINUE
      IF(NLOAD-3)99,65,65
C     AUGMENT COMMUNALITY VECTOR
   65 DO 66 J=1,NVAR
   66 H(J)=H(J)+C(J)*C(J)
C     STORE FACTOR LOADINGS ON DISC
      WRITE(2'L2)(C(J),J=1,NVAR)
      NFAC=NFAC+1
C     COMPUTE RESIDUAL CORRELATION MATRIX
      DO 68 J=1,NVAR
      DO 68 J2=1,NVAR
   68 A(J,J2)=A(J,J2)-C(J)*C(J2)
   75 CONTINUE
   99 NSTOP= NFAC
C     RETRIEVE AND PRINT MATRIX OF FACTOR LOADINGS
    8 FORMAT (5X,16H FACTOR LOADINGS,//)
      WRITE(3,8)
      L2=1
      DO 80 K=1,NFAC
      READ(2'L2)(C(J),J=1,NVAR)
      DO 79 J=1,NVAR
   79 WRITE(3,9)K,J,C(J)
    9 FORMAT(2I4,F12.3)
      ISKIP=1111
   80 WRITE(3,1)
C     LIST COMMUNALITIES
      WRITE(3,111)
  111 FORMAT (1X,14H COMMUNALITIES,/)
      DO 85 J=1,NVAR
   85 WRITE(3,112)J,H(J)
  112 FORMAT (I4,F10.4)
      CALL EXIT
      END
```

7 Oblique Primary- and Reference-factor Structures

7.1 Introduction

An orthogonal-factor analysis defines factor variates that are statistically uncorrelated. The factors are orthogonal if the matrix of intercorrelations or "cosines" among the factors is an identity matrix; i.e., all the off-diagonal elements are zero

$$D_1 \hat{W}_1' R \hat{W}_1 D_1 = \mathbf{\Sigma}_1 = I$$

or, for rotated factors,

$$D_2 T' D_1 \hat{W}_1' R \hat{W}_1 D_1 T D_2 = \mathbf{\Sigma}_2 = I$$

It has been emphasized that *factors* can be defined to be orthogonal even though the measurement variables within various different clusters are substantially related.

In cases where various clusters of measurement variables are not independent, orthogonal factors may provide a rather poor representation of the configuration or pattern of relationships among the original variables. For simplicity and meaningful interpretation, one would like each factor to represent a different homogeneous subset of variables so that the nature of the factor can be understood in terms of what the subset of variables measures in common.

Requiring that factors be independent, even though the various subsets of variables are not, produces a difficult problem.

The objective of simple structure, in which each variable tends to relate primarily to only one factor, can be achieved only by relaxing the requirement that all factors be mutually orthogonal. By permitting factor axes to become *oblique*, it is frequently possible to represent subsets or homogeneous clusters of variables in a more direct manner. A factor solution is said to be oblique if the matrix of factor cosines is not an identity matrix

$$D_1\hat{W}_1'R\hat{W}_1D_1 = \mathbf{\Sigma}_1 \quad \text{and} \quad \mathbf{\Sigma}_1 \neq I$$

or, for rotated factors,

$$D_2T'D_1\hat{W}_1'R\hat{W}_1D_1TD_2 = \mathbf{\Sigma}_2 \quad \text{and} \quad \mathbf{\Sigma}_2 \neq I$$

7.2 Primary Factors and Reference Factors

7.2.1 DEFINITIONS

Oblique-factor solutions are of two general types. Oblique *primary factors* are defined in such manner that each factor correlates as highly as possible with variables within a homogeneous subset or cluster *disregarding* relationships to variables in other clusters. Sometimes factors defined in this manner will be referred to as *cluster-centroid factors.* If the variables within different subsets tend to be positively related, then all will have positive correlations with the primary factors representing other clusters. In defining the oblique primary structure, the attempt is to maximize the correlations between each factor and the subset of variables that the factor represents. A problem is that the factor will tend to relate to a lesser (but perhaps substantial) degree to a large number of variables that are represented primarily by other factors. Although it may be obvious in oblique solutions which variables relate most highly to a factor, one has difficulty in specifying what the factor does *not* measure.

An alternative oblique solution is known as the *reference structure.* Reference factors are defined to represent that which is measured in common by a homogeneous subset of variables *independent* of that which is measured in common by variables in other homogeneous subsets or clusters. Reference factors can be conceived as residual or adjusted factors that result from partialing out the variance associated with all other *primary* factors. Reference factors thus tend to correlate substantially with variables from one homogeneous cluster and to have near-zero correlations with variables represented by other factors. The reference factors tend to define what Thurstone termed the simple-structure solution because each factor represents only one source of variance.

7.2.2 OBLIQUE CLUSTER-CENTROID FACTORS

Suppose that one has by some means already identified clusters of variables that tend to have high correlations. One way to do this is to complete a preliminary analysis such as the orthogonal powered-vector analysis or an orthogonal rotation of principal-axes factors. There are, of course, numerous other ways that the cluster configuration can be identified. Given that one has identified the cluster configuration, the primary-factor structure can be defined by applying a simple zero-one transformation matrix to the original matrix of intercorrelations

$$R_1A_1D_1 = F_p \tag{7.1}$$

where each column of A_1 is formed with unit elements, i.e., equal to 1.0, corresponding to variables in one cluster and zeros elsewhere. As usual, D_1 is a diagonal scaling matrix containing the reciprocals of the square roots of the diagonal elements in $V_1 = A_1'R_1A_1$. Note that Eq. (7.1) is identical to the general factor transformation defined in Eq. (4.1), except for use of a different symbol to represent the cluster-transformation matrix as a special case.

For example, suppose that three clusters have been identified among nine variables. The first cluster consists of variables 1, 2, and 6. The second cluster consists of variables 3, 5, and 7, while the third consists of variables 4, 8, and 9. The cluster-centroid transformation matrix which will define a primary-factor structure will appear as follows:

$$A_1 = \begin{bmatrix} 1 & 0 & 0 \\ 1 & 0 & 0 \\ 0 & 1 & 0 \\ 0 & 0 & 1 \\ 0 & 1 & 0 \\ 1 & 0 & 0 \\ 0 & 1 & 0 \\ 0 & 0 & 1 \\ 0 & 0 & 1 \end{bmatrix}$$

Where zero-one vectors are employed in the transformation matrix A_1, the factor loadings appearing as elements in columns of F_p will be the correlations between the original variables and new composite factor variables defined as the unweighted sums of *standard scores* for variables in the various subsets. The first vector of factor loadings obtained by applying the transformation matrix A_1 illustrated above will contain product-moment correlations between the nine variables and a new variable which is the sum of standard scores for variables 1, 2, and 6. If unities in the principal diagonal of R_1 have been replaced by communality estimates, the factor loadings will not be precisely product-moment correlation coefficients, but the general interpretation as indices of relationship is similar.

Cosines or correlations among the primary factors can be obtained as previously defined

$$D_1A_1'R_1A_1D_1 = \boldsymbol{\Sigma}_p \tag{7.2}$$

where D_1 is a diagonal scaling matrix containing the reciprocals of square roots of diagonal elements in $V_1 = A_1'R_1A_1$. Since the matrix V_1 will have been computed to obtain the scaling factors, the primary-factor cosine matrix can be easily obtained as follows:

$$D_1V_1D_1 = \mathbf{\Sigma}_p \tag{7.3}$$

7.2.3 OBLIQUE REFERENCE FACTORS

It is a simple matter to go directly from oblique primary or cluster-centroid factors to a solution in which each reference factor is orthogonal to all except one of the primary factors. This can be accomplished by using the inverse of the matrix of cosines among primary factors

$$F_p\mathbf{\Sigma}_p^{-1}D_2 = R_1A_1D_1\mathbf{\Sigma}_p^{-1}D_2 = F_r \tag{7.4}$$

where F_p is defined in Eq. (7.1), $\mathbf{\Sigma}_p^{-1}$ is the inverse of the primary-factor cosine matrix in Eq. (7.2) or (7.3), and D_2 is a diagonal scaling matrix containing the reciprocals of the square roots of diagonal elements in

$$V_2 = \mathbf{\Sigma}_p^{-1}D_1A_1'R_1A_1D_1\mathbf{\Sigma}_p^{-1} = \mathbf{\Sigma}_p^{-1}\mathbf{\Sigma}_p\mathbf{\Sigma}_p^{-1} = \mathbf{\Sigma}_p^{-1}$$

That only one of the original primary factors (cluster-centroid factors) will have nonzero correlation with each reference factor can readily be appreciated in terms of the definition of a matrix inverse. Consider $\mathbf{\Sigma}_p$ to be a matrix of intercorrelations between primary factors. The new reference factors are defined as weighted combinations of the original primary factors. As described in Sec. 4.2, the correlations between several original variables and any set of r new composite variables defined as linear functions of the original variables can be expressed in the general form of a factor transformation,

$$RWD = F$$

or, in this instance,

$$\mathbf{\Sigma}_p\mathbf{\Sigma}_p^{-1}D_2 = F^*$$

where $\mathbf{\Sigma}_p$ is the matrix of correlations among the original primary factors, $\mathbf{\Sigma}_p\mathbf{\Sigma}_p^{-1} = I$, and D_2 is a diagonal matrix. If the elements in F^* are correlations of the original primary factors with the new reference factors, it is obvious that only one of the original primary factors will have nonzero correlation with the new reference factors because $F^* = D_2$ is a diagonal matrix.

Cosines or correlations among the reference axes can be obtained from the inverse of the matrix of cosines among the primary factors

$$D_2\mathbf{\Sigma}_p^{-1}D_1A_1'R_1A_1D_1\mathbf{\Sigma}_p^{-1}D_2 = D_2\mathbf{\Sigma}_p^{-1}D_2 = \mathbf{\Sigma}_r \tag{7.5}$$

where D_2 contains the reciprocals of the square roots of diagonal elements in $\mathbf{\Sigma}_p^{-1}$, as previously defined.

7.3 Oblique Powered-vector Factor Analysis

The oblique factor structures based upon a preliminary orthogonal powered-vector factor analysis will be referred to as oblique powered-vector solutions. The sequence of computations involves preliminary calculation of factor loadings by the method of orthogonal powered-vector factor analysis (Chap. 6). Next a zero-one transformation matrix A_1 $(p \times r)$ is defined in which each row contains one and only one nonzero element. This matrix is obtained by setting the largest factor loading for each variable equal to 1.0 and the others equal to zero. Given the cluster-centroid transformation matrix A_1, the primary-factor structure can be computed using Eq. (7.1) and the reference-factor structure using Eq. (7.4).

To illustrate the oblique powered-vector solution, let us begin with the orthogonal powered-vector analysis for the 16 BPRS symptom rating variables derived in Sec. 6.6. The first step following the orthogonal powered-vector factor analysis is to define the oblique cluster-centroid transformation matrix as shown in Table 7.1. The factor loadings for each variable are replaced by 1.0, corresponding to the largest loading for that variable, and zeros elsewhere.

The next step is to compute the *unscaled* factor loadings by postmultiplying the original correlation matrix R_1 (presented previously in Table 6.1) by the zero-one cluster-centroid transformation matrix A_1. The product matrix R_1A_1 is presented in Table 7.2. Each element in the first column of the product matrix R_1A_1 is equal to the sum of *selected* elements from the corresponding row of R_1, as determined by the positions of nonzero elements in column 1 of the transformation matrix. For example, the first element in the first column of R_1A_1 is equal to the sum of elements 1, 2, 5, and 9 from the first row of R_1.

Table 7.1 Orthogonal Powered-vector Factor Loadings and Derived Zero-One Cluster-centroid Transformation Matrix

	I	*II*	*III*	$\mathbf{a}^{(1)}$	$\mathbf{a}^{(2)}$	$\mathbf{a}^{(3)}$
Somatic concern	.66	.12	.02	1	0	0
Anxiety	.76	.28	−.11	1	0	0
Emotional withdrawal	−.17	.06	.74	0	0	1
Conceptual disorganization	−.37	.24	.53	0	0	1
Guilt feelings	.86	.03	.05	1	0	0
Tension	.26	.34	−.01	0	1	0
Mannerisms and posturing	−.35	.16	.56	0	0	1
Grandiosity	−.40	.50	−.19	0	1	0
Depressive mood	.89	−.02	.02	1	0	0
Hostility	−.34	.72	.09	0	1	0
Suspiciousness	−.17	.78	−.08	0	1	0
Hallucinatory behavior	−.25	.57	.03	0	1	0
Motor retardation	.46	−.30	.53	0	0	1
Uncooperativeness	−.21	.34	.63	0	0	1
Unusual thought content	−.35	.50	.27	0	1	0
Blunted affect	−.27	−.07	.69	0	0	1

Table 7.2 The R_1A_1 Matrix (Unscaled) Formed by Postmultiplying the Correlation Matrix by the Zero-One Matrix

$Ra^{(1)}$	$Ra^{(2)}$	$Ra^{(3)}$
2.67	−.44	−.49
2.87	.04	−.96
−.54	.42	3.30
−1.03	1.63	2.89
2.93	−.88	−.55
.97	1.58	.02
−1.00	1.34	3.01
−1.05	2.79	−.07
3.00	−1.18	−.71
−.80	3.04	.89
−.22	3.18	.12
−.52	2.93	.70
1.32	−1.65	1.75
−.58	1.49	3.01
−.84	2.79	1.63
−.88	.06	3.09

The second element in the first column is equal to the sum of elements 1, 2, 5, and 9 from the second row of R_1. That multiplication by a zero-one transformation matrix results in summation of selected elements is obvious from consideration of the row-by-column rule for matrix multiplication (Sec. 2.4).

The final step in computing the matrix of factor loadings for the oblique *primary factors* is to scale the factors to unit variance. This requires calculation of the matrix product

$$V_1 = A_1'R_1A_1 = A_1'(R_1A_1)$$

Since the product matrix R_1A_1 has already been obtained, $V_1 = A_1'(R_1A_1)$ can be obtained by simply premultiplying the matrix R_1A_1 (Table 7.2) by the transpose of the zero-one cluster-centroid transformation matrix (Table 7.1). This is, of course, tantamount to forming a new matrix in which each element is the sum of *selected* elements in the *columns* of R_1A_1. The resulting primary-factor *covariance* matrix $V_1 = A_1'R_1A_1$ is presented in Table 7.3. Note that the first element in the first column of V_1 is equal to the sum of elements 1, 2, 5, and 9 from the first column of R_1A_1 in Table 7.2. The second element in the first column of V_1 is equal to the sum of elements 6, 8, 10, 11, 12, and 15 from the

Table 7.3 The Primary-factor Covariance Matrix $V_1 = A_1'R_1A_1$

$$\begin{bmatrix} 11.4659 & -2.4609 & -2.7239 \\ -2.4609 & 16.3199 & 3.2989 \\ -2.7239 & 3.2989 & 17.0479 \end{bmatrix}$$

Table 7.4 The Scaling Matrix D_1

$$\begin{bmatrix} .2953 & .0000 & .0000 \\ .0000 & .2475 & .0000 \\ .0000 & .0000 & .2422 \end{bmatrix}$$

first column of R_1A_1. These results follow from simple row-by-column multiplication with a zero-one matrix as the prefactor.

The matrix D_1 required to scale the primary factors to unit variance contains the reciprocals of the square roots of the variances of the unscaled factors that appear as diagonal elements of $V_1 = A_1'R_1A_1$. The diagonal matrix D_1 is presented in Table 7.4. The first element is the reciprocal of the square root of the first diagonal element in V_1; that is, $1/\sqrt{11.47} = .2953$. The second diagonal element is $1/\sqrt{16.32} = .2475$. Other elements were obtained in similar manner.

The scaled *primary-factor* loadings, which can be interpreted as correlations between the primary-factor variates and the original variables, are shown in Table 7.5. These were obtained by postmultiplying the product matrix R_1A_1 of Table 7.2 by the diagonal scaling matrix D_1 of Table 7.4

$$R_1A_1D_1 = (R_1A_1)D_1 = F_p$$

Postmultiplication by a diagonal matrix is equivalent to multiplying all elements in each column of the prefactor by a constant. Thus, elements in the first column of F_p are equal to the corresponding elements in the first column of

Table 7.5 Scaled Primary-factor Loadings
$R_1A_1D_1 = F_p$

	I	*II*	*III*
Somatic concern	.788	−.109	−.120
Anxiety	.848	.010	−.233
Emotional withdrawal	−.159	.105	.799
Conceptual disorganization	−.304	.404	.699
Guilt feelings	.864	−.218	−.134
Tension	.287	.392	.006
Mannerisms and posturing	−.297	.333	.728
Grandiosity	−.311	.690	−.018
Depressive mood	.886	−.291	−.173
Hostility	−.237	.753	.217
Suspiciousness	−.066	.788	.029
Hallucinatory behavior	−.152	.726	.171
Motor retardation	.389	−.407	.424
Uncooperativeness	−.172	.368	.730
Unusual thought content	−.247	.691	.395
Blunted affect	−.261	.014	.748

Table 7.6 The Primary-factor Cosine Matrix $D_1 V_1 D_1 = \mathbf{\Sigma}_p$

$$\begin{bmatrix} .9999 & -.1799 & -.1948 \\ -.1799 & 1.0000 & .1977 \\ -.1948 & .1977 & 1.0000 \end{bmatrix}$$

$R_1 A_1$ (Table 7.2) multiplied by the first element in D_1. Elements in the second column of F_p are equal to the elements in the second column of $R_1 A_1$ multiplied by the second element in D_1.

Examination of the primary-factor loadings reveals that the larger loadings on each factor became still larger in going from the orthogonal to the oblique primary-factor structure. In this sense, the oblique primary factors are better representations of the cluster configuration. It will also be noted that most variables correlate to some degree with each primary factor. This can pose a problem for interpretation since each factor represents something that is measured in part by almost all the original variables.

Cosines or correlations among the primary factors were defined in Eqs. (7.2) and (7.3). The latter equation is more convenient, since the matrix V_1 has already been computed. The matrix of cosines among the three primary factors can be obtained by pre- and postmultiplying $V_1 = A_1' R_1 A_1$ of Table 7.3 by the diagonal scaling matrix D_1 of Table 7.4, that is, $D_1 V_1 D_1 = \mathbf{\Sigma}_p$. The resulting primary cosine matrix is presented in Table 7.6. It should be noted that pre- and postmultiplication of a square matrix by the same diagonal matrix is equivalent to forming a new matrix in which each element is equal to the product of the corresponding element in the square matrix and two elements from the diagonal matrix, one associated with row and the other with column. The element σ_{ij} of the primary-factor cosine matrix $\mathbf{\Sigma}$ is equal to the corresponding element v_{ij} of the matrix $A_1' R_1 A_1 = V_1$ multiplied by the product of diagonal elements d_{ii} and d_{jj} from the matrix D_1

$$\sigma_{ij} = v_{ij} d_{ii} d_{jj}$$

For example, the element in the third row and second column of $\mathbf{\Sigma}_p$ is equal to the element in the third row and second column of $V_1 = A_1' R_1 A_1$ multiplied by the product of the second and third elements in the diagonal matrix D_1.

Examination of the primary cosine matrix reveals that the primary clusters are relatively independent in the original set of variables. Correlations between oblique factors of the order of .20 to .30 are generally considered to be quite acceptable. One should be concerned that each factor represents a reasonably distinct dimension of variation, and high correlations between primary factors, say above .50, should generally be rejected. One might consider collapsing such highly correlated factors into a single common factor.

If one wants to examine the reference-factor structure, a simple transformation of the primary-factor structure is required. The inverse of the primary-factor cosine matrix must be computed, and this can be done using the square-root

Table 7.7 Inverse of the Primary-factor Cosine Matrix

$$\begin{bmatrix} 1.0624 & .1563 & .1761 \\ .1563 & 1.0637 & -.1799 \\ .1761 & -.1799 & 1.0699 \end{bmatrix}$$

method described in Sec. 2.6. The inverse of the 3×3 primary-factor cosine matrix is presented in Table 7.7 for this example. Calculation of reference-factor loadings requires postmultiplication of the matrix of primary-factor loadings by the inverse of the primary-factor cosine matrix and subsequent rescaling

$$F_p \boldsymbol{\Sigma}_p^{-1} D_2$$

where D_2 is a diagonal matrix containing the reciprocals of the square roots of diagonal elements in $\boldsymbol{\Sigma}_p^{-1}$. The scaling matrix D_2 can be applied to $\boldsymbol{\Sigma}_p^{-1}$ before transforming the primaries. The matrix $\boldsymbol{\Sigma}_p^{-1} D_2$ is a new matrix in which elements in each column are equal to the corresponding elements in $\boldsymbol{\Sigma}_p^{-1}$ divided by the square root of the diagonal element in that column. The matrix $\boldsymbol{\Sigma}_p^{-1} D_2$ for this example is presented in Table 7.8. In each column, the elements were obtained by dividing elements in that column of $\boldsymbol{\Sigma}_p^{-1}$ by the square root of the diagonal entry for that column.

The final step in computing reference-factor loadings is to postmultiply the matrix of primary-factor loadings by the transformation matrix $\boldsymbol{\Sigma}_p^{-1} D_2$ of Table 7.8. The reference structure, or matrix of factor loadings on the oblique reference factors, is presented in Table 7.9. Each column in Table 7.9 was obtained by applying a separate set of weights, i.e., a particular column of $\boldsymbol{\Sigma}_p^{-1} D_2$, to the various rows of F_p. For example, the 16 elements in the first column were computed by applying the elements in the first column of the transformation matrix $\boldsymbol{\Sigma}_p^{-1} D_2$ to each of the 16 rows of F_p. Similarly, the 16 elements of the second column of F_r were obtained by applying weights in the second column of $\boldsymbol{\Sigma}_p^{-1} D_2$ to each of the 16 rows of F_p. Examination of the reference factor structure reveals that the variables having the largest relationship to each corresponding primary factor still have high loadings in the reference structure but that most other reference-factor loadings are close to zero. From the reference structure it is frequently easier to specify what each factor does and does not represent.

Table 7.8 Transformation Matrix $\boldsymbol{\Sigma}_p^{-1} D_2$ Used to Obtain Oblique Reference Factors

$$\begin{bmatrix} 1.0307 & .1515 & .1703 \\ .1516 & 1.0314 & -.1739 \\ .1709 & -.1744 & 1.0344 \end{bmatrix}$$

Table 7.9 Oblique Reference Factors
$F_p \Sigma_p^{-1} D_2 = F_r$

	I	*II*	*III*
Somatic concern	.775	.027	.029
Anxiety	.835	.179	−.098
Emotional withdrawal	−.011	−.055	.780
Conceptual disorganization	−.133	.248	.601
Guilt feelings	.835	−.070	.046
Tension	.356	.446	−.013
Mannerisms and posturing	−.131	.171	.644
Grandiosity	−.218	.667	−.191
Depressive mood	.839	−.135	.022
Hostility	−.093	.703	.052
Suspiciousness	.056	.797	−.118
Hallucinatory behavior	−.017	.695	.024
Motor retardation	.411	−.435	.575
Uncooperativeness	.003	.226	.661
Unusual thought content	−.082	.605	.246
Blunted affect	−.139	−.155	.726

The correlations or cosines among the reference factors can be obtained directly from the inverse of the primary-factor cosine matrix

$$D_2 \Sigma_p^{-1} D_2 = \Sigma_r$$

where D_2 is a diagonal matrix containing reciprocals of square roots of diagonal elements in Σ_p^{-1}. Thus, elements in Σ_r can be obtained by dividing each element in Σ_p^{-1} by the product of square roots of row and column diagonals

$$\sigma_{ij}^{(2)} = \frac{\sigma_{ij}^{(1)}}{\sqrt{\sigma_{ii}^{(1)} \sigma_{jj}^{(1)}}}$$

For example, the cosine between reference factors 2 and 3 can be obtained by dividing the element in second row and third column of Σ_p^{-1} (Table 7.7) by the product of the square roots of diagonal elements 2 and 3. The reference-factor cosine matrix is presented in Table 7.10.

Table 7.10 The Reference Cosine Matrix
$D_2 \Sigma_p^{-1} D_2 = \Sigma_r$

$$\begin{bmatrix} 1.000 & .147 & .165 \\ .147 & 1.000 & -.168 \\ .165 & -.168 & .999 \end{bmatrix}$$

7.4 Another Example

In the previous chapter, rotated principal-axes factor loadings were compared with orthogonal powered-vector factors before and after orthogonal rotation (Tables 6.6 and 6.7). It was concluded that with these data both solutions

Table 7.11 Scaled Primary-factor Loadings
$R_1 A_1 D_1 = F_p$

	I	*II*	*III*	*IV*
Somatic concern	−.035	.534	−.000	.074
Anxiety	−.083	.762	.038	.109
Emotional withdrawal	.823	.003	.250	.099
Conceptual disorganization	.398	−.060	.700	.098
Guilt feelings	.006	.627	.023	−.070
Tension	.159	.564	.212	.149
Mannerisms and posturing	.649	.022	.322	.130
Grandiosity	.021	.062	.549	.358
Depressive mood	.026	.703	−.160	.003
Hostility	.083	.494	.308	.824
Suspiciousness	.206	.285	.666	.604
Hallucinatory behavior	.245	.057	.706	−.003
Motor retardation	.620	.060	−.026	−.212
Uncooperativeness	.584	.119	.282	.680
Unusual thought content	.212	.049	.820	.146
Blunted affect	.711	−.060	.238	−.094

Table 7.12 The Primary-factor Cosine Matrix
$D_1 V_1 D_1 = \mathbf{\Sigma}_p$

$$\begin{bmatrix} 1.000 & .042 & .315 & .178 \\ .042 & 1.000 & .114 & .296 \\ .315 & .114 & 1.000 & .350 \\ .178 & .296 & .350 & 1.000 \end{bmatrix}$$

Table 7.13 Oblique Reference Factors
$F_p \mathbf{\Sigma}_p^{-1} D_2 = F_r$

	I	*II*	*III*	*IV*
Somatic concern	−.035	.536	−.023	−.068
Anxiety	−.096	.762	.016	−.103
Emotional withdrawal	.785	−.018	.006	−.041
Conceptual disorganization	.198	−.098	.630	−.138
Guilt feelings	.014	.677	.036	−.266
Tension	.099	.543	.130	−.079
Mannerisms and posturing	.577	−.012	.126	−.020
Grandiosity	−.175	−.053	.483	.197
Depressive mood	.087	.738	−.197	−.148
Hostility	−.069	.260	.036	.666
Suspiciousness	−.031	.104	.473	.353
Hallucinatory behavior	.045	.052	.713	−.273
Motor retardation	.682	.137	−.139	−.294
Uncooperativeness	.474	−.079	−.083	.582
Unusual thought content	−.036	−.004	.798	−.139
Blunted affect	.686	−.026	.090	−.225

Table 7.14 The Reference Cosine Matrix $D_2 \Sigma_p^{-1} D_2 = \Sigma_r$

$$\begin{bmatrix} 1.000 & .014 & -.274 & -.077 \\ .014 & 1.000 & -.015 & -.274 \\ -.274 & -.015 & .999 & -.297 \\ -.077 & -.274 & -.297 & 1.000 \end{bmatrix}$$

approximated the simple-structure criterion but left something to be desired. An oblique rotation might provide better definition of the powered-vector factors. It can be seen in Table 7.11 that the factor structure was considerably improved by oblique rotation. All four factors are clearly defined by a few high loadings with the remaining loadings much reduced, and there is a stronger tendency for each symptom construct to have a high loading on only a single factor. The primary cosine matrix of Table 7.12 shows that after oblique rotation, factor I is nearly orthogonal to factor II but the last three factors are no longer independent. Table 7.13 contains the oblique reference factors and Table 7.14 the reference cosine matrix.

7.5 General-purpose Orthogonal and Oblique Factor-analysis Program

Having discussed a variety of factor-analysis methods ranging from orthogonal principal-axes to oblique cluster-oriented factor solutions, we are now ready to consider a very general computer program that can be used to accomplish any of these types of analysis under control of only two parameter cards. In fact, since some of the parameters are continuously variable, the program can be used to explore *families* of related factor-analysis methods.

The program, which is written in Fortran IV for the IBM 1130 disk-oriented computer, can handle up to 159 variables on several hundred individuals. Since the raw data are read and stored on the disk, a certain latitude for trade-off between number of variables and number of individuals exists due to the possibility of redefining the disk storage files. As listed below, the files are defined to handle up to 159 variables and up to 350 individuals.

The general-purpose factor-analysis program consists of one mainline program plus ten subroutines. The program reads raw data, computes the matrix of intercorrelations, accomplishes an orthogonal-factor analysis and then the associated oblique cluster-oriented factor solution. The orthogonal-factor analysis can be a principal-axes, approximate principal-axes, powered-vector, or some related solution, depending on parameters provided in the control cards. The oblique primary and reference structures are then defined in terms of the cluster of variables having highest loadings on each orthogonal factor. Thus, if interest is in the oblique structure, a cluster-oriented orthogonal solution such as the powered vector should be used.

Two control cards must precede the raw data cards in the input sequence.

Entries in each control card are punched in successive four-column fields. The first control card defines the number of variables (NVAR), the number of individuals (NOBS), and a parameter NT which will cause a preliminary cluster-rotation transformation of the correlation matrix (Sec. 8.6). Ordinarily NT will be left blank or punched zero if a conventional factor analysis is desired.

The second control card includes three parameters MPOW, NPOW, and NIT. The parameter MPOW determines the method of selecting the column from the correlation matrix to be used in defining the hypothesis vector at each stage. For a principal-axes, or approximate principal-axes analysis, MPOW = 2. For a powered-vector analysis, MPOW = 4. The parameter NPOW is the exponent or power to which elements in the selected column of the correlation matrix are raised in defining the hypothesis vector. For principal axes, or approximate principal axes, the value of NPOW = 1. For powered-vector analysis, NPOW = 3. In general, the program will run and will yield meaningful results where MPOW is any *even* integer and NPOW is any *odd* integer; however, parameter values other than those mentioned above should be used only by people who have considerable insight into the general factor-analysis methodology discussed in the preceding chapters. The parameter NIT determines the number of iterations completed in defining the final hypothesis vector for each factor. For a principal-axes analysis, NIT should be a large positive integer, say NIT = 30. For a good approximate principal-axes analysis that requires much less computer time, let NIT = 2 or 3. For powered-vector analysis used in conjunction with MPOW = 4 and NPOW = 3, let NIT = 1. The standard control cards for three conventional types of factor analysis for 50 variables and 200 individuals are as follows.

Principal Axes

Card 1:	50	200	0
Card 2:	2	1	30

Approximate Principal Axes

Card 1:	50	200	0
Card 2:	2	1	3

Powered Vector

Card 1:	50	200	0
Card 2:	4	3	1

The printout resulting from use of the program includes (1) the complete matrix of intercorrelations partitioned into 16-column blocks, (2) the variance accounted for by each successively extracted orthogonal factor, and (3) the orthogonal, oblique primary, and oblique reference structures.

General-purpose Orthogonal and Oblique Factor-analysis Program

```
      DIMENSION A(159),B(159),Z(32,32)
      COMMON L3,L4,L5,L6,L7,L8,L9,L10
      DEFINE FILE 1(350,320,U,L3), 2(160,320,U,L4), 3(35,320,U,L5), 6(16
     10,320,U,L6), 7(35,320,U,L7), 8(35,320,U,L8), 9(35,70,U,L9), 10(35,
     2320,U,L10)
    1 FORMAT (3I4)
    2 FORMAT(32X,10F1.0,2X,7F4.0)
   18 FORMAT(/,' CORRELATION MATRIX VARIABLES',I4,'--',I4)
      READ (2,1) NVAR,NOBS,NT
      READ(2,1) MPOW,NPOW,NIT
      L3=1
      IF(NOBS-9999) 21,22,21
   21 DO 20 JX=1,NOBS
      READ(2,2)(A(J),J=1,NVAR)
   20 WRITE(1'L3)(A(K),K=1,NVAR)
      CALL DCOR (NVAR,NOBS)
      GO TO 23
   22 CALL DCOR2(NVAR)
   23 CALL CSQR(NVAR)
C     RETRIEVE AND PRINT CORRELATIONS
   13 FORMAT(1X,16F6.2)
      XNVAR=NVAR
      XNT=XNVAR/16.
      MT=XNT+.99
      DO 16I=1,MT
      JJ=16*I-15
      KK=16*I
      IF(KK-NVAR) 26,26,15
   15 KK=NVAR
   26 WRITE(3,18)JJ,KK
      DO 16 J=1,NVAR
      READ(2'J)(B(K),K=1,NVAR)
   16 WRITE(3,13)(B(K),K=JJ,KK)
   19 IF (NT) 7,8,7
    7 CALL DMULT (NVAR,NT)
    8 CALL FAC(NVAR,MPOW,NPOW,NIT,NFAC)
      CALL MCOS(NVAR,NFAC)
      CALL PRCOS(NVAR,NFAC)
      CALL PFAC(NVAR,NFAC)
      CALL JMINV (NFAC,Z)
      CALL RFCOS(NFAC,NVAR,Z)
      CALL EXIT
      END
```

Subroutine DCOR Computes Matrix of Intercorrelations and Stores on Disk

```
      SUBROUTINE DCOR (NVAR,NOBS)
      DIMENSION A(159),B(159),C(159),D(159)
      COMMON L3,L4,L5,L6,L7,L8,L9,L10
      FNOBS=NOBS
      DO 99 I=1,NVAR
      DO 95 J=1,NVAR
      B(J)=0.0
      C(J)=0.0
   95 D(J)=0.0
      L3 = 1
      DO 98 I2=1,NOBS
      READ (1'L3)(A(J),J=1,NVAR)
      DO 97 J=I,NVAR
      B(J)=B(J)+A(J)
      C(J)=C(J)+A(J)*A(I)
   97 D(J)=D(J)+A(J)*A(J)
   98 CONTINUE
      DO 94 J=I,NVAR
      C(J)=C(J)-B(I)*B(J)/FNOBS
      D(J)=D(J)-B(J)*B(J)/FNOBS
   94 C(J)=C(J)/SQRT(D(J)*D(I))
      WRITE(2'I)(C(J),J=1,NVAR)
   99 CONTINUE
      RETURN
      END
```

Subroutine DCOR2 Reads Previously Computed Correlation Matrix and Stores on Disk

```
      SUBROUTINE DCOR2 (NVAR)
      DIMENSION A(35,35)
      COMMON L3,L4,L5,L6,L7,L8,L9,L10
      L4=1
C     IF MATRIX IS FULL AND NOT JUST UPPER TRIANGLE,THEN READ STATEMENT
C     MUST BE CHANGED, AND THE THREE CARDS FOLLOWING IT TAKEN OUT
      DO 9 J=1,NVAR
    9 READ(2,3)(A(J,K),K=J,NVAR)
      DO 10 J=1,NVAR
      DO 10 K=J,NVAR
   10 A(K,J)=A(J,K)
      DO 11 J=1,NVAR
   11 WRITE(2'L4)(A(J,K),K=1,NVAR)
    3 FORMAT(13F6.3)
      RETURN
      END
```

Subroutine CSQR Completes the Square of Upper Triangular Matrix Stored on Disk

```
    SUBROUTINE CSQR (NVAR)
    DIMENSION A(159),B(159)
    COMMON L3,L4,L5,L6,L7,L8,L9,L10
    DO 9 J=1,NVAR
    READ(2'J)(A(I),I=1,NVAR)
    DO 9 K=J,NVAR
    READ(2'K)(B(I),I=1,NVAR)
    B(J)=A(K)
  9 WRITE(2'K)(B(I),I=1,NVAR)
    DO 96 J=1,NVAR
    READ(2'J)(B(I),I=1,NVAR)
    B(J)=0.0
    INDEX=1
    BIG=ABS(B(1))
    DO 14 JJ=2,NVAR
    IF(BIG-ABS(B(JJ))) 13,14,14
 13 BIG=ABS(B(JJ))
    INDEX=JJ
 14 CONTINUE
    B(J)=ABS(B(INDEX))
 96 WRITE(6'J)(B(I),I=1,NVAR)
    RETURN
    END
```

Subroutine DMULT Preconditions Correlation Matrix by Direct Cluster Rotation to Emphasize Cluster Structure

```
    SUBROUTINE DMULT (NVAR,NT)
    DIMENSION A(159),B(159),C(159),D(159)
    COMMON L3,L4,L5,L6,L7,L8,L9,L10
    DO 555 J5=1,NT
    DO 77 J=1,NVAR
    READ(2'J)(A(K),K=1,NVAR)
    A(J)=0.0
    DX=0.0
    DO 75 K=1,NVAR
    XSR=A(K)
    SR=XSR*XSR
    IF(SR-DX)75,75,74
 74 DX=SR
 75 A(K)=SR*A(K)
    A(J)=DX*SQRT(DX)
    WRITE(2'J)(A(K),K=1,NVAR)
 77 CONTINUE
    DO 15 J=1,NVAR
    READ(2'J)(A(K),K=1,NVAR)
    DO 14 I=J,NVAR
    READ(2'I)(B(K),K=1,NVAR)
    PROD=0.0
```

```
      DO 13 K=1,NVAR
   13 PROD=PROD+A(K)*B(K)
   14 C(I)=PROD
      D(J)=SQRT(1.0/C(J))
   15 WRITE(2'J)(C(K),K=1,NVAR)
C     COMPLETE SQUARE
      DO 100 J=1,NVAR
      READ(2'J)(A(I),I=1,NVAR)
      DO 90 K=J,NVAR
      READ(2'K)(B(I),I=1,NVAR)
      B(J)=A(K)
   90 WRITE(2'K)(B(I),I=1,NVAR)
  100 CONTINUE
C     SCALE TO UNIT DIAGONAL
      DO 200 J=1,NVAR
      READ(2'J)(A(I),I=1,NVAR)
      DO 201 I=1,NVAR
  201 A(I)=A(I)*D(I)*D(J)
      WRITE(3,66)(A(I),I=1,NVAR)
   66 FORMAT(1X,20F5.2)
  200 WRITE(2'J)(A(I),I=1,NVAR)
  555 CONTINUE
      RETURN
      END
```

Subroutine FAC Computes Orthogonal-factor Analysis

```
      SUBROUTINE FAC(NVAR,MPOW,NPOW,NIT,NFAC)
      DIMENSION A(159), B(159), C(159)
      COMMON L3,L4,L5,L6,L7,L8,L9,L10
      L7=1
C     FIND COLUMN WITH LARGEST SUM OF MPOW POWERS,WHERE
C     MPOW IS EQUAL TO 2 FOR APPROX PRINCIPAL AXIS AND EQUAL
C     TO 4 FOR POWERED VECTOR
      M=0
      M2=1
      FV=NVAR
   55 MAX=0
      FMAX=0.0
      L4 = 1
      DO 20 J=1,NVAR
      SUM=0.0
      SSQ=0.0
      READ(2' J)(A(K),K=1,NVAR)
      A(J)=0.0
      INDEX=1
      BIG=0.0
      DO 14 JJ=1,NVAR
      POW=A(JJ)**MPOW
      SUM=SUM+POW
      IF(BIG-POW)13,14,14
   13 BIG=POW
      INDEX=JJ
   14 CONTINUE
      A(J)=ABS(A(INDEX))
      WRITE(2' J)(A(K),K=1,NVAR)
      IF(SUM-FMAX)20,20,18
   18 MAX= J
      FMAX=SUM
   20 CONTINUE
C     FETCH HYPOTHESIS VECTOR
      READ(2'MAX)(B(J),J=1,NVAR)
      FMAX=0.0
      DO 21 J=1,NVAR
   21 B(J)=B(J)**NPOW
C     WHERE NPOW IS EQUAL TO 1 FOR APPROX PRINCIPAL AXIS AND
C     EQUAL TO 3 FOR POWERED VECTOR
C     ITERATE FACTOR TRANSFORMATION NIT TIMES
      DO 22 KK=1,NIT
      L4=1
      VAR=0.0
```

```
      DO 30 J=1,NVAR
      READ(2'L4)(A(K),K=1,NVAR)
      C(J)=0.0
      DO 29 K=1,NVAR
   29 C(J)=C(J)+B(K)*A(K)
   30 VAR=VAR+C(J)*B(J)
      SCL=1.0/SQRT(VAR)
      DO 35 J=1,NVAR
      C(J)=C(J)*SCL
   35 B(J)=C(J)**NPOW
   22 CONTINUE
C     TEST VARIANCE ACCOUNTED FOR
      MTES=0
      TEST=0.0
      DO 36 J=1,NVAR
      CSQR=C(J)*C(J)
      TEST= TEST+CSQR
      IF(CSQR-.09)36,136,136
  136 MTES=MTES+1
   36 CONTINUE
      IF(MPOW-4)199,137,199
  137 IF(MTES-3)99,199,199
  199 IF(TEST-1.0)99,37,37
   37 WRITE(7'L7)(C(J),J=1,NVAR)
   38 IF(M-32)60,99,99
   60 M=M+1
C     COMPUTE RESIDUAL MATRIX
      WRITE(3,236) M,TEST
  236 FORMAT(1X,'VARIANCE FOR FACTOR', I4,F10.5)
      DO 50 J=1,NVAR
      READ(2'J)(A(K),K=1,NVAR)
      DO 49 K=1,NVAR
   49 A(K)=A(K)-C(K)*C(J)
      WRITE(2'J)(A(K),K=1,NVAR)
   50 CONTINUE
      GO TO 55
   99 NFAC=M
      RETURN
      END
```

Subroutine MCOS Forms a Zero-One Cluster-centroid Transformation Matrix Based on Orthogonal Factor Structure

```
      SUBROUTINE MCOS (NVAR,MM)
      DIMENSION N(159),B(159),C(159)
      COMMON L3,L4,L5,L6,L7,L8,L9,L10
      DO 115 I=1,MM
      READ(7'I)(C(J),J=1,NVAR)
      SQR=0.0
      DO 13 J=1,NVAR
   13 SQR=SQR+C(J)*C(J)
      SQR=SQRT(SQR)
      DO 14 J=1,NVAR
   14 C(J)=C(J)/SQR
  115 WRITE(1'I)(C(J),J=1,NVAR)
      DO 5 K=1,NVAR
    5 B(K)=0.0
      DO 15 K=1,MM
      READ (1'K ) (C(J),J=1,NVAR)
      DO 15 J=1,NVAR
      IF(C(J)) 7,8,8
    7 KK= -1.0*K
      GO TO 9
    8 KK=K
    9 C(J)=ABS(C(J))
      IF (B(J)-C(J)) 12,12,15
   12 B(J)=C(J)
      N(J)=KK
   15 CONTINUE
      DO 20 K=1,MM
      DO 25 J=1,NVAR
```

```
   C(J)=0.0
   IF (N(J)) 17,18,18
17 KJ=K
   KN=IABS(N(J))
   IF(KN-KJ)25,26,25
26 C(J)=-1.0
   GO TO 25
18 IF (N(J)-K) 25,24,25
24 C(J)=1.0
25 CONTINUE
20 WRITE (1'K) (C(J),J=1,NVAR)
   RETURN
   END
```

Subroutine PRCOS Computes Cosines among Primary Axes Passed through Cluster Centroids

```
   SUBROUTINE PRCOS(NVAR,NFAC)
   DIMENSION A(159), B(159), C(159),D(159)
   COMMON L3,L4,L5,L6,L7,L8,L9,L10
   DO 25 K=1,NFAC
   READ(1'K)(A(I),I=1,NVAR)
   DO 24 J=1,NVAR
   C(J)=0.0
   READ(6'J)(B(I),I=1,NVAR)
   DO 22 I=1,NVAR
22 C(J)=C(J)+A(I)*B(I)
24 CONTINUE
   WRITE(8'K )(C(J),J=1,NVAR)
25 CONTINUE
   DO 35 J=1,NFAC
   READ(8'J)(C(I),I=1,NVAR)
   DO 34 K=1,NFAC
   READ(1'K)(B(I),I=1,NVAR)
   A(K)=0.0
   DO 32 I=1,NVAR
32 A(K)=A(K)+B(I)*C(I)
34 CONTINUE
   D(J)=1.0/SQRT(A(J))
35 WRITE(9'J)(A(K),K=1,NFAC)
   DO 45 J=1,NFAC
   READ(9'J)(A(K),K=1,NFAC)
   DO 44 K=1,NFAC
44 A(K)=A(K)*D(J)*D(K)
45 WRITE(9'J)(A(K),K=1,NFAC)
   RETURN
   END
```

Subroutine PFAC Computes Oblique Primary-factor Projections

```
   SUBROUTINE PFAC(NVAR,NFAC)
   DIMENSION A(159), B(159), C(159)
   COMMON L3,L4,L5,L6,L7,L8,L9,L10
   DO 25 K=1,NFAC
   READ(1'K)(A(I),I=1,NVAR)
   VAR=0.0
   DO 24 J=1,NVAR
   C(J)=0.0
   READ(6'J)(B(I),I=1,NVAR)
   DO 22 I=1,NVAR
22 C(J)=C(J)+A(I)*B(I)
24 VAR=VAR+C(J)*A(J)
   SCL=1.0/SQRT(VAR)
   DO 21 I=1,NVAR
21 C(I)=C(I)*SCL
25 WRITE(10'K)(C(I),I=1,NVAR)
   RETURN
   END
```

Subroutine JMINV Computes the Inverse of the Primary Cosine Matrix Stored on Disk

```
      SUBROUTINE JMINV(N,A)
      DIMENSION IPIVO(32),INDEX(32,2),PIVOT(32), A(32,32)
      COMMON L3,L4,L5,L6,L7,L8,L9,L10
      DO 284 K=1,N
      READ(9'K)(A(K,J),J=1,N)
  284 CONTINUE
      DET=1.0
      DO 20 J=1,N
   20 IPIVO (J)=0
      DO 550 I=1,N
      AMAX=0.0
      DO 105 J=1,N
      IF (IPIVO (J)-1) 60,105,60
   60 DO 100 K=1,N
      IF (IPIVO (K)-1) 80,100,740
   80 IF(ABS(AMAX)-ABS(A(J,K)))85,100,100
   85 IROW=J
      ICOLU =K
      AMAX=A(J,K)
  100 CONTINUE
  105 CONTINUE
      IF(ABS (AMAX)-2.0E-7) 800,800,801
  800 WRITE(3,666)
  666 FORMAT(20H DETERMINANT = ZERO )
      PAUSE
  801 IPIVO(ICOLU)=IPIVO(ICOLU)+1
      IF (IROW-ICOLU ) 140,260,140
  140 DET=-DET
      DO 200 L=1,N
      SWAP=A(IROW,L)
      A(IROW,L)=A(ICOLU ,L)
  200 A(ICOLU ,L)=SWAP
  260 INDEX(I,1)=IROW
      INDEX(I,2)=ICOLU
      PIVOT(I)=A(ICOLU ,ICOLU )
      DET=DET*PIVOT(I)
      A(ICOLU ,ICOLU )=1.0
      DO 350 L=1,N
  350 A(ICOLU ,L)=A(ICOLU ,L)/PIVOT(I)
      DO 550 L1=1,N
      IF(L1-ICOLU ) 400,550,400
  400 T=A(L1,ICOLU )
      A(L1,ICOLU )=0.0
      DO 450 L=1,N
  450 A(L1,L)=A(L1,L)-A(ICOLU ,L)*T
  550 CONTINUE
      DO 710 I=1,N
      L=N+1-I
      IF (INDEX(L,1)-INDEX(L,2)) 630,710,630
  630 IROW=INDEX(L,1)
      ICOLU =INDEX(L,2)
      DO 705 K=1,N
      SWAP=A(K,IROW)
      A(K,IROW)=A(K,ICOLU )
      A(K,ICOLU )=SWAP
  705 CONTINUE
  710 CONTINUE
  740 RETURN
      END
```

Subroutine RFCOS Computes the Oblique Reference-axis Projections and Prints All Three Solutions

```
      SUBROUTINE RFCOS(NFAC,NVAR,A)
      DIMENSION A(32,32), C(159), F(159), B(159)
      COMMON L3,L4,L5,L6,L7,L8,L9,L10
    1 FORMAT(12F10.3)
      DO 10 K=1,NFAC
   10 C(K)=1.0/ SQRT(A(K,K))
      L3=1
      WRITE(1'L3)(C(K),K=1,NFAC)
      DO 15 K=1,NFAC
      DO 15 J=1,NFAC
   15 A(J,K)=A(J,K)*C(K)
      DO 99 K1=1,NVAR
      DO 98 K2=1,NFAC
      READ(10'K2)(C(J),J=1,NVAR)
   98 F(K2)=C(K1)
      DO 95 K2=1,NFAC
      C(K2)=0.0
      DO 94 K3=1,NFAC
   94 C(K2)=C(K2)+F(K3)*A(K3,K2)
   95 CONTINUE
   99 WRITE(2'K1)(C(K2),K2=1,NFAC)
      DO 50 K=1,NFAC
      WRITE(3,11) K,K,K
      READ(7'K)(C(J),J=1,NVAR)
      READ(10'K)(F(J),J=1,NVAR)
      DO 50 J=1,NVAR
      READ(2'J)(B(I),I=1,NFAC)
   50 WRITE(3,13) J, C(J), F(J), B(K)
   11 FORMAT(' ORTHOGONAL LOADINGS',I3,20X,'PRIMARY LOADINGS',I3,20X,
     1'REFERENCE LOADINGS',I3)
   13 FORMAT(5X,I4,3(F10.5,25X))
      WRITE(3,4)
    4 FORMAT(//,' PRIMARY COSINE MATRIX')
      DO 21 J=1,NFAC
      READ(9'J)(B(K),K=1,NFAC)
   21 WRITE(3,1)(B(K),K=1,NFAC)
      L3=1
      READ(1'L3)(C(K),K=1,NFAC)
      DO 20 K=1,NFAC
      DO 20 J=1,NFAC
   20 A(J,K)=A(J,K)*C(J)
      WRITE(3,3)
      DO 2 J=1,NFAC
    2 WRITE(3,1)(A(J,K),K=1,NFAC)
    3 FORMAT(// ,' REFERENCE COSINE MATRIX')
      RETURN
      END
```

7.6 Interpretation of Primary and Reference Factors

Whether to seek an oblique primary structure or an oblique reference structure must be answered in terms of the objectives of the investigation. A geometric model will help tc illustrate the difference between the two. In Fig. 7.1 measurement variables have been plotted as points in a two-dimensional space using orthogonal factor loadings as cartesian coordinate values. Such plotting of factor loadings is helpful in visualizing relationships among variables. Primary factors tend to represent directly subsets of variables. That is, if a distinct cluster configuration is present, the oblique primary axes will tend to pass through the cluster centroids as illustrated by P_1 and P_2 in Fig. 7.1. The primary factors thus tend to measure that which is measured by a subgroup of variables *without regard for relationships to the other variables.* If one considers that the factor loadings represent perpendicular "projections" of the measurement

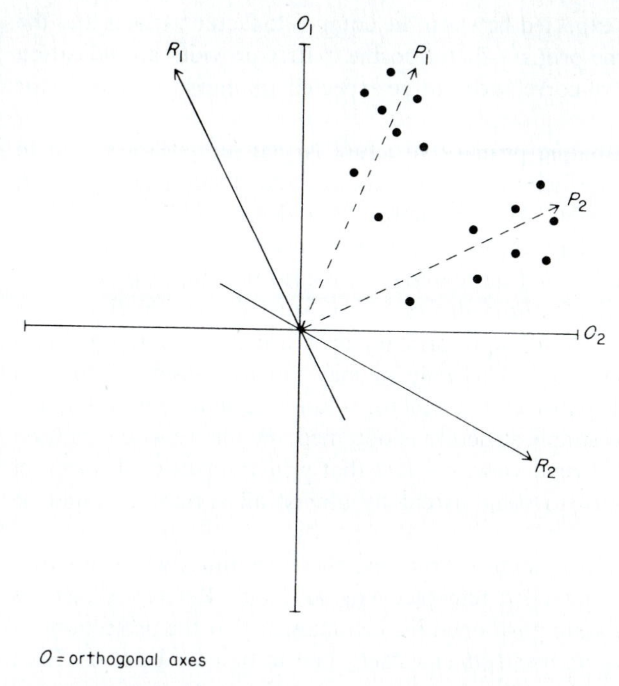

Figure 7.1 Geometric representation of relationships between primary and reference factors.

variables on each factor axis, it is clear in Fig. 7.1 that the variables which have greatest projection on primary factor P_1 also have substantial projection on P_2. Similarly, the variables that have greatest projection on primary factor P_2 also have substantial projection on P_1.

Each primary axis tends to be located in such a manner that the projections of a subset of variables are maximized without regard for the magnitudes of projections of other variables. The primary factor represents most directly that which is measured by a subset of the total collection of variables. The correlations, or cosines, between primary factors tend to approximate the average correlations of variables in one subset with variables in another. In fact, if one were to define factor scores as the simple sums of z-score values for variables having the greatest relationship to each factor, as is frequently done, the correlations between the factor scores should approximate very closely the oblique primary-axes cosines. In a sense, it appears more consistent to be concerned with the primary structure when the aim is to define factor scores as simple sums of selected test scores. The primary factor loadings provide a picture of the correlations to be expected between the composite factor variates and the original variables, and the primary-factor cosine matrix provides an indication of the degree and nature of correlations to be expected among the several factor scores.

A problem with the oblique primary structure is that all variables tend to project substantially on all factors. The matrix of factor loadings reveals that almost all variables correlate to some substantial degree with each factor. While one can say that each primary factor tends to measure that which is measured by the variables having highest loadings, it is also true that the source of variance represented by the factor is also present in variables which relate more highly to other factors. The oblique primary axes usually do not provide a good simple-structure solution. This may or may not be considered to be a problem. Many psychologists feel that a factor solution cannot be meaningfully interpreted unless a clear simple structure is obtained. While we would caution the student against such a rigid view, the fact that primary axes tend to reflect sources of variance shared to some extent by almost all variables should be noted.

Apart from the question of simple structure, there are times when the logic of a problem seems to demand a reference-axis solution. Reference axes, as shown in Fig. 7.1, usually are positioned in such manner that the projections of all variables, *except those in one distinct subset*, tend to be near to zero. Each reference axis is orthogonal to all primary axes except one. Of course, in this simplified geometric example only two primary axes and the two corresponding reference axes are shown. Thus, reference axis R_1 is orthogonal to primary axis P_2, and reference axis R_2 is orthogonal to primary axis P_1. This means that the variables which lie closest to P_2 will tend to have very low perpendicular projection on reference axis R_1, and variables which lie close to P_1 will have very low projections on R_2. If there were more than two factors, each reference axis would be positioned orthogonal to all except one of the oblique primary axes. Each reference factor tends to represent that variance which is shared in common by one subset of variables *independent* of the variance shared by other variables. In a very real sense, the transformation from primary structure to reference structure is analogous to calculating *partial correlations* of all variables with each primary factor after other primary factors are partialed out. In many investigations, the real concern centers about examination of the *independent* sources of variance and the ways in which different variables relate to them.

In spite of the fact that reference axes are properly conceived as representing independent sources of variance, it should be noted that the reference axes are not themselves mutually orthogonal. Each reference axis is orthogonal to all *primary* axes except the one with which it is identified; however, the oblique reference axes are not orthogonal to one another. If one were to calculate factor scores to represent precisely the *reference factors*, and not the associated primaries, the factor variates would represent variance shared in common by a relatively few tests after variance shared in common by other tests has been partialed out. The reference-factor variates would be correlated with one another.

7.7 Exercises

1 As described in Sec. 4.6, factor scores can be obtained by applying to the original data (*z*-score form) the same transformation matrix that is applied to the correlation matrix R in obtaining the matrix of factor loadings. Oblique primary-factor loadings are obtained by applying a transformation of the type $R_1 A_1 D_1 = F_p$, and reference-factor loadings are obtained by the additional transformation $R_1 A_1 D_1 \Sigma_p^{-1} D_2 = F_r$. If one wants factor variates which have correlations with the original BPRS variables such as those shown in Table 7.5, what weight should be given each of the 16 variables? If one wants factor variates which have correlations with the original BPRS variables such as those shown in Table 7.9, what weight should be given each of the 16 variables?

2 Assuming that the weights that you have specified are to be applied to z scores, what additional information would you need in order to define the same factors as weighted functions of the raw scores? Assume for purposes of this exercise that the individual BPRS variables have the variances shown in the principal diagonal of the matrix in Table 9.4 and that the means are the same as those for paranoid patients in Table 9.5. What would the raw-score factor-variate equations be? What would be the mean and variance of each factor variate, assuming means and variances from Tables 9.4 and 9.5? *Hint:* The variance of a linear composite is a function of the weighting coefficients and the variances and covariances among the original variables.

8

Empirical Methods for Developing Classification Typologies

8.1 Introduction

Recent years have witnessed increasing interest in the use of empirical methods for development of new taxonomies or typologies of mental illness. Other areas of social and medical research have also recognized a need to refine classification concepts through empirical analysis of similarity relationships. This chapter concerns empirical methods for classification grouping based on similarities among individuals or objects without regard for any previously existing classification scheme. This problem should be clearly distinguished from correct assignment of individuals among preestablished categories or diagnostic groups, methods for which will be considered in Chaps. 12 to 15.

Suppose a sample of individuals has been drawn at random from a mixed population composed of several distinct subpopulations. Several observations or measurements are taken on each individual. Solely on the basis of these multivariate observations, some individuals appear more alike and some more different from one another. Certain *modal patterns* tend to recur with substantial frequencies, and it is to be inferred that these patterns represent homogeneous subtypes. A relatively few prototype patterns can represent most individuals in the larger heterogeneous population. If the most frequently occurring patterns are, in fact, identified, then a majority of individuals in the total mixed population can be described as being like one of the modal types. Classification is the process of organizing individuals or objects according to similarities on specified variables.

The use of empirical methods to develop classification concepts, or typologies, raises several important considerations. The typology is derived entirely from internal relationships among individuals or objects without reference to any preexisting classification system, making it essential that the measurements or other observations used to define similarities among individuals be *relevant.* Variables that will define a typology must be carefully selected with reference to the purpose for which the typology is to be used. A typology of psychiatric patients *could* be derived from frequently occurring patterns of body measurements, hair and eye color, plus protein, carbohydrate, and fat breakdown of the individual's preferred diet; but if empirical methods are being used to define a psychiatric diagnostic classification system, the symptoms and background variables known to be associated with psychiatric pathology should be the starting point. If the empirical methods are employed to classify types of fat metabolism, measures of different lipid molecules should provide the starting point.

A second problem in using empirical methods to develop classification concepts is that the results from any analysis describe similarities and differences among individuals within the particular sample. One would, of course, like to use empirical clustering techniques to define a typology of general significance, but the methodology provides only a similarity grouping or cluster configuration for a sample of individuals. The authors have found profile cluster-analysis methods relatively sensitive to sampling variability, so that the modal types derived from one sample are often not replicated closely in the next. Most often certain modal patterns tend to reappear consistently, while, in different analyses, one or more unreproducible patterns will also appear. We recommend that research aimed at developing typologic concepts by means of empirical clustering techniques be replicated extensively in independent samples until the investigator feels confident that he can describe modal types representative of the majority of individuals in the population being studied. A strategy we have used to integrate the results of numerous replicated analyses will be mentioned in a later section of this chapter.

Regarding the problem of stability and replicability, different methods for defining modal clusters unquestionably differ in sensitivity to sampling vagaries in the data. The simple cluster-analysis methods that will be described first, primarily for heuristic purposes, seem especially sensitive to sampling variability because the starting point for each empirical group or cluster depends upon only two or three profiles out of the total sample. Other methods exist which, in a sense, take into account relationships among all individuals in the sample. The investigator should be sensitive to this problem and should give preference to a method that is minimally sensitive to relationships between single pairs of individuals.

A final consideration in classification research should be the validity and utility of the resulting grouping. Here classification, as opposed to mere empirical prediction of a specific criterion, offers a great advantage. For example, if one wishes to predict response to one particular drug, the most powerful and direct approach should be the general linear regression method described in Chap. 17. But there are hundreds of drugs, and clinicians will understandably resist evaluating hundreds of regression equations. A valid classification system provides the advantage of general utility for a number of different kinds of problems and, in some areas of research, may turn out to have fundamental significance for underlying psychological, social, or disease processes. Finally, scientists and laymen alike are accustomed to thinking and communicating in terms of classification concepts, and that alone recommends the effort to develop better ones.

8.2 General Nature of Classification Methods

Classification refers to the process of grouping together individuals that are highly similar in specified attributes. Such similar individuals can, for many practical purposes, be treated alike. Although the methods used to accomplish classification grouping vary, they have two major steps in common: (1) computing quantitative indices of multivariate similarity between all pairs of individuals or objects and (2) analyzing similarity indices to identify homogeneous subgroups.

Similarities between multivariate measurement profiles can be represented in terms of *distance-function* or *vector-product* coefficients. Although the methods of calculating these two types of coefficients appear quite different, their tendency to reflect the same information can be readily illustrated in a simple geometric model. The geometric interpretation of multivariate profile similarity arises from the fact that any profile (individual or object) can be located as a vector or point in multivariate space by using elements in the score vector (profile) as rectangular coordinate values. For a simplified example, suppose the score vector consists of only two elements. Accepting the two measurement variables as orthogonal coordinate axes, we can locate any individual profile in the simple two-dimensional space according to projections on the x and y axes, as shown in Fig. 8.1. The profile vector designated A was located in the model by going up three units on one axis and to the right two units on the other. The same principles hold for profiles containing more than two elements, although physical representation is more difficult.

The similarity between profiles A and B can be described in terms of the *distance* between the vector end points. According to the familiar pythagorean theorem, the square of the distance between the profiles equals the sum of squares of the differences in projections on the orthogonal coordinate axes, since the differences in projections are equal to the lengths of two sides of the

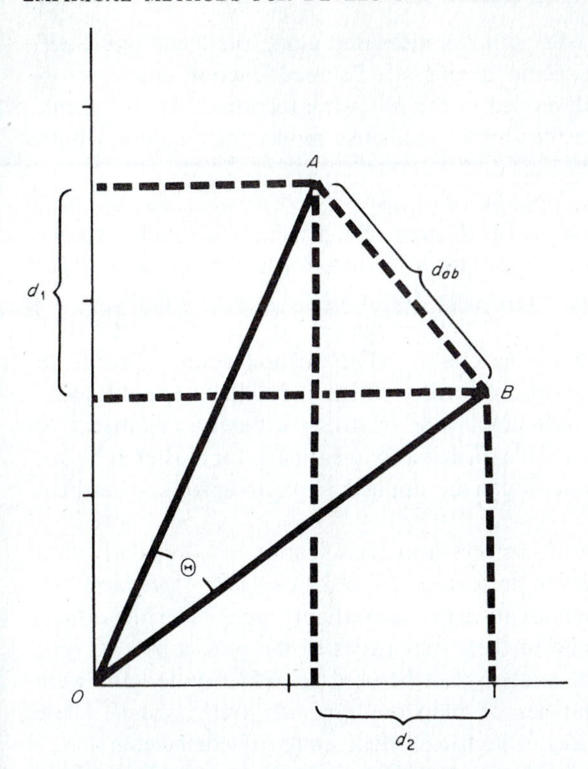

Figure 8.1 Geometric model illustrating relationship between distance-function and vector-product similarity coefficients.

right triangle indicated by the dotted lines. Because specific measurement variables are identified with the coordinate axes, the square of the distance between profiles can be calculated as the sum of squares of differences between corresponding scores in the two profiles. This same calculation can be generalized to profiles containing more than two elements, although physical representation becomes more difficult.

Inspection of Fig. 8.1 reveals that the distance between the vector end points is a function of the angle between the profile vectors and their respective lengths. In fact, the squared distance between the profiles can be calculated directly from knowledge of the vector lengths and the angle between them. If the profile vectors are scaled to unit length, the distance between their end points is a function solely of the angular separation between them. The cosine of the angle between two normalized profile vectors can be calculated as the sum of cross products between corresponding elements in the two profiles. This gives rise to *vector-product* indices of profile similarity. Obviously, the vector-product and distance-function indices contain much the same information since the distance between two vector end points is measured along the "side opposite" the angle between them. Specific details of distance-function and vector-product calculations will be discussed in the following sections. At this point, the student should appreciate the simple geometric model that underlies both types of multivariate similarity indices.

Given a matrix of distance-function or vector-product coefficients, the next problem is to identify subgroups, or clusters of individuals belonging to the same class. Cluster analysis or a variation on linear components analysis is most frequently used. Cluster analysis involves a logical and systematic search procedure in which a homogeneous cluster nucleus is identified and then similar individuals are added to the cluster. The method results directly in several mutually exclusive subgroups within which individuals are relatively similar and between which individuals are relatively different. Matrices of distance-function coefficients provide the most frequent basis for cluster analysis, although cluster-analysis methods can be applied to vector-product matrices as well.

The general linear-components approach to classification involves derivation of a set of prototype profiles representing hypothetical *pure types* and the evaluation of similarities of individuals to the derived pure types. The individuals can then be classified according to their similarities to the pure types. *Q*-type factor analysis or linear typal analysis can be used to define underlying pure types and to evaluate similarities of individuals to the pure types. These methods are applied to matrices of vector-product similarity coefficients.

Although the computational techniques of factor analysis serve effectively in classification research, the theoretical basis should be recognized as being quite different from that underlying the use of factor analysis for the study of relationships among measurement variables (Chaps. 5 to 7). Classification research is undertaken in the context of a specific set of measurement variables. The specified set of measurement variables constitutes the orthogonal reference frame within which relationships among individuals or objects are evaluated. Individuals are recognized to vary in degree of similarity, while the measurement variables constituting the frame of reference are accepted as distinct fixed dimensions of individual difference. There is really no way the orthogonality of the measurement reference frame can be justified on statistical grounds because statistical relationships among variables have geometric meaning only in terms of a reference frame in which persons are taken as the orthogonal coordinate axes. It is just as logical—or more logical—to accept specified measurement variables as a fixed orthogonal reference frame as it is to accept randomly selected persons as constituting such a reference frame. This view is quite distinct from that upon which traditional factor analysis is based and requires adoption of a distinctly different conceptual model, a fact that was difficult for one of the present authors to accept for several years (Overall, 1964). Ideally, variables should be chosen as the basis for classification research that are relevant, have reasonably comparable metrics, and tend a priori to represent different classification dimensions. Where this is not possible, statistical transformations discussed in the next section should be used with care.

8.3 Different Kinds of Quantitative Similarity Indices

8.3.1 INTRODUCTION

Two types of similarity indices used most frequently as the basis for classification research are the vector-product and the distance-function coefficients. As mentioned in the preceding section, these are highly related (Nunnally, 1962). Vector-product indices that can be used as the basis for cluster analysis include (1) product-moment correlation coefficients, (2) covariances, and (3) raw cross products between profiles. In general, a large vector-product coefficient represents greater similarity. The cluster analysis of a vector-product matrix proceeds through the identification of subsets of profiles that are interrelated by relatively *large* vector-product coefficients.

Distance-function indices that can be used as the basis for cluster analysis include (1) simple sums of squared differences between corresponding scores and (2) sums of squared differences between profiles of transformed scores, where the transformations are employed to equate variances or remove certain types of redundancies among the measurements. The distance-function indices are actually measures of dissimilarity, so that the *smaller* values represent greater similarity. If two profiles are identical, the distance between them will be zero. Cluster analysis of a matrix of distance-function indices proceeds by identification of subgroups of profiles that are interrelated by relatively *small* distance-function coefficients.

Before proceeding to a discussion of specific techniques for cluster analysis, it will be useful to discuss in more detail the problem of evaluating similarities between *person* profiles. Throughout the chapters concerned with factor analysis only the analyses of relationships among measurement variables were considered explicitly. Relationships between objects or individuals can be calculated in a similar fashion by turning the data matrix around and considering the individuals as variables. Other indices of multivariate profile similarity have no direct counterparts in the analysis of relationships among measurement variables.

8.3.2 DISTANCE-FUNCTION INDICES

The simplest and perhaps most generally useful index of multivariate similarity is the simple distance function, calculated as the sum of squares of differences between corresponding scores in two multivariate profiles. The geometric model illustrated in Fig. 8.1 shows the square of the distance between two profile points d_{ab} equal to the sum of squares of differences in projections d_1 and d_2 on the orthogonal reference axes. For simple distance-function calculations, each original measurement variable is associated with a distinct orthogonal axis.

Scores on these original measurements are then employed as cartesian coordinates to locate each profile as a point, or vector, in the multidimensional hyperspace. Differences between *scores* for two individuals or diseases can thus be conceived as differences in *projections* on orthogonal reference axes. According to a generalization of the pythagorean theorem, the square of the distance between two profile points in p-dimensional space is equal to the sum of squares of differences in projections on p orthogonal coordinate axes.

For simplicity, let $\mathbf{x}_i$ and $\mathbf{x}_j$ represent the score vectors for the ith and jth individuals regardless of any preliminary transformations that may have been effected on the data. The same calculations are required irrespective of prior transformations. The simple d^2 distance-function index can be represented in vector notation as

$$d_{ij}^2 = (\mathbf{x}_i - \mathbf{x}_j)'(\mathbf{x}_i - \mathbf{x}_j) = \mathbf{d}'\mathbf{d} \qquad (8.1)$$

If the scales of measurement for the several variables within the profiles differ quite obviously, they may need to be rescaled to a common metric. The within-class standard deviation is an acceptable and meaningful metric; however, the classes are usually undefined at the beginning of classification research. The purpose of the research is to define the classes. Nevertheless, external data are sometimes available for homogeneous groups of individuals. Alternatively, if the reliability coefficient is known for each measure within the profile, the *within-individual* variability can be derived as a meaningful scaling unit,

$$\sigma_e^2 = \sigma_t^2(1 - r_{tt})$$

If neither of these error variances can be estimated, one may have to resort to using as an initial approximation the standard deviations derived from a composite heterogeneous sample. In such event, it may be considered worthwhile to iterate the classification analysis using in a second analysis estimates of within-class variances derived from an initial tentative classification. The adjustment for differences in units of measurement among the several profile components can be represented in matrix notation as

$$d_{ij}^2 = (\mathbf{x}_i - \mathbf{x}_j)'V^{-1}(\mathbf{x}_i - \mathbf{x}_j) \qquad (8.2)$$

where V^{-1} is a diagonal matrix containing reciprocals of the within-class or the within-individual variances of the p profile elements.

If the several variables within the multiple-measurement profiles are substantially related in the sense that they tend to reflect the same underlying classification dimensions, then some type of orthogonal transformation may be useful to ensure a more equal consideration of all relevant dimensions. As previously stated, there seems to be little logic in requiring that profile elements be transformed to have precise statistical independence across all objects that are to be classified. An optimum approach defines an orthogonal transformation that tends to maximize differences between classification categories, but, once again, we note that the classification categories are not usually available at the outset of a classification analysis. An expedient solution, but one which requires considerable computation, iterates the classification analysis by starting with the simple d^2 of Eq. (8.1), progressing to the variance-corrected d^2 of Eq. (8.2), and finally using as reference axes canonical variates obtained from discrimination among tentatively defined classes in the final step. The orthogonal transformation can be represented in matrix notation as

$$d_{ij}^2 = (\mathbf{x}_i - \mathbf{x}_j)' W^{-1} (\mathbf{x}_i - \mathbf{x}_j) \tag{8.3}$$

where W^{-1} is the inverse of the *within-class* covariance matrix obtained by considering classification grouping from a previous iteration.

Where the number of elements within the profiles is large and the scores tend to be highly related, a canonical reduction to define a relatively small number of highly discriminating orthogonal dimensions will tend to provide more reliable and valid estimates of interprofile distance

$$d_{ij}^2 = (\mathbf{x}_i - \mathbf{x}_j)' A \Lambda^{-1} A' (\mathbf{x}_i - \mathbf{x}_j) \tag{8.4}$$

where A $(p \times r)$ is a matrix containing the first r solution vectors of the matrix equation $(B - \lambda_i W)\mathbf{a}_i = 0$ and Λ^{-1} is a diagonal matrix containing the reciprocals of the within-groups variances λ_i of the linear functions defined by the normalized solution vectors. The statistician frequently calls this canonical reduction a multiple discriminant analysis, and the method of accomplishing it is described in Chap. 10. The B and W matrices are between-groups and within-groups covariance matrices. The vectors in matrix A define linear combinations of the original variables that have maximum potential for discriminating between members of different subclasses. Once again, Eq. (8.4) cannot be used until a preliminary analysis has identified tentative classification groups. A logical progression involves use of Eq. (8.1) to define profile-similarity coefficients prior to any classification grouping. After classification grouping of the simple d^2 coefficients, Eq. (8.2) can be used to obtain refined estimates of interprofile distances. With a tentative grouping defined in the second iteration, a final set of interprofile distances can be computed using Eq. (8.3) or (8.4). The final classification grouping is obtained from the d_{ij}^2 values computed using these transformed variates.

8.3.3 VECTOR-PRODUCT INDICES OF PROFILE SIMILARITY

The pythagorean distance-function index of profile similarity has been represented as the squared length of one side of a triangle. Thus it depends upon the angular disparity between the profile vectors and their respective lengths, as shown in the schematic diagram of Fig. 8.1. It is possible to define another statistic, called a *vector-product index*, that also depends upon angular disparity between profile vectors and their respective lengths. Expanding the distance function of Eq. (8.1), we obtain

$$d_{ij}^2 = \mathbf{x}_i'\mathbf{x}_i + \mathbf{x}_j'\mathbf{x}_j - 2\mathbf{x}_i'\mathbf{x}_j \tag{8.5}$$

The first two terms are sums of squares representing total projections of the profile vectors on the orthogonal coordinate axes; thus, in terms of the geometric model, we conceive of them as the squared distances of the profile-vector end points from the origin or the squares of the profile-vector lengths. The cross-product term represents the cosine between the profile vectors multiplied by the product of the vector lengths

$$d_{ij}^2 = l_i^2 + l_j^2 - 2l_il_j \cos \theta \tag{8.6}$$

The raw vector-product index of profile similarity is simply the inner product of two score vectors. As described above, this raw vector product can be conceived as the cosine of angular separation between the two profile vectors multiplied by the product of their lengths

$$v_{ij} = \mathbf{x}_i'\mathbf{x}_j = l_il_j \cos \theta_{ij} \tag{8.7}$$

Obviously, the raw-score vector-product index has one disadvantage compared with the corresponding distance function. Whereas maximum possible similarity (identity) of two profiles results in $d^2 = 0$, the identity of two profiles results in a raw-score vector product equal to the square of the (common) vector length. Thus, the raw-score vector-product index can have different values for different pairs of *identical* profiles. The vector length, or sum of squares of profile elements, can be large for one pair of identical profiles and smaller for another. For this reason, the vector-product indices of profile similarity are almost universally computed using normalized profile vectors in which $\mathbf{x}_i'\mathbf{x}_i = \mathbf{x}_j'\mathbf{x}_j = 1.0$.

The simple d^2 index of profile similarity has the following expanded form when profile vectors have been normalized so that the sum of squares of elements in each is equal to unity:

$$d_{ij}^2 = 1.0 + 1.0 - 2\mathbf{x}_i'\mathbf{x}_j = 2(1 - \cos \theta) \tag{8.8}$$

Thus,

$$d_{ij}^2 = 2(1 - v_{ij}) \tag{8.9}$$

and

$$v_{ij} = \mathbf{x}_i'\mathbf{x}_j = 1 - \frac{d_{ij}^2}{2} \tag{8.10}$$

The normalized vector-product index has a maximum value of 1.0, corresponding to $d^2 = 0$, and the distance-function index has a maximum value of $d^2 = 4.0$, corresponding to $\mathbf{x}_i'\mathbf{x}_j = -1.0$. If the normalized profiles are not elevation-corrected and contain all positive scores, the maximum value for d^2 is 2.0 and the minimum vector product is $v = 0$.

For each transformed distance-function index there is a corresponding transformed vector-product index. In the face of incomparability of measurement units, a standard-score transformation may be required

$$d_{ij}^2 = (\mathbf{x}_i - \mathbf{x}_j)'V^{-1}(\mathbf{x}_i - \mathbf{x}_j)$$

so that

$$d_{ij}^2 = \mathbf{x}_i'V^{-1}\mathbf{x}_i + \mathbf{x}_j'V^{-1}\mathbf{x}_j - 2\mathbf{x}_i'V^{-1}\mathbf{x}_j \tag{8.11}$$

where V^{-1} is a diagonal matrix and contains the reciprocals of the within-class variances of the separate profile elements. The associated vector-product index is simply

$$v_{ij} = \mathbf{x}_i'V^{-1}\mathbf{x}_j \tag{8.12}$$

In a similar manner, it may be considered on a priori or statistical grounds that some type of orthogonal transformation is required. Such transformation can be inserted into the vector-product calculation in the same manner as it can be inserted into d^2 calculations

$$v_{ij} = \mathbf{x}_i'W^{-1}\mathbf{x}_j \tag{8.13}$$

or

$$V_{ij} = \mathbf{x}_i'A\Lambda^{-1}A'\mathbf{x}_j \tag{8.14}$$

Note that these equations represent the third terms from expansion of Eqs. (8.3) and (8.4). The matrices W^{-1} and $A\Lambda^{-1}A'$ are the orthogonal transformation matrices described with respect to the two related d^2 coefficients.

8.4 Cluster Analysis of d^2 Matrix

8.4.1 THE SIMPLE BASIC METHOD

Once coefficients of similarity have been computed, the logic of cluster analysis is quite simple: identify subsets of objects or individuals that tend to be relatively similar and group them together. The method of accomplishing this goal can be either naïvely simple or computationally laborious. In view of what we consider the most important potential application of cluster analysis, we shall describe a specific method for the grouping of person profiles.

If the number of profiles to be grouped is small, the clustering can be achieved by simple inspection of the matrix. If the number of profiles to be grouped is large, the calculation of average distances within and between clusters is laborious by hand, although it is reasonably efficient for computer applications. The general procedure involves identifying a *cluster nucleus* consisting of highly similar profiles and then adding to the cluster other profiles that are similar to the cluster members. New clusters are started when no acceptable candidates can be found for inclusion into existing clusters. The procedure terminates when no new cluster nucleus can be identified. The steps in cluster analysis of a matrix of d^2 coefficients can be summarized as follows:

1 Identify the single individual having the smallest average distance from two other individuals. This is accomplished by scanning columns of the d^2 matrix to identify the column having the two (on the average) smallest d^2

coefficients. The individual represented by the column plus the two other individuals from whom he has smallest distances become the nucleus for the first cluster.

2 For each individual not already in a cluster, calculate the ratio of average distance to the cluster members relative to average distance to all other individuals who are not in the particular cluster. This index will be designated as the ratio of cluster to noncluster distances.

3 Identify the single individual not already in a cluster who has the smallest ratio of average cluster to noncluster distances. This individual is the next candidate for admission to the cluster if the ratio is less than a prespecified critical value, usually $B \leq .6$. Initiate the search for another candidate for inclusion into the cluster. If the ratio of cluster to noncluster distances for the candidate exceeds $B = .6$, discontinue addition to the particular cluster and search for a new cluster nucleus.

4 The search for a new cluster nucleus resembles exactly the search for the first cluster nucleus except that only individuals not already in a cluster are considered.

5 When a new cluster nucleus has been identified, the process of identifying new candidates for admission is identical to that used for the first cluster. Consider only individuals not already in a cluster. Add individuals to the new cluster until the next candidate for inclusion does not have an adequately small ratio of cluster to noncluster distances. In this event, seek still another cluster nucleus.

6 The whole procedure terminates when no new cluster nucleus can be identified.

The computer program listed in the next section follows essentially this sequence of steps, except that a few additional tests and safeguards have been added. For example, a high degree of homogeneity within each initial cluster nucleus assumes importance. A test is included to ensure it. Specification of many minor details will probably hinder the student rather than help him to grasp the essential features of the profile-cluster method. In practice, quite effective classification grouping can be achieved by following the rules specified above.

8.4.2 COMPUTER PROGRAM FOR CLUSTER ANALYSIS OF INTERPROFILE DISTANCES

The computer program is written in Fortran IV for IBM 1130/1800 digital computers with disk storage. It is designed to calculate the interprofile distance between each pair of profiles and to cluster-analyze the d^2 matrix following the methods just described. The interprofile distance upon which similarity grouping is based is the simple sum of squares of differences in scores on corresponding profile elements [Eq. (8.1)]. The input sequence requires only one control card followed by the multivariate score vectors for up to 50 individuals or objects.

Input

1. One control card containing in first two four-column fields
   NPRO = number of profiles to be clustered
   NVAR = number of scores in each profile, not to exceed 60
2. Individual profiles, punched into four-column fields of one or more (consecutive) cards for each individual.

Output

1. Matrix of interprofile d^2 distances among all possible pairs of the NPRO profiles
2. Vector indicating profiles that have been grouped into each derived cluster
3. Mean profile for each derived cluster
4. Average within-cluster and between-cluster d^2 values for the derived cluster

8.4.3 CLUSTER CLASSIFICATION OF PSYCHIATRIC DIAGNOSTIC PROFILES

In giving a detailed example of cluster analysis we shall follow a procedure identical to that described in Sec. 8.4.1 except that the initial cluster nuclei are defined in terms of *pairs*, rather than triads, of similar profiles. This modification is desirable considering the limited size of the illustrative problem and the need for simplicity of required calculations.

This example illustrates an approach for evaluating the validity of cluster methods, as well as the calculations. Numerous profile-clustering procedures are available, but insufficient work has been accomplished to demonstrate their validity for identification of the true underlying group structure. The expectation in the use of profile-clustering techniques is that profiles of individuals who belong to the same underlying class will be grouped together. The experiment reported in this section was undertaken to test the validity of the cluster-analysis method against well-established clinical expectations.

In the area of psychiatric diagnosis, clinicians have expectations concerning relationships among diagnostic types. Although a finer subdivision may be possible, diagnostic groups are frequently classified under the broad rubrics of schizophrenia, paranoid reaction, and depression. It seems reasonable for a valid profile-clustering method to group together the various subtypes of depression, schizophrenia, and so on. If a cluster-analysis procedure can achieve such a clinically meaningful grouping in an area where the underlying structure is well agreed upon, it should be valid for defining classification groups in areas lacking a priori classification systems.

American diagnostic-prototype profiles representing 12 categories of functional psychiatric illness as described by 38 expert judges are presented in Table 8.1. The interprofile distances calculated as simple sums of squares of difference scores appear in Table 8.2. Cluster analysis of the d^2 matrix was accomplished in the following manner.

Program for Cluster Analysis of Distance-function Indices

```
C     PROGRAM FOR CLUSTER ANALYSIS OF INTERPROFILE DISTANCE MATRIX
      DIMENSION A(50,50),X(50),Y(50),Z(50),R(50)
      DEFINE FILE 1(50,120,U,L1),2(30,120,U,L2)
    1 FORMAT(3I4)
    2 FORMAT(18F4.2)
    6 FORMAT(/20H PROFILES IN CLUSTER,I4)
    7 FORMAT(30F4.0)
    8 FORMAT(1X,16F5.2)
    9 FORMAT(/10X,24H MEAN VECTOR FOR CLUSTER,I4)
   10 FORMAT(1X,10F6.1)
   11 FORMAT(/10X,38H DISTANCES WITHIN AND BETWEEN CLUSTERS)
      READ(2,1)NPRO,NVAR,NODAT
      FNPRO=NPRO
      T1=.6
      INDX=0
      L2=1
      DO 666 I=1,NPRO
  666 Z(I)=0.0
      IF(NODAT) 554,555,554
  554 DO 556 I=1,NPRO
      DO 556 J=I,NPRO
      IF(I-J)553,556,553
  553 READ(2,557) A(I,J)
  556 CONTINUE
  557 FORMAT(F6.3)
      DO 558 I=1,NPRO
      DO 559 J=I,NPRO
  559 A(J,I)=A(I,J)
  558 A(I,I)=0.0
      GO TO 65
  555 DO 12 I=1,NPRO
      READ(2,2)(X(J),J=1,NVAR)
C     AT THIS POINT TRANSFORM PROFILE SCORES IF NECESSARY, ALSO
C     AT THIS POINT REMOVE PROFILE ELEVATION AND VARIABILITY IF DESIRABLE
   12 WRITE(1' I)(X(J),J=1,NVAR)
C     BRING BACK PROFILE VECTORS AND COMPUTE D-SQR MATRIX
      DO 24 I=1,NPRO
      READ(1' I)(X(J),J=1,NVAR)
      DO 24 I2=I,NPRO
      READ(1'I2)(Y(J),J=1,NVAR)
      SSQ=0.0
      DO 23 J=1,NVAR
   23 SSQ=SSQ+(X(J)-Y(J))**2
      A(I,I2)=SSQ
   24 A(I2,I)=SSQ
C     FIND CLUSTER NUCLEUS CONSISTING OF THREE PROFILES
   65 NPR1=-1+NPRO
      NPR2=-2+NPRO
      XBAR=0.0
      DO 66 J=1,NPRO
      DO 66 J2=J,NPRO
   66 XBAR=XBAR+A(J,J2)
      XBAR=XBAR/(NPRO*NPR1/2)
   99 SML=9999.
      DO 70 J=1,NPRO
   70 X(J)=0.0
      DO 79 J1=1,NPR2
      IF(Z(J1))79,71,79
   71 J11=J1+1
      DO 78 J2=J11,NPR1
      IF(Z(J2))78,72,78
   72 J22=J2+1
      DO 77 J3=J22,NPRO
      IF(Z(J3))77,73,77
   73 DIST=A(J1,J2)+A(J1,J3)+A(J2,J3)
      IF(DIST-SML)74,77,77
   74 SML=DIST
      M1=J1
      M2=J2
      M3=J3
   77 CONTINUE
   78 CONTINUE
   79 CONTINUE
```

```
      RATIO=SML/(3.0*XBAR)
      IF(RATIO-.5)80,81,81
   80 X(M1)=1.0
      X(M2)=1.0
      X(M3)=1.0
      Z(M1)=1.0
      Z(M2)=1.0
      Z(M3)=1.0
      XN=3.0
      GO TO 88
C     FIND CLUSTER NUCLEUS CONSISTING OF TWO PROFILES
   81 SML=9999.
      DO 89 J1=1,NPR1
      IF(Z(J1))89,82,89
   82 J11=J1+1
      DO 87 J2=J11,NPRO
      IF(Z(J2))87,83,87
   83 IF(A(J1,J2)-SML)84,87,87
   84 SML=A(J1,J2)
      M1=J1
      M2=J2
   87 CONTINUE
   89 CONTINUE
      RATIO=SML/XBAR
      IF(RATIO-.5)90,900,900
   90 X(M1)=1.0
      X(M2)=1.0
      Z(M2)=1.0
      Z(M1)=1.0
      XN=2.0
C     COMPUTE RATIOS OF AVERAGE CLUSTER AND NON-CLUSTER DISTANCES
   88 DO 33 J=1,NPRO
      C2=0.0
      C3=0.0
      DO 34 I=1,NPRO
      IF(X(I))35,36,35
   35 C2=C2+ A(I,J)
      GO TO 34
   36 C3=C3+A(I,J)
   34 CONTINUE
      C2=C2/XN
      C3=C3/(FNPRO-XN-1)
   33 Y(J)=C2/C3
C     IDENTIFY PROFILE WITH SMALLEST CLUSTER RATIO
      MIN=1
      TEST =9999.
      NN=0
      DO 37 J=1,NPRO
      IF(Z(J))42,42,37
   42 NN=NN+1
      IF(Y(J)- TEST) 38,37,37
   38 MIN=J
      TEST= Y(J)
   37 CONTINUE
      IF(NN) 41,41,39
C     DETERMINE ADEQUACY OF NEW CLUSTER CANDIDATE
   39 IF(Y(MIN)- T1)40,40,41
   40 X(MIN)=1.0
      Z(MIN)=1.0
      XN=XN+1.0
      IF(XN-FNPRO-1)88,41,41
C     PRINT VECTOR IDENTIFYING PROFILES IN CLUSTER
   41 INDX=INDX+1
      WRITE(3,6)INDX
      WRITE(3,7)(X(J),J=1,NPRO)
C     STORE CLUSTER INDICATOR ON DISK
      WRITE(2'L2)(X(J),J=1,NPRO)
      GO TO 99
  900 CONTINUE
      IF(NODAT) 560,561,560
C     COMPUTE MEAN VECTOR FOR EACH CLUSTER
  561 DO 201 J=1,INDX
      READ(2' J)(X(I),I=1,NPRO)
      DO 200 I=1,NVAR
  200 Y(I)=0.0
      FNN=0
```

```
      DO 199 I=1,NPRO
      IF(X(I))199,199,198
  198 FNN=FNN+1
      READ(1' I)(R(K),K=1,NVAR)
      DO 197 K=1,NVAR
  197 Y(K)=Y(K)+R(K)
  199 CONTINUE
      DO 195 K=1,NVAR
  195 Y(K)=Y(K)/FNN
      WRITE(3,9)J
  201 WRITE(3, 8)(Y(K),K=1,NVAR)
C     COMPUTE AVERAGE WITHIN AND BETWEEN CLUSTER DISTANCES
  560 WRITE(3,11)
      DO 302 J=1,INDX
      READ(2' J)(X(I),I=1,NPRO)
      DO 301 J2=1,INDX
      READ(2'J2)(Y(I),I=1,NPRO)
      SUM=0.0
      FN=0.0
      DO 300 I=1,NPRO
      DO 300 I2=1,NPRO
      IF(I-I2) 298,300,298
  298 XX=X(I)*Y(I2)
      IF(XX)299,300,299
  299 SUM=SUM+A(I,I2)
      FN=FN+1.0
  300 CONTINUE
      IF(FN) 309,309,312
  309 R(J2)=0.0
      GO TO 301
  312 R(J2)=SUM/FN
  301 CONTINUE
  302 WRITE(3,10)(R(J2),J2=1,INDX)
      CALL EXIT
       END
```

The first cluster nucleus was defined to include profiles 11 and 12 because those two profiles had the smallest distance, $d^2 = 3.65$. In considering other profiles for inclusion in the first cluster, we found the ratio of average cluster distance to noncluster distance smallest for profile 10. This profile was not accepted because the B ratio exceeded the arbitrarily prespecified minimum value of 0.6. Thus, the first cluster was defined to contain only profiles 11 and 12.

Profiles 1 and 2 were taken as the nucleus for the second cluster because of the small interprofile distance $d^2 = 7.40$. The ratio of average cluster distance to noncluster distance was found to be smallest for profile 3, which was then added to the second cluster. Next, the ratio of average cluster distance to noncluster distance was found to be smallest for profile 13, and that profile was added to the cluster. The next candidate for inclusion into the second cluster was profile 8, but in this case the ratio of average cluster distance to noncluster distance was too large. Cluster 2 was closed with profiles 1, 2, 3, and 13 included.

The nucleus for the third cluster was identified as profiles 7 and 8. Other profiles were added to the third cluster in the order 9, 6, and then 5. After that, the next candidate for inclusion into the third cluster was profile 4, but the associated B ratio did not pass the test. Cluster 3 was closed with profiles 5, 6, 7, 8, and 9 included. The pair of profiles 4 and 10 formed a final dyad cluster with acceptably small interprofile distance. At this point, each profile had been included in a cluster and the analysis was terminated.

Let us examine the cluster-analysis results in view of clinical expectations concerning relationships among the diagnostic groups. Cluster 1 contains profiles for manic-depressive and psychotic depressive reactions. Cluster 2 contains profiles representing paranoia, paranoid state, paranoid schizophrenia, and manic state. Reference to Figs. 3.2 to 3.7 confirms a tendency for the manic-type profile to be grouped with the paranoid profiles in other sets of BPRS data analyzed by other means. (Because of this, an excitement scale has been added to the BPRS to provide more power in differentiating manic and related clinical states.) Cluster 3 contains profiles for catatonic, hebephrenic, simple, chronic undifferentiated, and residual schizophrenic reactions. Note that the prototypes for these groups have marked elevation of the primary signs of emotional withdrawal and blunting of affect, with little manifest affect, anxiety, and tension. Cluster 4 contains prototypes for acute undifferentiated and schizo-affective types. These are both mixed types, and the prototype profiles emphasize anxiety, tension, and affect, as well as the primary schizophrenic symptoms. Mean BPRS vectors for the four empirically derived groups are presented in Table 8.3. The results conform well to clinical expectation and represent what might well have been an a priori classification of major psycho-pathologies. The four mean profiles are descriptive of the higher-order classification concepts of primary schizophrenia, paranoid states, depression, and acute undifferentiated or mixed schizophrenia as conceived by a number of American psychiatrists and psychologists. These four major diagnostic-prototype profiles can be used as the basis for classification of individual patients using methods described in Chaps. 14 and 15.

Results from cluster analyses are often presented in graphic form. The average within-cluster and between-cluster distances can be calculated from the d^2 matrix. In the American psychiatric diagnostic-prototype example, average within-cluster and between-cluster d^2 values were calculated with the following results:

$$
\begin{array}{c} \\ C1 \\ C2 \\ C3 \\ C4 \end{array}
\begin{array}{c} \begin{array}{cccc} C1 & C2 & C3 & C4 \end{array} \\
\begin{bmatrix} 3.6 & 105.4 & 76.5 & 38.7 \\ 105.4 & 25.0 & 61.4 & 40.7 \\ 76.5 & 61.4 & 31.6 & 34.1 \\ 38.7 & 40.7 & 34.1 & 10.3 \end{bmatrix} \end{array}
$$

The diagonal entries represent average within-cluster distances. The within- and between-cluster distances were used to position the four clusters and to relate them in a reduced geometric model, as shown in Fig. 8.2.

A second example provides an extended empirical basis for judging the validity of cluster-analysis methodology in defining meaningful relationships among multivariate profile patterns. The German diagnostic-stereotype data were used.

Table 8.1 Prototype Profiles for 13 Functional Psychotic Classes Described in the American Psychiatric Association Standard Nomenclature

	1	*2*	*3*	*4*	*5*	*6*	*7*	*8*	*9*	*10*	*11*	*12*	*13*	*14*	*15*	*16*
1. Paranoia	1.86	1.92	1.94	1.17	.72	2.44	.61	4.31	.75	4.19	4.94	.61	.19	2.89	3.92	1.56
2. Paranoid state	2.18	3.00	2.13	2.71	1.39	3.29	.71	3.42	1.42	4.03	4.87	1.55	.42	3.47	3.97	1.61
3. Paranoid schizophrenic	3.32	3.16	3.24	3.49	1.30	3.57	1.43	4.30	1.41	4.84	5.49	3.62	.68	4.08	4.78	2.38
4. Acute undifferentiated schizophrenic	3.24	4.37	3.68	4.76	2.45	4.24	2.84	2.05	2.18	3.13	3.61	4.00	1.53	3.37	4.63	2.76
5. Catatonic schizophrenic	1.66	2.79	5.42	4.68	2.47	3.68	5.53	1.61	1.76	3.13	2.97	4.13	5.16	4.87	4.58	4.18
6. Hebephrenic schizophrenic	2.21	1.74	4.89	5.50	1.18	2.71	5.00	2.32	1.18	2.26	2.58	4.74	1.39	3.47	5.29	4.71
7. Simple schizophrenic	1.92	1.37	4.47	3.39	1.05	1.50	2.18	.84	1.13	1.39	1.97	1.50	2.03	2.32	2.58	4.61
8. Chronic undifferentiated schizophrenic	2.47	2.16	3.21	2.79	1.58	2.13	1.71	1.29	1.37	1.79	2.45	1.29	1.24	1.74	2.32	2.95
9. Residual schizophrenic	2.89	2.16	4.32	4.11	1.92	2.53	2.97	2.18	1.61	2.50	3.21	3.18	1.95	2.84	4.03	4.16
10. Schizo-affective	2.92	3.68	2.65	3.68	3.30	4.00	1.81	2.16	3.49	2.86	3.00	2.68	1.86	2.59	3.92	2.03
11. Psychotic depressive	4.53	4.79	3.34	3.29	5.00	3.82	1.95	.39	5.47	2.76	2.29	1.82	4.13	3.03	3.47	2.05
12. Manic-depressive, depressive	4.55	4.84	3.66	2.42	5.37	3.37	1.71	.45	5.53	2.00	2.21	1.21	5.03	2.74	2.87	1.63
13. Manic-depressive, manic	.68	.63	1.61	3.82	.58	4.74	2.32	5.26	.42	3.76	2.13	1.16	.00	3.50	2.79	.92

Source: J. E. Overall and L. E. Hollister, Computer Procedures for Psychiatric Classification, *J. Amer. Med. Assoc.*, **195**: 946–948 (1966). By permission.

Table 8.2 Interprofile Distances among American Diagnostic Prototypes

	1	*2*	*3*	*4*	*5*	*6*	*7*	*8*	*9*	*10*	*11*	*12*	*13*
1	.00	7.40	26.31	60.53	115.66	91.23	59.53	35.55	50.04	46.60	108.00	118.89	29.91
2	7.40	.00	11.55	30.02	85.44	67.88	51.11	28.00	32.50	22.62	73.70	88.29	28.88
3	26.31	11.55	.00	20.68	68.11	50.42	67.16	49.64	28.85	29.52	82.60	103.69	46.11
4	60.53	30.02	20.68	.00	33.81	26.31	46.65	35.89	13.40	10.31	40.52	59.79	62.91
5	115.66	85.44	68.11	33.81	.00	24.32	55.06	69.46	28.05	52.30	69.13	83.77	103.25
6	91.23	67.88	50.42	26.31	24.32	.00	37.18	49.90	13.28	49.12	96.17	119.11	75.85
7	59.53	51.11	67.16	46.65	55.06	37.18	.00	8.25	14.74	41.66	74.21	81.10	66.15
8	35.55	28.00	49.64	35.89	69.46	49.90	8.25	.00	16.61	23.22	57.05	63.27	46.99
9	50.04	32.50	28.85	13.40	28.05	13.28	14.74	16.61	.00	19.27	53.85	68.15	55.76
10	46.60	22.62	29.52	10.31	52.30	49.12	41.66	23.22	19.27	.00	21.25	33.24	53.33
11	108.00	73.70	82.60	40.52	69.13	96.17	74.21	57.05	53.85	21.25	.00	3.65	125.61
12	118.89	88.29	103.69	59.79	83.77	119.11	81.10	63.27	68.15	33.24	3.65	.00	142.80
13	29.91	28.88	46.11	62.91	103.25	75.25	66.15	46.99	55.76	53.33	125.61	142.80	.00

Table 8.3 Mean BPRS Profiles of Four Clusters Formed from American Diagnostic-stereotype Data

	Cluster			
	C1	*C2*	*C3*	*C4*
Somatic concern	4.5	2.0	2.2	3.0
Anxiety	4.8	2.1	2.0	4.0
Emotional withdrawal	3.5	2.2	4.4	3.1
Conceptual disorganization	2.8	2.7	4.0	4.2
Guilt feelings	5.1	.9	1.6	2.8
Tension	3.5	3.5	2.5	4.1
Mannerisms and posturing	1.8	1.2	3.4	2.3
Grandiosity	.4	4.3	1.6	2.1
Depressive mood	5.5	.9	1.4	2.8
Hostility	2.3	4.2	2.2	2.9
Suspiciousness	2.2	4.3	2.6	3.3
Hallucinatory behavior	1.5	1.7	2.9	3.3
Motor retardation	4.5	.3	2.3	1.6
Uncooperativeness	2.8	3.4	3.0	2.9
Unusual thought content	3.1	3.8	3.7	4.2
Blunted affect	1.8	1.6	4.1	2.3

From each of the diagnostic groups labeled schubförmige paranoide Schizophrenie, Schizophrenie simplex, and endogene Depression, 10 profiles were selected *at random*. This example was designed to simulate the situation where a random sample from a mixed population consists of three distinct subpopulations. Previous discriminant analyses had already shown that the larger groups from which these samples were drawn differed significantly. If the empirical clustering technique is valid and useful for identification of underlying latent classes, it should result in the grouping of individual profiles according to previously posited diagnostic classification.

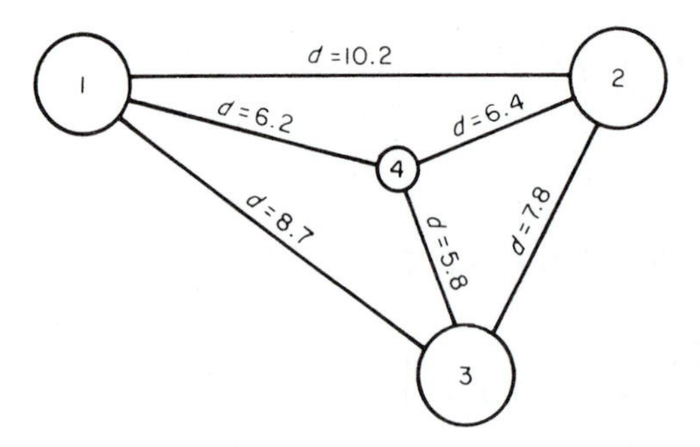

Figure 8.2 Schematic diagram of cluster-analysis-derived classification of major psychiatric disorders: (1) depressive, (2) paranoid, (3) schizophrenic, (4) mixed types.

Table 8.4 Number of Prototype Profiles of Each Diagnostic Type Appearing in Each Cluster

	Diagnostic type		
Cluster	*Paranoid*	*Simple schizophrenia*	*Depression*
1	0	0	10
2	0	7	0
3	0	3	0
4	9	0	0
5	1	0	0
Total	10	10	10

The 30 diagnostic prototype profiles were subjected to computer analysis using the program listed in Sec. 8.4.2. The complete matrix of d^2 coefficients relating all possible pairs of profiles was computed, and the cluster-analysis procedure was executed. The results appear in Table 8.4 cross-tabulated against the original expert grouping.

Table 8.4 shows that although no misclassification occurred, such as a paranoid profile being grouped with profiles for schizophrenic types, the clustering method divided the schizophrenic group into two clusters and refused admission of one paranoid profile into the same cluster with the other nine. Examination of the mean BPRS profiles for the five clusters shown in Table 8.5 provides a basis for understanding these results. The three schizophrenic profiles (cluster 3) not included with the other seven in cluster 2 place much greater emphasis on affective elements of somatic concern, anxiety, guilt feelings, and depressive

Table 8.5 Mean BPRS Prototype Profiles of Five Clusters Formed from German Diagnostic-stereotype Data

	Cluster				
	1	*2*	*3*	*4*	*5*
Somatic concern	4.1	1.2	4.0	2.1	.0
Anxiety	5.3	.4	4.0	4.0	2.0
Emotional withdrawal	3.7	4.8	5.3	3.6	4.0
Conceptual disorganization	.4	1.7	2.3	2.6	1.0
Guilt feelings	4.8	.1	2.6	1.6	.0
Tension	1.7	.8	2.3	3.5	2.0
Mannerisms and posturing	.0	2.2	2.6	1.8	1.0
Grandiosity	.0	.7	1.3	2.2	4.0
Depressive mood	5.5	.7	3.3	.4	.0
Hostility	.3	1.0	2.3	3.3	.0
Suspiciousness	.8	.5	2.3	5.6	1.0
Hallucinatory behavior	.0	.0	1.0	2.6	1.0
Motor retardation	5.0	4.0	4.3	.7	1.0
Uncooperativeness	1.1	2.2	4.0	3.2	5.0
Unusual thought content	2.3	1.5	2.3	4.5	5.0
Blunted affect	1.2	4.5	4.3	1.2	1.0

mood. Apparently, German physicians disagreed among themselves whether these symptoms were prominent in schizophrenia. Clearly, the one paranoid profile (cluster 5) was rejected by the clustering method because of a surprising underemphasis on hostility and suspiciousness as key elements. One clinician conceived of the typical paranoid state in terms of uncooperativeness, unusual thought content, emotional withdrawal, and grandiosity, while all others (cluster 4) felt the paranoid prototype implied elevated levels of hostility and suspiciousness. The example has thus illustrated another implication of profile-cluster analysis, i.e., the identification of outliers or misfits in clinically constituted groups.

8.5 Clustering Based on Vector-product Indices

The method for grouping variables or profiles on the basis of vector-product coefficients of similarity is identical to that for grouping based on distance indices except that larger values rather than smaller values are emphasized. Initially, the first cluster identifies the pair of profiles with largest vector product (correlation or cross product). The profile having the highest average vector product with those already in the cluster is identified as the next candidate for admission to the cluster. The new candidate is added to the cluster provided it does not have a higher vector product with some other individual not in the cluster. When the next candidate is excluded because of higher relationship to some other individual, a new cluster is started. When no new cluster nucleus can be identified, any remaining variables are added to the clusters with which they relate most highly.

The method of Holzinger and Harmon (1941), applied to a matrix of profile correlations or other vector products, involves the following steps. The nucleus for the first cluster is identified as the pair of profiles having the highest correlation. The profile having the highest average correlation with those in the cluster is identified. The ratio of the average correlation with cluster profiles relative to the average correlation with all profiles not in the particular cluster is formed. If the ratio B is greater than 1.3 (or some other arbitrary prespecified coefficient), the variable is added to the cluster. If the ratio B is less than 1.3, the variable is not added and a new cluster is started.

Obviously, an almost infinite number of variations on the basic clustering scheme can be developed. Some possible modifications become important under certain circumstances. For example, in very large matrices the ratio of cluster correlations to correlations with *all* profiles not in the particular cluster may be relatively large; yet, the profile under consideration can actually have a substantially larger average correlation with a small subset of the remaining variables than it has with the cluster. Assuming a minimum of three variables to form a completed cluster, it may be more meaningful to compute the B coefficient as the ratio of the average correlation with variables in a cluster to the average correlation with any other *triad* of variables. This suggestion is intended simply to lead the student to regard alternatives that may provide better insight into the basic problems and objectives of cluster analysis.

The matrix of Q-type product-moment correlation coefficients relating the 13 American diagnostic profiles from Table 8.1 appears in Table 8.6. Rather than illustrating the clustering procedure step by step, we shall leave the actual clustering as an exercise for the student. These are the same prototype profiles that were cluster-analyzed using the d^2 index of profile distance in the preceding section. Results from cluster analysis of the vector-product (correlation) matrix can thus be compared with those previously obtained from analysis of the distance-function matrix. In comparing results, one should remember that profiles are adjusted for differences in both elevation and variability in the process of calculating Q-type correlation coefficients. Similar adjustment was not made in computing the d^2 coefficients previously presented in Table 8.2.

Before we discuss another method for clustering vector-product indices of profile similarity, note that alternative methods for defining initial cluster nuclei can be specified. Although most traditional cluster-analysis methods start with the simple pair of profiles having greatest similarity, this procedure has a serious disadvantage in that the cluster nucleus is determined by the relationship between only two profiles. A single pair of profiles can quite possibly be identical by chance, even though neither is central to an underlying population. An alternative method for defining each initial cluster nucleus in terms of relationships among several profiles would appear superior from this point of view. The factor-analysis and direct cluster-rotation analysis methods described in the next sections of this chapter appear to offer an advantage in this regard, although the more involved computation makes these methods less useful in providing intuitive understanding of what cluster analysis is and how it can be accomplished.

8.6 Use of Factor Analysis for Identifying Clusters of Individuals

8.6.1 INTRODUCTION

As previously mentioned, standard factor-analysis methods can be used to analyze matrices of Q-type correlations between persons. In such analyses, the persons replace the variables, and correlations between pairs of persons are computed across the measurement variables. In all other respects, the "obverse," or Q-type, factor analysis is mechanically just like any other factor analysis. The reader should be forewarned, however, that this type of application of factor-analysis methodology has had a stormy and controversial history. From a practical point of view, we have found the application of cluster-oriented factor methods to the study of natural groupings among individuals so useful that we recommend it as a preferred approach.

The major objections to the use of factor-analysis methods for identifying clusters or natural groupings among persons have come from individuals who see the linear factor model as reflecting the partitioning of variance into factor

Table 8.6 Matrix of Q-type Correlations among 13 American Diagnostic Profiles

	1	*2*	*3*	*4*	*5*	*6*	*7*	*8*	*9*	*10*	*11*	*12*	*13*
1	1.000	.927	.876	.226	—.259	—.015	—.112	.173	.142	.169	—.409	—.439	.608
2	.927	1.000	.928	.501	—.231	.027	—.113	.259	.183	.449	—.236	—.335	.614
3	.876	.928	1.000	.569	—.128	.250	—.007	.289	.369	.304	—.435	—.539	.606
4	.226	.501	.569	1.000	.245	.550	.247	.539	.569	.661	.028	—.167	.287
5	—.259	—.231	—.128	.245	1.000	.685	.625	.256	.561	—.278	—.205	—.211	.036
6	—.015	.027	.250	.550	.685	1.000	.695	.509	.890	—.081	—.483	—.587	.261
7	—.112	—.113	—.007	.247	.625	.695	1.000	.802	.856	—.205	—.156	—.175	.097
8	.173	.259	.289	.539	.256	.509	.802	1.000	.796	.216	.026	—.049	.015
9	.142	.183	.369	.569	.561	.890	.856	.796	1.000	—.004	—.359	—.447	.146
10	.169	.449	.304	.661	—.278	—.081	—.205	.216	—.004	1.000	.549	.371	.146
11	—.409	—.236	—.435	.028	—.205	—.483	—.156	.026	—.359	.549	1.000	.961	—.542
12	—.439	—.335	—.539	—.167	—.211	—.587	—.175	—.049	—.447	.371	.961	1.000	—.623
13	.608	.614	.606	.287	.036	.261	—.097	.015	.146	.146	—.542	—.623	1.000

components. In this context, the partitioning of individuals does not appear at first blush to make much sense. Within the context of the linear factor model, however, it is quite reasonable to conceive of "person factors" as *ideal types* and the factor loadings as indices of relationship of individuals to the several ideal types. Given a reasonably good simple-structure solution, most individuals will relate primarily to only one ideal type (Q-type factor), although some individuals will be recognized as complex. Whereas factor variates in the usual R-type analysis are defined as weighted functions of the original multiple measurements, by direct analogy the "ideal types" can be defined as weighted averages of profiles for individuals.

From a practical point of view, the use of cluster-oriented factor analysis has two distinct advantages over simple cluster-analysis routines. Most important, relationships among all or at least a substantial number of individuals are considered simultaneously in defining each cluster. Perhaps due to this fact, factor-analysis methods and the direct cluster-rotation method described in Sec. 8.7 have, in our experience, tended to yield more consistent and reproducible results in classification research. Second, Q-type factor analysis tends to define factors, and thus ideal types, that are more distinct from one another than cluster nuclei derived from other methods. The reason for this can easily be demonstrated geometrically, but it is sufficient to point out that the extraction of all "variance" associated with each previously defined ideal type leaves only clusters in the matrix that are located at a distance from any previously defined ideal types.

Factor analysis has been discussed and illustrated in considerable detail. The methods of Chaps. 4 to 7 can be applied to the study of relationships among individuals. Algebraic definitions related to Q-type analyses of relationships between person profiles may be helpful in defining the procedure more explicitly.

8.6.2 ALGEBRAIC DEFINITIONS RELATED TO FACTOR ANALYSES OF PERSON PROFILES

Most frequently Q-type factor analyses have been accomplished using a matrix of simple product-moment correlations among persons computed across profile elements. Let S be an $n \times p$ data matrix with the scores in each row corrected so that the row mean is zero. The Q-type correlation matrix is then

$$\underset{n\times n}{D}\ \underset{n\times p}{S}\ \underset{p\times n}{S'}\ \underset{n\times n}{D} = \underset{n\times n}{R} \tag{8.15}$$

where D is a diagonal scaling matrix containing the reciprocals of the square roots of diagonal elements in $C = SS'$. Inclusion of the scaling matrix D is tantamount to normalizing the mean-corrected profiles in the data matrix before computing $R = \hat{S}\hat{S}'$.

One problem with simple Q-type profile correlations is that the different elements within the profile may have substantially different means and variances across all individuals. Suppose, for example, that the p-element profile for each individual contains variables measured in quite different units, such as height in feet and weight in pounds. In all profiles, the height (in foot units) will be of smaller numerical value than the weight (in pound units), and this will tend to build in a positive correlation among all profiles which has nothing to do with individual differences. Such built-in correlations tend to result in a general factor unless prior score transformations are undertaken. Some investigators refer to such a general factor as a "species" factor and prefer to partial it out as the first factor and then disregard it.

Another way of handling this problem is to subtract out mean values and standardize measurements *across all individuals* before calculating the Q-type correlations. Let Z represent an $n \times p$ matrix in which variables within each of the p *columns* are first standardized to zero mean and unit variance. Subsequently, the elements in each of n rows are transformed by subtracting out the row mean. The *double-centered* Q-type correlation matrix is then defined as before

$$\underset{n\times n}{D}\;\underset{n\times p}{Z}\;\underset{p\times n}{Z'}\;\underset{n\times n}{D} = \underset{n\times n}{R} \tag{8.16}$$

Where mean values for the different profile elements differ substantially, equating variable means by subtracting an appropriate constant from each score seems desirable before calculating profile correlations. The problem of what constant to subtract from each profile element to remove general factors and yet not eliminate important group factors is discussed in Sec. 8.11. As previously mentioned, some investigators prefer to accomplish the same thing by extracting the first complete centroid factor of R ($n \times n$) defined in Eq. (8.15) and disregarding it in further calculations (Tucker, 1968).

In addition to the problem created by profile elements with grossly different mean values, correlations among profile elements affect the results from a Q-type factor analysis and thus affect the apparent clustering of individuals. Assuming that the different variables in the profile have roughly comparable mean values or that the variable means have been adjusted by subtracting a constant from each score, one still may be faced with different variances and correlations among profile elements. These differences can be taken into consideration, and adjusted for, by using any one of several orthogonal-transformation techniques. Let S be the $n \times p$ score matrix and C^{-1} be the $p \times p$ inverse of the matrix of variances and covariances for the p profile elements. An adjusted Q-type correlation matrix can be computed as follows:

$$\underset{n\times n}{D}\;\underset{n\times p}{S}\;\underset{p\times p}{C^{-1}}\;\underset{p\times n}{S'}\;\underset{n\times n}{D} = \underset{n\times n}{R_0} \tag{8.17}$$

where the n rows of S are corrected to zero mean, C^{-1} is the inverse of $C = (1/n)S'S$, and D is a diagonal scaling matrix required to bring diagonal elements of R_0 ($n \times n$) to unity.

As noted in Sec. 8.2, use of the inverse matrix for an orthogonal transformation is tantamount to factor-analyzing the matrix and defining all possible factors ($r = p$). This can tend to emphasize measurement errors in cases where profile elements are substantially correlated. A preferable orthogonal transformation would seem to involve the use of only the major factors of the correlation or covariance matrix. Let W be the raw-score factor-score transformation matrix

$$\underset{n\times n}{D}\ \underset{n\times p}{S}\ \underset{p\times r}{W}\ \underset{r\times p}{W'}\ \underset{p\times n}{S'}\ \underset{n\times n}{D} = \underset{n\times n}{R_0} \tag{8.18}$$

where the rows of S are adjusted to zero mean and D is a diagonal scaling matrix which in effect normalizes the n rows of SW so that the diagonal elements in R_0 are scaled to unity.

So far, we have been concerned solely with alternative ways of defining the Q-type product-moment matrix that is to be factor-analyzed for purposes of defining ideal types (Q-type factors) and identifying clusters of individuals who relate most highly to each ideal type. It is unlikely that any of the transformations should, in practice, be carried out as specified in the defining equations. These transformations simply amount to subtracting out mean values, dividing by standard deviations in order to standardize scores, or applying weighting coefficients to the original scores to obtain new transformed scores. They can be accomplished on the raw data before computing the Q-type matrix. These kinds of transformations should be considered in appropriate circumstances, and they can result in overcoming objections that some people have voiced with regard to the use of factor analysis for studying relationships among persons. The point is that these preliminary transformations have nothing to do with the subsequent factor analysis.

The Q-type factor analysis is accomplished like any other factor analysis described in Chap. 4

$$\underset{n\times n}{R}\ \underset{n\times r}{W} = \underset{n\times r}{F} \tag{8.19}$$

where the matrix F ($n \times r$) contains coefficients relating each of the n individuals to r ideal types. It is important in the factor analysis of a Q-type matrix to achieve a good simple-structure solution. If a good simple structure is obtained in the matrix F ($n \times r$), most of the original individuals will be recognized to be highly similar to one or another of the ideal types. If the ideal types are taken as classification prototypes, individuals can be assigned to groups according to their similarity to the prototypes.

In the discussion of linear typal analysis (Sec. 8.10), methods for estimating ideal-type profiles using the factor-transformation matrix will be discussed. In the case of more general Q-type factor analysis, prototype profiles representing the ideal types should be calculated as the simple unweighted means of the individual profiles that are most highly related to each ideal type. Let $\hat{W}$ represent a zero-one factor-transformation matrix in which a single nonzero element in each row corresponds to the factor with which a particular individual has highest relationship, and let $\hat{D}$ be a diagonal scaling matrix containing the reciprocals of the column sums in matrix $\hat{W}$. The columns of $\hat{W}\hat{D}$ will be recognized to sum to zero, and the vectors in $\hat{Z}$ are simple means of vectors in the original data matrix from which $R = DSS'D$ was computed

$$S'\hat{W}\hat{D} = \hat{Z} \tag{8.20}$$

8.6.3 EXAMPLE OF USE OF Q-TYPE FACTOR ANALYSIS TO DEFINE CLASSIFICATION GROUPS

For comparison with results obtained from cluster analysis of simple d^2 coefficients relating American diagnostic prototypes, the powered-vector method of factor analysis (Chap. 6) was applied to the matrix of Q-type profile intercorrelations shown in Table 8.6. The powered-vector analysis resulted in only three factors with substantial relationship to the diagnostic profiles. The Q-type factor loadings are presented in Table 8.7. The factor loadings were interpreted as coefficients of relationship between the 13 diagnostic profiles and three ideal types. BPRS profiles for the three ideal types were calculated by first normalizing the factor loadings within each of the three columns and then assigning each individual profile to the factor cluster which this (normalized) index indicated to be most appropriate. This was done in lieu of using the factor transformation to compute weighted matrices as in Eq. (8.20) and is tantamount to defining a zero-one transformation matrix. The mean vectors for profiles assigned to each factor cluster are displayed in Table 8.8. Clearly, the Q-type powered-vector factor analysis resulted in three ideal types which correspond closely to the general clinical concepts of paranoid, schizophrenic, and depressive syndromes.

Table 8.7 Powered-vector Factor Loadings Relating 13 American Diagnostic Prototypes to Three Q-type Factors

	I	*II*	*III*
Paranoia	.94	−.09	−.10
Paranoid state	.97	−.06	.13
Paranoid schizophrenic	.96	.09	−.06
Acute undifferentiated schizophrenic	.49	.43	.39
Catatonic schizophrenic	−.19	.66	−.17
Hebephrenic schizophrenic	.14	.82	−.28
Simple schizophrenic	−.07	.94	−.00
Chronic undifferentiated schizophrenic	.27	.81	.34
Residual schizophrenic	.27	.94	−.06
Schizo-affective	.36	−.11	.74
Psychotic depressive	−.37	−.21	.87
Manic-depressive, depressive	−.47	−.26	.82
Manic-depressive, manic	.66	−.02	−.32

Table 8.8 BPRS Profiles for Three Ideal Types Derived from *Q*-type Powered-vector Factor Analysis

	Paranoid	*Schizophrenic*	*Depressive*
Somatic concern	2.45	2.37	3.99
Anxiety	2.69	1.85	4.43
Emotional withdrawal	2.43	4.22	3.21
Conceptual disorganization	2.45	3.94	3.12
Guilt feelings	1.13	1.43	4.55
Tension	3.09	2.21	3.73
Mannerisms and posturing	.91	2.96	1.82
Grandiosity	4.00	1.65	.99
Depressive mood	1.19	1.32	4.82
Hostility	4.35	1.98	2.53
Suspiciousness	5.10	2.55	2.50
Hallucinatory behavior	1.92	2.67	1.90
Motor retardation	.42	1.65	3.67
Uncooperativeness	3.48	2.59	2.78
Unusual thought content	4.22	3.55	3.41
Blunted affect	1.84	4.10	1.90

8.7 Direct Cluster-rotation Analysis

8.7.1 INTRODUCTION

One method of conditioning a matrix of product-moment coefficients so that the multiple cluster structure appears in sharp relief is known as *direct cluster-rotation analysis*. This method differs substantially from other standard cluster-analysis methods both in computation and conception. *In direct cluster-rotation analysis, each original profile vector is rotated toward the vectors with which it has greatest relationship and away from vectors with which its relationship is relatively weaker.* A matrix results in which profiles in each cluster are related by coefficients approaching $\pm$ 1.0, while profiles in different clusters are related by coefficients of near-zero value. The cluster configuration is readily apparent in such a matrix.

The clusters resulting from direct cluster rotation do not always resemble those identified through the use of factor analysis or other methods of cluster analysis because each profile is progressively transformed through a series of steps that alter its relationships with the other profiles. A profile may move toward a second profile with which it has substantial relationship and end up in a cluster with several other profiles. The validity of such a method for grouping profiles depends upon the meaningfulness of results obtained. The results from direct cluster-rotation analysis should be meaningful in terms of what is known from other sources about groupings of individuals in a measurement domain.

8.7.2 COMPUTATIONAL PROCEDURE

The rotation of measurement vectors to display the cluster structure can be accomplished easily by simple matrix transformations. Let R_1 be the original matrix of intercorrelations among p variables

$$D_1 \mathbf{\Sigma}_1^{(3)\prime} \mathbf{\Sigma}_1^{(3)} D_1 = \mathbf{\Sigma}_2 \tag{8.21}$$

where in this notation $\mathbf{\Sigma}_1^{(3)}$ indicates a matrix in which each element is equal to the third power of the corresponding element in the original correlation matrix R_1, and where D_1 is a diagonal scaling matrix containing reciprocals of the square roots of diagonal elements in the matrix $V_1 = \mathbf{\Sigma}_1^{(3)\prime} \mathbf{\Sigma}_1^{(3)}$. The transformation process can be contained through any number of iterations, at each stage operating on the transformed matrix obtained from the previous iteration

$$\begin{aligned} D_2 \mathbf{\Sigma}_2^{(3)\prime} \mathbf{\Sigma}_2^{(3)} D_2 &= \mathbf{\Sigma}_3 \\ D_3 \mathbf{\Sigma}_3^{(3)\prime} \mathbf{\Sigma}_3^{(3)} D_3 &= \mathbf{\Sigma}_4 \\ \cdots\cdots\cdots\cdots \\ D_i \mathbf{\Sigma}_i^{(3)\prime} \mathbf{\Sigma}_i^{(3)} D_i &= \mathbf{\Sigma}_{i+1} \end{aligned} \tag{8.22}$$

where $\mathbf{\Sigma}_i^{(3)}$ contains the cubes of elements in $\mathbf{\Sigma}_i$ obtained from the last preceding transformation and D_i is a diagonal matrix containing the reciprocals of the square roots of diagonal elements in $V_i = \mathbf{\Sigma}_i^{(3)\prime} \mathbf{\Sigma}_i^{(3)}$.

It is interesting that this cluster-transformation procedure will result in a unique number of clusters for any matrix. Thus, one is not faced with arbitrary decisions concerning how many clusters to define. In our experience, only four or five iterations have invariably been found adequate to bring the cluster structure into sharp relief. The general cluster-oriented factor-analysis program for digital computers presented in Chap. 7 includes a subroutine which accomplishes cluster rotation under control of the entry NT in the control card. If one wants to sharpen the cluster structure prior to the preliminary factor analysis, the integer NT = 4 or NT = 5 can be inserted in the proper field of the control card. If one does not want the program to go through the cluster rotation prior to the factor analysis, NT = 0 should be specified.

8.7.3 CLUSTER MEMBERSHIP COEFFICIENTS DERIVED FROM ANALYSIS OF TRANSFORMED MATRIX

The method of direct cluster-rotation analysis, described in Sec. 8.7.2, is merely a method for transforming an $n \times n$ correlation matrix in such manner that the cluster structure is emphasized. The powered-vector method of factor analysis discussed in Chap. 6 can be applied to the resulting matrix to yield vectors of *cluster-membership coefficients*. Since the cluster-rotation transformation results in a matrix the elements of which are essentially zero or one, the powered-vector factor loadings obtained from such a matrix (which we choose to call cluster-membership coefficients) will also be essentially zero or one values. From these cluster-membership coefficients, individuals belonging to each cluster are easily identified.

To illustrate this method, let us start with the matrix of Q-type product-moment correlation coefficients among the 13 American diagnostic prototypes

Table 8.9 Matrix of Cosines among American Diagnostic Prototypes Following Five Iterations of Direct Cluster Rotation

	1	*2*	*3*	*4*	*5*	*6*	*7*	*8*	*9*	*10*	*11*	*12*	*13*
1. Paranoia	.99	.99	.99	.00	.00	.00	.00	.00	.00	.00	.00	.00	.99
2. Paranoid state	.99	.99	.99	.00	.00	.00	.00	.00	.00	.00	.00	.00	.99
3. Paranoid schizophrenia	.99	.99	.99	.00	.00	.00	.00	.00	.00	.00	.00	.00	.99
4. Acute undifferentiated schizophrenia	.00	.00	.00	.99	.00	.00	.00	.00	.00	.99	.00	.00	.00
5. Catatonic schizophrenia	.00	.00	.00	.00	.99	.99	.99	.99	.99	.00	.00	.00	.00
6. Hebephrenic schizophrenia	.00	.00	.00	.00	.99	.99	.99	.99	.99	.00	.00	.00	.00
7. Simple schizophrenia	.00	.00	.00	.00	.99	.99	.99	.99	.99	.00	.00	.00	.00
8. Chronic undifferentiated schizophrenia	.00	.00	.00	.00	.99	.99	.99	.99	.99	.00	.00	.00	.00
9. Residual schizophrenia	.00	.00	.00	.00	.99	.99	.99	.99	.99	.00	.00	.00	.00
10. Schizo-affective	.00	.00	.00	.99	.00	.00	.00	.00	.00	.99	.00	.00	.00
11. Psychotic depressive reaction	.00	.00	.00	.00	.00	.00	.00	.00	.00	.00	.99	.99	.00
12. Manic-depressive, depressed	.00	.00	.00	.00	.00	.00	.00	.00	.00	.00	.99	.99	.00
13. Manic-depressive, manic	.99	.99	.99	.00	.00	.00	.00	.00	.00	.00	.00	.00	.99

presented in Table 8.6. Four direct rotational transformations yielded the cluster structure shown in Table 8.9. The procedure that produced the matrix, described in Sec. 8.7.2, follows:

$$\mathbf{\Sigma}_2 = D_1 \mathbf{\Sigma}_1^{(3)\prime} \mathbf{\Sigma}_1^{(3)} D_1$$

where $\mathbf{\Sigma}_1^{(3)}$ contained the cubes of the corresponding elements in the original correlation matrix R_1;

$$\mathbf{\Sigma}_3 = D_2 \mathbf{\Sigma}_2^{(3)\prime} \mathbf{\Sigma}_2^{(3)} D_2$$

where $\mathbf{\Sigma}_2^{(3)\prime}$ contained the cubes of the corresponding elements in the matrix $\mathbf{\Sigma}_2$;

$$\mathbf{\Sigma}_4 = D_3 \mathbf{\Sigma}_3^{(3)\prime} \mathbf{\Sigma}_3^{(3)} D_3$$

where $\mathbf{\Sigma}_3^{(3)\prime}$ contained the cubes of the corresponding elements in the matrix $\mathbf{\Sigma}_3$;

$$\mathbf{\Sigma}_5 = D_4 \mathbf{\Sigma}_4^{(3)\prime} \mathbf{\Sigma}_4^{(3)} D_4$$

It is the matrix $\mathbf{\Sigma}_5$ that is shown in Table 8.9.

A simple orthogonal powered-vector factor analysis (Chap. 6) was applied to the matrix in Table 8.9 to obtain the cluster-membership coefficients shown in Table 8.10. In the case of the American diagnostic-prototype profiles, the direct cluster-rotation analysis based on Q-type profile correlations yielded exactly the same cluster configuration as the cluster analysis of simple d_2 distance indices described in Sec. 8.2. Reference to the transformed cluster-rotation matrix in Table 8.9 will verify that the profiles in each cluster are related by essentially unit cosines in the transformed matrix. The direct cluster-rotation analysis clearly indicated that only four higher-order diagnostic concepts should be considered and these can be described as primary schizophrenia, acute mixed schizophrenia, paranoid states, and depression. The results from the cluster-rotation analysis are satisfyingly in agreement with clinical expectation and

Table 8.10 Cluster-membership Coefficients Resulting from Application of Powered-vector Analysis to Direct Cluster-rotation Matrix of Table 8.9

	C1	*C2*	*C3*	*C4*
1. Paranoia	.00	.99	.00	.00
2. Paranoid state	.00	.99	.00	.00
3. Paranoid schizophrenia	.00	.99	.00	.00
4. Acute undifferentiated schizophrenia	.00	.00	.99	.00
5. Catatonic schizophrenia	.99	.00	.00	.00
6. Hebephrenic schizophrenia	.99	.00	.00	.00
7. Simple schizophrenia	.99	.00	.00	.00
8. Chronic undifferentiated schizophrenia	.99	.00	.00	.00
9. Residual schizophrenia	.99	.00	.00	.00
10. Schizo-affective	.00	.00	.99	.00
11. Psychotic depressive reaction	.00	.00	.00	.99
12. Manic-depressive, depressed	.00	.00	.00	.99
13. Manic-depressive, manic	.00	.99	.00	.00

common clinical parlance. They appear to lend confirmation to the validity of the empirical clustering technique as a method for identifying underlying populations or higher-order classification concepts.

8.7.4 THE FACTOR-ANALYSIS MODEL AND CLUSTER ROTATION

In Sec. 8.6.2, we discussed how the factor-analysis model can be conceived as defining ideal types in the analysis of Q-type profile relationships. To this point, the method of direct cluster-rotation analysis has been treated as cluster analysis without reference to the basic factor model. This is the manner in which we would prefer to view it, but academic interest leads us to examine the relationship to the general oblique-factor model.

Let $\mathbf{\Sigma}^*$ represent the final matrix obtained by iteration of the cluster-rotation procedure [Eq. (8.22)]. The powered-vector factor analysis of the transformed $\mathbf{\Sigma}^*$ matrix can be represented, according to Eq. (4.22), as

$$\mathbf{\Sigma}^* HT = F_1 \tag{8.23}$$

where F_1 contains cluster-membership coefficients similar to those shown in Table 8.10 for the diagnostic-prototype problem.

The matrix F_1 is essentially a zero-one group-centroid type of transformation matrix (Sec. 6.2). To obtain the Q-type factor loadings for the n original profiles on the r oblique primary factors in the original measurement space, we can use the matrix $\mathbf{\Sigma}^* HT = F_1 = W$ as a factor-transformation matrix [Eq. (4.22)]. *This reduces the direct cluster-rotation analysis to a factor-transformation matrix*, which we have repeatedly stressed can be accomplished by a variety of (sometimes arbitrary) techniques. Projections of the n person vectors (profiles) on the r oblique primary axes are then given by

$$R\mathbf{\Sigma}^* HTD_1 = RF_1D_1 = F_p \tag{8.24}$$

where R is the $n \times n$ matrix of Q-type person correlations, F_1 is the $n \times r$ matrix of cluster-membership coefficients derived from powered-vector factor analysis of the cluster-transformed matrix $\mathbf{\Sigma}^*$, and D_1 is a diagonal scaling matrix containing the reciprocals of the square roots of the diagonal elements in $V_1 = F_1' R F_1$.

To obtain ideal (factor) types, the elements of the transformation matrix should be rescaled so that the coefficients in each of the r columns of F_1 sum to unity. This can be accomplished by defining a new diagonal scaling matrix D_2 containing the reciprocals of column sums for F_1. Ideal types, analogous to factor scores and representing weighted means of person profiles most highly related to each Q-type factor, can be obtained by applying

$$\underset{n\times n}{\mathbf{\Sigma}^*}\ \underset{n\times r}{H}\ \underset{r\times r}{D_2} = \underset{n\times r}{F_1}\ \underset{r\times r}{D_2}$$

to the original score matrix standardized within rows

$$\underset{p\times n}{S'}\ \underset{n\times n}{\Sigma^*}\ \underset{n\times r}{H}\ \underset{r\times r}{D_2} = \underset{p\times n}{S'}\ \underset{n\times r}{F_1}\ \underset{r\times r}{D_2} = \underset{p\times r}{P} \tag{8.25}$$

where S' is the original score matrix standardized (z-score form) by rows for each of the p variables, F_1 is the matrix of cluster-membership coefficients derived from the direct cluster-rotation analysis, and D_2 is a diagonal scaling matrix defined above. The matrix P will contain the standard-score profiles for r ideal types, each of which is a weighted mean of the profiles of the n individual standard-score profiles. (In practice, we would recommend defining F^* to contain zero-one coefficients and the matrix D_2 to contain elements which are the reciprocals of the sums of nonzero elements in each column. This is equivalent to defining ideal types as the unweighted means of profiles for individuals most highly related to each Q-type factor.) To get from the standard-score profiles for the r ideal types to the raw-score profiles, one must simply multiply by the standard deviation and add the mean for each variable.

8.8 Empirically Derived Classification of Psychiatric Depressions

In the introduction to this chapter we emphasized the problem of sampling variability in profile cluster-analysis results and the need to replicate empirical findings in numerous independent samples before using cluster-analysis results as the basis for postulating a classification system. Our purpose in this section is to illustrate the kinds of results that can be obtained, to display the degree of variability that we feel must be tolerated, and to describe an empirically derived classification of psychiatric depressions that has proved useful in clinical research. The analyses upon which this tripartite classification is based have been undertaken over a period of half a dozen years on samples drawn from inpatient and outpatient populations. The methods of profile clustering have varied, although most frequently Q-type factor analysis or direct cluster-rotation analysis has been the method employed.

The first series of analyses undertaken involved BPRS profiles for inpatients in Veterans Administration studies of chemotherapy for depression. Patients were selected as being appropriate for treatment with an antidepressant drug, although the basic study involved a double-blind comparison of a tricyclic (antidepressant) with a phenothiazine (major tranquilizer). Q-type powered-vector factor analysis resulted in the identification of three major ideal types of profile clusters. The bar-graph profiles defining the three depressive subtypes are shown in Fig. 8.3. These three empirically derived depressive subtypes were described as anxious, hostile, and retarded depressions.

Numerous subsequent profile-clustering analyses were accomplished on larger and smaller samples of data representing BPRS rating profiles for clinically depressed patients. A single sample of 96 patients was analyzed by the direct cluster-rotation method to yield the three cluster prototypes (series A) shown in the left-hand section of Table 8.11. In another investigation (Overall and

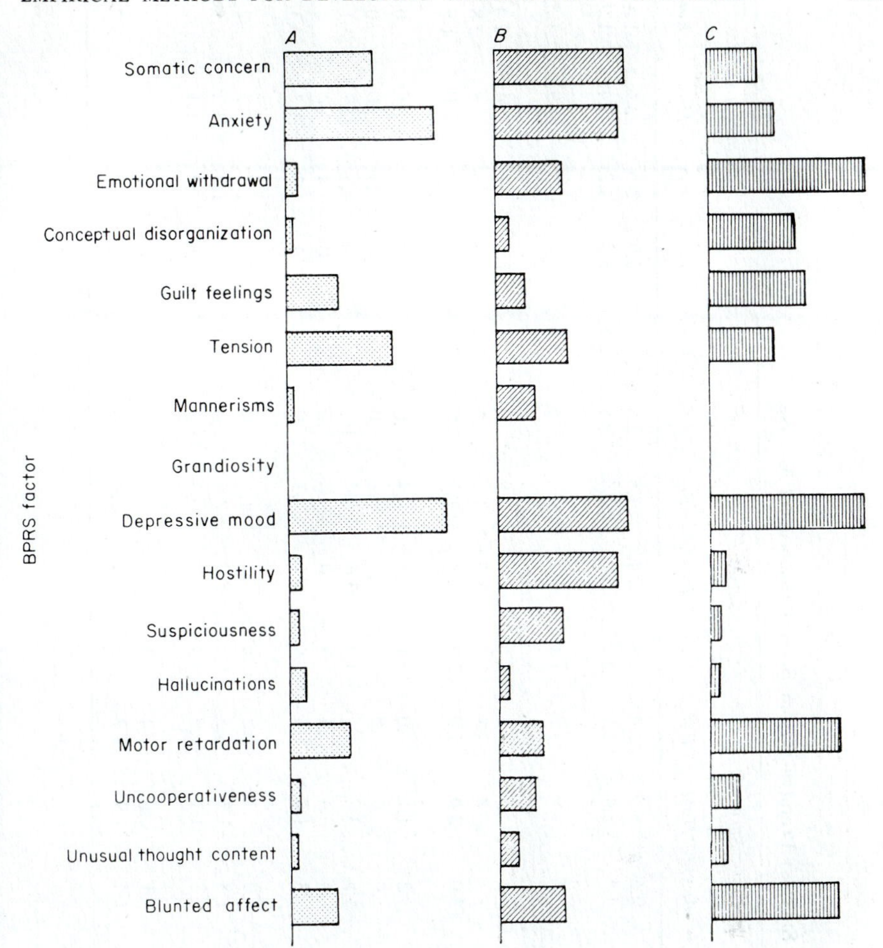

Figure 8.3 Profiles for three replicating modal types derived from four analyses of 40×40 matrices of profile correlations: (*a*) anxious depression, (*b*) hostile depression, (*c*) withdrawn-retarded depression. [*From J. E. Overall, L. E. Hollister, M. Johnson, and V. Pennington, Nosology of Depression and Differential Response to Drugs, J. Amer. Med. Assoc.*, **195:** 946–948 (1966). *By permission.*]

Hollister, 1967), six different subsamples of $n = 40$ were analyzed separately to yield six different sets of results consisting of three to five clusters each. The resulting 24 cluster (mean) profiles from the six preliminary analyses were then analyzed by the direct cluster-rotation method to identify the derived patterns that appeared similar enough to be grouped together. The resulting three higher-order clus er prototypes are shown in the section of Table 8.11 identified as series B. Other profile-clustering analyses of later samples of data for clinically depressed patients yielded still another set of three depressive classification prototypes (series C) that were used in a later drug study (Overall et al.,

Table 8.11 Empirically Derived Depression Subclass Prototypes from Several Independent Analyses

	Series A			*Series B*			*Series C*			*Series D*			*Current studies*		
	Anx.	*Host.*	*Ret.*	*Anx.*	*Host.*	*Ret.*	*Anx.*	*Host.*	*Ret.*	*Anx.*	*Host.*	*Ret.*	*Anx.*	*Host.*	*Ret.*
Somatic concern	3.1	1.2	3.9	2.5	3.0	1.4	1.8	2.4	1.5	2.8	2.3	3.0	2.5	2.6	1.4
Anxiety	3.8	3.2	1.5	4.3	3.4	1.9	3.7	3.2	2.4	2.8	2.8	3.0	4.3	3.4	1.9
Emotional withdrawal	1.0	3.7	4.2	.4	1.9	4.2	1.1	1.7	2.7	.4	1.0	3.1	.4	1.9	4.2
Conceptual disorganization	.6	2.2	1.5	.1	.4	2.4	.5	.5	.9	.0	.5	1.1	.1	.4	2.4
Guilt feelings	1.8	2.1	1.9	1.5	.8	2.8	2.3	1.5	1.7	.2	1.3	.5	2.5	.8	2.8
Tension	3.2	3.9	1.5	3.0	2.0	1.9	2.7	2.8	1.1	2.2	2.3	1.1	3.0	2.0	1.9
Mannerisms and posturing	.2	.6	.3	.2	1.1	.0	.1	.2	.4	.0	.0	.0	.2	1.1	.0
Grandiosity	.0	.0	.3	.0	.0	.0	.2	.2	.1	.0	.5	.0	.0	.0	.0
Depressive mood	4.5	4.0	4.1	4.5	3.7	4.1	3.9	3.4	3.4	1.8	2.5	3.4	4.5	3.7	4.1
Hostility	1.9	3.0	.8	.4	3.4	.4	.9	3.0	.4	.0	2.2	.0	.4	3.4	.4
Suspiciousness	.4	2.9	.3	.3	1.8	.3	.5	1.1	.3	.0	.8	.0	.3	1.8	.3
Hallucinatory behavior	.0	1.8	.4	.0	.3	.2	.1	.7	1.1	.0	.0	.0	.0	.3	.2
Motor retardation	1.0	2.0	4.2	1.8	1.3	3.5	1.2	1.0	2.4	.0	.0	2.4	1.8	1.3	3.5
Uncooperativeness	.3	1.9	.7	.2	1.1	.9	.3	1.0	.6	.0	1.3	.3	.2	1.1	.9
Unusual thought content	.1	2.9	.5	.1	.6	.4	.3	.5	.3	.0	.0	.0	.1	.6	.4
Blunted affect	.6	3.0	3.3	1.4	1.9	3.7	.5	.8	2.8	.0	.5	3.1	1.4	1.9	3.7

1969). A similar series of analyses was run on rating profiles for outpatients seen at the University of Texas Medical Branch clinic and reported at an annual American Psychiatric Association convention (Vanderpool et al., 1967). The three depression profile patterns resulting from these analyses are shown in the section of Table 8.11 marked series D. Note the same three patterns are distinguishable but the general level of pathology was lower in the outpatient samples. Currently, we are conducting a series of clinical drug studies using a still later set of empirically derived depression prototypes. These appear in the right-hand section of Table 8.11. From a clinical research point of view, we recommend these final (right-hand section) profiles for investigators interested in classifying psychiatric depression using methods described in Chap. 15.

The point of this discussion has been to illustrate the general degree of consistency and yet obvious variability that occur in results from empirical profile-clustering analyses. The investigator using these types of analyses to establish a fixed and final typology should be aware of the degree of variability he is likely to encounter. With one exception, the results presented in this section involved the integration of cluster profiles obtained from several preliminary analyses. One must be prepared to tolerate a certain degree of ambiguity in comparing results from different analyses and be prepared to reach the final solution by an inductive process of integration and arbitrary termination. While we feel these methods are extremely valuable, it would be improper to present them without emphasizing their limitations.

8.9 Most Representative Profile Patterns in the General Psychiatric Population

With the aid of a computer, an extensive series of profile-clustering analyses was undertaken using a random sample of 2,000 BPRS profiles from the available collection of 6,400. The 6,400-patient sample was a composite drawn from a variety of hospital and outpatient settings. The purpose of the investigation was to define the most representative or most frequently occurring patterns of manifest psychopathology in the general psychiatric population. The plan of the research was as follows.

First, the sample of 2,000 cases was divided into 50 samples of $n = 40$. A 40×40 Q-type correlation matrix was calculated to represent the interrelationships among profiles in each of the 50 samples. A cluster-rotation analysis (Sec. 8.7) was undertaken on each of the 50 matrices to group together similar BPRS profiles. These analyses tended to result in four to six clusters from each matrix. The mean of BPRS profiles for patients falling into each empirically derived cluster was computed to yield 250 cluster means from the 50 analyses.

The second step was to perform empirical cluster-rotation analyses on the *cluster means* derived from the 50 initial analyses. These analyses were accomplished in order to group together, empirically and objectively, similar cluster means derived from different preliminary analyses. Six cluster prototypes emerged from the higher-order analysis which combined results from the 50 preliminary analyses. The six profiles derived directly from the cluster analysis are shown in Table 8.12. The first three profile types appear to represent rather precisely the three depression subtypes previously identified in analyses of samples of clinically diagnosed depressions (Sec. 8.8). The remaining three profile types can be described as a paranoid hostile-suspiciousness syndrome, a withdrawn-disorganized schizophrenic syndrome, and an acute florid-thinking-disorder syndrome.

The six cluster prototype profiles shown in Table 8.12 were modified slightly in a subsequent series of analyses to improve their representativeness. A set of classification prototypes is representative to the extent that most patients in the population can be recognized to have a symptom profile highly similar to one of the prototype profiles. An index of representativeness is the average distance [Eq. (8.1)] of patient profiles from the closest prototype profile. A computer program was written to evaluate the distance of each patient profile from each of the prototype profiles and to assign the patient to the class from which his profile had least distance. The program was used to classify a random sample of 3,000 profiles according to their similarity to the cluster prototypes shown in Table 8.12. New mean vectors were then calculated for each of the six classes. The process was repeated, classifying the 3,000 profiles according to similarity to the new derived mean profiles, and still another set of classification prototypes was obtained. At each stage, the average distance of patient profiles from the nearest prototype profile was checked to ensure that the newly derived set was more "representative" than the previous set. The process was repeated until no further improvement in the representativeness coefficient could be achieved. The resulting prototype profiles, useful for classifying patients from a general psychiatric population, are presented in Table 8.13.

The major change in prototype profiles that resulted from the iterative re-classification procedure was a reduced general level of symptomatology in the simple anxious-depression prototype. This reduction was no doubt due to the substantial number of patients in the sample derived from outpatient treatment settings where relatively mild anxious depression is not uncommon.

8.10 The Theory and Rationale of Linear Typal Analysis

The use of factor analysis to study similarities and differences between individuals has generally not enjoyed popularity among adherents of Thurstonian factor analysis. The present authors were once counted among the critics because the theory and rationale of factor analysis, as originally developed, concerns the partitioning of test variance into multiple independent components. A somewhat different theoretical approach seems required for analysis of relationships among individuals. Therefore, in this section, we abandon the terminology and the theoretical constructs of factor analysis and consider *linear typal analysis*.

Linear typal analysis is a method of studying relationships of observed individuals to underlying pure types. From a mechanical point of view, it is

Table 8.12 Cluster Prototypes from the General Psychiatric Population

	Anxious depression	*Hostile depression*	*Retarded depression*	*Paranoid hostile-suspicious*	*Withdrawn-disorganized schizophrenia*	*Florid thinking disorder*
Somatic concern	2.6	.6	1.4	1.4	.7	.7
Anxiety	2.8	2.7	1.7	1.5	.8	1.3
Emotional withdrawal	1.1	1.1	3.0	1.0	3.1	2.4
Conceptual disorganization	.5	1.1	1.2	1.4	3.4	3.9
Guilt feelings	.8	2.0	.7	.4	.1	.2
Tension	1.8	1.8	1.1	1.4	1.1	2.0
Mannerisms and posturing	.2	.3	.6	.4	1.3	1.5
Grandiosity	.2	.3	.1	1.0	.2	1.4
Depressive mood	2.5	2.5	3.4	.5	.5	.8
Hostility	.8	2.9	.5	3.4	.4	1.4
Suspiciousness	.4	2.2	.5	2.6	1.0	3.0
Hallucinatory behavior	.1	.2	.3	.1	1.5	3.5
Motor retardation	1.0	.5	2.2	.4	1.8	.7
Uncooperativeness	.3	1.0	.8	1.6	1.2	1.6
Unusual thought content	.4	.7	.4	1.2	2.2	4.2
Blunted affect	1.0	.7	2.7	.7	3.6	2.6

Table 8.13 Prototype Profiles after Several Iterative Reclassifications

	Simple anxious depression	*Hostile depression*	*Retarded depression*	*Paranoid hostile-suspicious*	*Withdrawn-disorganized schizophrenia*	*Florid thinking disorder*
Somatic concern	1.4	2.2	3.0	1.8	.5	1.5
Anxiety	1.5	4.1	2.6	1.4	.6	2.5
Emotional withdrawal	.6	1.2	2.7	1.2	3.8	3.3
Conceptual disorganization	.5	1.4	1.4	2.5	3.1	4.1
Guilt feelings	.4	2.3	1.6	.6	.2	1.0
Tension	1.0	2.9	1.7	1.3	1.5	2.7
Mannerisms and posturing	.1	.5	.8	.5	1.8	2.1
Grandiosity	.2	.3	.7	2.3	.2	1.6
Depressive mood	1.3	3.9	3.8	1.0	.6	1.4
Hostility	.4	2.6	1.3	2.2	.5	2.0
Suspiciousness	.2	2.1	.6	2.0	.8	3.6
Hallucinatory behavior	.2	.4	.4	.7	.6	3.9
Motor retardation	.5	.8	2.3	.3	2.0	1.3
Uncooperativeness	.2	1.0	.9	1.4	1.5	1.9
Unusual thought content	.3	1.3	.6	2.2	1.6	4.4
Blunted affect	.8	1.0	2.9	1.2	3.6	3.1

much like factor analysis or linear components analysis, except for minor scaling differences. From a theoretical point of view, one starts by assuming that underlying any heterogeneous group of individuals are a relatively few basic *pure types*. Any one individual can be a mixture of several types, although he is likely to resemble most closely one particular pure type. The situation can be likened to the study of genetic mixtures, where several highly inbred pure strains have been somewhat diluted by the introduction of foreign genes. Most individuals will have the general appearance of one of the pure strains, but particular individuals will tend to manifest alien characteristics to greater or lesser degree. Without knowing the number or nature of the underlying inbred strains, we must determine how many are present and what they look like. This can be accomplished from the analysis of similarities and differences among individuals randomly sampled from a heterogeneous population.

Linear typal analysis aims at discovering the nature and number of underlying pure types, defining a prototype score vector to represent each pure type, and then determining the similarity of each individual to each of the hypothetical pure types. If most individuals can be recognized as being highly similar to one and only one of the underlying pure types, a classification grouping can be accomplished by assigning each individual to the class represented by the pure type having a profile most similar to his own.

The model for linear typal analysis is thus a simple linear model in which the observed profile for any individual is conceived as a weighted average of pure-type profiles, plus an error component

$$\mathbf{x} = b_1\mathbf{z}_1 + b_2\mathbf{z}_2 + \cdots + b_r\mathbf{z}_r + b_0\boldsymbol{\epsilon} \tag{8.26}$$

where the $\mathbf{z}_i$ are prototype profiles representing r pure types and the b_i are scalar weighting coefficients. Disregarding the random-error component, the observed score profiles are approximated as linear combinations of the pure types

$$\begin{bmatrix} x_{i1} \\ x_{i2} \\ \cdot \\ \cdot \\ \cdot \\ x_{ip} \end{bmatrix} \approx b_{i1}\begin{bmatrix} z_{11} \\ z_{21} \\ \cdot \\ \cdot \\ \cdot \\ z_{p1} \end{bmatrix} + b_{i2}\begin{bmatrix} z_{12} \\ z_{22} \\ \cdot \\ \cdot \\ \cdot \\ z_{p2} \end{bmatrix} + \cdots + b_{ir}\begin{bmatrix} z_{1r} \\ z_{2r} \\ \cdot \\ \cdot \\ \cdot \\ z_{pr} \end{bmatrix}$$

or in matrix notation

$$\underset{n\times p}{X} \approx \underset{n\times r}{B}\;\underset{r\times p}{Z'} \tag{8.27}$$

Since the observed profile vectors can be represented, apart from an error component, as the weighted average of pure-type profiles, the reciprocal relationship requires that it be possible for pure-type profiles to be estimated as simple weighted averages of the individual profile vectors. The pure-type profiles can be approximated by a simple weighted function of the individual profiles even though no individuals in the population are actually pure types themselves

$$\mathbf{z} = a_1\mathbf{x}_1 + a_2\mathbf{x}_2 + \cdots + a_n\mathbf{x}_n \tag{8.28}$$

or in matrix notation

$$\underset{p\times r}{Z} = \underset{p\times n}{X'}\ \underset{n\times r}{A} \tag{8.29}$$

In these equations, the $\mathbf{x}_i$ which taken together form the matrix X are observed score vectors for n individuals, and the a_i are scalar weighting coefficients. Note a separate and distinct set of a_i coefficients defining each of the r pure types. The problem in linear typal analysis is to determine how many pure-type profiles are required to account for the observed score vectors and to determine the appropriate values for weighting coefficients used to define the prototype profiles as linear combinations of the observed score vectors.

Linear typal analysis is applied to vector-product indices of profile similarity. A complete matrix of vector-product indices relating all pairs of individuals can be defined as

$$\underset{n\times n}{Q} = \underset{n\times p}{X}\ \underset{p\times n}{X'} \tag{8.30}$$

where X $(n \times p)$ contains profiles consisting of p elements for each of n individuals.

Various types of transformations can be undertaken on the profiles before calculating Q. For example, the mean of each row can be subtracted from the p elements in the row. As previously noted, this has the effect of removing the elevation component. The elements in the score matrix can be adjusted by subtracting out both the row and column means. Some investigators consider analysis of double-centered matrices most appropriate when interest really lies in studying only individual differences (Gollob, 1968). Other investigators favor use of raw scores in generating the vector-product matrix (Nunnally, 1962; Tucker, 1968). Many investigators in practical research applications choose to subtract out the row means and then to normalize the resulting deviation scores within rows of X. The vector product between elevation-corrected and normalized score vectors is a product-moment correlation matrix. Thus, if one simply computes the product-moment correlation coefficients between all pairs of profiles, the resulting matrix can be taken as a special case of the Q matrix of Eq. (8.30) where elements in each row of X have been mean-corrected and normalized. Common Q-type matrices contain product-moment correlations, covariances, or raw cross products between pairs of profiles. For simplicity in this discussion, assume the data matrix X contains profiles that have been normalized so that the sum of squares across all p elements is equal to unity. In the next section, problems related to adjustment of mean levels will be discussed in more detail.

Assume for a moment that several homogeneous subsets of profiles have been identified. Each pure-type profile is approximated as a weighted combination of the observed profiles [Eq. (8.28)]. Given a Q-type matrix of vector-product similarity coefficients, the vector-product similarity indices relating each of the original profiles to the underlying pure types can easily be obtained. Let A $(n \times r)$ be a matrix containing r columns in each of which nonzero weights are associated with a different subset of individuals. The matrix of vector products relating each of the original profiles to each of the r new hypothetical pure types can be defined as

$$QAS = F \tag{8.31}$$

where the matrix A $(n \times r)$ has been chosen to emphasize the cluster configuration and S $(r \times r)$ is a diagonal scaling matrix containing the reciprocals of the square roots of the diagonal elements in $V = A'QA$. The matrix S is necessary for the pure-type profiles to be normalized consistent with the original profiles in the data matrix X.

The weighting coefficients defining the pure-type composition of the original (normalized) profiles are obtained by applying an orthogonalizing transformation to the matrix F

$$QAS(S'A'QAS)^{-1} = F(S'A'QAS)^{-1} = B \tag{8.32}$$

The matrix B $(n \times r)$ contains a distinct set of weighting coefficients $b_1, b_2, \ldots, b_r$ that specify the relative contributions of the r pure types to each of the n individuals [Eq. (8.26)]. The original data matrix consisting of n normalized profiles can be approximately reproduced according to the basic model

$$X \approx BZ'$$

The theory of types suggests that most individuals should be predominantly of one type. This means the pattern of weighting coefficients in the matrix B should reveal, for each individual, a relatively large contribution of one pure type. Each pure type should tend to determine the profile pattern for a distinct subgroup of individuals and should contribute little to the profiles of most other individuals. This type of pattern for weighting coefficients comes very close to Thurstone's concept of simple structure. The theory underlying linear typal analysis implies that each individual will *tend* to be relatively simple with regard to his pure-type composition. The computational problem in linear typal analysis thus becomes that of defining a transformation matrix A such that the pattern of weighting coefficients in the matrix B will approach the simple-structure ideal.

The prototype profiles representing r pure types are obtained by applying the cluster-oriented transformation AS, as defined in connection with Eq. (8.31), to the original data matrix

$$X'AS = Z \tag{8.33}$$

When the diagonal matrix S $(r \times r)$ is defined to contain the reciprocals of the square roots of the diagonal elements in $V = A'QA$, the pure-type profiles in Z $(p \times r)$ will be normalized in the same manner as the original profiles in X $(n \times p)$.

Numerous methods are available for defining appropriate cluster transformations, pure-type profiles, and weighting coefficients. The method of oblique powered-vector factor analysis (Sec. 7.3) can be applied to the matrix Q to obtain a cluster-centroid transformation matrix, as illustrated in Table 7.1. The matrix A of Eqs. (8.31) to (8.33) can thus be defined as a zero-one cluster-centroid transformation matrix. Direct cluster-rotation analysis (Sec. 8.7) can be employed to obtain a zero-one matrix of cluster-membership coefficients, as illustrated in Table 8.10. Such a zero-one matrix represents an appropriate A matrix for linear typal analysis. Given an appropriate cluster-oriented transformation matrix, pure-type profiles and weighting coefficients can be obtained using equations presented in this section.

Section 8.13 describes a computer program that uses powered-vector factor analysis to obtain the simple-structure matrix. It defines prototype vectors as the means of raw-score profile vectors for individuals most highly related to each pure type. Thus, it is a classification program which describes the classification groups in terms of prototypes which are mean vectors. Cluster means are the columns of the Z matrix defined in Eq. (8.33) except that they are in raw-score form.

8.11 Determination of Optimum Profile Elevation

The term *elevation* indicates the average level of scores within a profile. In psychological and social science, the zero points for scales of measurement are usually rather arbitrary, which means that the raw-score profile elevation is arbitrary. Cronbach and Gleser (1953) have discussed the elimination of arbitrary elevation components by subtracting the profile mean from each score. An *average* elevation of zero across all profiles can also be obtained by subtracting from each variable the mean for that particular variable. Some investigators recommend subtracting out both person and test means to remove average effects entirely in the study of individual differences in pattern (Gollob, 1968). These types of corrections and adjustments are not without implications for linear typal analysis.

If each variable within all profiles is corrected to zero mean across all individuals, the origin in the multivariate space is moved to the person centroid. This means that the rank of the Q matrix is effectively reduced by 1 and the number of pure types defined by linear typal analysis or factor analysis will be, in general, 1 less than the number of homogeneous subclasses represented in the data. On the other hand, if raw scores are analyzed and the arbitrary origin lies far from the center of the person space, the analysis will tend to result in a large general factor plus $r - 1$ pattern factors (Tucker, 1968). It is frequently possible to rotate the orthogonal linear components to a reasonable pure typal structure, but the orthogonal axes cannot be made to pass through the cluster centroids in the presence of a large general factor. Examples of raw-score profile analysis presented by Nunnally (1962) reveal this problem.

Correction of all variables within the profiles to zero mean is equivalent to subtracting a constant (mean) profile from all individual profiles. Although the constant profile used most frequently consists of the variable means, no reason exists why another constant might not be used to adjust the average profile elevation in a perhaps more meaningful way. Discussions in the literature have tended to polarize around either analysis of raw scores or analysis of mean-corrected scores.

Following Gollob (1968), we propose that the meaningful pattern differences are exhibited in a data matrix from which specific row and column effects have been removed. The proper origin for linear typal analysis does not lie, however, in the center of the person space. A constant should be added to all mean-corrected scores to yield a vector-product matrix Q that will result in a maximally simple pure-type structure. Three possible origins are illustrated in Fig. 8.4. The uncorrected raw-score origin is too far from the center of the person configuration and results in elements in $Q = XX'$ being uniformly large. The mean-corrected origin is in the middle of the person space and yields both positive and negative vector products. The ideal or optimal origin is located somewhat below the center of gravity for most profiles and results in a $Q = XX'$ matrix in which elements are predominantly positive, although some minor negative coefficients may be present. In a Q matrix of this type, the cluster configuration is most evident. A linear typal analysis applied to such a matrix will tend to yield a pure-type profile for each homogeneous cluster of individuals, and the structure of the relationships will reveal a good pure-type pattern.

A simple and satisfactory manner of determining the optimum correction for profile elevation involves double-correcting the original score matrix by subtracting out both row and column means. This can be done in one step as follows:

$$\hat{x}_{ij} = x_{ij} - \bar{x}_{i\cdot} - \bar{x}_{\cdot j} + \bar{x}_{\cdot\cdot} \tag{8.34}$$

Next, the matrix $\hat{Q} = \hat{X}\hat{X}'$ is computed using the double-centered data matrix. The elements in $\hat{Q}$ will be both positive and negative, and the columns will sum to zero since the origin for the profiles in the data matrix $\hat{X}$ is at the center of the person space.

As discussed in Sec. 8.10, a good solution in linear typal analysis requires a simple pattern in the matrix of coefficients relating individual profiles to pure types [Eqs. (8.31) and (8.32)]. If the matrix $Q = XX'$ has uniformly large positive elements due to an arbitrary origin that is too far from the center of the person configuration, any transformation $QAS = F$ in which A is a simple cluster-centroid type of matrix will yield an F in which elements are also uniformly large and positive. On the other hand, if the origin has been shifted to the center of the person space by subtracting the mean for each variable, the

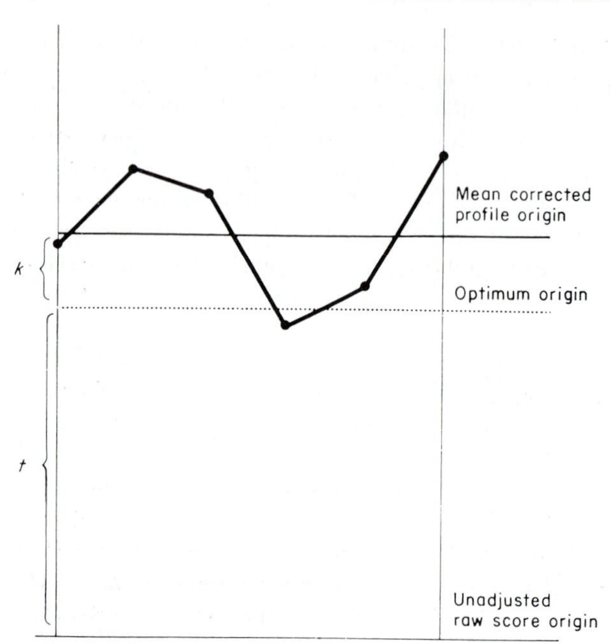

Figure 8.4 Schematic diagram showing multivariate profile in relation to three different origins.

matrix $\hat{Q}$ (columns sum to zero) will yield $\hat{Q}AS = F$, in which elements are both positive and negative. It would be desirable to adjust the average value in $\hat{Q}$ so that the transformation $\hat{Q}AS = F$ results in a good simple structure. This suggests that a constant of some magnitude should be added to each element in the matrix $\hat{Q}$, the constant being chosen to maximize the simplicity of the pattern in F.

The matrix $\hat{\hat{Q}}$ that best reveals the cluster configuration is one in which most elements are either positive or near zero in value. Such a matrix can be obtained either by adding a constant c to all elements in the double-centered $\hat{Q} = \hat{X}\hat{X}'$ or by adding a constant elevation correction to each double-corrected element in $\hat{X}$. Given a matrix $\hat{Q} = \hat{X}\hat{X}'$ computed from a double-centered data matrix, the product matrix $\hat{\hat{Q}}$ can be obtained as follows:

$$\hat{\hat{Q}} = \hat{Q} + \mathbf{cc}' \tag{8.35}$$

where $\mathbf{c}$ is a vector containing uniform positive elements of magnitude sufficient to raise most of the negative values in $\hat{Q}$ to near zero. It is difficult to specify how the value of c can be calculated in advance without inspection of the elements in $\hat{Q}$. Thus, if one is going to transform the origin by operating on $\hat{Q} = \hat{X}\hat{X}'$, it is suggested that the elements in $\mathbf{c}$ should be equal to the square root of the absolute value of the (average) larger negative elements in $\hat{Q}$.

The matrix $\hat{\hat{Q}}$, calculated by adding a constant to each element in $\hat{Q} = \hat{X}\hat{X}'$, can be computed directly from vector products of double-corrected profiles to which a constant k has been added. Let $\hat{X}$ be a double-corrected data matrix in which rows and columns sum to zero, and let $\hat{\hat{X}}$ be the double-corrected data matrix after a constant k has been added to each element. Then the vector-product matrix that adequately reveals the cluster configuration can be computed in the usual way, $\hat{\hat{Q}} = \hat{\hat{X}}\hat{\hat{X}}'$. The value of k is represented by the difference between the mean corrected origin and the optimum origin in Fig. 8.4. The value of k that will result in the vector-product matrix $\hat{\hat{Q}} = \hat{Q} + \mathbf{cc}'$ is related to the value of elements in $\mathbf{c}$ as follows:

$$k = \sqrt{\frac{c}{p}} \tag{8.36}$$

where c is the general element in $\mathbf{c}$ and p is the number of elements in each profile. Again, the value c is the absolute value of the larger (averaged) negative elements in $\hat{Q} = \hat{X}\hat{X}'$, as computed from a double-corrected data matrix without a constant k added.

The effect on profile elevation of adding a constant k to each element is illustrated in Fig. 8.4. The transformed profiles have differential person and test means removed, but a constant elevation factor has been added. Elements in the transformed data matrix $\hat{\hat{X}}$ are obtained as follows:

$$\hat{\hat{x}}_{ij} = x_{ij} - \bar{x}_{i.} - \bar{x}_{.j} + \bar{x}_{..} + k \tag{8.37}$$

In Q-type factor analysis and in linear typal analysis, it is usually desirable to standardize scores across individuals in order to eliminate arbitrary and disparate units of measurement. The magnitudes of negative elements in $\hat{Q} = \hat{X}\hat{X}'$, computed from a *standardized* and then double-corrected data matrix without a constant k added, will depend on the number of groups or pure types required to account for individual differences. A good rule of thumb that has some mathematical basis leads to the following estimate of the constant k to be added to standardized, double-corrected score vectors

$$k = \frac{1}{\sqrt{\hat{r} - 1}} \tag{8.38}$$

where $\hat{r}$ is the expected number of pure types or groups. In practice, the constant $k = .7$ provides an adequate general elevation correction for most double-centered matrices in which scores are initially standardized to z-score form across individuals.

It must be admitted that the mathematical rules for determining in advance an optimum value for k are very crude. One must frequently resort to using an estimate, such as $k = 1/\sqrt{\hat{r} - 1}$ or even $k = .7$, in a preliminary analysis and then refining the estimate on the basis of the results obtained. If k is too large, projections on one or more of the orthogonal factors will tend to be uniformly positive and large. If k is too small, projections on one or more orthogonal factors derived from $\hat{\hat{Q}} = \hat{\hat{X}}\hat{\hat{X}}'$ will tend to be equally positive and negative. An optimum choice of k will yield good orthogonal simple structure in which some profiles have large positive projections and the remainder have near-zero projections on each factor.

Another way to evaluate the adequacy of the constant k is by inspection of the matrix $\hat{\hat{Q}} = \hat{\hat{X}}\hat{\hat{X}}'$. If k is too large, all elements in $\hat{\hat{Q}}$ will be positive. If k is too small, the number of positive and negative elements will tend to be equal. Optimum choice of k will lead to a matrix $\hat{\hat{Q}} = \hat{\hat{X}}\hat{\hat{X}}'$ in which 80 to 90 percent of the elements are positive and the remainder are small negative values. If an initial choice of k does not result in such a matrix, the value should be revised up or down to produce a more desirable cluster configuration.

At the time of this writing, the notion of shifting the multivariate origin to provide optimum display of the cluster configuration is a new one. In Q-type factor analysis and in linear typal analysis, location of the origin at the center of the person configuration by double correction of the data matrix has serious drawbacks, the major one being loss of a factor or dimension. Similarly, the raw-score origin has disadvantages, particularly since different arbitrary units of measurement may be involved in the multivariate profile. Where scores within the profiles have been standardized, there is need to shift the origin away from the center of the person configuration. The problem is to specify how far to move it. We have attempted to provide some suggestions. The student is encouraged to try various alternatives in order to grasp the problem more fully and perhaps to arrive at a better solution than that offered here.

Adjustment of profile elevation by adding a constant k to all scores will usually result in some negative elements, as illustrated in Fig. 8.4. Following a Q-type factor analysis or linear typal analysis, it may be desirable to place the resulting cluster means or pure-type profiles back in original raw-score form. A reverse transformation is required in which the constant k is subtracted from each score and then the standard scores are transformed to raw scores through multiplying by the standard deviation and adding the mean value of each variable. Alternatively, one can use the analysis merely to identify the profiles that constitute various clusters and then go back to the original raw scores to compute mean profiles to represent the pure types.

8.12 An Example of Linear Typal Analysis

In this section, the computations and results obtained from linear typal analysis will be illustrated using purely fictitious data. The departure from our usual practice of employing clinically relevant data for examples is motivated by the desire to limit the size of the problem so that intermediate computational steps can be shown in detail. A raw-data matrix consisting of score vectors

Table 8.14 Raw-data and Corrected-data Matrix

	Raw-score data matrix						*Standardized and double-centered with origin adjusted*					
	1	*2*	*3*	*4*	*5*	*6*	*1*	*2*	*3*	*4*	*5*	*6*
1	95	16	20	35	55	19	1.82	.03	−.59	1.41	1.74	−.17
2	93	20	22	34	53	16	1.94	.38	−.19	1.17	1.10	−.17
3	90	19	23	36	55	18	1.46	.04	−.41	1.81	1.67	−.32
4	50	16	35	35	55	50	−.94	−.15	.48	1.22	1.55	2.07
5	55	15	37	30	53	48	−.16	.30	1.15	−.65	1.18	2.41
6	53	18	38	35	54	46	−.72	−.05	.78	1.26	1.16	1.80
7	85	85	50	35	50	30	.78	1.68	1.48	.94	−.89	.22
8	83	86	52	33	48	27	1.00	2.04	1.98	.32	−1.43	.32
9	87	88	49	30	52	26	1.18	2.06	1.68	−1.14	.25	.19

for nine hypothetical individuals is shown in the left-hand portion of Table 8.14. The score vectors were purposely constructed so that successive sets of three are highly similar.

The first step in the analysis was to standardize the $p = 6$ variables across all individuals and then to subtract out the profile means (elevations) and add the constant $k = 1/\sqrt{\hat{r} - 1} = .71$. This resulted in the double-centered matrix shown in the right-hand section of Table 8.14. This double-centered matrix, computed from scores that were first standardized across individuals, contains the elements of $\hat{\hat{x}}_{ij}$ of Eq. 8.37. The constant $k = 1/\sqrt{2}$ was added to each of the double-corrected scores in the expectation that the true number of pure types is $\hat{r} = 3$. In fact, it would not have affected the results adversely if the expected number of clusters had been estimated to be two, three, or four. With only nine profiles and six elements in each, it is unlikely that anyone would expect more than four pure types. If there is no other basis for estimating the expected number of clusters, a fairly adequate estimate of $\hat{r}$ is the square root of the number of profiles entered in the analysis or the number of variables within each profile, whichever is smaller.

The next step in the analysis involved calculation of the vector-product matrix $\hat{\hat{Q}} = \hat{\hat{X}}\hat{\hat{X}}'$, where $\hat{\hat{X}}$ is the double-corrected matrix shown in the right-hand section of Table 8.14. The resulting Q-type matrix is shown in Table 8.15. It will be noted that the matrix $\hat{\hat{Q}} = \hat{\hat{X}}\hat{\hat{X}}'$ does not have unities in the principal diagonal because the scores were not standardized *within* profiles. Elements in $\hat{\hat{Q}} = \hat{\hat{X}}\hat{\hat{X}}'$ tend to be positive, although a few rather small negative values are present.

An orthogonal powered-vector analysis (Chap. 6) was applied to the matrix Q to obtain the orthogonal pure-type projections shown in the matrix F which appears in the left-hand section of Table 8.16. The projections on the orthogonal pure-type axis are not restricted to the range ± 1 because the $\hat{\hat{Q}}$ matrix was not

Table 8.15 Vector-product Matrix Computed from Double-corrected Data Matrix with Origin Adjusted

	1	*2*	*3*	*4*	*5*	*6*	*7*	*8*	*9*
1	8.75	7.29	8.44	2.06	−.25	1.70	.34	−1.39	.02
2	7.29	6.59	6.97	.80	−.29	.87	1.97	1.08	1.69
3	8.44	6.97	8.48	2.54	−.70	2.25	.75	−1.18	−.57
4	2.06	.80	2.54	9.38	6.72	8.19	−.04	−1.46	−1.20
5	−.25	−.29	−.70	6.72	9.10	5.91	.97	1.61	3.90
6	1.70	.87	2.25	8.19	5.91	7.37	1.07	.04	−.43
7	.34	1.97	.75	−.04	.97	1.07	7.42	8.85	5.64
8	−1.39	1.08	−1.18	−1.46	1.61	.04	8.85	11.38	8.05
9	.02	1.69	−.57	−1.20	3.90	−.43	5.64	8.05	9.89

a correlation matrix. Examination of the projections obtained from the orthogonal powered-vector analysis reveals that one subset of three similar profiles projects highly on each pure-type factor axis. The matrix A, required to define the pure-type profiles [Eq. (8.31)] and the weighting coefficients descriptive of the pure-type composition of each individual B [Eq. (8.32)], was obtained by setting the element corresponding to the largest projection in each row of F equal to 1.0 and then scaling by dividing all elements in each column by the sum of the elements. The resulting typal-transformation matrix A is shown in the right-hand section of Table 8.16. Note that elements in each column sum to unity.

The matrix Z $(p \times r)$, which contains the pure-type profiles in standard form, was computed by applying the matrix A to the corrected-data matrix $\hat{\hat{X}}'$. The standard pure-type profiles shown in the left section of Table 8.17 were converted back to raw-score form using the following reverse transformation. The constant $k = 1/\sqrt{2}$ was *subtracted* from each element, the difference was multiplied by the standard deviation of the original variable, and the mean of the

Table 8.16 Powered-vector Factor Loadings and Linear Typal Transformations

	Factor loadings			*Transformations*		
	I	*II*	*III*	*I*	*II*	*III*
1	−.23	2.94	−.01	.00	.33	.00
2	.47	2.50	−.24	.00	.33	.00
3	−.19	2.88	.11	.00	.33	.00
4	−.35	.70	2.94	.00	.00	.33
5	.63	−.12	2.58	.00	.00	.33
6	.07	.63	2.59	.00	.00	.33
7	2.61	.36	.13	.33	.00	.00
8	3.35	−.20	−.08	.33	.00	.00
9	2.59	.12	.05	.33	.00	.00

Table 8.17 Pure-type Profiles

	Standard-score form			*Raw-score form*		
	Type I (7,8,9)	*Type II (1,2,3)*	*Type III (4,5,6)*	*Type I (7,8,9)*	*Type II (1,2,3)*	*Type III (4,5,6)*
1	.99	1.74	−.61	81.71	94.84	53.77
2	1.93	.15	.03	80.20	22.39	18.39
3	1.71	−.40	.80	48.11	23.13	37.41
4	.04	1.46	.61	32.27	35.26	33.46
5	−.68	1.50	1.30	49.56	54.62	54.14
6	.24	−.22	2.09	25.27	19.25	48.80

original variable was added to each element. The pure-type profiles in raw-score form are shown in the right-hand section of Table 8.17.

The weighting coefficients defining the original (standardized and double-corrected) score profiles as linear functions of the r pure types were computed as follows:

$$QA(A'QA)^{-1} = B$$

where the matrices Q and A are shown in Tables 8.15 and 8.16. The matrix B, containing three weighting coefficients for each individual, is presented in Table 8.18. Due to the tight cluster configuration that was purposely built into the data matrix, each individual is highly similar to one of the pure-type profiles. Thus, a coefficient of approximately 1.00 represents the weight given one of the pure type profiles in reproducing each individual profile. The other coefficients tend to have near-zero values.

This example has illustrated in detail how linear typal analysis can be accomplished. The results include prototype profiles representing the pure types that

Table 8.18 Weighting Coefficients Defining Pure-type Composition of Each Individual

	Type I	*Type II*	*Type III*
1	−.09	1.06	.03
2	.16	.91	−.07
3	−.09	1.02	.06
4	−.18	.10	1.09
5	.23	−.19	.96
6	−.04	.07	.97
7	.90	.08	.02
8	1.16	−.11	−.06
9	.94	.01	.05

are necessary to account for differences in individual profiles and weighting coefficients that indicate how similar each individual is to each pure type. Individuals can readily be classified according to predominant pure-type likeness. The computer program listed in Sec. 8.13 accomplishes a linear typal analysis similar to that described in this section with the exception that the transformed profiles are normalized so that the diagonal elements of $\hat{\hat{Q}}$ will be equal to unity. This profile normalizing has advantages with regard to interpretation of projections on pure-type axes, but it adds a needless complication in a computational example like the one considered here.

8.13 Computer Program for Linear Typal Analysis

The computer program presented here accomplishes linear typal analysis as described in Sec. 8.10 with adjustment of origin as described in Sec. 8.11. The analysis differs from the example in Sec. 8.12 in that the profiles are normalized so that the diagonal elements in the matrix $\hat{\hat{Q}}$ are equal to unity. The pure-type profiles are converted back to raw-score units with original origin for each variable. The program consists of a mainline program plus seven subroutines.

The computer program reads raw data, transforms variables to z-score form across individuals, corrects profiles to zero mean, adds constant k, and then normalizes the resulting transformed profiles to yield $\hat{\hat{Q}} = \hat{\hat{X}}\hat{\hat{X}}'$ with unity in principal diagonal. The Q-type vector products are subjected to an orthogonal powered-vector factor analysis to obtain the projections of the n profiles on r pure-type factors. Each individual is assigned to the cluster corresponding to the pure-type factor on which his (double-corrected and normalized) profile has the largest positive projection. A mean profile is computed to represent the pure type associated with each orthogonal component of the $\hat{\hat{Q}}$ matrix.

The setup for the linear typal analysis is quite simple. It requires only one control card followed by data cards containing raw-score profiles for up to 200 persons. A single profile can contain up to 150 elements.

Input

1 Control card containing NPER, NSCOR, MSTP, NCL in successive four-column fields.
NPER = number of profiles
NSCOR = number of scores per profile
MSTP = maximum number of pure types that the program will be permitted to define. (Set at 999 if no limit is desired.)
NCL = minimum number of orthogonal projections that must exceed 0.50 on each pure-type factor. (Usually set at 3 or 4.)

2 Data cards with profile elements punched into successive four-column fields. (Format 110 in subroutine STD.)

Output

1 The double-corrected and origin-adjusted $\hat{\hat{Q}} = \hat{\hat{X}}\hat{\hat{X}}$ vector-product matrix.
2 Value of constant k that was added to elements in each double-corrected profile.
3 Pure-type profiles computed as means of profiles for individuals most highly related to each pure type.
4 Projections of individual profiles on orthogonal and oblique pure-type axes and weighting coefficients defining individual profiles as linear functions of pure types.

Mainline Program for Linear Typal Analysis

```
 DIMENSION A(200)
 COMMON L,L2,L3,L4,L8,L9,L10
 DEFINE FILE 1(200,300,U,L), 2(200,300,U,L2), 3(200,300,U,L3),
14(200,300,U,L4),8(35,320,U,L8),9(35,70,U,L9),10(35,320,U,L10)
 CALL STD(NVAR,NSCOR,MSTP,NCL)
 CALL QCORR(NVAR,NSCOR)
 CALL VFAC(NVAR,A,MSTP,M, NCL)
 CALL MEAN(M,A,NSCOR,NVAR)
 CALL MMCOS (NVAR,M)
 CALL PRCOS (NVAR,M)
 CALL PFAC (NVAR,M)
 CALL JMINV (M,Z)
 CALL QRCOS (M,NVAR,Z,NSCOR)
 CALL EXIT
 END
```

Subroutine to Read Profile Vectors

```
      SUBROUTINE STD (NVAR, NSCOR, MSTP, NCL)
      DIMENSION XBR(150), SD(150),A(200)
      COMMON L,L2,L3,L4,L8,L9,L10
  100 FORMAT (4I4)
  110 FORMAT (14F2.1)
      READ (2,100) NPER,NSCOR,MSTP,NCL
      NVAR = NPER
      FSCOR=NSCOR
      FNPER=NPER
      DO 120 K=1,NSCOR
      XBR(K) = 0.0
      SD(K) = 0.0
  120 CONTINUE
C     READ PROFILE VECTORS
      DO 140 J=1,NPER
      READ (2,110) (A(K),K=1,NSCOR)
      WRITE(3'J)(A(K),K=1,NSCOR)
      DO 130 K=1,NSCOR
      XBR(K) = XBR(K) + A(K)
      SD(K) = SD(K)+A(K)*A(K)
  130 CONTINUE
  140 CONTINUE
C     CALCULATE MEAN AND STANDARD DEVIATION FOR EACH VARIABLE
      DO 150 K=1,NSCOR
      SD(K)=SQRT((SD(K)-XBR(K)*XBR(K)/FNPER)/(FNPER-1))
      XBR(K) = XBR(K) / FNPER
  150 CONTINUE
      JJ=NPER + 1
      WRITE (3'JJ) (XBR(K), K=1,NSCOR)
      JJ = NPER + 2
      WRITE (3'JJ) (SD(K), K=1,NSCOR)
      RETURN
      END
```

Subroutine to Compute *Q*-type Vector-product Matrix

```
      SUBROUTINE QCORR(NVAR,NSCOR)
      DIMENSION A(200),B(200),R(200),XBR(150),SD(150)
      COMMON L,L2,L3,L4,L8,L9,L10
      NPER = NVAR
      FNPER=NPER
      FSCOR=NSCOR
      CONST = 1.0/FSCOR**.25
  100 FORMAT (1X,'CONSTANT ADDED IS ', F6.4)
      WRITE (3,100) CONST
C     CONVERT TO STANDARD SCORES
      JJ=NPER + 1
      READ  (3'JJ) (XBR(K), K=1,NSCOR)
      JJ = NPER + 2
      READ  (3'JJ) (SD(K), K=1,NSCOR)
      DO 140 J=1,NPER
      READ (3'J) (A(K),K=1,NSCOR)
      DO 110 K=1,NSCOR
      A(K)=(A(K)-XBR(K))/SD(K)
  110 CONTINUE
C     NORMALIZE PROFILE VECTORS AND ADD CONSTANT
      SSQ=0.0
      SUM=0.0
      DO 120 K=1,NSCOR
      SUM=SUM+A(K)
  120 SSQ=SSQ+A(K)*A(K)
      SSQ=SSQ-SUM*SUM/FSCOR
      XBAR=SUM/FSCOR
      SSQ = SQRT(SSQ/FSCOR)
      DO 130 K=1,NSCOR
  130 A(K)=(A(K)-XBAR)/SSQ+ CONST
  140 WRITE(1'J)(A(K),K=1,NSCOR)
C     COMPUTE PROFILE CORRELATIONS
      DO 170 J=1,NPER
      READ(1'J)(A(K),K=1,NSCOR)
      DO 160 J2 = J, NPER
      READ(1'J2)(B(K),K=1,NSCOR)
      COR=0.0
      DO 150 K=1,NSCOR
  150 COR=COR+A(K)*B(K)
  160 R(J2) = COR/ (FSCOR * (1.0+CONST**2.0))
  170 WRITE(2'J)(R(J2),J2=1,NPER)
C     COMPLETE SQUARE OF LARGE SYMMETRIC MATRIX
      DO 180 J=1,NVAR
      READ(2'J)(A(I),I=1,NVAR)
      DO 180 K=J,NVAR
      READ(2'K)(B(I),I=1,NVAR)
      B(J) =A(K)
      WRITE(1'K)(B(I),I=1,NVAR)
  180 WRITE(2'K)(B(I),I=1,NVAR)
      RETURN
      END
```

Subroutine to Compute Orthogonal Powered-vector Analysis of *Q*-type Vector-product Matrix

```
      SUBROUTINE VFAC(NVAR,A,MSTP,M,NCL)
      DIMENSION A(200), B(200), C(200)
      COMMON L,L2,L3,L4,L8,L9,L10
      L4 = 1
      M=0
      FV=NVAR
C     FIND COLUMN WITH LARGEST VARIANCE OF SQUARED ELEMENT
  100 MAX=0
      MTES = 0
      FMAX=0.0
      DO 130 J=1,NVAR
      SUM=0.0
      SSQ=0.0
      READ(1'J)(A(K),K=1,NVAR)
      DO 110 K=1,NVAR
      X=A(K)*A(K)
      SUM=SUM+X
  110 SSQ=SSQ+X*X
      SSQ=SSQ-SUM*SUM/FV
      IF(SSQ-FMAX)130,130,120
  120 MAX=J
      FMAX=SSQ
  130 CONTINUE
C     FETCH HYPOTHESIS VECTOR
      READ(1'MAX)(B(J),J=1,NVAR)
      DO 140 J=1,NVAR
      X=B(J)
  140 B(J)=X*X*X
C     COMPUTE FACTOR LOADINGS
      VAR=0.0
      DO 170 J=1,NVAR
      READ(1'J)(A(K),K=1,NVAR)
      C(J)=0.0
      A(J)=0.0
      FMAX=0.0
      DO 160 K=1,NVAR
      C(J)=C(J)+B(K)*A(K)
      X=A(K)*A(K)
      IF (X-FMAX)160,160,150
  150 FMAX=X
  160 CONTINUE
      A(J)=SQRT(FMAX)
      C(J)=C(J)+B(J)*A(J)
  170 VAR=VAR+C(J)*B(J)
      SCL=1.0/SQRT(VAR)
      TEST=0.0
      DO 190 J=1,NVAR
      C(J)=C(J)*SCL
      CSQ = C(J)*C(J)
      IF (CSQ -.25) 190,190,180
  180 MTES = MTES + 1
  190 TEST=TEST+C(J)*C(J)
      IF (MTES - NCL) 250,250,200
  200 STOP = 1.0 + .05 * FV
      IF (TEST-STOP) 250,250,210
  210 CONTINUE
      M=M+1
      WRITE(4'L4) (C(J),J=1,NVAR)
C     COMPUTE RESIDUAL MATRIX
      IF(M-MSTP)220,250,250
  220 DO 240 J=1,NVAR
      READ(1'J)(A(K),K=1,NVAR)
      DO 230 K=1,NVAR
  230 A(K)=A(K)-C(K)*C(J)
  240 WRITE(1'J)(A(K),K=1,NVAR)
      GO TO 100
  250 CONTINUE
      RETURN
      END
```

Subroutine to Calculate Pure-type Profiles

```
      SUBROUTINE MEAN(MM,A,NSCOR,NVAR)
      DIMENSION A(200),B(200),C(200),N(200)
      COMMON L,L2,L3,L4,L8,L9,L10
  100 FORMAT (//1X,  'INPUT PROFILES, LISTED IN ORDER, WERE ASSIGNED TO
     2THE INDICATED TYPES')
  110 FORMAT(20I4)
  120 FORMAT (//)
  130 FORMAT (' PURE TYPE PROFILE',I4,4X,'N=', F4.0)
  140 FORMAT(1X,20F6.2)
      DO 150 K=1,NVAR
      N(K)=0
  150 B(K)=.05
      DO 170 K=1,MM
      READ(4'K)(C(J),J=1,NVAR)
      DO 170 J=1,NVAR
      IF(B(J)-C(J)) 160,160,170
  160 B(J)=C(J)
      N(J)=K
  170 CONTINUE
      WRITE (3,100)
      WRITE (3,110) (N(J), J=1,NVAR)
      WRITE (3,120)
      DO 230 I=1,MM
      COUNT=0.0
      DO 180 K=1,NSCOR
  180 B(K) = 0.0
C     COMPUTE MEAN VECTOR FOR EACH CLUSTER
      DO 210 J=1,NVAR
      IF(N(J)-I) 210,190,210
  190 READ(3'J)(A(K),K=1,NSCOR)
      COUNT = COUNT + 1.0
      DO 200 K=1,NSCOR
  200 B(K) = B(K) + A(K)
  210 CONTINUE
      DO 220 K=1,NSCOR
  220 B(K) = B(K)/COUNT
      WRITE (3,130) I, COUNT
  230 WRITE (3,140) (B(K),K=1,NSCOR)
      RETURN
      END
```

Subroutine to Compute Pure-type Transformation Matrix *A*

```
      SUBROUTINE MMCOS (NVAR,MM)
      DIMENSION N(159),B(159),C(159)
      COMMON L,L2,L3,L4,L8,L9,L10
      DO 120 I=1,MM
      READ(4'I)(C(J),J=1,NVAR)
      SQR=0.0
      DO 100 J = 1, NVAR
  100 SQR=SQR+C(J)*C(J)
      SQR=SQRT(SQR)
      DO 110 J = 1, NVAR
  110 C(J)=C(J)/SQR
  120 WRITE(1'I)(C(J),J=1,NVAR)
      DO 130 K = 1, NVAR
  130 B(K)=0.0
      DO 180 K = 1, MM
      READ(1'K)(C(J),J=1,NVAR)
      DO 180 J = 1, NVAR
      IF (C(J)) 140, 150, 150
  140 KK=-1.0*K
      GO TO 160
  150 KK=K
  160 C(J)=ABS(C(J))
      IF(B(J)-C(J)) 170, 170, 180
  170 B(J)=C(J)
      N(J)=KK
  180 CONTINUE
  190 FORMAT (1X, 16F7.2)
  200 FORMAT (//,' TRANSFORMATION MATRIX A')
      WRITE (3, 200)
      DO 280 K = 1, MM
      SUM=0.0
      DO 250 J = 1, NVAR
      C(J)=0.0
      IF (N(J)) 210, 230, 230
  210 KJ = K
      KN=IABS(N(J))
      IF (KN - KJ) 250, 220, 250
  220 C(J)=0.0
      GO TO 250
  230 IF (N(J) -K) 250,240,250
  240 C(J)=1.0
      SUM=SUM+1.0
  250 CONTINUE
      DO 260 J=1,NVAR
  260 C(J)=C(J)/SUM
      WRITE (3,190)(C(J), J = 1,NVAR)
  280 WRITE(1'K)(C(J),J=1,NVAR)
  290 FORMAT (//)
      WRITE (3, 290)
      RETURN
      END
```

Subroutine to Compute Vector Products between Pure Types

```
C     COMPUTE COSINES AMONG CLUSTER CENTROIDS
      SUBROUTINE PRCOS(NVAR,NFAC)
      DIMENSION A(159), B(159), C(159),D(159)
      COMMON L,L2,L3,L4,L8,L9,L10
      DO 120 K = 1, NFAC
      READ(1'K)(A(I),I=1,NVAR)
      DO 110 J = 1, NVAR
      C(J)=0.0
      READ(2'J)(B(I),I=1,NVAR)
      DO 100 I = 1, NVAR
  100 C(J)=C(J)+A(I)*B(I)
  110 CONTINUE
      WRITE(8'K )(C(J),J=1,NVAR)
  120 CONTINUE
      DO 150 J = 1, NFAC
      READ(8'J)(C(I),I=1,NVAR)
      DO 140 K = 1, NFAC
      READ(1'K)(B(I),I=1,NVAR)
      A(K)=0.0
      DO 130 I = 1, NVAR
  130 A(K)=A(K)+B(I)*C(I)
  140 CONTINUE
      D(J)=1.0/SQRT(A(J))
  150 WRITE(9'J)(A(K),K=1,NFAC)
      DO 170 J=1,NFAC
      READ(9'J)(A(K),K=1,NFAC)
      DO 160 K=1,NFAC
  160 A(K)=A(K)*D(J)*D(K)
  170 WRITE(9'J)(A(K),K=1,NFAC)
      RETURN
      END
```

Subroutine to Compute Vector Products of Individual Profiles with Pure Types

```
C     COMPUTE PRIMARY FACTOR LOADINGS
      SUBROUTINE PFAC(NVAR,NFAC)
      DIMENSION A(159), B(159), C(159)
      COMMON L,L2,L3,L4,L8,L9,L10
      DO 130 K = 1, NFAC
      READ(1'K)(A(I),I=1,NVAR)
      VAR=0.0
      DO 110 J = 1, NVAR
      C(J)=0.0
      READ(2'J)(B(I),I=1,NVAR)
      DO 100 I = 1, NVAR
  100 C(J)=C(J)+A(I)*B(I)
  110 VAR=VAR+C(J)*A(J)
      SCL=1.0/SQRT(VAR)
      DO 120 I = 1, NVAR
  120 C(I)=C(I)*SCL
  130 WRITE(10'K)(C(I),I=1,NVAR)
      RETURN
      END
```

Matrix-inversion Subroutine Used in Linear Typal Analysis

```
C       COMPUTE MATRIX INVERSE
        SUBROUTINE JMINV(N,A)
        DIMENSION IPIVO(32),INDEX(32,2),PIVOT(32), A(32,32)
        COMMON L,L2,L3,L4,L8,L9,L10
        DO 100 K = 1, N
        READ(9'K)(A(K,J),J=1,N)
  100   CONTINUE
        DET=1.0
        DO 110 J = 1, N
  110   IPIVO (J) = 0
        DO 260 I = 1, N
        AMAX=0.0
        DO 160 J = 1, N
        IF (IPIVO (J)-1) 120, 160, 120
  120   DO 150 K = 1, N
        IF (IPIVO (K)-1) 130, 150, 300
  130   IF(ABS(AMAX)-ABS(A(J,K))) 140, 150, 150
  140   IROW=J
        ICOLU =K
        AMAX=A(J,K)
  150   CONTINUE
  160   CONTINUE
        IF(ABS (AMAX)-2.0E-7) 170, 170, 190
  170   WRITE (3,180)
  180   FORMAT(20H DETERMINANT = ZERO )
        PAUSE
  190   IPIVO(ICOLU)=IPIVO(ICOLU)+1
        IF (IROW-ICOLU ) 200, 220, 200
  200   DET=-DET
        DO 210 L = 1, N
        SWAP=A(IROW,L)
        A(IROW,L)=A(ICOLU ,L)
  210   A(ICOLU ,L)=SWAP
  220   INDEX(I,1)=IROW
        INDEX(I,2)=ICOLU
        PIVOT(I)=A(ICOLU ,ICOLU )
        DET=DET*PIVOT(I)
        A(ICOLU ,ICOLU )=1.0
        DO 230 L = 1, N
  230   A(ICOLU ,L)=A(ICOLU ,L)/PIVOT(I)
        DO 260 L1 = 1, N
        IF (L1-ICOLU) 240, 260, 240
  240   T=A(L1,ICOLU )
        A(L1,ICOLU )=0.0
        DO 250 L = 1, N
  250   A(L1,L)=A(L1,L)-A(ICOLU ,L)*T
  260   CONTINUE
        DO 290 I = 1, N
        L=N+1-I
        IF (INDEX(L,1)-INDEX(L,2)) 270, 290, 270
  270   IROW=INDEX(L,1)
        ICOLU =INDEX(L,2)
        DO 280 K = 1, N
        SWAP=A(K,IROW)
        A(K,IROW)=A(K,ICOLU )
        A(K,ICOLU )=SWAP
  280   CONTINUE
  290   CONTINUE
  300   RETURN
        END
```

Subroutine to Compute Weighting Coefficients *B* Revealing Pure-type Composition of Individuals

```
      SUBROUTINE QRCOS (NFAC, NVAR, A, NSCOR)
      DIMENSION A(32,32), C(159), F(159), B(159)
      COMMON L,L2,L3,L4,L8,L9,L10
100   FORMAT(12F10.3)
      DO 110 K=1,NFAC
110   C(K)=1.0/ SQRT(A(K,K))
      L3=1
      WRITE(1'L3)(C(K),K=1,NFAC)
      DO 120 K=1,NFAC
      DO 120 J=1,NFAC
120   A(J,K)=A(J,K)*C(K)
      DO 160 K1 = 1, NVAR
      DO 130 K2 = 1, NFAC
      READ(10'K2)(C(J),J=1,NVAR)
130   F(K2)=C(K1)
      DO 150 K2=1,NFAC
      C(K2)=0.0
      DO 140 K3=1,NFAC
140   C(K2)=C(K2)+F(K3)*A(K3,K2)
150   CONTINUE
160   WRITE(2'K1)(C(K2),K2=1,NFAC)
170   FORMAT(' ORTHOGONAL LOADINGS',I3,20X,'PRIMARY LOADINGS',I3,20X,
     1'MATRIX B--COLUMN',I3)
180   FORMAT(5X,I4,3(F10.5,25X))
      DO 190 K=1,NFAC
      WRITE (3, 170) K,K,K
      READ(4'K)(C(J),J=1,NVAR)
      READ(10'K)(F(J),J=1,NVAR)
      DO 190 J=1,NVAR
      READ(2'J)(B(I),I=1,NFAC)
190   WRITE(3,180)J, C(J), F(J), B(K)
      WRITE (3, 200)
200   FORMAT(//,' PRIMARY COSINE MATRIX')
      DO 210 J=1,NFAC
      READ(9'J)(B(K),K=1,NFAC)
210   WRITE (3,100) (B(K), K = 1, NFAC)
      L3=1
      READ(1'L3)(C(K),K=1,NFAC)
      DO 220 K=1,NFAC
      DO 220 J=1,NFAC
220   A(J,K)=A(J,K)*C(J)
      WRITE (3, 230)
230   FORMAT(// ,' REFERENCE COSINE MATRIX')
      DO 240 J = 1, NFAC
240   WRITE (3,100) (A(J,K),K=1,NFAC)
      RETURN
      END
```

8.14 References

Cronbach, L. J., and G. C. Gleser: Assessing Similarities between Profiles, *Psychol. Bull.*, **50**:456–473 (1953).

Gollob, H. F.: Confounding of Sources of Variation in Factor-analytic Techniques, *Psychol. Bull.*, **70**:330–344 (1968).

Holzinger, K. J., and H. H. Harmon: "Factor Analysis," University of Chicago Press, Chicago, 1941.

Nunnally, J.: The Analysis of Profile Data, *Psychol. Bull.*, **59**:311–319 (1962).

Overall, J. E.: Note on Multivariate Methods for Profile Analysis, *Psychol. Bull.*, **61**:195–198 (1964).

———, and L. E. Hollister: Differential Drug Responses of Different Depressive Syndromes, in H. Brill (ed.), "Neuro-psychopharmacology: Proceedings of the Fifth International Congress of the Collegium Internationale Neuro-psycho-Pharmacologicum," Exerpta Medica, Amsterdam, 1967.

———, and ———: Studies of Quantitative Approaches to Psychiatric Classification, in M. Katz, J. O. Cole, and W. E. Barton (eds.), "The Role and Methodology of Classification in Psychiatry and Psychopathology," Government Printing Office, Washington, 1968.

———, ———, M. Johnson, and V. Pennington: Nosology of Depression and Differential Response to Drugs, *J. Amer. Med. Assoc.*, **195**:162–164 (1966).

———, ———, J. Shelton, I. Kimbell, Jr., and V. Pennington: Broad Spectrum Screening of Psychotherapeutic Drugs: Thiothixene as an Antipsychotic and Antidepressant, *Clin. Pharmacol. Therap.*, **10**:36–43 (1969).

Tucker, L. R.: Comments on "Confounding of Sources of Variation in Factor-analytic Techniques," *Psychol. Bull.*, **70**:345–354 (1968).

Vanderpool, J. P., J. E. Overall, and B. W. Henry: Self-improving Computer-based Outpatient Treatment Program, "Scientific Proceedings in Summary Form: 123*d* Annual Meeting of the APA," American Psychiatric Association, Washington, 1967.

PART THREE

DISCRIMINATION OF GROUPS

9 Discriminant Function for Two Groups

9.1 Introduction

It has been emphasized that the reduction of multiple measurements to a single weighted composite is the key to much of multivariate analysis. Nowhere is this truer than in discriminant-function analysis. By assigning appropriate weighting coefficients, several scores can be transformed to a single score which has maximum potential for distinguishing between members of two groups. In this manner, the multivariate problem is actually reduced to a simple univariate problem, and assignment of individuals between two groups depends upon the value of a single variable. Under appropriate assumptions, the new composite score can be conceived as having a normal distribution with estimable mean and variance for each group. Thus, one can utilize probability tables for the unit-normal distribution to determine probabilities of misclassification and the likelihood with which an individual case belongs to each group. Having reduced the several correlated measurements for each individual to a single weighted composite score, it is possible to determine a critical value or cutting point which will minimize errors of misclassification or which will yield known, but unequal, probabilities of error within the two groups. Under appropriate assumptions, tests of significance of multivariate mean differences can be applied to discriminant-analysis results.

The clinical problems in which simple discriminant function analysis is useful involve questions of differences between two groups of individuals and appropriate assignment of individuals between the groups. A familiar example would be the classification of patients as neurotic or psychotic on the basis of Minnesota Multiphasic Personality Inventory (MMPI) scores. For each individual, p different measurements are available. The several measurements can be correlated or uncorrelated, and completely different scale units may be involved. It is considered that the several measurements available for each individual are quantitative scores having (multivariate) normal distributions within the two populations.

9.2 Definitions

The solution to a discriminant-function problem involves determining the weight to be given each of the p original measurements in order that the resulting composite score will have maximum utility in distinguishing between members of the two groups. We begin by assuming that some unknown set of weighting coefficients exists which will define a composite score providing maximum discrimination between the two groups. The desired discriminant function is thus of the form

$$y = a_1x_1 + a_2x_2 + \cdots + a_px_p \tag{9.1}$$

where $a_1, a_2, \ldots, a_p$ are the weighting coefficients to be applied to the p original scores for each individual. The problem is to determine optimal values for the weighting coefficients such that the difference between mean scores for the two groups will be maximized relative to the variation within groups. This is equivalent to saying that weighting coefficients are to be derived such that the t statistic or F ratio between groups will be maximum. The function to be maximized, as first defined by R. A. Fisher, is the ratio of the between-groups variance to the within-groups variance

$$\begin{aligned} f(a_1, a_2, \ldots, a_p) &= \frac{n_1n_2}{n_1 + n_2} \frac{(a_1d_1 + a_2d_2 + \cdots + a_pd_p)^2}{\sum\sum c_{ij}a_ia_j} \\ &= \frac{n_1n_2}{n_1 + n_2} \frac{\mathbf{a}'\mathbf{dd}'\mathbf{a}}{\mathbf{a}'C\mathbf{a}} \end{aligned} \tag{9.2}$$

where $\mathbf{d}' = [d_1 \quad d_2 \quad \cdots \quad d_p]$ is the vector of mean differences on the p original measures and C is the within-groups covariance matrix. The above function is not used in calculation but simply defines the criterion function which is maximized when one computes optimal values for the unknown weighting coefficients, $a_1, a_2, \ldots, a_p$. The denominator of the criterion function $\sum\sum c_{ij}a_ia_j = \mathbf{a}'C\mathbf{a}$ is the within-groups variance of a linear combination in which the a_i are weighting coefficients. This quadratic form is one of the familiar forms discussed in Chap. 2. The numerator of the criterion function contains the square of the mean difference on the composite function and is thus proportional to the between-groups variance on that composite.

The problem is to solve for a set of weighting coefficients that will maximize the criterion function given that one has estimates of the mean vectors and within-groups variance-covariance matrix obtained from reasonably large samples. The general form of the solution is obtained by applying calculus to $f(a_i)$ and maximizing the function with regard to the a_i $(i = 1, 2, \ldots, p)$. This yields a set of p equations in p unknowns that can be solved simultaneously to obtain values for $a_1, a_2, \ldots, a_p$

$$\begin{aligned} a_1c_{11} + a_2c_{12} + \cdots + a_pc_{1p} &= d_1 \\ a_1c_{21} + a_2c_{22} + \cdots + a_pc_{2p} &= d_2 \\ \cdots\cdots\cdots\cdots\cdots\cdots\cdots\cdots \\ a_1c_{p1} + a_2c_{p2} + \cdots + a_pc_{pp} &= d_p \end{aligned}$$

In these equations, the c_{ij} are elements of the within-groups variance-covariance matrix among the p original measurements. In this solution, it is assumed that the variances and covariances are the same in the two populations from which the samples were drawn. The within-groups covariance matrix is computed by calculating separately the corrected sums of squares and cross products within each group, summing the two SP matrices, and then dividing by degrees of freedom $n_1 + n_2 - 2$, as described in Sec. 2.7.6. The vector of mean differences $[d_1 \quad d_2 \quad \cdots \quad d_p]$ is computed by calculating the means of the two groups on the original p measurements and then taking the differences between them.

9.3 Computing the Weighting Coefficients and Related Values

There are many ways to solve systems of simultaneous equations. If p is small, say 2 or 3, the equations can be solved directly by substitution and elimination, as described in elementary algebra texts. When p is larger than 3, direct solutions are not practical and matrix techniques become important. Recalling the row-by-column rule for matrix multiplication, the system of equations can be written in matrix form as

$$\begin{bmatrix} c_{11} & c_{12} & \cdots & c_{1p} \\ c_{21} & c_{22} & \cdots & c_{2p} \\ \cdots & \cdots & \cdots & \cdots \\ c_{p1} & c_{p2} & \cdots & c_{pp} \end{bmatrix} \begin{bmatrix} a_1 \\ a_2 \\ \cdot \\ a_p \end{bmatrix} = \begin{bmatrix} d_1 \\ d_2 \\ \cdot \\ d_p \end{bmatrix}$$

or in matrix notation

$$\mathbf{C}\mathbf{a} = \mathbf{d}$$

Premultiplication of both sides by C^{-1} yields the equation from which the vector $\mathbf{a}$ can be obtained by simple matrix-vector multiplication

$$\mathbf{a} = C^{-1}\mathbf{d} \tag{9.3}$$

where C^{-1} is the inverse of the within-groups variance-covariance matrix and $\mathbf{d}$ is the vector of mean differences.

The vector **a** contains weighting coefficients defining the most discriminating linear combination of the original x_i scores

$$y = a_1x_1 + a_2x_2 + \cdots + a_px_p \tag{9.1}$$

The mean value of the discriminant function for group I can be obtained by applying the weighting coefficients to the mean scores for group I on the original variables. The mean value for the discriminant function for group II can be obtained in similar manner

$$\bar{y}^{(1)} = a_1\bar{x}_1^{(1)} + a_2\bar{x}_2^{(1)} + \cdots + a_p\bar{x}_p^{(1)} \tag{9.4}$$

and

$$\bar{y}^{(2)} = a_1\bar{x}_1^{(2)} + a_2\bar{x}_2^{(2)} + \cdots + a_p\bar{x}_p^{(2)} \tag{9.5}$$

The variance of the discriminant function within each group, assumed to be identical as a result of the preliminary assumption of equal variance-covariance matrices for the original variables, is given by the quadratic form

$$V(y) = \mathbf{a}'C\mathbf{a} = \mathbf{a}'CC^{-1}C\mathbf{a} = \mathbf{d}'C^{-1}\mathbf{d} \tag{9.6}$$

If it is further assumed that the original x_i have a multivariate normal distribution within groups, it becomes apparent that the discriminant-function variate can be considered as having a normal distribution within groups, with mean values $\bar{y}^{(1)}$ and $\bar{y}^{(2)}$ and standard deviation $\sigma = \sqrt{\mathbf{d}'\mathbf{a}}$. This means that the deviation of an individual discriminant-function score from each of the group means can be regarded as a unit-normal deviate or z score

$$z_y = \frac{y - \bar{y}^{(i)}}{\sqrt{V(y)}} \tag{9.7}$$

Thus, for any particular discriminant-function score, say y_c, the z-score deviation from each group mean can be computed. The unit-normal distribution in Appendix A.1 can be used to obtain an estimate of the probability of a deviation from each group mean as large as that represented by any particular z-score value. A particular discriminant-function score value y_c can be accepted as a cutting point for classifying individuals in the two groups. The proportion of misclassifications can be read from the unit-normal distribution tables after transforming y_c to z-score form using Eq. (9.7). By varying the location of the cutting point, the proportion of misclassifications within the two groups can be changed.

Let y_c be a particular discriminant-function value falling between the two group means $\bar{y}^{(1)}$ and $\bar{y}^{(2)}$ as shown in Fig. 9.1. If every individual having a discriminant-function value less than y_c were classified into group I, the proportion of individuals actually belonging to group II who would be *misclassified* by being assigned to group I would be represented by the area under G_2 to the left of y_c. To determine this proportion, transform y_c to z-score form

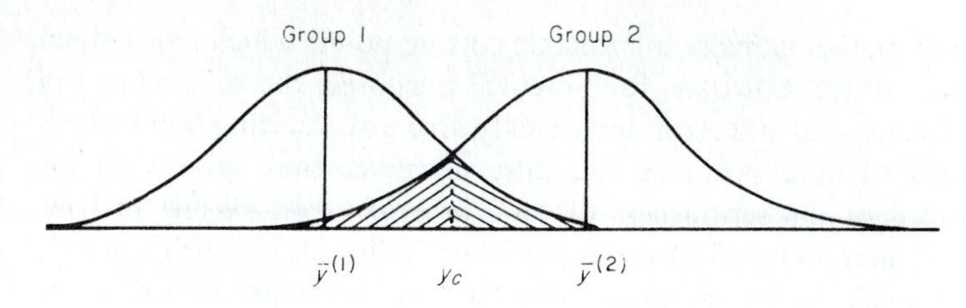

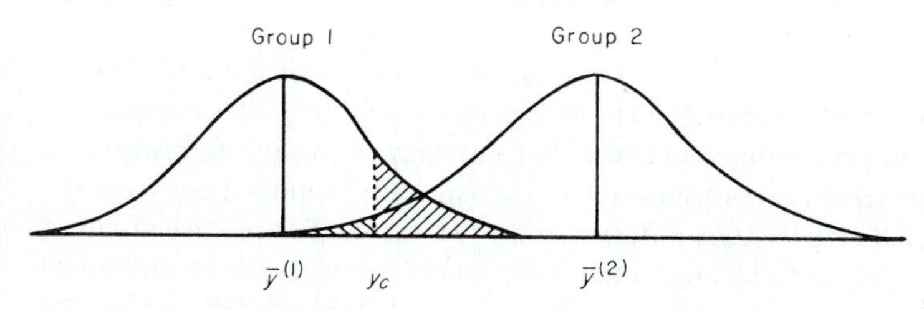

Figure 9.1 Schematic diagram of distributions of discriminant-function scores in two groups showing errors of misclassifications.

using Eq. (9.7)

$$z_y = \frac{y_c - \bar{y}^{(2)}}{\sqrt{V(y)}}$$

and then look up the area in the smaller portion of the unit-normal curve corresponding to z_y. This will provide an estimate of the proportion of individuals from group II who would be incorrectly classified as belonging to group I. If $p_\varepsilon^{(2)}$ represents the probability of misclassification for individuals in group II, then the probability of correctly classifying an individual from group II by using y_c as the cutting point will be $1 - p_\varepsilon^{(2)}$, or the area in the larger portion of the curve corresponding to $z_y = (y_c - \bar{y}^{(2)})/\sqrt{V(y)}$.

In a similar manner the probability of misclassifying an individual from group I using the cutting y_c can be calculated. Transform y_c to a z-score deviation about the mean for group I

$$z_y = \frac{y_c - \bar{y}^{(1)}}{\sqrt{V(y)}}$$

Look up z_y in Appendix A.1. The probability of misclassifying an individual from group I using y_c as a cutting score is the area under the smaller portion of the curve G_1; call this probability of misclassification $p_\varepsilon^{(1)}$. The probability of correctly classifying an individual from group I is equal to the area under the larger portion of the curve, or $1 - p_\varepsilon^{(1)}$.

It is not difficult mathematically to calculate cutting points which are optimal in one sense or another; however, for practical purposes, the researcher can vary the value of y_c by trial and error and quickly arrive at a cutting point which yields proportions of false positives and false negatives that agree with his subjective appraisal of the seriousness of each type of error. While in later sections of this chapter some of these mathematical solutions will be explored, for practical purposes, the writers suggest the trial-and-error procedure. What is important is that optimal discriminant-function weighting coefficients can be obtained, the mean values and variances of the derived composite can be estimated, and probabilities of error in classification for members of each group can be evaluated appropriately.

It is not always the best solution to choose a cutting point for classification which equates the probabilities of error in assignment of individuals between the two groups. The probabilities of error that one may be willing to accept will depend upon the relative seriousness of misclassification of individuals from the two groups, which may be quite different in the two cases. The relative numbers of individuals expected to belong to the two *populations* may also be an important consideration, since the actual numbers of individuals misclassified will be equal to the probabilities of misclassification times the relative numbers in the two populations. Such considerations tend to be highly subjective, so that an estimate of the proportions correctly and incorrectly classified from each group should be available for any particular cutting point chosen by an investigator.

A useful mechanism for arriving at an appropriate cutting point for assigning individuals between two groups is to construct frequency histograms of the discriminant-function scores for available cases in each group. Unless the samples are quite large, it is unnecessary to attempt to place cutting points so as to absolutely minimize errors of classification in the particular *samples*, because this involves too much emphasis on a few extreme cases. The general shape of the distributions can be appreciated from the sample distributions and an appropriate cutting point selected.

9.4 A Computational Example

Before proceeding to consider available tests of significance and other problems related to discriminant-function analysis, it should be worthwhile to examine step by step the calculation of optimal discriminant-function coefficients and selection of a cutting point which tends to minimize errors of classification in a problem of some psychological and psychiatric significance. In a series of clinical drug studies conducted in five Veterans Administration hospitals, patients were selected by psychiatrists to be placed in two different types of studies: investigations of efficacy of major tranquilizers in treatment of "newly admitted schizophrenics" and investigations of efficacy of antidepressant (tricyclic) drugs in treatment of "depression." Patients were placed in one type of study or the other depending on psychiatrists' clinical judgments concerning major therapeutic requirement. We shall refer to the two groups as "depressive" and "schizophrenics."

Each patient was interviewed soon after admission by two experienced professionals, psychiatrists or psychologists, and symptom profiles were recorded using the BPRS. The question of current interest is the extent to which patients assigned to the two kinds of drug studies (depressive and schizophrenics) can be distinguished on the basis of four factor scores derived from the BPRS ratings. These four factors are described as thinking disturbance, withdrawal-retardation, hostile-suspiciousness, and anxious depression. Assuming that discrimination is possible, do the results suggest that effective assignment between the two types of treatments could be made on the basis of symptom ratings alone?

A total of 480 rating profiles were available for depressive patients and 999 for schizophrenic patients. Mean scores on the four higher-order symptom factors are presented in Table 9.1 for the two groups. The pooled within-groups covariance matrix, estimated on $n_1 + n_2 - 2 = 1{,}477$ degrees of freedom, is also presented in Table 9.1. This within-groups variance-covariance matrix was computed as described in Sec. 2.7.6 by first computing separately the SP matrix for each group, adding the two matrices, and then dividing by degrees of freedom. The inverse of the pooled within-groups variance-covariance

Table 9.1 Preliminary Statistics for Discriminant-function Analysis between Depressive and Schizophrenic Groups
$n_1 = 480$, $n_2 = 999$

MEAN SCORES ON FOUR SYMPTOM FACTORS

	Depression	*Schizophrenia*	*Difference*
Thinking disturbance	2.7166	7.4434	−4.7268
Withdrawal-retardation	5.0104	5.9599	−.9495
Hostile-suspiciousness	2.9625	5.4584	−2.4959
Anxious depression	10.0395	5.1991	4.8404

POOLED WITHIN-GROUPS COVARIANCE MATRIX

$df = 1{,}477$	*TD*	*W-R*	*H-S*	*AD*
Thinking disturbance	14.1016	8.2064	2.9642	−1.5214
Withdrawal-retardation	8.2064	14.6254	1.6509	.1900
Hostile-suspiciousness	2.9642	1.6509	15.0074	1.5341
Anxious depression	−1.5214	.1900	1.5341	14.2972

INVERSE OF POOLED WITHIN-GROUPS COVARIANCE MATRIX
FOR DEPRESSIVE AND SCHIZOPHRENIC GROUPS

	TD	*W-R*	*H-S*	*AD*
Thinking disturbance	.1113	−.0607	−.0167	.0144
Withdrawal-retardation	−.0607	.1024	.0015	−.0079
Hostile-suspiciousness	−.0167	.0015	.0707	−.0093
Anxious depression	.0144	−.0079	−.0093	.0725

matrix (Table 9.1) was computed by the square-root method described in Chap. 2.

The optimal weighting coefficients providing maximum discrimination between the clinically depressed and clinically schizophrenic groups can be obtained using Eq. (9.3) by multiplying the inverse of the covariance matrix by the vector of mean differences. As will be recalled from the row-by-column rule for matrix multiplication, this is equivalent to setting each unknown a_i equal to a weighted sum of the elements in one row of the inverse matrix, each element being weighted by the appropriate mean difference

$$a_i = d_1 c^{i1} + d_2 c^{i2} + \cdots + d_j c^{ij} + \cdots + d_p c^{ip}$$

where c^{ij} is the element in the ith row and jth column of the matrix inverse and d_j is the difference between mean values for the two groups on the jth variable

$$a_1 = -4.7268(.1113) + -.9495(-.0607) + -2.4959(-.0167) + 4.8404(.0144)$$

$$a_2 = -4.7268(-.0607) + -.9495(.1024) + -2.4959(.0015) + 4.8404(-.0079)$$

$$a_3 = -4.7268(-.0167) + -.9495(.0015) + -2.4959(.0707) + 4.8404(-.0093)$$

$$a_4 = -4.7268(.0144) + -.9495(-.0079) + -2.4959(-.0093) + 4.8404(-.0725)$$

Solution of these equations yields the values

$$a_1 = -.3568 \qquad a_2 = .1474 \qquad a_3 = -.1441 \qquad a_4 = .3140$$

These are the weighting coefficients which should be applied to the four symptom factor scores in order to obtain a single composite score having maximum average difference in value for clinically depressed and clinically schizophrenic patients. It will be noted immediately that the distinction between the two clinical groups depends most strongly on the contrast or difference between only two of the symptom factors, i.e., thinking disturbance and anxious depression. The factors of withdrawal-retardation and hostile-suspiciousness are assigned relatively little weight. The weights should be interpreted relative to the within-groups standard deviations for the particular variables, but in this case the relative similarities of the σ make consideration of them less critical.

Mean discriminant-function values for the depressive and schizophrenic groups can be computed by applying the weighting coefficients to the group means on the original symptom factor scores as previously shown in Eqs. (9.4) and (9.5)

$$\bar{y}^{(1)} = -.3568(2.7166) + .1474(5.0104) - .1441(2.9625) + .3140(10.0395) = 2.4948$$

$$\bar{y}^{(2)} = -.3568(7.4434) + .1474(5.9599) - .1441(5.4584) + .3140(5.1991) = -.9314$$

The variance of the weighted composite within the two diagnostic groups, assuming equal variance-covariance matrices, can be obtained most easily by applying the weighting coefficients to the vector of mean differences as given in Eq. (9.6)

$$V(y) = \mathbf{d}'C^{-1}\mathbf{d} = \mathbf{d}'\mathbf{a} = 3.4267$$

It will be noted that the within-groups variance of the discriminant function is precisely equal to the mean difference $\bar{y}^{(1)} - \bar{y}^{(2)}$. Thus, separate calculation of $V(y)$ is not necessary if the group means on the discriminant function have already been calculated.

Appropriate multivariate tests of significance of mean differences will be discussed in the next section; however, an *approximate* univariate test can be used to judge quickly the effectiveness of discrimination. Remembering that the discriminant-function analysis has reduced the p original variables to a single weighted score which can be considered to be normally distributed with variance $V(y)$ within groups, we can express the mean difference $\bar{y}^{(1)} - \bar{y}^{(2)}$ relative to the standard error of the mean difference and refer the derived statistic to the table of the normal curve in Appendix A.1. The approximate standard error of mean difference is

$$\sigma_{\bar{y}_1 - \bar{y}_2} = \sqrt{\frac{V(y)}{n_1} + \frac{V(y)}{n_2}} \tag{9.8}$$

where n_1 and n_2 are sample sizes. For purposes of quick evaluation, the statistic

$$\tilde{z} = \frac{\bar{y}^{(1)} - \bar{y}^{(2)}}{\sqrt{V(y)/n_1 + V(y)/n_2}} \tag{9.9}$$

can be treated as a z-score unit-normal deviate. The probability of a chance (plus or minus) deviation as great as the observed can be read from the table for areas under the normal curve. This probability is equal to twice the area in the smaller portion of the curve, since the z score can be either positive or negative depending only on which group is called group I. In the present example,

$$\tilde{z} = \frac{3.4267}{\sqrt{.007139 + .003430}} = \frac{3.4267}{\sqrt{.010569}} = 33.33$$

Since z-score values of this magnitude have such low probability under the null hypothesis that they are not included in standard normal-probability tables, it is to be expected on the basis of this quick approximate test that a true symptom-profile difference does exist between the depressive and schizophrenic groups.

It should be stressed that the simple z-score transformation provides an extremely crude approximate test subject to considerable bias. No cognizance is taken of the number of variables which have been combined in the attempt to achieve maximum separation between group means. The value of $\tilde{z}$ tends to be inflated as the number of variables included in the discriminant function is increased. The degree of inflation is approximately $.25(p - 1)$. Thus, one can take $t = \tilde{z} - .25(p - 1)$ as a t statistic with $n = n_1 + n_2 - p - 1$ degrees of freedom. There is actually little reason to consider such an approximate test, however, since an exact F test is available. The primary motivation for discussing the z-score transformation here has been to emphasize that the discriminant-function analysis actually transforms the original multivariate problem into a simple univariate problem. The function which is maximized in deriving the optimal weights is precisely proportional to the between-groups z score as defined in Eq. (9.9). The discriminant-function weights are computed to yield a composite variable having maximum mean difference relative to variability within groups.

More adequate statistical tests of significance for differences in mean discriminant-function scores are readily available and are, in fact, computationally simple. The within-groups variance $V(y) = \mathbf{d}'C^{-1}\mathbf{d}$ can be used as the Mahalanobis D^2, which can be related to the F distribution, under the assumption that the several original measurements have a multivariate normal distribution within the populations from which samples were drawn and that the variance-covariance matrices are equal for the two populations. These assumptions are obviously multivariate generalizations of the usual parametric assumptions of normality and homogeneity of variance in the univariate analysis of variance. Given these assumptions,

$$F = \frac{n_1 n_2 (n_1 + n_2 - p - 1)}{p(n_1 + n_2)(n_1 + n_2 - 2)} D^2 \tag{9.10}$$

where $D^2 = V(y)$, n_1 and n_2 are sample sizes, and p is the number of variables entering into the discriminant function. The statistic F can be referred to probability tables for the F distribution with p and $n_1 + n_2 - p - 1$ degrees of freedom (Table A.2).

A question of more concern than mere statistical significance in most discriminant-function analyses is the utility of the derived function for assignment of individuals between the two groups. In this discussion, it will be assumed that each individual is certain to belong to one of the two groups; no other alternatives are available. For example, every patient is assumed to be either depressed or schizophrenic. The utility of the discriminant function can then be judged by assessing the probabilities of misclassification.

If there is no basis for assuming that one group should contain substantially more individuals than the other, and if the seriousness of errors of misclassification are about equal, the logical cutting point lies halfway between the two group means $\bar{y}^{(1)}$ and $\bar{y}^{(2)}$. Defining the cutting point as

$$y_c = \frac{\bar{y}^{(1)} + \bar{y}^{(2)}}{2} \tag{9.11}$$

a discriminant-function value is defined above which individuals should be assigned to group II and below which individuals should be assigned to group I (or vice versa if the mean discriminant-function score is greater for group I). It is convenient to add the constant $-y_c$ to the discriminant-function equation. The decision rule then becomes: assign to group II if the individual's discriminant score is positive and to group I if negative (or vice versa depending on which group has the largest discriminant-function mean). This is known as a *maximum-likelihood procedure* because each individual is assigned to the group in which the occurrence of his particular discriminant score is most likely.

In the problem of discrimination between clinically depressed and schizophrenic cases, the discriminant-function value lying halfway between the two group means is

$$y_c = \frac{\bar{y}^{(1)} + \bar{y}^{(2)}}{2} = .7817$$

Since the within-groups variance of the discriminant scores was estimated to be $V(y) = 3.4267$, the probability of misclassification can be obtained by computing the z score of y_c as follows:

$$z^{(1)} = \frac{y_c - \bar{y}^{(1)}}{\sqrt{V(y)}} = -.92 \qquad \text{and} \qquad z^{(2)} = \frac{y_c - \bar{y}^{(2)}}{\sqrt{V(y)}} = .92$$

Since the cutting point y_c was calculated to lie equidistant between $\bar{y}^{(1)}$ and $\bar{y}^{(2)}$, the two z-score values are identical except for sign. Reference to Table A.1 for the unit-normal distribution indicates that 17.88 percent of the area is located in the smaller portion of the curve. Thus, the probability of misclassification is $p_\varepsilon = .1788$ for members of each of the two clinical groups. The term "misclassification" in this context means that assignment of individuals to depressive or schizophrenic diagnostic groups on the basis of symptom ratings should be expected to disagree with clinical diagnostic classification made by psychiatrists in approximately 18 percent of the cases. It might be of interest to examine further the therapeutic responses of the cases for which discriminant-function classification did not agree with clinical classification.

9.5 Estimating Probabilities of Error in Assignment of Individuals

While it is important in the development of a discriminant-function classification procedure to evaluate the long-run probabilities of misclassification for all members of each group, it is recognized that such probabilities are not uniform for all possible discriminant-function scores. Individuals whose discriminant-function scores lie near the selected cutting point are more likely to be placed in the wrong group than those having extremely large or small discriminant-score values. Therefore, it becomes a matter of considerable interest to estimate the probability of misclassification for each individual.

Even without the theory and mathematics of probability-density functions, it can be appreciated from examination of Fig. 9.1 that the probability that an individual selected at random will belong to group I is proportional to the height of the normal curve G_1 at a point coinciding with his discriminant score. Similarly, the probability of the individual's belonging to group II is proportional to the height of G_2 at the same point. Thus, the relative probabilities of belonging to groups I or II are a function of the relative heights (densities) of the two normal curves at the point corresponding to the individual's discriminant-function score.

Mathematically, the height, or ordinate value, of a normal probability distribution is given by

$$f(x) = \frac{1}{\sqrt{2\pi\sigma^2}} e^{-(x-\mu)^2/2\sigma^2} \tag{9.12}$$

Fortunately, the normal-probability-density function has been evaluated for a wide range of z-score values, and ordinates are presented in Table A.1. The equation for the normal-probability-density function is introduced now because it will be required in later chapters dealing with more complex multivariate distribution problems and it is worthwhile to recognize the similarity of the table look-up procedure described here to the more complex calculations described later. In each case, the probability-density function $f(y)$ is related to the probability of occurrence of a particular score value (or set of values) in some particular theoretical distribution.

Let $f(y^{(1)})$ represent the probability density, or the ordinate, of the normal curve for a particular discriminant-function score value y in group I. Let $f(y^{(2)})$ represent the corresponding value of the probability density for the particular discriminant-function score value y in group II. If the total number of individuals in group I is equal to the total number of individuals in group II, the probability of belonging to group I for an individual having the particular score y can be estimated from the ratio

$$p^{(1)} = \frac{f(y^{(1)})}{f(y^{(1)}) + f(y^{(2)})} \tag{9.13}$$

In the case of the unit-normal distribution, the z-score deviation of the individual discriminant-function score y from each group mean, $\bar{y}^{(1)}$ and $\bar{y}^{(2)}$, can be calculated, and the associated $f(y^{(i)})$ values can be found as ordinates of the normal curve in Table A.1.

Returning to the problem of assigning individuals to depressive or schizophrenic diagnostic groups on the basis of four symptom factor scores, it will be recalled that the best linear discriminant function was computed to be

$$y = -.3568x_1 + .1474x_2 - .1441x_3 + .3140x_4$$

Mean discriminant-function scores for the two diagnostic groups were $\bar{y}^{(1)} = 2.4948$ and $\bar{y}^{(2)} = -.9314$. The within-groups variance of the discriminant-function scores was $V(y) = 3.4267$.

Suppose there are three individual cases to assign. Their symptom factor scores x_1, x_2, x_3, and x_4 are as follows:

Patient	*Thinking disturbance*	*Withdrawal-retardation*	*Hostile-suspiciousness*	*Anxious depression*
A	8	6	2	3
B	6	2	4	9
C	1	3	4	9

The discriminant-function scores computed by applying appropriate weights to the symptom factor scores for the three patients are

$$y_a = -.3568(8) + .1474(6) - .1441(2) + .3140(3) = -1.32$$
$$y_b = -.3568(6) + .1474(2) - .1441(4) + .3140(9) = .40$$
$$y_c = -.3568(1) + .1474(3) - .1441(4) + .3140(9) = 2.34$$

The simple discriminant-function scores are transformed to z-score deviations about the two diagnostic group means

$$z_a^{(1)} = \frac{y_a - \bar{y}^{(1)}}{\sqrt{V(y)}} = \frac{-3.81}{\sqrt{3.4267}} = 2.06$$

$$z_a^{(2)} = \frac{y_a - \bar{y}^{(2)}}{\sqrt{V(y)}} = \frac{-.39}{\sqrt{3.4267}} = -.21$$

$$z_b^{(1)} = \frac{y_b - \bar{y}^{(1)}}{\sqrt{V(y)}} = \frac{-2.09}{\sqrt{3.4267}} = -1.13$$

$$z_b^{(2)} = \frac{y_b - \bar{y}^{(2)}}{\sqrt{V(y)}} = \frac{1.33}{\sqrt{3.4267}} = .72$$

$$z_c^{(1)} = \frac{y_c - \bar{y}^{(1)}}{\sqrt{V(y)}} = \frac{-.15}{\sqrt{3.4267}} = -.08$$

$$z_c^{(2)} = \frac{y_c - \bar{y}^{(2)}}{\sqrt{V(y)}} = \frac{3.26}{\sqrt{3.4267}} = 1.76$$

Having obtained z-score deviations of the individual discriminant-function scores about the two group means, it is a simple matter to look up the associated probability-density values $f(y^{(i)})$ as ordinates of the normal curve using Table A.1. The ratios of these ordinate values provide the required estimates of probability of the individual's belonging to each diagnostic group.

Patient A: $f(y_a^{(1)}) = .0478$ and $f(y_a^{(2)}) = .3902$

$$p^{(1)} = \frac{f(y_a^{(1)})}{f(y_a^{(1)}) + f(y_a^{(2)})} = .11$$

$$p^{(2)} = \frac{f(y_a^{(2)})}{f(y_a^{(1)}) + f(y_a^{(2)})} = .89$$

Thus, for patient A, the probability of belonging to group I (depression) is estimated to be $p^{(1)} = .11$, and the probability of belonging to group II (schizophrenic) is estimated to be $p^{(2)} = .89$. Following the rule of assigning the patient to the more probable group, the probability that he will be incorrectly assigned is $p_\varepsilon = p^{(1)} = .11$.

Patient B: $f(y_b^{(1)}) = .2107$ and $f(y_b^{(2)}) = .3079$

$$p^{(1)} = \frac{f(y_b^{(1)})}{f(y_b^{(1)}) + f(y_b^{(2)})} = .41$$

$$p^{(2)} = 1.0 - p^{(1)} = .59$$

Assigning patient B to the more probable alternative yields an estimated probability of incorrect classification of $p_\varepsilon = p^{(1)} = .41$. Patient B is a difficult case to assign, and one would not feel very confident about the correctness of the decision made on the basis of symptom evaluation.

Patient C: $f(y_c^{(1)}) = .3977$ and $f(y_c^{(2)}) = .0848$

$$p^{(1)} = .82$$

$$p^{(2)} = 1.0 - p^{(1)} = .18$$

The probability that patient C belongs in the depressive group is clearly higher than the probability that he belongs in the schizophrenic group. Assigning patient C to the depressive group results in an estimated probability of error of $p_\varepsilon = .18$.

These three examples were contrived to illustrate that the discriminant-function approach to classification of individuals can lead to results having widely different probabilities of error. For some individual cases, the decision can be made with confidence, while in other cases the best decision will be subject to great uncertainty. The probabilities of error can be evaluated for each individual case.

An alternative decision rule is to permit one of three decisions: (1) classify the individual into group I if the probability of his belonging to group II is smaller than some prespecified value, say $p_\varepsilon^{(2)} < .10$, (2) classify the individual into group II if the probability of his belonging to group I is smaller than some prespecified value, say $p_\varepsilon^{(1)} < .5$, or (3) refrain from making a definite classification pending examination of additional information. This approach appears especially reasonable in clinical diagnostic decision making, where the expense of additional tests, for example, means that they should be employed only if necessary.

9.6 Bayesian Considerations in Classification Procedures

Up to this point it has been tacitly assumed that there was no basis for expecting that more individuals should actually belong to one group than to the other. The relative frequencies of occurrence of individuals in the two populations has little or no implication for computation of the best linear discriminant function so long as group means are based on a reasonable n. Frequently, solutions for discriminant-function coefficients are based on two samples of approximately equal size although the two populations from which samples were drawn may contain very different numbers of individuals. However, when it comes to use of the discriminant function for classification of individuals, the relative frequencies in the two *populations* become important considerations. This is the familiar base-rate problem.

In deriving optimal discriminant functions, it is good practice to obtain a substantial sample from each population to ensure stable estimates of mean values. Frequently, an approximately equal number of individuals from each of two populations is selected. For example, in the study of suicide proneness, 100 suicides and 100 controls might be selected for use in developing a discriminant function. It has been found repeatedly that significant discrimination is possible in such studies. A problem arises when the derived function is used for identification of potential suicides in a population where there are 100 nonsuicides for every suicide. For any realistic value of the discriminant function, the probability will be greater that a randomly selected individual will not become a suicide!

Let π_1 be the proportion of individuals in the population from which group I was drawn, and let π_2 be the proportion of individuals in the population from which group II was drawn. Then to estimate the probability that an individual selected at random belongs to group I, consider both the probability of his score pattern in the distribution of scores for group I *and* the probability that he belongs to group I regardless of score pattern. The probability of belonging in each of the two groups is a joint function of the base rates and pattern probabilities

$$p^{(1)} = \frac{\pi_1 f(y^{(1)})}{\pi_1 f(y^{(1)}) + \pi_2 f(y^{(2)})} \tag{9.14}$$

and

$$p^{(2)} = \frac{\pi_2 f(y^{(2)})}{\pi_1 f(y^{(1)}) + \pi_2 f(y^{(2)})} \tag{9.15}$$

where $\pi_1 + \pi_2 = 1.0$ are the relative proportions in populations I and II, respectively.

It will be recognized that in this formulation the problem of estimating $p^{(1)}$ and $p^{(2)}$ is identical to Eq. (9.13), except that the probability densities (normal-curve ordinate values) have each been weighted by the corresponding base rate or a priori probability.

To illustrate the importance of considering base rates in clinical classification problems, a diagnostic-classification problem involving two populations that are expected to contain different numbers of patients will be examined. This problem also provides an opportunity to examine the utility of discriminant-function methods where the original scores do not actually have the assumed multivariate normal distributions. This will permit an examination of practical ways of evaluating the utility of a derived function and of choosing a critical cutting point by examination of the distributions of discriminant-function scores in the two sample groups. Such an approach is a practical expedient that should not be overlooked when distribution assumptions are in doubt.

In a series of parallel studies conducted in France, Germany, Czechoslovakia, and Italy, psychiatrists were asked to provide BPRS profiles descriptive of patients belonging to various diagnostic groups. Some of the resulting data will be used to examine the feasibility of discriminating between patients diagnosed as paranoid or paranoid schizophrenic and those diagnosed as manic. This is a clinical problem of practical significance because paranoid-type patients are frequently active, tense, agitated, and grandiose—as the manics are—and in certain cases a diagnostic decision may be difficult.

Although the available data provided by psychiatrists in the four countries included approximately one-half as many "typical" manic profiles as "typical" paranoid-type profiles, it will be assumed that in the general clinical population paranoid-type patients (including paranoid schizophrenics) outnumber true manics by approximately 9 to 1. Accordingly, we shall set $\pi_1 = .1$ and $\pi_2 = .9$. The two actual sample sizes were $n_1 = 363$ and $n_2 = 751$, frequencies which bear no necessary relationship to the estimated population proportions π_1 and π_2.

The data consisted of BPRS symptom-rating profiles for prototypical cases belonging to the two populations and did not include difficult to classify atypical cases. It is reasoned that if a valid discriminant function can be derived to distinguish effectively between typical cases, that same discriminant function should have utility for classifying atypical cases.

Rather than relying on the four higher-order factor scores, which do tend to be normally distributed within groups, it was felt necessary to use the detailed 16-variable rating profiles for this problem. In spite of the fact that the individual 7-point ratings are recognized not to be normally distributed within different diagnostic groups, the detailed symptom description was considered necessary for discrimination. This fact will also provide the opportunity to examine the distribution of composite discriminant-function scores in a problem

Table 9.2 Within-groups Covariance Matrix among 16 BPRS Symptom Constructs for Manic-Paranoid Problem

	1	*2*	*3*	*4*	*5*	*6*	*7*	*8*	*9*	*10*	*11*	*12*	*13*	*14*	*15*	*16*
1	1.916	.723	−.051	−.167	.357	.168	−.028	.412	.463	.316	.489	.163	.015	.082	−.023	−.096
2	.723	2.183	.258	.051	.462	.688	.064	−.224	.508	.442	.724	.337	.144	.197	.244	−.214
3	−.051	.258	2.572	.970	.254	.436	.986	−.626	.215	.251	−.004	.312	.373	.948	.410	1.125
4	−.167	.051	.970	3.703	.270	.374	1.731	−.501	.008	−.435	−.683	1.483	.443	.351	1.505	1.043
5	.357	.462	.254	.270	1.074	.229	.204	−.088	.452	−.057	.055	.227	.244	.022	.233	.065
6	.168	.688	.436	.374	.229	2.580	.731	−.349	.249	.458	.222	−.002	.129	.517	.131	.112
7	−.028	.064	.986	1.731	.204	.731	3.817	−.514	.087	−.002	−.580	.898	.462	.640	1.270	1.150
8	.412	−.224	−.626	−.501	−.088	−.349	−.514	3.223	−.004	.665	.548	−.112	−.457	−.139	.218	−.231
9	.463	.508	.215	.008	.452	.249	.087	−.004	1.109	.092	.162	.095	.246	.179	.061	.186
10	.316	.442	.251	−.435	−.057	.458	−.002	.665	.092	2.715	1.500	−.322	−.247	1.086	−.266	.091
11	.489	.724	−.004	−.683	.055	.222	−.580	.548	.162	1.500	2.373	.015	−.187	.493	−.113	−.201
12	.163	.337	.312	1.483	.227	−.002	.898	−.112	.095	−.322	.015	3.644	.303	−.009	1.411	.500
13	.015	.144	.373	.443	.244	.129	.462	−.457	.246	−.247	−.187	.303	1.204	.237	.189	.486
14	.082	.197	.948	.351	.022	.517	.640	−.139	.179	1.086	.493	−.009	.237	2.848	−.011	.589
15	−.023	.244	.410	1.505	.233	.131	1.270	.218	.061	−.266	−.113	1.411	.189	−.011	3.336	.676
16	−.096	−.214	1.125	1.043	.065	.112	1.150	−.231	.186	.091	−.201	.500	.486	.589	.676	2.878

where multivariate normal-distribution assumptions with regard to the original variables do not hold. As will be seen, the composite scores tend to be normally distributed in spite of lack of multivariate normality of the original scores.

The 16 $\times$ 16 within-groups variance-covariance matrix is presented in Table 9.2. Mean vectors for manic and paranoid groups are presented in Table 9.3, and the inverse of the within-groups covariance matrix is presented in Table 9.4. These are the basic sample statistics required to complete the discriminant-function analysis. The discriminant-function weighting coefficients obtained by multiplying the matrix inverse by the vector of mean differences, $\mathbf{a} = C^{-1}\mathbf{d}$, are as follows:

$$y = -.405x_1 - .377x_2 - .576x_3 + .342x_4 + .134x_5 + .909x_6 + .208x_7 + .794x_8 - .394x_9 + .024x_{10} - 1.375x_{11} - .537x_{12} - .287x_{13} - .017x_{14} - .400x_{15} - .361x_{16}$$

where $x_1, x_2, \ldots, x_{16}$ are the 16 BPRS rating variables in the order listed in Chap. 1.

Mean scores on the composite discriminant-function variable for the two groups were calculated to be

$$\text{Manic } \bar{y}^{(1)} = 2.78 \qquad \text{and} \qquad \text{paranoid } \bar{y}^{(2)} = -10.04$$

The within-groups variance of the derived function should be $V(y) = \bar{y}^{(1)} - \bar{y}^{(2)} = 12.82$. For purposes of evaluating probabilities of misclassification using the previously described procedures for statistical inference, it is important that

Table 9.3 Mean Vectors among 16 BPRS Symptom Constructs for Manic-Paranoid Problem

	Manic	*Paranoid*
1	1.3691	2.8215
2	1.4820	3.6378
3	1.9834	4.0053
4	4.1322	4.0199
5	1.1845	1.7976
6	5.2286	3.9587
7	3.6528	3.3515
8	6.1074	4.4234
9	1.1487	2.2210
10	3.4462	4.8308
11	2.2424	5.5539
12	1.6969	4.2902
13	1.0440	1.8934
14	3.0275	4.1065
15	3.6969	5.1784
16	1.9586	3.4873

Table 9.4 Inverse of Within-groups Covariance Matrix for Manic-Paranoid Problem

	1	*2*	*3*	*4*	*5*	*6*	*7*	*8*	*9*	*10*	*11*	*12*	*13*	*14*	*15*	*16*
1	.681	−.165	.031	.015	−.105	.016	−.031	−.089	−.170	.008	−.061	−.025	.022	−.003	.042	−.002
2	−.165	.702	−.043	−.003	−.117	−.124	.018	.085	−.152	−.027	−.157	−.035	−.028	.025	−.052	.089
3	.031	−.043	.578	−.071	−.078	−.013	−.038	.073	−.017	−.014	−.014	.018	−.008	−.125	.002	−.155
4	.015	−.003	−.071	.460	−.054	−.026	−.088	.015	.057	.010	.093	−.110	−.023	−.020	−.101	−.047
5	−.105	−.117	−.078	−.054	1.250	−.025	−.005	−.006	−.379	.038	.000	−.003	−.133	.048	−.035	.057
6	.016	−.124	−.013	−.026	−.025	.473	−.080	.033	−.037	−.056	.004	.033	.010	−.033	.011	.024
7	−.031	.018	−.038	−.088	−.005	−.080	.412	.029	.023	−.065	.111	−.021	−.044	−.033	−.090	−.065
8	−.089	.085	.073	.015	−.006	.033	.029	.395	−.027	−.099	−.026	.003	.092	.016	−.073	−.011
9	−.170	−.152	−.017	.057	−.379	−.037	.023	−.027	1.254	.016	.016	.006	−.145	−.041	.009	−.083
10	.008	−.027	−.014	.010	.038	−.056	−.065	−.099	.016	.719	−.389	.050	.102	−.185	.053	−.032
11	−.061	−.157	−.014	.093	.000	.004	.111	−.026	.016	−.389	.787	−.068	.001	−.011	−.046	.002
12	−.025	−.035	.018	−.110	−.003	.033	−.021	.003	.006	.050	−.068	.373	−.033	.007	−.096	−.008
13	.022	−.028	−.008	−.023	−.133	.010	−.044	.092	−.145	.102	.001	−.033	1.039	−.072	.022	−.116
14	−.003	.025	−.125	−.020	.048	−.033	−.033	.016	−.041	−.185	−.011	.007	−.072	.491	.023	−.015
15	.042	−.052	.002	−.101	−.035	.011	−.090	−.073	.009	.053	−.046	−.096	.022	.023	.440	−.037
16	−.002	.089	−.155	−.047	.057	.024	−.065	−.011	−.083	−.032	.002	−.008	−.116	−.015	−.037	.494

this estimate of within-groups variance be adequate and that the composite function have a normal distribution within groups. If the normality assumption holds, an equal probability of misclassification for patients within *each* population separately should result from selection of a cutting point lying halfway between the two mean values

$$y_c = \frac{\bar{y}^{(1)} + \bar{y}^{(2)}}{2} = -3.63$$

The z-score deviation of this cutting point from the group means should be

$$z^{(1)} = \frac{y_c - \bar{y}^{(1)}}{\sqrt{V(y)}} = \frac{-6.41}{3.58} = -1.79$$

and

$$z^{(2)} = \frac{y_c - \bar{y}^{(2)}}{\sqrt{V(y)}} = \frac{6.41}{3.58} = 1.79$$

Reference to the unit-normal distribution (Table A.1) indicates 3.67 percent of the area in the smaller portion of the curve beyond $z = 1.79$. The estimated probability of misclassification, which is equal for the two groups, is $p_\varepsilon = .0367$.

If it is assumed that the true population proportions are $\pi_1 = 0.1$ and $\pi_2 = 0.9$, the total number of errors of misclassification will not be minimum using a cutting point which results in an equal proportion of misclassifications in each group. This is intuitively obvious since the number of paranoid patients misclassified will be nine times the number of manic patients misclassified. The total number of misclassifications can be minimized using a cutting point calculated as follows:

$$y_c = \frac{\bar{y}^{(1)} + \bar{y}^{(2)}}{2} + \log_e \pi_2 - \log_e \pi_1 \tag{9.16}$$

where $\bar{y}^{(1)}$ is the larger of the two discriminant-function means and π_1 is the population ratio associated with the group having largest mean value

$$\begin{aligned} y_c &= -3.63 + \log_e .9 - \log_e .1 \\ &= -3.63 - .1054 + 2.3026 = -1.4328 \end{aligned}$$

Use of the revised cutting point which takes the population base rates into consideration results in proportions of misclassification associated with the z-score deviations

$$z^{(1)} = \frac{y_c - \bar{y}^{(1)}}{\sqrt{V(y)}} = \frac{-1.43 - 2.78}{3.58} = -1.18$$

and

$$z^{(2)} = \frac{y_c - \bar{y}^{(2)}}{\sqrt{V(y)}} = \frac{-1.43 + 10.04}{3.58} = 2.41$$

The areas in the smaller portions of the two curves are .119 and .008, respectively. The respective proportions of the groups that should be misclassified using the new cutting point are

$$p_\varepsilon^{(1)} = \pi_1(.119) = .0119 \qquad \text{and} \qquad p_\varepsilon^{(2)} = \pi_2(.008) = .0072$$

This yields a total estimated error $p_\varepsilon^{(1)} + p_\varepsilon^{(2)} = .0191$. The total probability of error obtained by taking base rates into consideration therefore is substantially smaller than that resulting from using a cutting point placed halfway between the two group means, but it should be noted that the proportion of manics that will be misclassified is substantially greater than the proportion of paranoid types that will be misclassified: 12 percent vs. less than 1 percent for paranoids. Whether or not the Bayesian solution which minimizes total number of errors at the expense of misclassifying a higher proportion from the smaller population is chosen will depend on the objectives and values involved. A cutting point can be chosen to accomplish either objective.

9.7 An Empirical Method for Determining the Critical Value to Minimize Total Classification Errors

Rather than using the logarithmic conversion of Eq. (9.16) to define y_c, successive probabilities of misclassification for trial values can be read from normal-probability tables by starting with a z-score value lying halfway between the two means and moving the trial cutting point toward the mean of the group having the smaller population proportion.

First calculate the z-score value of the point lying halfway between the two means

$$z^{(1)} = \frac{y_c - \bar{y}^{(1)}}{\sqrt{V(y)}} \qquad \text{where } y_c = \frac{\bar{y}^{(1)} + \bar{y}^{(2)}}{2}$$

and

$$z^{(2)} = -z^{(1)} = \frac{y_c - \bar{y}^{(2)}}{\sqrt{V(y)}}$$

If $\bar{y}^{(1)} > \bar{y}^{(2)}$, compute successive trial values by *adding* increments of .10 to $z^{(1)}$ and also to $z^{(2)}$. If $\bar{y}^{(1)} < \bar{y}^{(2)}$, compute successive trial values by *subtracting* increments of .10 from both $z^{(1)}$ and $z^{(2)}$.

For each successive pair of trial values $z^{(1)} + k$ and $z^{(2)} + k$, look up the corresponding area in the smaller portion of the normal curve and multiply by the associated population proportion or base rate. Let $p_\varepsilon^{(1)}$ be the area in the smaller portion of the curve determined from $z^{(1)} + k$, and let $p_\varepsilon^{(2)}$ be the area in the smaller portion of the curve determined from $z^{(2)} + k$. Multiply each $p_\varepsilon^{(i)}$ by the associated π_i to obtain the joint probability that a randomly selected individual will both belong to group I and have a discriminant-function score which falls beyond the cutting point $z^{(i)} + k$. For successive values of $z^{(1)}$ and

$z^{(2)}$ in the series, the total probability of error is

$$\pi_1 p_\varepsilon^{(1)} + \pi_2 p_\varepsilon^{(2)} = p_\varepsilon \tag{9.17}$$

Identify the minimum value of p_ε in the series and associated with this value the $z^{(1)}$ and $z^{(2)}$ representing z-score deviations of the optimal cutting point about the two group means. Since

$$z^{(1)} = \frac{y_c - \bar{y}^{(1)}}{\sqrt{V(y)}}$$

the discriminant-function score value for the cutting point y_c is simply

$$y_c = z^{(1)}\sqrt{V(y)} + \bar{y}^{(1)} \tag{9.18}$$

or

$$y_c = z^{(2)}\sqrt{V(y)} + \bar{y}^{(2)} \tag{9.19}$$

As an example of the empirical method for identifying the optimal cutting point taking base rates into consideration, the method can be applied to the manic-paranoid discrimination problem. The midpoint between manic $\bar{y}^{(1)}$ and paranoid $\bar{y}^{(2)}$ is -3.58 on the discriminant-function scale, yielding z scores

$$z^{(1)} = \frac{y_c - \bar{y}^{(1)}}{\sqrt{V(y)}} = \frac{-3.63 - 2.78}{3.58} = -1.79$$

and

$$z^{(2)} = \frac{y_c - \bar{y}^{(2)}}{\sqrt{V(y)}} = \frac{-3.63 + 10.04}{3.58} = +1.79$$

Since $\bar{y}^{(1)} > \bar{y}^{(2)}$, increments of $k = .1$ are added to obtain successive trial values for $z^{(1)}$ and $z^{(2)}$. Results are shown in Table 9.5. It is apparent that the total probability of error decreases as the cutting point is moved toward the mean for the smaller population and that it reaches a minimum value of $p_\varepsilon = .0193$ after the sixth increment has been added. The z-score values at this point are $z^{(1)} = -1.19$ and $z^{(2)} = 2.39$. The optimal cutting point on the discriminant-function scale can be calculated to be

$$y_c = z^{(1)}\sqrt{V(y)} + \bar{y}^{(1)} = -1.19(3.58) + 2.78 = -1.48$$

or

$$y_c = z^{(2)}\sqrt{V(y)} + \bar{y}^{(2)} = 2.39(3.58) - 10.04 = -1.48$$

Both the estimated total probability of error and the optimal cutting point arrived at by the systematic trial-and-error method agree closely with the corresponding values calculated by Eq. (9.16).

Table 9.5 Empirical Evaluation of Optimal Cutting Point for Discriminant-function Classification

	$z^{(1)} + k$	$z^{(2)} + k$	$\pi_1 p_1$	$\pi_2 p_2$	p_ε
1	−1.69	1.89	.00455	.02646	.0310
2	−1.59	1.99	.00559	.02097	.0266
3	−1.49	2.09	.00681	.01647	.0233
4	−1.39	2.19	.00823	.01287	.0211
5	−1.29	2.29	.00985	.00990	.0198
6	−1.19	2.39	.01170	.00756	.0193
7	−1.09	2.49	.01379	.00576	.0196
8	−.99	2.59	.01611	.00432	.0204
9	−.89	2.69	.01867	.00324	.0219
10	−.79	2.79	.02148	.00234	.0238
11	−.69	2.89	.02451	.00171	.0262
12	−.59	2.99	.02776	.00126	.0290
13	−.49	3.09	.03121	.00090	.0321
14	−.39	3.19	.03483	.00063	.0355
15	−.29	3.29	.03859	.00045	.0390
16	−.19	3.39	.04247	.00027	.0427
17	−.09	3.49	.04641	.00018	.0466
18	+.01	3.59	.04960	.00018	.0498
19	+.11	3.69	.04562	.00009	.0457
20	+.21	3.79	.04168	.00009	.0418

9.8 A Graphic Method of Determining an Appropriate Cutting Point

The simple graphic method of examining distribution of discriminant-function scores is important because normality and equality of variances can be assessed and an appropriate cutting point chosen even if these assumptions do not appear valid. Since the graphic method concentrates on the distributions of scores within the particular samples, they should be large enough to provide stable estimates of the corresponding population distributions.

Computer programs for discriminant-function analysis often follow calculation of the discriminant-function equation with an evaluation of the distribution of discriminant-function scores within each sample. For each individual, a discriminant-function score is computed, and the relative frequency or percentage of scores within class intervals is tabulated for each group. The results from this type of empirical approach applied to the manic-paranoid discrimination problem are shown graphically in Fig. 9.2. From examination of the distributions of scores within the two samples, it appears that assumptions of normality and equality of variances were reasonably satisfied, although some degree of skewness is present. From inspection, it further appears that an approximately equal proportion of subjects from each group will be misclassified if a cutting point of −3.0 on the discriminant-function scale is used. This value is similar to $y_c = (\bar{y}^{(1)} + \bar{y}^{(2)})/2 = -3.6$, which is the halfway point calculated assuming normal distributions with equal variances.

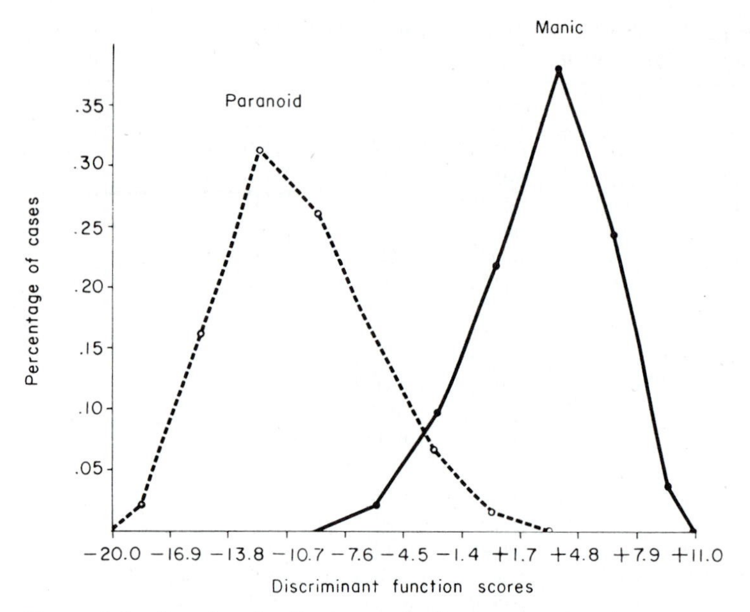

Figure 9.2 Sample distributions of discriminant-function scores derived from paranoid and manic profiles.

Unequal base rates, π_1 and π_2, can be taken into consideration in the graphic method by multiplying the relative frequencies in various class intervals by the appropriate π_i value for that population. The results obtained by multiplying the frequencies in various intervals of the manic distribution by $\pi_1 = .1$ and frequencies in the corresponding intervals of the paranoid distribution by $\pi_2 = .9$ are shown in Fig. 9.3. It is obvious that a dividing point that falls halfway between the two group means will result in many more paranoid types being misclassified. The two curves appear to intersect at a point having a discriminant-function scale value of approximately $-.5$. A cutting point somewhere to the left of the intersection of the two curves in Fig. 9.3 appears required to minimize total errors of misclassification. A cutting point of -1.0 would probably be chosen from inspection of the curves alone, without any other information, and this value is satisfactorily close to the value of -1.4 that was calculated as the optimal cutting point using normal-distribution assumptions [Eq. (9.16)]. Discriminant-function scores were actually calculated for each profile, and the number falling above and below -1.0 were counted. The proportion of the manic group falling below -1.0 was 10 percent, and the proportion of paranoid profiles falling above -1.0 was only 2 percent. Assuming a 9:1 ratio of paranoids to manics in the population, the total proportion of errors should be $p_\varepsilon = .9(.02) + .1(.10) = .028$. This value can be compared with the $p_\varepsilon = .019$ estimate of total probability of error derived from the

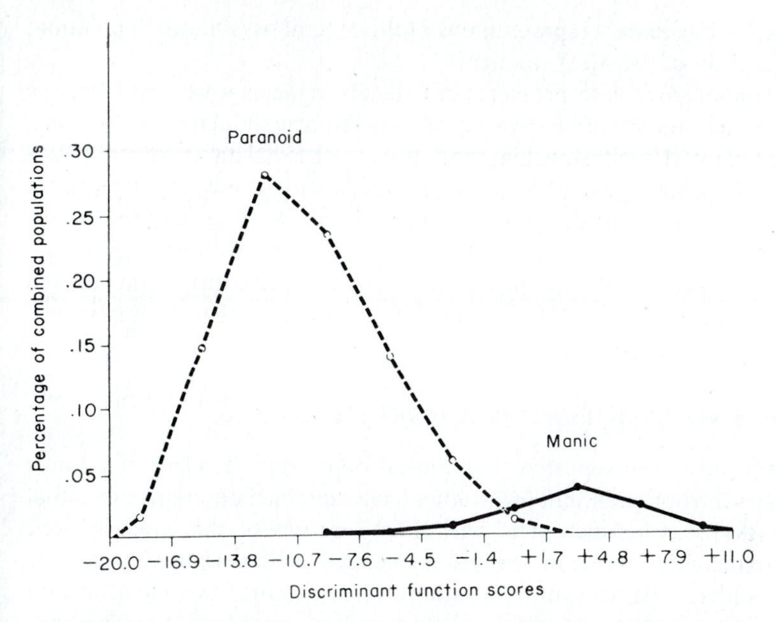

Figure 9.3 Sample distributions adjusted for differences in base rates.

theoretical normal-distribution model. We feel that this emphasizes the practical utility and general comparability of the different methods for deriving appropriate cutting points for patient classification. In practice, the mathematical, empirical, or graphic method can be used to define classification regions. Some advantage does result from being able to examine actual sampling distribution by the graphic method.

9.9 Some Practical Applications

9.9.1 INTRODUCTION

In this section we present for computational purposes several problems of clinical significance for which discriminant-function analyses appear relevant. Assuming that the student of this methodology can compute means and within-groups covariance matrices, these tedious preliminary statistics have been calculated. Similarly, the inverse of the covariance matrix has been computed for each problem. These preliminary statistics provide all the necessary information for discriminant-function analyses and related tests of significance. Although the problems have been chosen as clinically significant, complete solutions are not provided but are left for the student.

The data utilized for discriminant-function analyses included BPRS symptom ratings and basic demographic data on a total sample of 3,498 cases. The patients in the sample were drawn from Veterans Administration hospitals, state hospitals, university hospitals, and outpatient clinics. The total sample is considered to be adequately representative of the general psychiatric population, with the exception of private outpatients.

Other data concerned with problems of clinical diagnosis were drawn from a series of parallel studies of diagnostic nomenclature conducted in America, France, Germany, Czechoslovakia, and Italy. These data involved BPRS symptom-rating profiles provided by qualified experts for hypothetical "typical" cases representing different diagnostic groups. From these stereotype ratings, discriminant functions representing differences among ideal "pure types" can be derived. It appears a reasonable clinical application to consider the derived functions as appropriate for classification of real patients that may have less clearly prototypic symptom profiles.

9.9.2 AGE AND MANIFEST PSYCHOPATHOLOGY

It has frequently been reported that clinical depressions tend to occur at a later age than schizophrenia, but few studies have examined differences in actual symptom patterns as a function of age. For purposes of this investigation, patients were included in one of two groups defined as "less than 40 years" or "40 years or older." BPRS symptom ratings were combined to yield scores on four higher-order factors of thinking disturbance, withdrawal-retardation,

Table 9.6 Preliminary Statistics for Discriminant-function Analysis between Age Groups
$n_1 = 1{,}563,\ n_2 = 1{,}917$

MEANS

Factor	Age group < 40 years	> 40 years
Thinking disturbance	5.604606	5.143543
Withdrawal-retardation	5.714331	5.766302
Hostile-suspiciousness	6.438260	5.369849
Anxious depression	7.343890	7.134585

WITHIN-GROUPS COVARIANCE MATRIX

$$\begin{bmatrix} 17.425048 & 7.702976 & 8.344831 & 3.220274 \\ 7.702976 & 16.449260 & 5.768603 & 6.039369 \\ 8.344831 & 5.768603 & 17.726661 & 7.096571 \\ 3.220274 & 6.039369 & 7.096571 & 18.497192 \end{bmatrix}$$

INVERSE OF WITHIN-GROUPS COVARIANCE MATRIX

$$\begin{bmatrix} .085785 & -.031447 & -.033413 & .008152 \\ -.031447 & .084609 & -.004562 & -.020399 \\ -.033413 & -.004562 & .083530 & -.024740 \\ .008152 & -.020399 & -.024740 & .068795 \end{bmatrix}$$

hostile-suspiciousness and anxious depression. Mean scores for the younger and older patient groups plus the within-groups covariance matrix and its inverse are presented in Table 9.6. The sample sizes were 1,563 and 1,917, respectively.

9.9.3 RACE AND MANIFEST PSYCHOPATHOLOGY

Of the total available sample of 3,498 cases, 2,627 were Anglo and 585 were Negro. The remaining 286 were predominantly Latin American. In recent years, there have been several reports of differences in the relative frequencies of depressive and schizophrenic diagnoses among Anglo and Negro populations. It is of interest to know whether the two ethnic groups can be distinguished in terms of symptom-rating factors, and if so, what weighted combination of the symptom measures is most discriminating. The necessary data appear in Table 9.7.

9.9.4 SEX AND MANIFEST PSYCHOPATHOLOGY

In specific types of hospital populations, sex differences in the frequencies of various clinical diagnoses are apparent. Since the nature of these differences has not always been found to be consistent from one type of treatment setting to another, it is of more than incidental interest to know whether consistent differences in symptom characteristics distinguish male and female patients in a heterogeneous population drawn from a variety of types of treatment settings.

Table 9.7 Preliminary Statistics for Discriminant-function Analysis between Race Groups
$n_1 = 2{,}627$, $n_2 = 585$

MEANS

Factor	*Anglo*	*Negro*
Thinking disturbance	5.298629	5.354701
Withdrawal-retardation	5.710506	5.564103
Hostile-suspiciousness	6.095737	5.047009
Anxious depression	7.524362	5.991453

WITHIN-GROUPS COVARIANCE MATRIX

$$\begin{bmatrix} 17.180652 & 7.559639 & 8.565032 & 3.361519 \\ 7.559639 & 16.171505 & 5.750782 & 6.225525 \\ 8.565032 & 5.750782 & 18.155651 & 7.112420 \\ 3.361519 & 6.225525 & 7.112420 & 18.583187 \end{bmatrix}$$

INVERSE OF WITHIN-GROUPS COVARIANCE MATRIX

$$\begin{bmatrix} .087860 & -.031884 & -.034476 & .007984 \\ -.031884 & .086472 & -.003833 & -.021733 \\ -.034476 & -.003833 & .081890 & -.023821 \\ .007984 & -.021733 & -.023821 & .068766 \end{bmatrix}$$

Table 9.8 Preliminary Statistics for Discriminant-function Analysis between Sex Groups
$n_1 = 1{,}906$, $n_2 = 1{,}579$

MEANS

Factor	*Male*	*Female*
Thinking disturbance	5.237933	5.459468
Withdrawal-retardation	5.667104	5.843572
Hostile-suspiciousness	5.417891	6.353388
Anxious depression	6.894019	7.615896

WITHIN-GROUPS COVARIANCE MATRIX

$$\begin{bmatrix} 17.351745 & 7.669935 & 8.404842 & 3.197005 \\ 7.669935 & 16.450531 & 5.673948 & 5.971371 \\ 8.404842 & 5.673948 & 17.785270 & 6.968650 \\ 3.197005 & 5.971371 & 6.968650 & 18.351863 \end{bmatrix}$$

INVERSE OF WITHIN-GROUPS COVARIANCE MATRIX

$$\begin{bmatrix} .086569 & -.031573 & -.034014 & .008108 \\ -.031573 & .084290 & -.003969 & -.020419 \\ -.034014 & -.003969 & .083103 & -.024339 \\ .008108 & -.020419 & -.024339 & .068964 \end{bmatrix}$$

If significant differences exist, what single weighted combination of symptom variables best represents the multivariate differences? The necessary preliminary statistics required for complete discriminant-function analysis are presented in Table 9.8. The sample sizes involved were 1,906 male and 1,579 female.

9.9.5 MARITAL STATUS AND MANIFEST PSYCHOPATHOLOGY

Higher incidence of clinical diagnoses of schizophrenia has frequently been reported among single (never married) psychiatric patients. In the total sample of 3,498 patients, 750 were classified as single and 1,534 as currently married. Divorced, separated, and widowed patients were excluded for purposes of the present analysis. Preliminary calculations needed for complete discriminant-function analysis are presented in Table 9.9. Is the best linear discriminant function consistent with the notion that schizophrenia is relatively more prevalent among single patients and depression relatively more prevalent among married patients? What is the probability that the observed difference in multivariate mean profiles is due to chance sampling variability?

9.9.6 EDUCATIONAL ACHIEVEMENT AND MANIFEST PSYCHOPATHOLOGY

It has been frequently observed that schizophrenia has higher incidence among lower social classes, while depression has been reported to have higher incidence in middle and upper social-class groups. Paranoid-type reactions

Table 9.9 Preliminary Statistics for Discriminant-function Analysis between Marital-status Groups
$n_1 = 750$, $n_2 = 1,534$

MEANS

Factor	*Single*	*Married*
Thinking disturbance	6.628666	4.769557
Withdrawal-retardation	6.828666	5.406780
Hostile-suspiciousness	6.363333	5.839961
Anxious depression	6.674667	7.588983

WITHIN-GROUPS COVARIANCE MATRIX

$$\begin{bmatrix} 16.870266 & 7.068760 & 8.641761 & 3.764628 \\ 7.068760 & 15.998180 & 5.753693 & 6.236890 \\ 8.641761 & 5.753693 & 18.243072 & 7.225377 \\ 3.764628 & 6.236890 & 7.225377 & 17.811367 \end{bmatrix}$$

INVERSE OF WITHIN-GROUPS COVARIANCE MATRIX

$$\begin{bmatrix} .087780 & -.028419 & -.034803 & .005516 \\ -.028419 & .085330 & -.004759 & -.021941 \\ -.034803 & -.004759 & .082481 & -.024436 \\ .005516 & -.021941 & -.024436 & .072574 \end{bmatrix}$$

have been reported to occur more frequently among the better educated, while primary-process-type schizophrenia has been associated with low levels of accomplishment. Discriminant-function analysis can be employed to determine whether educational groups differ in patterns of manifest psychopathology and, if so, to evaluate whether the nature of the difference is consistent with the previously reported differences in clinical diagnoses. The mean vectors and within-groups covariance matrix derived from data on 1,579 patients with less than high school graduation and 1,577 patients who completed high school or beyond are presented in Table 9.10.

9.9.7 WORK ACHIEVEMENT AND MANIFEST PSYCHOPATHOLOGY

Patients in the total sample of 3,498 cases were grouped according to work achievement with 1,654 falling into the "never worked" or "unskilled" group and 1,396 falling into the "skilled," "clerical," and "managerial or professional" groups. Housewives and students were excluded from the samples for purposes of this investigation. Mean vectors and the covariance matrix are presented in Table 9.11. Do the two groups differ significantly in multivariate mean profiles? Is the most discriminating linear combination of symptom variables consistent with the notion that schizophrenia has higher incidence in low-work-achievement groups, while depression has higher incidence among

Table 9.10 Preliminary Statistics for Discriminant-function Analysis between Educational-achievement Groups
$n_1 = 1{,}579$, $n_2 = 1{,}577$

MEANS

Factor	Non-high-school graduates	High school graduates and above
Thinking disturbance	5.108930	5.535193
Withdrawal-retardation	5.759658	5.574192
Hostile-suspiciousness	5.382520	6.513317
Anxious depression	6.851172	7.870006

WITHIN-GROUPS COVARIANCE MATRIX

$$\begin{bmatrix} 17.652957 & 7.803820 & 8.693702 & 3.469159 \\ 7.803820 & 16.491977 & 5.907439 & 6.483780 \\ 8.693702 & 5.907439 & 18.292236 & 7.121626 \\ 3.469159 & 6.483780 & 7.121626 & 18.910213 \end{bmatrix}$$

INVERSE OF WITHIN-GROUPS COVARIANCE MATRIX

$$\begin{bmatrix} .085559 & -.031527 & -.033489 & .007725 \\ -.031527 & .085694 & -.004105 & -.022052 \\ -.033489 & -.004105 & .080820 & -.022885 \\ .007725 & -.022052 & -.022885 & .067644 \end{bmatrix}$$

Table 9.11 Preliminary Statistics for Discriminant-function Analysis between Work-level Groups
$n_1 = 1{,}654$, $n_2 = 1{,}396$

MEANS

Factor	Never worked and unskilled	Skilled and managerial
Thinking disturbance	5.450120	5.082378
Withdrawal-retardation	5.664148	5.683739
Hostile-suspiciousness	5.373640	6.194126
Anxious depression	6.498187	7.924785

WITHIN-GROUPS COVARIANCE MATRIX

$$\begin{bmatrix} 17.618087 & 7.936758 & 8.749113 & 3.627100 \\ 7.936758 & 16.964191 & 6.005671 & 6.018490 \\ 8.749113 & 6.005671 & 18.210735 & 7.428151 \\ 3.627100 & 6.018490 & 7.428151 & 18.378200 \end{bmatrix}$$

INVERSE OF WITHIN-GROUPS COVARIANCE MATRIX

$$\begin{bmatrix} .086048 & -.030627 & -.034011 & .006794 \\ -.030627 & .081554 & -.004493 & -.018846 \\ -.034011 & -.004493 & .083096 & -.025402 \\ .006794 & -.018846 & -.025402 & .069510 \end{bmatrix}$$

patients with higher work achievement? Not knowing the work achievement of patients, what portion could one expect to classify correctly using symptom factor scores alone? What critical value on the discriminant-function scale should be used as the cutting point for maximally effective classification?

9.9.8 RÉSUMÉ

While classification of individuals may not be a practical problem with regard to criterion groups defined by age, sex, race, marital status, and the like, discriminant-function analysis may nevertheless have practically important implications. First of all, the analysis provides the basis for a multivariate test of significance of difference in mean profiles. Second, the analysis provides a description of the combination of characteristics which best separates members of the two groups. Comparison of the empirically derived discriminant function with a priori functions derived from other considerations, such as diagnostic-group differences, permits one to evaluate whether or not the a priori function is adequate to account for all the differences between the two groups. For example, is the empirically derived discriminant function between single and married groups consistent with the notion that the symptom difference is totally explained in terms of a previously derived contrast between schizophrenia and depression? Calculation of the proportion of misclassification provides a meaningful index of the degree of separation between the two groups. The student is encouraged to select one of the problems presented in this section and to work through it, as if it were a personal research project, using the techniques presented in this chapter. Just as important as the mere computational techniques is a comprehension of what can and cannot be achieved in understanding data from use of the techniques.

9.9.9 DIAGNOSTIC PROBLEMS

Of more practical interest from the point of view of the classification of individuals are problems related to diagnosis. In the present section, several such problems will be presented using diagnostic-stereotype data derived from America, France, Germany, Czechoslovakia, and Italy. The diagnostic stereotypes were grouped into three major classes: depressive; primary, or core, schizophrenic; and paranoid (including paranoid schizophrenia). This grouping provides the basis for discriminant-function analyses contrasting (1) depression and primary schizophrenia, (2) depression and paranoid types, and (3) primary schizophrenia and paranoid types. The required preliminary calculations are presented in Tables 9.12 to 9.14.

For each of these problems, it is relevant to ask the following questions: Do the groups of diagnostic stereotype profiles differ significantly? What is the weighted combination of symptom factor scores which best characterizes the difference between groups? If one neglects the possibility of differential base rates, what cutting point on the discriminant-function scale should lead to minimum number of misclassifications? If primary schizophrenia is assumed to

Table 9.12 Preliminary Statistics for Discriminant-function Analysis between Core-schizophrenic Patients and Depressed Patients
$n_1 = 1{,}419$, $n_2 = 650$

MEANS

Factor	*Schizophrenics*	*Depressives*
Thinking disturbance	13.358703	7.886154
Withdrawal-retardation	13.881607	12.055385
Hostile-suspiciousness	11.390417	8.058462
Anxious depression	7.679351	17.947692

WITHIN-GROUPS COVARIANCE MATRIX

$$\begin{bmatrix} 13.242288 & 8.993814 & 6.797463 & .568898 \\ 8.993814 & 17.022060 & 4.377313 & -.378113 \\ 6.797463 & 4.377313 & 13.278839 & 2.070281 \\ .568898 & -.378113 & 2.070281 & 10.610044 \end{bmatrix}$$

INVERSE OF WITHIN-GROUPS COVARIANCE MATRIX

$$\begin{bmatrix} .146299 & -.063385 & -.054065 & .000446 \\ -.063385 & .092104 & .001076 & .006471 \\ -.054065 & .001076 & .105376 & -.017624 \\ .000446 & .006471 & -.017624 & .097895 \end{bmatrix}$$

Table 9.13 Preliminary Statistics for Discriminant-function Analysis between Depressed Patients and Paranoid or Paranoid-schizophrenic Patients
$n_1 = 650$, $n_2 = 888$

MEANS

Factor	*Depressives*	*Paranoid types*
Thinking disturbance	7.886154	12.534910
Withdrawal-retardation	12.055385	9.031532
Hostile-suspiciousness	8.058462	13.890766
Anxious depression	17.947692	8.762388

WITHIN-GROUPS COVARIANCE MATRIX

$$\begin{bmatrix} 14.605951 & 8.618133 & 4.560335 & 1.034421 \\ 8.618133 & 12.723989 & 4.650770 & 2.113172 \\ 4.560335 & 4.650770 & 13.589275 & 1.301300 \\ 1.034421 & 2.113172 & 1.301300 & 9.617801 \end{bmatrix}$$

INVERSE OF WITHIN-GROUPS COVARIANCE MATRIX

$$\begin{bmatrix} .116568 & -.074767 & -.014085 & .005796 \\ -.074767 & .140213 & -.020987 & -.019925 \\ -.014085 & -.020987 & .086024 & -.005513 \\ .005796 & -.019925 & -.005513 & .108474 \end{bmatrix}$$

Table 9.14 Preliminary Statistics for Discriminant-function Analysis between Core-schizophrenic Patients and Paranoid or Paranoid-schizophrenic Patients

$n_1 = 1,419$, $n_2 = 888$

MEANS

Factor	*Core schizophrenics*	*Paranoid types*
Thinking disturbance	13.358703	12.583333
Withdrawal-retardation	13.881607	9.063064
Hostile-suspiciousness	11.390417	13.938064
Anxious depression	7.679351	8.798423

WITHIN-GROUPS COVARIANCE MATRIX

$$\begin{bmatrix} 12.789344 & 7.719555 & 4.190029 & .949338 \\ 7.719555 & 14.883285 & 3.252998 & -.220359 \\ 4.190029 & 3.252998 & 13.558361 & 2.620621 \\ .949338 & -.220359 & 2.620621 & 11.732202 \end{bmatrix}$$

INVERSE OF WITHIN-GROUPS COVARIANCE MATRIX

$$\begin{bmatrix} .120898 & -.057913 & -.022330 & -.005992 \\ -.057913 & .098974 & -.007435 & .008205 \\ -.022330 & -.007435 & .085823 & -.017503 \\ -.005882 & .008205 & -.017503 & .089775 \end{bmatrix}$$

occur with twice the frequency of either depression or paranoid disorders, what should the optimal cutting points be to minimize total errors of misclassification between relevant pairs of groups? If one wanted to divide the total discriminant-function scale into three regions such that the two extreme regions result in 95 percent accuracy of classification and patients falling in the central region remained unclassified, where should the two cutting points be placed? Assuming normal distribution, diagrams can be prepared to represent these various solutions. Given individual patterns of scores on the four symptom factors, probabilities of the individual belonging in each group can be calculated.

9.10 Computer Program for Simple Discriminant-function Analysis

This program computes optimal weighting coefficients that provide maximum separation between two groups. The input to the program can be either raw-data cards or previously computed means and within-groups covariance matrix. The printout includes (1) group means on the original variables, (2) the within-groups covariance matrix, (3) the inverse of the within-groups covariance matrix, (4) discriminant-function coefficients, (5) group means on the discriminant function, (6) the Mahalanobis D^2 and associated F-ratio test statistics, (7) a detailed frequency distribution of discriminant-function scores in the two groups, and (8) a condensed decile frequency distribution. The means and covariance matrices are printed 10 fields per line, with continuation on the next line if more than 10 variables are involved.

Input

1 *One control card* NVAR NG(1) NG(2) NTAG

NVAR = number of variables
NG(1) = number in group 1
NG(2) = number in group 2
NTAG = code given nonzero value only if previously computed covariance matrix is to be read

2 Data cards organized by group and read according to format 567

Printout

1 Group means on NVAR original variables
2 Within-groups covariance matrix
3 Inverse of within-groups covariance matrix
4 Discriminant-function weighting coefficients
5 Group means on discriminant function
6 D^2 and F-ratio test statistics
7 Frequency distribution of scores on the discriminant function for each group

Computer Program for Simple Discriminant Function

```
      DIMENSION ASUM(18),A(18,18),WCOV(18,18),DATA(36),XBAR(2,18),NG(2),
     1DIFF(18),W(18),XMEAN(2),Y(3500),NSUM(101),MSUM(101)
      DEFINE FILE 1( 1500,80,U,L1)
    3 FORMAT(4I4)
    4 FORMAT(4X,'GROUP MEANS ON DISC. FUNCTION')
    5 FORMAT(1X,'FREQUENCY DISTRIBUTION OF DISCRIMINANT FUNCTION SCORES'
     1/,6X,'GP. 1',8X,'GP. 2', 4X,'SCORE')
    6 FORMAT(10X,'MEANS ON ORIGINAL VARIABLES')
    7 FORMAT(1X,'DECILE FREQUENCIES AND PROPORTIONS',/,10X,'GROUP 1',15X
     1,'GROUP 2')
    9 FORMAT(11F7.4,/,5F7.4)
   10 FORMAT(13F6.3,/,3F6.3)
   11 FORMAT(20X,'COVARIANCE MATRIX')
   12 FORMAT(15X,'INVERSE OF COVARIANCE MATRIX')
   13 FORMAT(2X,'D-SQUARE',3X,'F')
   14 FORMAT(10X,'DISCRIMINANT FUNCTION COEFFICIENTS')
      L1=1
      READ(2,3) NVAR,(NG(I),I=1,2),NTAG
C     COMPUTE THE WITHIN GROUPS COVARIANCE MATRIX
      TOT=0.0
      DO 20 M=1,NVAR
      W(M)=0.0
      ASUM(M)=0.0
      DO 20 M2=1,NVAR
      A(M,M2)=0.0
   20 WCOV(M,M2)=0.0
      FN1=NG(1)
      FN2=NG(2)
      DF=FN1+FN2-2.0
      NGPS=2
      IF(NTAG)1,2,1
    1 READ(2,10)((WCOV(M,M2),M2=1,NVAR),M=1,NVAR)
      READ(2,9)((XBAR(M,M2),M2=1,NVAR),M=1,2)
      GO TO 15
```

```
    2 DO 126 K=1,NGPS
      DO 526 J5=1,NVAR
  526 ASUM(J5)=0.0
      NX=NG(K)
      FNX=NX
      TOT=TOT+FNX
      FFNX=FNX-2
      DO 25 J=1,NX
  139 FORMAT(10F10.3)
      NVAR2=NVAR*2
  567 FORMAT(18F3.1)
      READ(2,567)(DATA(M5),M5=1,NVAR)
      WRITE(1'L1)(DATA(M5),M5=1,NVAR)
      DO 25 M=1,NVAR
      ASUM(M)=ASUM(M)+DATA(M)
      DO 25 M2=M,NVAR
      XDAT=DATA(M)*DATA(M2)
      A(M,M2)=A(M,M2)+XDAT
   25 CONTINUE
      DO 26 M=1,NVAR
      XBAR(K,M)=ASUM(M)/FNX
      DO 26 M2=M,NVAR
      WCOV(M,M2)=WCOV(M,M2)+A(M,M2)-ASUM(M)*ASUM(M2)/FFNX
   26 A(M,M2)=0.0
  126 CONTINUE
      WRITE(3,6)
      WRITE(3,139)(XBAR(1,M),M=1,NVAR)
      WRITE(3,139)(XBAR(2,M),M=1,NVAR)
      WRITE(3,11)
      DO 227 M=1,NVAR
      DO 227 M2=M,NVAR
      WCOV(M,M2)=WCOV(M,M2)/DF
  227 WCOV(M2,M)=WCOV(M,M2)
   15 DO 16 M=1,NVAR
   16 WRITE(3,139)(WCOV(M,M2),M2=1,NVAR)
C     TAKE INVERSE OF WITHIN GROUPS COVARIANCE MATRIX
      CALL RMINV(WCOV,NVAR)
      WRITE(3,12)
      DO 17 M=1,NVAR
   17 WRITE(3,139)(WCOV(M,M2),M2=1,NVAR)
      DO 127 M=1,NVAR
      DIFF(M)=XBAR(1,M)-XBAR(2,M)
  127 CONTINUE
      DO 130 M=1,NVAR
      DO 130 M2=1,NVAR
  130 W(M)=W(M)+WCOV(M,M2)*DIFF(M2)
      WRITE(3,14)
      WRITE(3,139)( W   (M),M=1,NVAR)
      DSQR=0.0
      DO 142 M=1,NVAR
  142 DSQR=DSQR+W(M)*DIFF(M)
      FNVAR=NVAR
      FVAL=(FN1*FN2*(FN1+FN2-1.-FNVAR)*DSQR)/((FN1+FN2)*(FN1+FN2-2.)*FNV
     1AR)
      WRITE(3,13)
      WRITE(3,139) DSQR,FVAL
      IF(NTAG) 42,41,42
   41 NTOT=TOT
      DO 60 II=1,2
      XMEAN(II)=0.0
      DO 60 KK=1,NVAR
   60 XMEAN(II)=XMEAN(II)+W(KK)*XBAR(II,KK)
      WRITE(3,4)
      WRITE(3,139) (XMEAN(II),II=1,2)
      L1=1
      DO 40 I=1,NTOT
      Y(I)=0.0
      READ(1'L1)(DATA(M5),M5=1,NVAR)
      DO 40 J=1,NVAR
   40 Y(I)=Y(I)+W(J)*DATA(J)
      DO 161 M=1,101
      NSUM(M)=0
  161 MSUM(M)=0
      XLARG=Y(1)
      SMALL=Y(1)
```

```
      DO 140 I=1,NTOT
      IF(SMALL-Y(I)) 141,141,242
  242 SMALL=Y(I)
      GO TO 140
  141 IF(XLARG-Y(I)) 143,140,140
  143 XLARG=Y(I)
  140 CONTINUE
      XXLAR=100./(XLARG-SMALL)
      N1=NG(1)
      N2=N1+1
      DO 150 I=1,N1
      Y(I)=XXLAR*(Y(I)-SMALL)
      NN=Y(I)+1.0
  150 NSUM(NN)=NSUM(NN)+1
      DO 160 I=N2,NTOT
      Y(I)=XXLAR*(Y(I)-SMALL)
      NN=Y(I)+1.
  160 MSUM(NN)=MSUM(NN)+1
      WRITE(3,5)
      DO 162 I=1,101
      X=I-1
      YMEAN=X*((XLARG-SMALL)/100.) + SMALL
  162 WRITE(3,163) NSUM(I),MSUM(I),YMEAN
  163 FORMAT(2I10,F10.4)
      WRITE (3,31)
      K=1
      WRITE(3,7)
      DO 164 I=1,10
      NN=0
      MM=0
      DO 165 J=1,10
      NN=NN+NSUM(K)
      MM=MM+MSUM(K)
  165 K=K+1
      ZN=NN
      ZM=MM
      ZN=ZN/FN1
      ZM=ZM/FN2
  164 WRITE(3,131) NN,ZN,MM,ZM
   31 FORMAT(///)
  131 FORMAT(5X,I4,F10.4,5X,I4,F10.4)
   42 CALL EXIT
      END
```

Matrix-inversion Subroutine Required by Simple Discriminant-function Program

```
      SUBROUTINE RMINV(A,N)
      DIMENSION IPIVO(18),INDEX(18,2),PIVOT(18), A(18,18)
      DET=1.0
      DO 20 J=1,N
   20 IPIVO (J)=0
      DO 550 I=1,N
      AMAX=0.0
      DO 105 J=1,N
      IF (IPIVO (J)-1) 60,105,60
   60 DO 100 K=1,N
      IF (IPIVO (K)-1) 80,100,740
   80 IF(ABS(AMAX)-ABS(A(J,K)))85,100,100
   85 IROW=J
      ICOLU =K
      AMAX=A(J,K)
  100 CONTINUE
  105 CONTINUE
      IF(ABS (AMAX)-2.0E-7) 800,800,801
  800 WRITE (3,666)
  666 FORMAT(20H DETERMINANT = ZERO )
      DET =0.0
      PAUSE
  801 CONTINUE
      IPIVO (ICOLU )=IPIVO (ICOLU )+1
      IF (IROW-ICOLU ) 140,260,140
  140 DET=-DET
      DO 200 L=1,N
      SWAP=A(IROW,L)
```

```
    A(IROW,L)=A(ICOLU ,L)
200 A(ICOLU ,L)=SWAP
260 INDEX(I,1)=IROW
    INDEX(I,2)=ICOLU
    PIVOT(I)=A(ICOLU ,ICOLU )
    DET=DET*PIVOT(I)
    A(ICOLU ,ICOLU )=1.0
    DO 350 L=1,N
350 A(ICOLU ,L)=A(ICOLU ,L)/PIVOT(I)
    DO 550 L1=1,N
    IF(L1-ICOLU ) 400,550,400
400 T=A(L1,ICOLU )
    A(L1,ICOLU )=0.0
    DO 450 L=1,N
450 A(L1,L)=A(L1,L)-A(ICOLU ,L)*T
550 CONTINUE
    DO 710 I=1,N
    L=N+1-I
    IF (INDEX(L,1)-INDEX(L,2)) 630,710,630
630 IROW=INDEX(L,1)
    ICOLU =INDEX(L,2)
    DO 705 K=1,N
    SWAP=A(K,IROW)
    A(K,IROW)=A(K,ICOLU )
    A(K,ICOLU )=SWAP
705 CONTINUE
710 CONTINUE
740 RETURN
    END
```

10 Multiple Discriminant Analysis

10.1 Introduction

Multiple discriminant analysis is a generalization of the method of discriminant-function analysis appropriate for only two groups. It is useful primarily as a method of studying relationships among several groups or populations, but it also can provide a basis for classification of individuals among several groups. It is appropriate where samples of individuals have been drawn from several different populations and where p different quantitative scores are available for each individual. The p measurements are assumed to have a multivariate normal distribution with equal variance-covariance matrices within the several populations.

The method of multiple discriminant analysis results in reduction of the multiple measurements to *one or more* weighted combinations having maximum potential for distinguishing among members of the different groups. The first *canonical variate*, or discriminant function, is that single weighted composite which of all possible weighted composites provides maximum average separation between the groups relative to variability within the groups. More precisely, the first canonical variate is that particular artificial composite variable on which the sum of squared differences among group means is maximally great relative to the within-groups variance for the same weighted composite. In the case of only two groups, a single optimal combination of the multiple measures will account for all differences; however, in the case of several groups, one weighted combination of the scores may distinguish well between certain groups but not between others. In such instance, a second or even a third composite may be required to distinguish between groups that were not well separated by the first discriminant function. The second canonical variate, or discriminant function, is that weighted composite which of all possible weighted composites uncorrelated with the first (within groups) provides for maximum average separation among the groups. The third canonical variate, or discriminant function, is that weighted composite which of all possible weighted composites uncorrelated with either of the first two provides for maximum average separation between the groups. The maximum number of potential discriminant functions in any problem is equal to the number of variables p or to 1 less than the number of groups, $k - 1$, whichever is smaller. Tests of significance are available, and one does not usually need to consider all the possible discriminant functions to account for all significant differences.

As stated, the primary importance of multiple discriminant analysis in clinical research is for the study of relationships among several groups in terms of the multiple measurements. Mean scores on the two or more most discriminating composite functions can be taken as coordinate values to locate the groups in a geometric space of minimum dimensionality. In addition, the results from multiple discriminant analysis can provide the basis for classification of individuals among the several groups. The approach provides tests of significance for certain important hypotheses concerning relationships among several groups, e.g., that a single composite score accounts for all significant differences among the groups.

10.2 Definitions

The first discriminant function is that single weighted combination of measurements which has maximum variance between groups relative to the variance within groups. Once again, the problem is one of combining multiple measurements to obtain one (or possibly more than one in the present case) composite variable having maximum utility for distinguishing between different groups. The criterion function to be maximized can be written as the ratio of between-groups to within-groups variance for a weighted linear function of several variables. As described in Sec. 2.7.8, the variance of a linear combination of p variables can be obtained by pre- and postmultiplying the matrix of covariances among the p variables by the vector of weighting coefficients

$$V(y) = \mathbf{a}'C\mathbf{a}$$

The within-groups and between-groups variances of a linear composite can be defined in terms of the within- and between-groups covariance matrices in a similar manner. In multiple discriminant analysis, the problem is to choose a set of compounding coefficients that will define a function having maximum variance between groups relative to the variance within groups. The criterion function to be maximized is

$$f(\mathbf{a}) = \frac{\mathbf{a}'B\mathbf{a}}{\mathbf{a}'W\mathbf{a}} \tag{10.1}$$

Maximizing the criterion function with the introduction of a *convenience coefficient* gives the matrix equation

$$(B - \lambda W)\mathbf{a} = 0 \tag{10.2}$$

where λ is the scalar coefficient known as a *Lagrange multiplier*. In advanced calculus, the Lagrange multiplier is employed to take care of scaling or proportionality problems, but, as we shall see, in the solution of certain statistical problems such as the present one the value of λ has familiar meaning. Equation (10.2) has a solution where $|B - \lambda W| = 0$. Subject to the restriction that the within-groups variance be scaled to unity, $\mathbf{a}'W\mathbf{a} = 1$, the value of λ is the portion of between-groups variance (adjusted for within-groups variance) accounted for by the single best discriminant function. Obviously, any number of (proportional) vectors will yield an identical ratio of between- to within-groups variance [Eq. (10.1)]. The vector $\mathbf{a}$ is uniquely defined only under the constraint that $\mathbf{a}'W\mathbf{a} = 1$.

Let $\mathbf{a}^{(1)}$ contain the weighting coefficients defining the first discriminant function, i.e., that linear combination of the multiple measurements which accounts for the largest discriminant variance λ_1. It is quite possible, in the case of several groups, that true group differences will be present that are not adequately represented in the first discriminant function. The first discriminant function is defined in a manner to reflect maximum *average* difference between group means, but some particular groups that are really quite different in terms of the original multiple measurements may appear similar on the first discriminant function. For example, in Fig. 10.1 (presented later in this chapter) one can readily appreciate that single and married groups differ substantially in terms of scores on one weighted function $y^{(1)}$. It is equally obvious that the widowed and divorced groups have almost identical scores on $y^{(1)}$; however, this does not mean that the widowed and divorced groups do not differ in terms of mean scores on the original variables. They simply do not differ on the particular weighted function of those scores that defines the dimension of maximum average difference between all groups. A second set of weighting coefficients applied to the same original variables defines another quite different composite function $y^{(2)}$, and the widowed and divorced groups are seen to differ markedly on this second function.

The second discriminant function is that weighted combination of the p variables which of all possible weighted combinations independent of the first discriminant function accounts for a maximum of the remaining group differences. The second weighted composite is defined by the vector $\mathbf{a}^{(2)}$ associated with the largest remaining between-groups variance λ_2 after all variance associated with the first discriminant function has been partialed out. The values λ_1

and λ_2 are the two largest roots of the determinantal equation $|B - \lambda W| = 0$. By definition the vectors $\mathbf{a}^{(1)}$ and $\mathbf{a}^{(2)}$ associated with two different roots define functions that are statistically independent. Thus,

$$\mathbf{a}^{(1)\prime} W \mathbf{a}^{(1)} = \mathbf{a}^{(2)\prime} W \mathbf{a}^{(2)} = 1$$

and

$$\mathbf{a}^{(1)\prime} W \mathbf{a}^{(2)} = 0$$

It will be recalled from Sec. 2.7.8 that pre- and postmultiplication of a covariance matrix by a single vector of weighting coefficients yields a scalar quantity which is the variance of the composite variable obtained by applying the weights to the original scores from which the covariance matrix was derived. Similarly, premultiplication by one vector of weighting coefficients and postmultiplication by another yields a scalar quantity which is the covariance between the two weighted functions. Successive discriminant functions are defined as the vectors associated with the remaining roots $\lambda_3, \lambda_4, \ldots, \lambda_r$. The maximum number of possible discriminant functions is equal to the number of nonzero roots of the determinantal equation, which is the number of original variables p or 1 less than the number of groups $k - 1$, whichever is smaller. Each discriminant function is scaled to unit variance within groups, $\mathbf{a}^{(i)\prime} W \mathbf{a}^{(i)} = 1$, and each is statistically independent of each other discriminant function within the groups, $\mathbf{a}^{(i)\prime} W \mathbf{a}^{(j)} = 0$, where $i \neq j$.

10.3 Computing the Discriminant Functions

To compute the sets of weighting coefficients $\mathbf{a}^{(1)}, \mathbf{a}^{(2)}, \ldots, \mathbf{a}^{(r)}$ defining the first r best linear functions for use in discriminating among $k > r$ groups, we begin by computing the within-groups and between-groups SP matrices as described in Chap. 2. Although the discussion in Sec. 10.2 implied the use of within- and between-groups *covariance* matrices in calculating discriminant-function coefficients, it will be recalled that SP matrices and associated covariance matrices differ only by a scalar constant which is the degree of freedom. From a scaling point of view and in order that the various roots $\lambda_1, \lambda_2, \ldots, \lambda_r$ will have approximate chi-square distributions without further calculation, it is most convenient to work with the between-groups SP matrix and the within-groups covariance matrix. Thus, in the remainder of this section (which is concerned with actual computation of discriminant-function coefficients and with tests of significance for multivariate mean differences) let B represent the between-groups SP matrix and let W represent the within-groups covariance matrix.

In order to use the simple iterative method described in Chap. 3 to solve for the roots and vectors satisfying Eq. (10.2), we must manipulate the expression to put it in the standard form $(C - \lambda I)\mathbf{a} = 0$. It is not convenient simply to multiply Eq. (10.2) on the left by W^{-1} because the matrix $W^{-1}B$ will generally turn out to be nonsymmetric

$$(B - \lambda W)\mathbf{a} = (W^{-1}B - \lambda I)\mathbf{a} = 0 \tag{10.3}$$

A computational device which does result in a symmetric matrix involves computing the square-root inverse of W, as obtained from the square-root method of matrix inversion. Referring back to Sec. 2.6.2 in which the square-root method of matrix inversion was discussed, let the original matrix be W and the lower triangular square root be V^{-1}. Using this notation, the format for the square-root inverse calculations is

W	I
V'	V^{-1}

The matrix V' is an upper triangular factor of W such that $VV' = W$, and V^{-1} is a lower triangular square-root inverse such that $V^{-1}V = V'V'^{-1} = I$. Then $V^{-1}VV'V'^{-1} = V^{-1}WV'^{-1} = I$, where $V'^{-1} = \nabla$ is the upper triangular inverse of $V' = \nabla$ and V^{-1} is the lower triangular inverse of $V = \triangle$.

Starting with the matrix equation

$$(B - \lambda W)\mathbf{a} = B\mathbf{a} - \lambda W\mathbf{a} = 0$$

and recalling that W can be expressed as the product of a triangular square-root matrix and its transpose,

$$B\mathbf{a} - \lambda VV'\mathbf{a} = 0 \tag{10.4}$$

Multiplying on the left by V^{-1}, we obtain

$$V^{-1}B\mathbf{a} - \lambda V'\mathbf{a} = 0 \tag{10.5}$$

Next, the identity matrix $V'^{-1}V' = I$ is inserted into the first term of the equation

$$V^{-1}BV'^{-1}V'\mathbf{a} - \lambda V'\mathbf{a} = 0$$

Factoring the product $V'\mathbf{a}$ on the right gives

$$(V^{-1}BV'^{-1} - \lambda I)V'\mathbf{a} = 0 \tag{10.6}$$

The result is recognized as the familiar form

$$(C - \lambda I)\mathbf{z} = 0 \tag{10.7}$$

where $C = V^{-1}BV'^{-1}$ and $\mathbf{z} = V'\mathbf{a}$.

Since the matrix $C = V^{-1}BV'^{-1}$ is symmetric, the standard iterative method described in Chap. 3 can be used to compute successive roots λ_i and vectors $\mathbf{z}^{(i)}$. The vectors $\mathbf{z}^{(i)}$ are the principal components of $V^{-1}BV'^{-1}$; thus the reader can refer to Chap. 3 for further insight into the method. It will be noted that the solution vectors $\mathbf{z}^{(i)}$ are not the vectors $\mathbf{a}^{(i)}$ that are needed to maximize the original criterion function; however, the defined relationship $\mathbf{z}^{(i)} = V'\mathbf{a}^{(i)}$ makes solution from this point relatively easy

$$\mathbf{a}^{(i)} = V'^{-1}\mathbf{z}^{(i)} \tag{10.8}$$

Computation of multiple discriminant functions is not a simple matter, but the essential steps can now be outlined in sequence.

1 Compute the within-groups and between-groups SP matrices as described in Chap. 2. Divide the elements in the within-groups SP matrix by degrees of freedom $k(n_i - 1) = n_t - k$ to obtain the within-groups covariance matrix W.
2 Compute the triangular square-root inverse of the matrix W using the square-root method described in Chap. 2. The matrix V^{-1} is the lower triangular matrix that results from this computation.
3 Pre- and postmultiply the between-groups SP matrix by the triangular square root V^{-1} and its transpose.
4 Compute the characteristic roots and vectors of the symmetric matrix $V^{-1}BV'^{-1}$ to obtain roots $\lambda_1, \lambda_2, \ldots, \lambda_r$ and associated vectors $\mathbf{z}^{(1)}, \mathbf{z}^{(2)}, \ldots, \mathbf{z}^{(r)}$ using the iterative method described in Chap. 3.
5 Obtain the vectors of discriminant-function coefficients by premultiplying each $\mathbf{z}^{(i)}$ by the triangular square-root matrix V'^{-1}.

10.4 A Computational Example

For many readers, an example of clinical significance will help to clarify the computational procedure. The study of relationships among marital-status groups in terms of psychiatric symptom factors provides such a problem. Do patients belonging to several marital-status groups differ significantly in total multivariate symptom profiles? It has frequently been reported that schizophrenia has higher incidence among single (never married) individuals, while married individuals are relatively more prone to depression. Can all significant symptom-profile differences among the several marital-status groups be accounted for by a single dimension which represents the contrast between schizophrenia and depression? What are the relationships among the several marital-status groups as far as manifest psychopathology is concerned? Which are more similar and which more different?

The sample of 3,482 psychiatric patients was divided into single, married, divorced (or separated), and widowed. For each patient, factor scores for thinking disturbance, withdrawal-retardation, hostile-suspiciousness, and anxious depression were computed from BPRS symptom ratings. The first step in the analysis was to compute the within-groups and between-groups SP matrices as described in Chap. 2. This was accomplished by computing separately the SP matrix for each group and summing them to obtain the combined within-groups SP matrix SP(W). The total SP matrix SP(T) was computed across all individuals disregarding marital status. The between-groups SP matrix was computed by subtracting the within-groups matrix from the total SP matrix, SP(B) = SP(T) − SP(W). Mean values for the four symptom factors, the within-groups matrix SP(W) estimated on 3,478 degrees of freedom,

Table 10.1 Preliminary Statistics for Marital-status Example

MEAN FACTOR SCORES FOR FOUR MARITAL-STATUS GROUPS

Factor	*Single*	*Married*	*Divorced*	*Widowed*
Thinking disturbance	6.63	4.77	5.11	5.75
Withdrawal-retardation	6.83	5.41	5.25	6.32
Hostile-suspiciousness	6.36	5.84	5.60	5.23
Anxious depression	6.66	7.59	6.85	7.93

WITHIN-GROUPS SP MATRIX FOR FOUR MARITAL-STATUS GROUPS, df = 3,478

58,711.8946	25,138.8785	28,823.1572	12,022.3335
	56,007.8017	19,438.3496	21,441.8967
		62,240.9957	25,280.9798
			63,853.5802

BETWEEN-GROUPS SP MATRIX FOR FOUR MARITAL-STATUS GROUPS, df = 3

1,840.2739	1,516.9771	481.0301	−707.7481
	1,384.1647	437.2358	−367.3215
		356.7078	−253.2305
			703.6728

and the between-groups matrix SP(B) estimated on 3 degrees of freedom, are shown in Table 10.1 for the four marital-status groups.

The next step in the computation was to calculate the triangular square-root inverse V^{-1} of the matrix W, where W is the within-groups *covariance* matrix obtained by dividing elements in SP(W) by degrees of freedom for error, $n_t - k$. This was done using the square-root method described in Chap. 2. The calculations are presented in Table 10.2. The upper triangle of the symmetric covariance matrix appears at the upper left, and the triangular matrix V^{-1} resulting from these calculations appears in the lower right-hand section of the table, as outlined. Note that this *lower* triangular matrix is V^{-1}, which appears on the left in Eq. (10.6).

Table 10.2 Computation of Triangular Square-root Inverse V^{-1} of Within-groups Covariance Matrix

16.880935	7.227969	8.287279	3.456680	1.000000	.000000	.000000	.000000
	16.103450	5.588944	6.165007	.000000	1.000000	.000000	.000000
		17.895628	7.268826	.000000	.000000	1.000000	.000000
			18.359281	.000000	.000000	.000000	1.000000
4.108641	1.759211	2.017036	.841319	.243389	.000000	.000000	.000000
7.227969	3.606747	.565759	1.298940	−.118714	.277258	.000000	.000000
8.287279	5.588944	3.675201	1.316109	−.115302	−.042681	.272093	.000000
3.456680	6.165007	7.268826	3.772542	.026821	−.080574	−.094924	.265073

Table 10.3 The Symmetric Matrix $V^{-1}BV'^{-1}$ Obtained by Pre- and Postmultiplication of the Between-groups SP Matrix by the Triangular Square-root Inverse

$$\begin{bmatrix} 109.014 & 49.195 & -35.546 & -74.510 \\ 49.195 & 32.477 & -14.552 & -21.801 \\ -35.546 & -14.552 & 27.988 & 10.696 \\ -74.510 & -21.801 & 10.696 & 79.019 \end{bmatrix}$$

The third step in the computation was to pre- and postmultiply the between-groups matrix SP(B) by the triangular square-root matrix V^{-1} and its transpose to obtain $V^{-1}BV'^{-1}$. The arrangement of these matrices in forming the symmetric product is as follows: ◺ □ ◹ $= V^{-1}BV'^{-1}$. The symmetric product matrix $V^{-1}BV'^{-1}$ is presented in Table 10.3.

The characteristic roots and vectors of the matrix $V^{-1}BV'^{-1}$ provide the latent roots $\lambda_1, \lambda_2, \ldots, \lambda_r$ and the vectors $\mathbf{z}^{(1)}, \mathbf{z}^{(2)}, \ldots, \mathbf{z}^{(r)}$. Characteristic roots and vectors have been described in Chap. 2. It has been noted that the principal-components analysis of Chap. 3 is tantamount to computing the characteristic roots and vectors of a covariance matrix. The lengthy iterative procedure will not be illustrated in detail again in this chapter. The reader is reminded that the solution vectors $\mathbf{z}^{(1)}, \mathbf{z}^{(2)}, \ldots, \mathbf{z}^{(r)}$ for the matrix $V^{-1}BV'^{-1}$ are the principal-component vectors computed exactly as illustrated in Chap. 3. Results of the iterative solution for characteristic roots and vectors of $V^{-1}BV'^{-1}$ are presented in Table 10.4. The student can verify these results by applying the iterative computational procedure previously described.

The final step in computation of the discriminant functions requires multiplication of the $\mathbf{z}^{(i)}$ vectors by the triangular square-root inverse V'^{-1}. This transformation for the first principal vector is illustrated in Table 10.5. Note that

Table 10.4 Characteristic Roots and Vectors of Symmetric Matrix $(V^{-1}BV'^{-1} - \lambda I)\mathbf{z} = 0$

Factor	*I*	*II*
Thinking disturbance	.7408	.1969
Withdrawal-retardation	.3162	.4014
Hostile-suspiciousness	−.2196	−.5371
Anxious depression	−.5504	.7153
	$\lambda_1 = 195.94$	$\lambda_2 = 37.80$

Table 10.5 Obtaining Discriminant-function Coefficients from Characteristic Vectors of Symmetric Matrix $V'^{-1}\mathbf{z} = \mathbf{a}$

$$\begin{bmatrix} .2434 & -.1187 & -.1153 & .0268 \\ & .2772 & -.0427 & -.0806 \\ & & .2721 & -.0949 \\ & & & .2651 \end{bmatrix} \begin{bmatrix} .7408 \\ .3162 \\ -.2196 \\ -.5504 \end{bmatrix} = \begin{bmatrix} .1533 \\ .1413 \\ -.0075 \\ -.1459 \end{bmatrix}$$

Table 10.6 Coefficients for the First Two Discriminant Functions

$$y^{(1)} = .1533x_1 + .1413x_2 - .0075x_3 - .1459x_4$$
$$y^{(2)} = .0813x_1 + .0765x_2 - .2140x_3 + .1896x_4$$

V'^{-1} at this stage is in upper triangular form. The remaining discriminant-function vectors can be computed in a similar manner. The coefficients for the two significant discriminant functions obtained in this analysis are presented in Table 10.6.

The first question that one may want to ask of the results is whether the several groups differ significantly in multivariate mean profiles. This test is analogous to the overall test of significance of differences between means for k groups in the univariate analysis of variance. Unless there is some good a priori reason for expecting that the several groups should "line up" along a single dimension in the multivariate measurement space, one should perhaps require that this test be significant before going on to examine the significance of individual discriminant functions. In cases where the multiple measurements are highly correlated and the groups do tend to order themselves along one, or even a few, of the total possible discriminant functions, the overall test is not powerful. This overall test of significance of differences between groups on all discriminant functions is accomplished by taking the sum of the diagonal elements in $V^{-1}BV'^{-1}$ (Table 10.3) as a chi-square statistic with $p(k - 1)$ degrees of freedom, where p is the number of variables and k is the number of groups. It is worthy noting that the sum of the diagonal elements in the matrix $V^{-1}BV'^{-1}$, which is distributed as chi square with $p(k - 1)$ degrees of freedom, is precisely equal to the sum of all latent roots $\lambda_1, \lambda_2, \ldots, \lambda_r$ of the symmetric matrix $V^{-1}BV'^{-1}$. As will be seen, this total chi square can be partitioned into several independent components for testing the significance of specific discriminant functions. For the present problem, the sum of elements in the principal diagonal of matrix $V^{-1}BV'^{-1}$ (Table 10.3) is 248.5. As a chi square with $p(k - 1) = 4(3)$ degrees of freedom, this value is highly significant (Table A.3). It can be concluded that patients from different marital-status groups have different symptom characteristics. Alternative tests of overall multivariate mean differences will be discussed in the next chapter.

The next question that should be asked concerns the significance of successive individual discriminant functions. The total degrees of freedom associated with multivariate mean differences are divided as follows among the several discriminant functions:

$$p(k - 1) = (p + k - 2) + (p + k - 4) \cdots \tag{10.9}$$

The first characteristic root λ_1 of the matrix $V^{-1}BV'^{-1}$ is approximately distributed as chi square with $p + k - 2 = 6$ degrees of freedom. In this case, the value $\lambda_1 = 195.9$ is highly significant, indicating that the several marital-status groups differ significantly in terms of the composite score defined as the first discriminant function. One would like to continue, perhaps, to test each of the separate roots $\lambda_2, \lambda_3, \ldots, \lambda_r$, but the statistical distributions of the individual roots have not been defined precisely enough for this purpose. It appears better to test the successive composite residuals to determine whether any significant dimensions of group difference remain, in addition to those already accepted as significant.

The strategy favored by most statisticians for determining the number of significant discriminant functions is the following. The total chi square, which is the sum of diagonal elements in $V^{-1}BV'^{-1}$ (or the sum of all latent roots $\lambda_1, \lambda_2, \ldots, \lambda_r$ for that matrix), can be partitioned into r independent components with degrees of freedom $p(k-1) = (p+k-2) + (p+k-4) + \ldots$, as described above. Rather than testing the significance of each root separately, the strategy is to test the total discrimination as a chi square with $p(k-1)$ degrees of freedom. If significant, it is accepted that *at least* one discriminant function is significant, and if any is significant, it should be the one with largest associated variance λ_1. Next the first root λ_1 is subtracted from the total of all roots or from the sum of the diagonal elements in $V^{-1}BV'^{-1}$, and the residual is tested as a chi square with $p(k-1) - (p+k-2)$ degrees of freedom. If this test is significant, it is concluded that *at least* one discriminant function in addition to the first is significant. Next, the residual remaining after subtracting out λ_1 and λ_2 is tested as a chi square with $p(k-1) - (p+k-2) - (p+k-4)$ degrees of freedom to determine whether there is any evidence to suggest that one or more additional dimensions of group separation remain.

The advantage of this approach is that it is consistent in providing tests of consecutive residual matrices. When a residual is not judged to be significant, the testing stops. When an investigator tests individual discriminant functions, for example, χ_1^2 with $p + k - 2$ degrees of freedom, it is possible to obtain indication of significance even though the overall chi square with $p(k-1)$ degrees of freedom fails to suggest a significant systematic difference between groups. The procedure of testing successive residuals appears to be more conservative; yet there are problems in which an ordering of groups may be present such that the groups tend to fall along a single continuum in the measurement space. For example, age groups tend to fall along a single continuum. Although a technique such as multiple-regression analysis should probably be used in the study of age relationships, one can readily appreciate that such ordered relationships may exist among groups constituted in other ways. In such cases, the first discriminant function may be highly significant, and yet the sum of all roots may not appear significant. Similarly, the measurements chosen for group discrimination may be highly correlated, such as the subtests of an IQ test. In this case, any group differences that exist are likely to be in terms of the single common factor, and other dimensions will be essentially error. Where multiple measurements are highly correlated, the largest-root test would seem more appropriate.

Table 10.7 Group Means on the First Two Discriminant Functions

	Single	*Married*	*Divorced*	*Widowed*
$\bar{y}^{(1)}$	.96	.34	.48	.57
$\bar{y}^{(2)}$	.96	.99	.92	1.33

In the analysis of differences among marital-status groups, the general conclusions are the same whether one tests individual roots or residuals. The total discriminable variance of 248.5 is distributed as chi square with 12 degrees of freedom, indicating at least one significant dimension of difference among the marital-status groups. The residual after subtracting out the first root is 248.5 − 195.9 = 52.6, which can be taken as a chi square with 12 − 6 = 6 degrees of freedom. This test indicates that there is at least one significant dimension of group difference *in addition to the first*. The first two roots can be subtracted out and the residual tested as a chi square with 12 − 6 − 4 = 2 degrees of freedom. In this particular example, of course, only three discriminant functions exist among the four marital-status groups, so that the final residual is precisely the third and last root λ_3. Thus, the conclusions resulting from the strategy of testing successive residual variances are the same as those resulting from testing the significance of individual roots in this example. The third discriminant function actually appears to provide significant additional group discrimination; however, since it accounts for less than 10 percent of the sum of all three roots (total discriminable variance), differences between the marital-status groups can be represented quite well in a simple two-dimensional model.

Given that the major portion of multivariate differences among the four marital-status groups can be represented in a two-dimensional space, what is the resulting configuration? How do the groups relate? Group means on the two principal discriminant functions shown in Table 10.7 were computed by applying the discriminant-function weights (Table 10.6) to the group means on the original variables (Table 10.1)

Single:
$$y^{(1)} = .1533(6.63) + .1413(6.83) - .0075(6.36) - .1459(6.66) = .9621$$

$$y^{(2)} = .0813(6.63) + .0765(6.83) - .2140(6.36) + .1896(6.66) = .9632$$

Married:
$$y^{(1)} = .1533(4.77) + .1413(5.41) - .0075(5.84) - .1459(7.59) = .3445$$

$$y^{(2)} = .0813(4.77) + .0765(5.41) - .2140(5.84) + .1896(7.59) = .9910$$

Divorced: $y^{(1)} = .1533(5.11) + .1413(5.25) - .0075(5.60) - .1459(6.85)$
$= .4838$

$y^{(2)} = .0813(5.11) + .0765(5.25) - .2140(5.60) + .1896(6.85)$
$= .9174$

Widowed: $y^{(1)} = .1533(5.75) + .1413(6.32) - .0075(5.23) - .1459(7.93)$
$= .5783$

$y^{(2)} = .0813(5.75) + .0765(6.32) - .2140(5.23) + .1896(7.93)$
$= 1.3353$

The mean values for $y^{(1)}$ and $y^{(2)}$ were used as cartesian coordinates to plot the locations of marital-status groups in the two-dimensional space. It will be recalled that the functions $y^{(1)}$ and $y^{(2)}$ are statistically uncorrelated within groups, so that representation as orthogonal reference axes is appropriate. Results are shown in Fig. 10.1. It is evident that the single group differs substantially from all others and occupies an extreme position in the measurement domain. The separation of single from married and divorced groups is primarily along the axis of the first discriminant function. The widowed group lies close to the divorced on the axis of the first discriminant function, but the widowed group tends to separate from the other groups along the axis of the second discriminant function. Before consideration of the particular nature of symptom-profile differences, it can be concluded that the primary dimension of separation among marital-status groups represents the difference between single

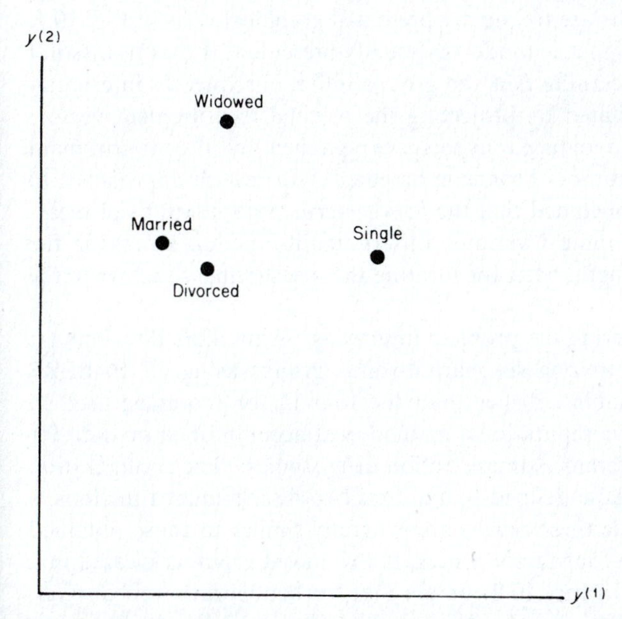

Figure 10.1 Configuration of marital-status groups derived from analysis of four symptom factor scores.

and married patients. The secondary dimension of group difference separates the widowed from other groups.

10.5 Interpretation

The problem of interpreting the nature of discriminant-function dimensions deserves serious consideration. One important way to characterize the dimensions is in terms of the groups they contrast or separate most, as was done in the preceding paragraph. In the past, investigators have attempted to define or describe the nature of the discriminant functions by examining relative magnitudes of the weighting coefficients. This can be hazardous because the magnitudes of the coefficients are dependent upon the units of measurement, which may be different for the different original measures. The effect of differences in units of measurement can largely be removed by *multiplying* each discriminant-function coefficient by the standard deviation of the particular variable to which the weight is applied. Following multiplication by the standard deviations, the relative magnitudes of coefficients can be compared to determine which variables contribute most to definition of the composite function.

10.6 Inserting Measurement Vectors into the Geometric Model

Multiple discriminant analysis provides a geometric model of the similarities and differences among groups in a reduced measurement space. The discriminant functions are accepted as orthogonal coordinate axes, and groups are located in the model by using mean scores on the discriminant functions as coordinate values. Results are frequently presented graphically, as in Fig. 10.1.

A problem with the geometric model, as usually presented, is that it does not provide information concerning *how* the groups differ. Frequently interpretation can be greatly facilitated by projecting the original measurement vectors (say symptoms) into the reduced hyperspace spanned by the discriminant functions. The present authors have experimented with several approaches to this problem and have concluded that the *between-groups* correlations of measurement vectors with canonical variates (discriminant-function scores) is the most conceptually meaningful basis for locating the measurement vectors in the geometric model.

Let us begin by considering the problem intuitively. A multiple discriminant analysis of relationships among six marital-status groups using all 16 BPRS symptom ratings as variables, rather than the four factor scores as used in Sec. 10.4 to illustrate the computational method, was accomplished on data for patients in a series of Veterans Administration drug studies. The configuration of group means in the plane defined by the first two discriminant functions is shown in Fig. 10.2. While these results are generally similar to those obtained previously using only the four factor scores, the widowed group is located in a more extreme position. In Fig. 10.2, we see that the configuration of marital-status groups tends to form a triangle with single, various divorced groups, and widowed at the vertices and the married group somewhere toward the

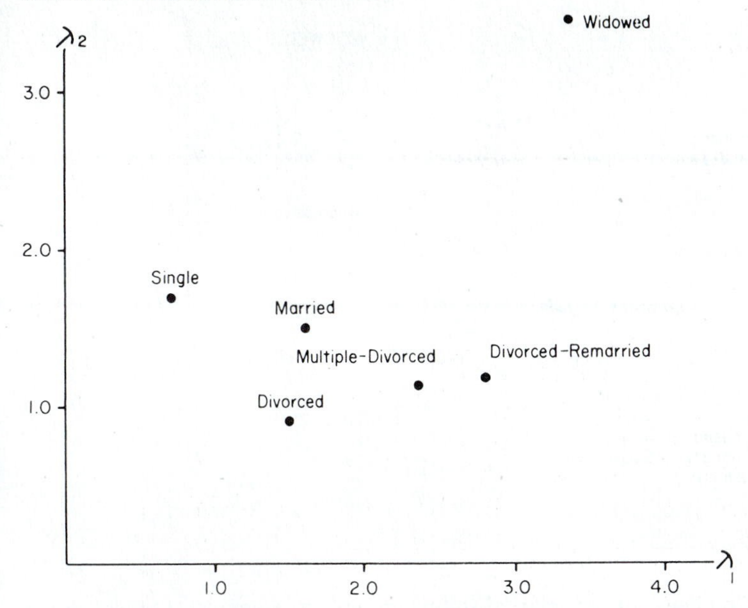

Figure 10.2 Configuration of marital-status groups in Veterans Administration sample derived from analysis of 16 symptom-rating variables.

center. As noted, the geometric representation fails to convey information concerning how the groups differ on the original variables, since the axes are composite canonical variates.

We would like to insert symptom vectors into the model in such manner that they tend to point toward groups having highest mean levels and away from groups having lowest mean levels. More specifically, we would like to position symptom vectors so that the projections of group means on each reflects the *relative* prominence of the symptom in that group. The length of the symptom vector can be used to represent its potency as a discriminator among the groups.

To locate symptom vectors in the geometric model in a manner which satisfies these requirements, one need simply calculate the product-moment correlation coefficient of group means on the original variables with group means on the derived canonical variates. Accepting as the origin for measurement vectors the unweighted centroid (mean) for all groups in the discriminant space, plot the symptom vectors with reference to vertical and horizontal dimensions just as one might plot factor loadings. That is, the *between-groups* correlation of a symptom variable with a canonical variate is taken as the cosine of the angle between the symptom vector and the axis (passing through centroid origin) representing each discriminant function. The *length* of the symptom vector is determined by multiplying the simple between-groups correlations (cosines) by the ratio of between-groups variance to within-groups variance for the particular symptom variable.

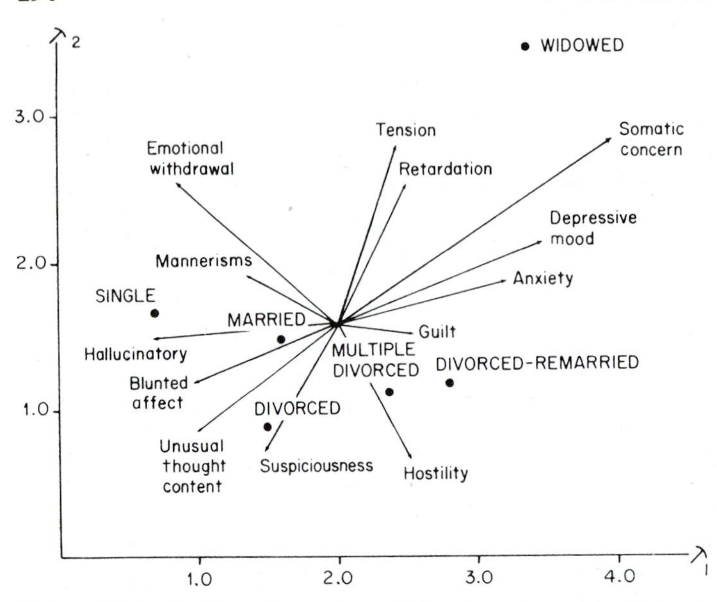

Figure 10.3 Configuration of marital-status groups in Veterans Administration sample with symptom vectors projected into model.

The geometric model obtained by inserting symptom vectors into the configuration of marital-status groups is shown in Fig. 10.3. The correlation between group means on the individual variables and group means on the first two discriminant functions was computed to define the cosines between symptom vectors and discriminant-function axes. These cosines were then multiplied by the simple F ratio associated with group differences on the individual symptom variables. The resulting values were plotted with reference to vertical and horizontal axes (not shown), which passed through the unweighted group means on the discriminant functions.

From the resulting picture, we see that the single group differed from the other groups, particularly the widowed, divorced and remarried, and multiply divorced, by having relatively high levels of emotional withdrawal, hallucinatory behavior, and blunted affect. Conversely, the widowed and divorced-and-remarried groups evidenced *relatively* less of this type of symptomatology. The widowed group evidenced most tension, motor retardation, somatic concern, depressive mood, and anxiety. The divorced group tended to evidence least of the types of symptomatology most prominent in the widowed group. The various divorced groups tended to manifest relatively more suspiciousness and hostility. It should be stressed that, in each instance, the prominence of a symptom is represented relative to the prominence of that same symptom in other groups and *not relative to the prominence of other symptoms* in the same group. Failure to keep this fact constantly in mind can lead to fallacious interpretations. For example, somatic concern may not be very prominent in any group profile, including the profile for the widowed group. It appears as a long vector because the prominence of this symptom is relatively great in the widowed group as compared with other groups. This does not mean that the widowed group should be characterized as having higher levels of somatic concern than of other symptoms.

Because of the manner in which measurement vectors are projected into the model, there is no reason for concern about equality of units of measurement. This can be important in understanding the nature of group differences. It is well known that profiles containing different kinds of measurements, with grossly different units of measurement, are difficult to compare visually. Translation to this type of geometric model eliminates that problem. Obviously, if no more than two or three groups are involved, the approach is of no value.

10.7 Additional Examples

10.7.1 INTRODUCTION

In this section the preliminary statistics required for complete multiple discriminant analyses are provided for a number of clinically relevant problems. As in other chapters, the solutions are omitted so that the student can get a better feel for the actual application of the methodology. Assuming that any reader feels confident that he can obtain basic dispersion matrices and mean vectors, these preliminary results are provided to conserve space and to eliminate the need for tedious routine calculations. Similarly, the triangular square-root inverse has been computed for each within-groups covariance matrix on the assumption that this represents only tedious routine computation. Given these basic data, the discriminant-function analysis and interpretation are left for the reader to accomplish. It should be noted that the preliminary statistics presented are exactly those which result from use of the general multivariate-analysis computer program described in Chap. 2.

10.7.2 DRUG ASSIGNMENT AND MANIFEST PSYCHOPATHOLOGY

The first problem to be considered involves an examination of the multivariate psychiatric symptom differences among 670 patients treated clinically in four different doctor's-choice treatment groups. Each patient was interviewed, and symptomatology was rated on the BPRS by one of three clinical psychologists. Independently of the symptom evaluations which were accomplished for research purposes, the treating psychiatrist assigned each patient to treatment with antidepressant, major tranquilizer, minor tranquilizer, or no drug.

Mean scores on the BPRS factors of thinking disturbance, withdrawal-retardation, hostile-suspiciousness and anxious depression for the four treatment groups and the within-groups and between-groups SP matrices are presented in Table 10.8. The triangular square-root inverse of the within-groups covariance

Table 10.8 Preliminary Statistics for Drug-assignment Example

MEAN FACTOR SCORES FOR PATIENTS IN FOUR TREATMENT GROUPS

	Treatment group			
Factor	*Antidepressant*	*Major Tranquilizer*	*Minor Tranquilizer*	*No Drug*
Thinking disturbance	2.5	5.5	2.3	3.9
Withdrawal-retardation	3.5	3.6	3.7	3.8
Hostile-suspiciousness	5.8	7.9	5.4	6.5
Anxious depression	9.2	7.8	9.2	8.3

WITHIN-GROUPS SP MATRIX FOR FOUR TREATMENT GROUPS, df = 666

$$\begin{bmatrix} 9{,}348.3540 & 1{,}785.3644 & 3{,}852.6332 & -34.5704 \\ & 8{,}528.6288 & 115.9970 & 772.7017 \\ & & 9{,}370.7477 & -738.5452 \\ & & & 7{,}614.0004 \end{bmatrix}$$

BETWEEN-GROUPS SP MATRIX FOR FOUR TREATMENT GROUPS, df = 3

$$\begin{bmatrix} 1{,}055.0205 & -2.1793 & 779.0861 & -482.4399 \\ & 7.5741 & -9.8209 & -5.5196 \\ & & 584.2433 & -349.3114 \\ & & & 227.2070 \end{bmatrix}$$

matrix is provided in Table 10.9. These preliminary statistics were obtained using the general multivariate-analysis program presented in Chap. 2. From these basic statistics a complete multiple discriminant analysis can be calculated.

What can be said concerning differences between the drug-treatment groups? Are all the significant differences adequately represented by a single linear function of the four factor scores which corresponds roughly to the difference between depressive and schizophrenic symptomatology? Are patients who are treated with a minor tranquilizer more like those treated with a major tranquilizer than they are like those treated with an antidepressant? Which drug-treatment group differs most from the no-drug group? Does it appear feasible to use symptom-rating profiles as a basis for treatment assignment?

Table 10.9 Computation of Triangular Square-root Inverse of Within-groups Covariance Matrix for Drug-treatment Problem

14.036567	2.680727	5.784734	−.051907	1.000000	.000000	.000000	.000000
	12.805749	.174169	1.160212	.000000	1.000000	.000000	.000000
		14.070191	−1.108926	.000000	.000000	1.000000	.000000
			11.432433	.000000	.000000	.000000	1.000000
3.746540	.715520	1.544020	−.013854	.266912	.000000	.000000	.000000
2.680727	3.506248	−.265414	.333725	−.054468	.285205	.000000	.000000
5.784734	.174169	3.408188	−.293105	−.125162	.022210	.293411	.000000
−.051907	1.160212	−1.108926	3.351858	−.004418	−.026454	.025657	.298341

Table 10.10 Preliminary Statistics for Ethnic-group Example

MEAN SYMPTOM FACTOR SCORES FOR THREE ETHNIC GROUPS

Factor	*Anglo*	*Negro*	*Latin*
Thinking disturbance	5.2	5.6	5.7
Withdrawal-retardation	5.6	5.9	6.5
Hostile-suspiciousness	6.2	4.6	5.0
Anxious depression	7.6	5.4	6.8

WITHIN-GROUPS SP MATRIX FOR THREE ETHNIC GROUPS, df = 3,416

$$\begin{bmatrix} 59{,}376.7905 & 26{,}345.2230 & 29{,}409.3125 & 12{,}055.6954 \\ & 56{,}527.5977 & 20{,}035.6856 & 21{,}300.0880 \\ & & 60{,}693.7119 & 23{,}409.8517 \\ & & & 61{,}605.9657 \end{bmatrix}$$

BETWEEN-GROUPS SP MATRIX FOR THREE ETHNIC GROUPS, df = 2

$$\begin{bmatrix} 74.9785 & 88.6087 & -306.8132 & -373.7507 \\ & 143.3824 & -305.6538 & -303.6471 \\ & & 1{,}339.3139 & 1{,}732.6587 \\ & & & 2{,}355.9196 \end{bmatrix}$$

10.7.3 RACE AND MANIFEST PSYCHOPATHOLOGY

In the chapter on simple discriminant function, differences between Anglo and Negro ethnic groups were considered. A third ethnic subgroup composed almost entirely of Latin American patients was not included in the previous analysis. Does the third group fall along the continuum separating the Anglo and Negro groups in the psychiatric symptom space? What can be concluded about the symptom relationships among Anglo, Negro, and Latin American groups in the psychiatric population? The four higher-order syndrome factors derived from BPRS ratings provide the basic data.

Mean vectors plus within-groups and between-groups SP matrices are presented in Table 10.10. The triangular square-root inverse of the within-groups covariance matrix is presented in Table 10.11. These statistics provide

Table 10.11 Computation of Triangular Square-root Inverse V^{-1} of Within-groups Covariance Matrix for Ethnic-group Problem

17.381964	7.712301	8.609283	3.529184	1.000000	.000000	.000000	.000000
	16.547891	5.865247	6.235388	.000000	1.000000	.000000	.000000
		17.767480	6.853001	.000000	.000000	1.000000	.000000
			18.034533	.000000	.000000	.000000	1.000000
4.169168	1.849841	2.064988	.846496	.239855	.000000	.000000	.000000
7.712301	3.622979	.564548	1.288857	−.122467	.276015	.000000	.000000
8.609283	5.865247	3.631058	1.205536	−.117365	−.042914	.275401	.000000
3.529184	6.235388	6.853001	3.768753	.025550	−.080666	−.088094	.265339

the basis for a multiple discriminant analysis. Note that the computer program presented in Chap. 2 provides these same basic statistics for any multivariate analysis involving several groups.

10.7.4 PARANOID, SCHIZOPHRENIC, AND DEPRESSIVE DIAGNOSTIC CLASSES

Diagnostic-prototype profiles drawn from the cross-cultural study of diagnostic nomenclature were employed for this investigation of multivariate relationships. The total collection of diagnostic prototypes was divided into major classifications of paranoid (including paranoid schizophrenia), schizophrenic, and depressive. Factor scores for thinking disturbance, withdrawal-retardation, hostile-suspiciousness and anxious depression were computed. Mean profiles and within-groups and between-groups SP matrices are presented in Table 10.12. The triangular square-root inverse of the within-groups covariance matrix is presented in Table 10.13. What can be said about the relationships among the three groups in the symptom space? Is the paranoid group more like the schizophrenic group or the depressive group?

10.7.5 PARANOID SCHIZOPHRENIA IN FOUR DIFFERENT COUNTRIES

The cross-cultural study of diagnostic classification concepts yielded prototype profiles representing paranoid schizophrenia in each of four European countries. The diagnostic terminology is similar enough so that one might infer that

Table 10.12 Preliminary Statistics for Diagnostic Example

MEAN SYMPTOM FACTOR SCORES FOR THREE DIAGNOSTIC-PROTOTYPE PROFILES

Factor	*Schizophrenic*	*Depressive*	*Paranoid*
Thinking disturbance	13.2	7.9	14.9
Withdrawal-retardation	13.9	12.1	9.1
Hostile-suspiciousness	11.4	8.1	13.9
Anxious depression	7.7	18.0	8.8

WITHIN-GROUPS SP MATRIX FOR THREE DIAGNOSTIC-PROTOTYPE PROFILES, df = 2,954

$$\begin{bmatrix} 47{,}569.2774 & 13{,}858.1307 & 17{,}875.9055 & 6{,}679.3140 \\ & 45{,}120.6343 & 12{,}400.1460 & 1{,}495.0423 \\ & & 40{,}434.0939 & 6{,}669.2591 \\ & & & 32{,}432.3752 \end{bmatrix}$$

BETWEEN-GROUPS SP MATRIX FOR THREE DIAGNOSTIC-PROTOTYPE PROFILES, df = 2

$$\begin{bmatrix} 19{,}452.4749 & -4{,}591.5289 & 15{,}263.6535 & -28{,}696.6104 \\ & 12{,}721.8471 & -6{,}786.2988 & -2{,}674.8792 \\ & & 12{,}847.6532 & -19{,}932.6717 \\ & & & 50{,}004.3850 \end{bmatrix}$$

Table 10.13 Computation of Triangular Square-root Inverse V^{-1} of Within-groups Covariance Matrix for the Paranoid, Schizophrenic, and Depressive Stereotypes

16.103343	4.691310	6.051423	2.261108	1.000000	.000000	.000000	.000000
	15.274419	4.197747	.506107	.000000	1.000000	.000000	.000000
		13.687912	2.257704	.000000	.000000	1.000000	.000000
			10.979138	.000000	.000000	.000000	1.000000
4.012897	1.169058	1.507993	.563460	.249196	.000000	.000000	.000000
4.691310	3.729305	.652886	−.040921	−.078117	.268146	.000000	.000000
6.051423	4.197747	3.314755	.432830	−.097981	−.052815	.301681	.000000
2.261108	.506107	2.257704	3.236144	−.031271	.010454	−.040349	.309009

schubförmige paranoide Schizophrenie (German), schizophrenia paranoidní (Czechoslovak), schizophrénie paranoïde (French), and schizophrenia paranoide (Italian) all represent the same diagnostic classification.

Factor scores for thinking disturbance, withdrawal-retardation, hostile-suspiciousness, and anxious depression were computed for each diagnostic profile. Mean vectors and SP matrices are presented in Table 10.14. The triangular square root of the within-groups covariance matrix is presented in Table 10.15. Do the diagnostic prototypes for paranoid schizophrenia differ significantly in the various countries? In which country is the diagnostic concept most different? What is the configuration of the four national samples in the two-dimensional space defined by the first two discriminant functions?

Table 10.14 Preliminary Statistics for Paranoid Stereotype Example

MEAN FACTOR SCORES FOR PARANOID STEREOTYPES FROM FOUR COUNTRIES

Factor	*German*	*Czech*	*French*	*Italian*
Thinking disturbance	11.7129	12.4888	18.4710	17.5111
Withdrawal-retardation	8.9537	8.8111	14.5785	11.0222
Hostile-suspiciousness	14.5000	15.8000	13.9090	16.4444
Anxious depression	8.7592	10.0888	8.3388	9.0666

WITHIN-GROUPS SP MATRIX FOR PARANOID STEREOTYPES FROM FOUR COUNTRIES, df = 360

$$\begin{bmatrix} 2{,}994.9839 & 695.3896 & 986.2596 & 468.7785 \\ & 2{,}971.0393 & 597.0191 & 297.5217 \\ & & 3{,}678.5111 & 448.9939 \\ & & & 2{,}995.9839 \end{bmatrix}$$

BETWEEN-GROUPS SP MATRIX FOR PARANOID STEREOTYPES FROM FOUR COUNTRIES, df = 3

$$\begin{bmatrix} 3{,}469.6864 & 2{,}728.7202 & -304.7211 & -431.4378 \\ & 2{,}432.2574 & -515.8653 & -447.3019 \\ & & 315.9827 & 182.3329 \\ & & & 165.9942 \end{bmatrix}$$

Table 10.15 Computation of Triangular Square-root Inverse V^{-1} of Within-groups Covariance Matrix for the Paranoid Diagnostic Stereotypes from Four European Countries

8.322177	1.931637	2.739610	1.302162	1.000000	.000000	.000000	.000000
	8.252886	1.658386	.826449	.000000	1.000000	.000000	.000000
		10.218086	1.247205	.000000	.000000	1.000000	.000000
			9.847047	.000000	.000000	.000000	1.000000
2.884818	.669587	.949664	.451384	.346642	.000000	.000000	.000000
1.931637	2.793660	.366008	.187641	−.083083	.357953	.000000	.000000
2.739610	1.658386	3.030224	.247461	−.098601	−.043235	.330008	.000000
1.302162	.826449	1.247205	3.089798	−.037697	−.018275	−.026430	.323945

10.8 Computer Program for Multiple Discriminant Analysis

Although multiple-discriminant-analysis programs are available in many computer centers, the program presented here is unique in some respects and provides results in a form consistent with considerations regarding interpretation discussed in this chapter. The program is written in Fortran IV for the IBM 1130/1800 disk-oriented computer system and is designed to complete in a single pass through the computer numerous discriminant-function analyses involving the same multivariate measurements but different classification variables. The program then completes a separate multiple discriminant analysis between categories or groups for each classification variable. The raw data are stored on the disk and retrieved for the successive discriminant analyses.

Input

The setup for use of the discriminant-analysis program involves one general control card and data cards followed by three control cards for each discriminant analysis that is to be accomplished.

1 One general control card containing three parameters: number of individuals NOBS, total number of classification variables to be used as basis for different analyses NHIST, and number of variables in multivariate score profile NVAR punched in successive four-column fields.

2 Raw-data cards containing (group) classification variables coded 0 to 9 and multiple measurement variables that are to be the basis for discrimination. Data are read in format 600.

3 Three control cards for *each* analysis to be completed.

- **a** Name of classification variable to be used for identification purposes.
- **b** Card containing ordinal position among the NHIST classification variables of the group classification with which the particular analysis is concerned and the number of groups NGPS to be considered in the analysis. These entries are punched in successive four-column fields.
- **c** A *regrouping control card* with integer group designation punched into columns 1 to 10. The number in each column represents the analysis group into which individuals scored with that ordinal position code should

be placed. The purpose of this card is to provide a means for flexible regrouping at the time of analysis without requiring recoding. For example, suppose that marital status is the classification variable to be analyzed and that it is coded 1 = single, 2 = married, 3 = separated, 4 = divorced, 5 = widowed. For the analysis, suppose that separated and divorced groups are to be combined so that the control card should appear as follows: 0123340000.

Output

1. Identification of classification variable and reproduction of the regrouping control card
2. Frequencies in groups
3. Within- and between-groups SP matrices
4. Total chi square derived as sum of diagonal elements in $V^{-1}BV'^{-1}$
5. Discriminant-function coefficients in raw-score and standard form
6. The latent root associated with each discriminant function
7. The group means on the canonical variates or discriminant functions
8. The Mahalanobis D^2 values between all pairs of groups
9. Group means on the original variables
10. Test vector projections on the discriminant functions (if as many as four groups are involved)

Mainline Program for Multiple Discriminant Analysis

```
      DIMENSION A(18,18 ),B(18,18),W(18,36),ASUM(18),BSUM(18),NG(10),DAT
     1A(20),HIST(31),N(10),XBAR(18)
      COMMON L,L2,L3,L4
      DEFINE FILE 1(100,100,U,L), 2(100,100,U,L2), 3(1500,100,U,L3), 4(
     110,100,U,L4)
    1 FORMAT(20I4)
    2 FORMAT (1H0,5I6)
    3 FORMAT(10I1)
   12 FORMAT(     40H
      NLAT=0
      READ(2,1)NOBS,NHIST,NVAR
      WRITE(3,2)NOBS,NHIST,NVAR
      NT=NHIST
      L3=1
  600 FORMAT(15X,16F1.0,29X,F1.0)
      DO 11 I=1,NOBS
      READ(2,600)(DATA(M5),M5=1,NVAR),(HIST(N5),N5=1,NHIST)
   11 WRITE(3'L3)(DATA(M5),M5=1,NVAR),(HIST(N5),N5=1,NHIST)
      DO 500 KKKK = 1,NT
      READ(2,12)
      WRITE(3,12)
      READ(2,1) NITEM,NGPS
      READ(2,3) (N(J),J=1,10)
      WRITE(3,51)(N(J),J=1,10)
   51 FORMAT(' EXPANDED-' 10I1)
      CALL SPMAT(NVAR,NGPS,ASUM,BSUM,A,B,W,NG,MGPS,NOBS,NITEM,N,NHIST)
      IF (MGPS-1) 5,500,5
    5 CONTINUE
      CALL TRISQ(NVAR,W,A)
      CALL PDMTX (W,B,NVAR,A)
      CALL ROOT(NVAR,A,B,W,ASUM,BSUM,NLAT,MGPS,NOBS)
      CALL DSQ(NVAR,NLAT,MGPS,XBAR,W,A)
      IF(MGPS-4)500,400,400
  400 CALL VPRO(NVAR,NLAT,MGPS,W)
  500 CONTINUE
      CALL EXIT
      END
```

Subroutine to Calculate Within- and Between-groups SP Matrices

```
      SUBROUTINE SPMAT(NVAR,NGPS,ASUM,BSUM,A,B,W,NG,MGPS,NOBS,NITEM,N,
     1NHIST)
      DIMENSION ASUM(18),BSUM(18),A(18,18),B(18,18),W(18,36),
     1DATA(20),XBAR(18),NG(10),HIST(31),N(10)
      COMMON L,L2,L3,L4
      L4=1
      TOT=0.0
      DO 20 M=1,NVAR
      ASUM(M)=0.0
      BSUM(M)=0.0
      DO 20 M2=1,NVAR
      A(M,M2)=0.0
      B(M,M2)=0.0
   20 W(M,M2)=0.0
      L=1
      MGPS=NGPS
      DO 126 K=1,NGPS
      DO 526 J5=1,NVAR
  526 ASUM(J5)=0.0
      NX = 0
      L3 = 1
      DO 25 J = 1,NOBS
      READ (3'L3)(DATA(M5),M5=1,NVAR),(HIST(N5),N5=1,NHIST)
      JJ = HIST(NITEM)
      JJ =JJ +1
      IF (N(JJ)-K) 25,24,25
   24 NX=NX+1
      DO 25 M=1,NVAR
      ASUM(M)=ASUM(M)+DATA(M)
      BSUM(M)=BSUM(M)+DATA(M)
      DO 25 M2=M,NVAR
      XDAT=DATA(M)*DATA(M2)
      A(M,M2)=A(M,M2)+XDAT
      B(M,M2)=B(M,M2)+XDAT
   25 CONTINUE
      FNX = NX
      TOT = TOT + FNX
      IF (NX-8) 30,30,31
   30 MGPS=MGPS-1
      DO 32 M = 1,NVAR
      BSUM(M) = BSUM(M) - ASUM(M)
      DO 32 M2 = M,NVAR
      B(M,M2) = B(M,M2)-A(M,M2)
   32 A(M,M2)= 0.0
      TOT =TOT - FNX
      GO TO 126
   31 DO 26 M =1,NVAR
      XBAR(M)=ASUM(M)/FNX
      DO 26 M2=M,NVAR
      W(M,M2)=W(M,M2)+A(M,M2)-ASUM(M)*ASUM(M2)/FNX
   26 A(M,M2)=0.0
      WRITE (1'L)(XBAR(M),M=1,NVAR)
      NG(K)=NX
      WRITE(3,503) K,NG(K)
  503 FORMAT(2X,I4,4X,I5)
  126 CONTINUE
      WRITE(3,502)
      DO 27 M=1,NVAR
      DO 27 M2=M,NVAR
      W(M2,M)=W(M,M2)
      B(M,M2)=B(M,M2)-BSUM(M)*BSUM(M2)/TOT
      B(M,M2)=B(M,M2)-W(M,M2)
      B(M2,M)=B(M,M2)
   27 WRITE(3,501)W(M2,M),B(M2,M),M2,M
      FNGPS=MGPS
      DFTOT=TOT-FNGPS
      DO 202 J=1,NVAR
  202 W(J,36)=(B(J,J)*(DFTOT))/(W(J,J)*(MGPS-1))
      WRITE (4'L4) (W(M,36),M=1,NVAR)
      L4=10
      WRITE(4'L4)(W(J,J),J=1,NVAR)
      DO 23 M2=1,NVAR
      DO 23 M=1,NVAR
      W(M,M2)=W(M,M2)/DFTOT
```

```
 23 A(M,M2)=W(M,M2)
501 FORMAT(1X,2F18.4,2I5)
502 FORMAT(1X,'ELEMENTS OF WITHIN AND BETWEEN MATRICES')
    RETURN
    END
```

Subroutine to Calculate Triangular Square-root Inverse

```
    SUBROUTINE TRISQ(NVAR,W,A)
    DIMENSION W(18,36),A(18,18)
    COMMON L,L2,L3,L4
    N2=2*NVAR
    N1=NVAR+1
    DO 411 J=N1,N2
    DO 411 I=1,NVAR
411 W(I,J)=0.0
    DO 412 J=N1,N2
    I=J-NVAR
412 W(I,J)=1.0
    W(1,1)=SQRT(W(1,1))
    X=1.0/W(1,1)
    DO 413 J = 2,N2
413 W(1,J)=W(1,J)*X
    DO 415 K=2,NVAR
    L=K-1
    DO 414 M=1,L
414 W(K,K)=W(K,K)-W(M,K)*W(M,K)
    W(K,K)=SQRT(W(K,K))
    L2=K+1
    DO 416 I= L2,N2
    DO 417 M=1,L
417 W(K,I)=W(K,I)-W(M,K)*W(M,I)
416 W(K,I)=W(K,I)/W(K,K)
415 CONTINUE
    DO 425 J=1,NVAR
    DO 425 J1=1,NVAR
425 W(J,J1)=0.0
    DO 420 J=N1,N2
    DO 420 I=1,NVAR
    K=J-NVAR
    W(I,K)=W(I,J)
420 W(I,J) = A(I,K)
    RETURN
    END
```

Subroutine to Form Symmetric Matrix $V^{-1}BV'^{-1}$

```
    SUBROUTINE PDMTX(W,B,NVAR,A)
    DIMENSION W(18,36),A(18,18),B(18,18)
    COMMON L,L2,L3,L4
    DO 101 M=1,NVAR
    DO 101 M2=1,NVAR
    A(M,M2)=0.0
    DO 101 M3=1,NVAR
101 A(M,M2)=A(M,M2)+W(M,M3)*B(M3,M2)
    DO 102 M=1,NVAR
    DO 102 M2=1,NVAR
    B(M,M2)=0.0
    DO 103 M3=1,NVAR
103 B(M,M2)=B(M,M2)+A(M,M3)*W(M2,M3)
102 CONTINUE
    RETURN
    END
```

Subroutine to Compute Latent Roots and Vectors of Symmetric Matrix $V^{-1}BV'^{-1}$

```
SUBROUTINE ROOT(NVAR,A,B,W,ASUM,BSUM,NLAT,NGPS,NOBS)
DIMENSION ASUM(18),BSUM(18),B(18,18),W(18,36),A(18,18),CSUM(18)
```

```
      COMMON L,L2,L3,L4
    5 FORMAT(/1X,'DISCRIMINANT FUNCTION COEFFICIENTS')
    6 FORMAT(/1X,'LATENT ROOT NUMBER',I3,F12.2)
    7 FORMAT(1X,F10.4,2I5,F10.4)
      SUMRT=0.0
      TACT=0.0
      DFTR=NVAR*(NGPS-1)
      SAM=0.0
      DO 600 J=1,NVAR
  600 SAM=SAM+B(J,J)
      XSQ=SAM*0.90
      WRITE (3,802)SAM,DFTR
      NDFL=NVAR+NGPS-2
      WRITE (3,5)
      L2=0.0
  130 DO 110 M=1,NVAR
  110 ASUM(M)=1.0
      L2=L2+1
      L4=10
  120 VAR=0.0
      DO 112 M=1,NVAR
      BSUM(M)=0.0
      DO 111 M2=1,NVAR
  111 BSUM(M)=BSUM(M)+ASUM(M2)*B(M,M2)
  112 VAR=VAR+BSUM(M)*ASUM(M)
      SCALE=SQRT(VAR)
      TEST= .0001
      K=0
      DO 113 M=1,NVAR
      BSUM(M)=BSUM(M)/SCALE
      TEST2=(ASUM(M)-BSUM(M))**2
      IF (TEST2-TEST) 113,113,114
  114 K=K+1
  113 CONTINUE
      IF (K) 115,115,116
  116 DO 117 M=1,NVAR
  117 ASUM(M)=BSUM(M)
      GO TO 120
  115 SSQ=0.0
      DO 121 M=1,NVAR
      DO 119 M2=1,NVAR
  119 B(M,M2)=B(M,M2)-ASUM(M)*ASUM(M2)
  121 SSQ=SSQ+ASUM(M)*ASUM(M)
      FLAM=SSQ
      TACT=TACT+FLAM
      SSQ=SQRT(SSQ)
  199 VAR2=0.0
      DO 280 M=1,NVAR
      BSUM(M)=0.0
      DO 280 M2=1,NVAR
  280 BSUM(M)=BSUM(M)+ASUM(M2)*W(M2,M)
      DO 281 M=1,NVAR
  281 ASUM(M)=BSUM(M)
      DO 903 N2=1,NVAR
      K=N2+NVAR
      DO 903 N3=1,NVAR
  903 VAR2=VAR2+W(N3,K)*ASUM(N3)*ASUM(N2)
      SCL=SQRT(VAR2)
      READ(4'L4) (CSUM(N2),N2=1,NVAR)
      DO 905 N2=1,NVAR
      ASUM(N2)=ASUM(N2)/SCL
      A(L2,N2)=ASUM(N2)
    9 CSUM(N2)=SQRT(CSUM(N2)/(NOBS-NGPS))
      CSUM(N2)=ASUM(N2)*CSUM(N2)
  905 WRITE (3,7)ASUM(N2),L2,N2,CSUM(N2)
  802 FORMAT(/1X,'TOTAL VARIATION',E18.8,3X,'CHI SQ WITH DF',F5.0)
      WRITE (3,6)L2,FLAM
      SUMRT=FLAM+ SUMRT
      IF(SUMRT-XSQ)775,775,777
  775 NLAT=L2
      NDFL=NDFL-2
      GO TO 130
  777 CONTINUE
      NLAT= L2
      RETURN
      END
```

Subroutine to Calculate Group Means on Canonical Variates and the Mahalanobis D^2 Between Groups

```
      COMPUTE CANONICAL VARIATE MEANS AND D-SQUARES
      SUBROUTINE DSQ(NVAR,NLAT,NGPS,XBAR,W,A)
      DIMENSION XBAR(18),W(18,36),A(18,18),ASUM(18),BSUM(18)
      COMMON L,L2,L3,L4
      WRITE(3,8)
    8 FORMAT(/1X,'CANONICAL VARIATE MEANS')
      L=1
      DO 4 K5 =1,NVAR
      BSUM(K5)=0.0
    4 ASUM(K5)=0.0
      GPS=NGPS
      DO 212 M=1,NGPS
      READ (1'L)(XBAR(K5),K5=1,NVAR)
      DO 213 K4 = 1,NVAR
  213 ASUM(K4) = ASUM(K4)+XBAR(K4)
      DO 211 M2=1,NLAT
      W(M2,M)=0.0
      DO 211 M3=1,NVAR
  211 W(M2,M)=W(M2,M)+XBAR(M3)*A(M2,M3)
      WRITE (3,10)
  212 WRITE(3,9)(W(M2,M),M2=1,NLAT)
      DO 214 K4 = 1,NVAR
  214 ASUM(K4) = ASUM(K4)/GPS
      DO 111 M2 = 1,NLAT
      DO 110 N= 1,NGPS
  110 BSUM(M2)=BSUM(M2)+W(M2,N)
  111 BSUM(M2)=BSUM(M2)/GPS
   10 FORMAT(1H )
    9 FORMAT(1X,8F10.4)
C     COMPUTE D-SQUARE COEFFICIENTS
      WRITE(3,11)
   11 FORMAT(/1X,'D SQUARE COEFFICIENTS BETWEEN GROUPS')
      NGP2=NGPS-1
      DO 310 M=1,NGP2
      M3=M+1
      DO 310 M2=M3,NGPS
      SSDIF=0.0
      DO 311 J=1,NLAT
  311 SSDIF=SSDIF+(W(J,M)-W(J,M2))**2
  310 WRITE(3,7)SSDIF,M,M2
    7 FORMAT(1X,F10.4,2I5)
C     GROUP MEANS ON ORIGINAL VARIABLES
      L=1
      WRITE(3,666)
  666 FORMAT(/1X,'GROUP MEANS ON ORIGINAL VARIABLES')
      DO 670 M=1,NGPS
      L8=L
      READ(1'L) (XBAR(K5),K5=1,NVAR)
      WRITE(3,667)M
  667 FORMAT(1X,'GROUP',I7)
  673 WRITE(3,9)(XBAR(K5),K5=1,NVAR)
  670 CONTINUE
      RETURN
      END
```

Subroutine to Calculate Test-vector Projections in Discriminant-function Space

```
      SUBROUTINE VPRO(NVAR,NLAT,NGPS,W)
      DIMENSION A(18,18),B(18,18),W(18,36),XSUM(18),XSQR(18),YSUM(18),YS
     1QR(18),SCLE(18)
      COMMON L,L2,L3,L4
      L=1
      DO 10 M=1,NGPS
   10 READ(1'L)(B(M,K5),K5=1,NVAR)
      DO 11 M=1,NVAR
      YSUM(M)=0.0
   11 YSQR(M)=0.0
      DO 12 M=1,NLAT
      XSUM(M)=0.0
   12 XSQR(M)=0.0
      DO 14 N=1,NGPS
      DO 13 M2=1,NLAT
      XSUM(M2)=XSUM(M2)+W(M2,N)
   13 XSQR(M2)=XSQR(M2)+W(M2,N)*W(M2,N)
      DO 14 M2=1,NVAR
      YSUM(M2)=YSUM(M2)+B(N,M2)
   14 YSQR(M2)=YSQR(M2)+B(N,M2)*B(N,M2)
      L4=1
      READ(4'L4)(SCLE(J),J=1,NVAR)
      DO 300 M=1,NLAT
      DO 300 M2=1,NVAR
      A(M,M2)=0.0
      DO 300 M3=1,NGPS
  300 A(M,M2)=A(M,M2)+W(M,M3)*B(M3,M2)
      DO 400 I=1,NLAT
      DO 400 J=1,NVAR
  400 B(I,J)=(A(I,J)-(XSUM(I)*YSUM(J))/NGPS)/((SQRT(XSQR(I)-( XSUM(I)**2
     1)/NGPS))*(SQRT(YSQR(J)- (YSUM(J)**2)/NGPS)))
      WRITE(3,5)
    5 FORMAT (1X,'TEST VECTOR PROJECTIONS SCALED UNIT LENGTH')
      DO 500 J=1,NVAR
  500 WRITE(3,501)(B(I,J),I=1,NLAT)
  203 FORMAT(16F6.1)
      WRITE(3,203)(SCLE(J),J=1,NVAR)
      WRITE(3,6)
    6 FORMAT (1X,'TEST VECTOR PROJECTIONS SCALED DISCRIMINANT LENGTH')
      DO 504 J=1,NVAR
      DO 502 I=1,NLAT
  502 B(I,J)=B(I,J)*SCLE(J)
  504 WRITE(3,501)(B(I,J),I=1,NLAT)
  501 FORMAT (10F12.4)
      RETURN
      END
```

11
Tests of Multivariate Hypotheses

11.1 Introduction

In the preceding chapters, tests were described for determining the significance of multivariate mean differences in discriminant-function analysis. In the present chapter, several other multivariate tests of significance which appear to have special implications for clinical research will be summarized. Certain of these tests of significance are equivalent to those discussed in connection with discrimination problems, while others are different. The complexity of multivariate hypotheses has resulted in an array of alternative tests of significance. While it is frequently possible to use any one of several different methods to test a particular hypothesis, the alternatives may not have the same power or appropriateness. In this chapter, some of the more useful multivariate tests will be described briefly, and an attempt will be made to illustrate how they can be used in clinical research. For a technical discussion of the distribution theory associated with these tests, the reader should consult a text on mathematical statistics, such as Anderson (1958).

11.2 Multivariate Generalization of *t* Test

11.2.1 DEFINITION

Hotelling (1931) defined the distribution for the statistic T^2, which is a multivariate generalization of the univariate Student's t. When hypotheses are multivariate analogs of univariate hypotheses that can be tested using the familiar t test, they can be tested using the Hotelling T^2 statistic. Such hypotheses include (1) equivalence of multivariate mean vectors derived from two independent samples, (2) equivalence of multivariate mean vectors for paired observations such as derived from a test-retest situation, (3) equivalence of sample mean vector to hypothesized population mean vector, plus (4) several types of tests that are peculiar to the multivariate situation. In general, if one has several different variables to consider, the T^2 statistic can be used for an overall multivariate test comparable to any univariate t test that might be undertaken for each variable separately.

The Student's t statistic for testing the significance of the deviation of a sample mean $\bar{x}$ about the hypothesized population mean μ can be written

$$t = \frac{\bar{x} - \mu}{\sqrt{s^2/n}} = \sqrt{n}\,\frac{\bar{x} - \mu}{s}$$

Hotelling's T^2 statistic can be written in similar form by substituting mean *vectors* and *covariance matrix* for the univariate means and variances. As in the preceding chapter (Sec. 10.3), let V' be the upper triangular square-root factor of the within-groups covariance matrix $\mathbf{\Sigma}$ such that $VV' = \mathbf{\Sigma}$, and let V^{-1} and V'^{-1} be the inverses of these respective matrices. Then $V'^{-1}V^{-1} = \mathbf{\Sigma}^{-1}$. By analogy to the univariate t statistic, we define

$$T = \sqrt{n}\,\frac{(\bar{\mathbf{x}} - \mathbf{u})'}{V'} = \sqrt{n}(\bar{\mathbf{x}} - \mathbf{u})'V'^{-1}$$

and

$$T^2 = n(\bar{\mathbf{x}} - \mathbf{u})'\mathbf{\Sigma}^{-1}(\bar{\mathbf{x}} - \mathbf{u}) \tag{11.1}$$

where $\bar{\mathbf{x}}$ is the mean vector for the sample, $\mathbf{u}$ is the hypothesized population mean vector, and $\mathbf{\Sigma}^{-1} = V'^{-1}V^{-1}$ is the inverse of the sample covariance matrix (Sec. 2.7.6).

11.2.2 TEST OF THE HYPOTHESIS THAT THE SAMPLE WAS DRAWN FROM POPULATION HAVING ZERO MEAN VECTOR

One of the problems for which the univariate t test is frequently employed is to test the hypothesis that the mean in the population from which a random sample was drawn is $\mu = 0$. This hypothesis is frequently tested where concern is with simple difference scores or change scores. If there has been no systematic change between pre- and posttreatment measurements, the mean *difference* score should be zero. A multivariate analogy is the T^2 test of the hypothesis that the multivariate mean *vector* in the population is a null vector, $\boldsymbol{\mu}_0' = [0 \;\; 0 \;\; 0 \;\; \cdots \;\; 0]$.

Suppose that one obtains p different measurements, or different kinds of measurements, on n subjects prior to treatment and then obtains the same p

measurements after treatment. If there has been no systematic effect, the difference between pre- and postmeasures should be zero (within chance limits) for each variable separately and thus the mean change-score vector should be zero. Assuming the p variates to have a multivariate normal distribution, Hotelling's T^2 statistic can be used to test the hypothesis that no systematic change is reflected in the complete p-variate vector of change-score means.

Since the hypothesized mean vector is a null vector $\boldsymbol{\mu}_0' = [0 \quad 0 \quad 0 \quad \cdots \quad 0]$, the T^2 statistic is defined in terms of the observed vector of change-score means and the matrix of variances and covariances among the change scores

$$T^2 = n\bar{\mathbf{x}}'\boldsymbol{\Sigma}^{-1}\bar{\mathbf{x}} \tag{11.2}$$

where $\bar{\mathbf{x}}'$ is the sample mean vector containing mean difference scores for p variates and $\boldsymbol{\Sigma}^{-1}$ is the inverse of the matrix of covariances among the difference scores.

The T^2 statistic can be referred to the F distribution for evaluating the significance of deviation from the hypothesis. In this instance, the distribution is that of T^2 with $n - 1$ degrees of freedom and the statistic

$$F = \frac{T^2(n - p)}{(n - 1)p} \tag{11.3}$$

is distributed as F with p and $n - p$ degrees of freedom.

A random sample of 670 inpatients in a university hospital setting was interviewed soon after admission and again after 4 weeks of treatment. BPRS symptom ratings were completed following each interview. The interviewers were clinical psychologists employed in the research program and were in no way involved in the treatment of the patients. Scores on the four BPRS factors representing thinking disturbance, withdrawal-retardation, hostile-suspiciousness, and anxious depression were computed for each patient, as was the difference between the prescores and follow-up scores on each factor. The mean difference-score vector, the variance-covariance matrix computed from *difference scores* for the 670 patients, and the inverse of the matrix of covariances among the difference scores are shown in Table 11.1. The value of $T^2 = 362.04$ was obtained by pre- and postmultiplying the inverse matrix by the vector of mean change scores as in Eq. (11.2). The statistic

$$F = \frac{T^2(n - p)}{(n - 1)p} = \frac{362.04(670 - 4)}{(670 - 1)4} = 90.10$$

was found to be highly significant when evaluated as an F ratio with 4 and 666 degrees of freedom. The hypothesis that there was no change in the symptomatology evidenced by the patients can be rejected. Following this general multivariate test one may want to examine specific variables in more detail.

The paired-difference T^2 test suffers from the same limitations that the simple paired-difference t test does. It may be useful to be able to state with confidence

Table 11.1 Statistics Needed to Evaluate Change during Treatment

MEAN DIFFERENCE-SCORE VECTOR

Thinking disturbance	*Withdrawal-retardation*	*Hostile-suspiciousness*	*Anxious depression*
1.6104	1.0492	1.8552	3.5567

VARIANCE-COVARIANCE MATRIX Σ

$$\begin{bmatrix} 12.14847 & 1.28528 & 5.09000 & 2.14395 \\ 1.28528 & 12.82118 & 1.15960 & 1.59137 \\ 5.09000 & 1.15960 & 15.60980 & 2.02989 \\ 2.14395 & 1.59137 & 2.02989 & 14.76434 \end{bmatrix}$$

INVERSE OF VARIANCE-COVARIANCE MATRIX Σ^{-1}

$$\begin{bmatrix} -.09716 & -.00586 & -.03003 & -.00934 \\ -.00586 & .07976 & -.00306 & -.00732 \\ -.03003 & -.00306 & .07481 & -.00559 \\ -.00934 & -.00732 & -.00559 & .07064 \end{bmatrix}$$

that change did occur in the symptom ratings for the patients, but one can only guess at the reason for the change. If the patients in the sample had been treated with one particular treatment, there would be no basis in the absence of a control group for concluding that the significant change was due to that treatment. Unless one is working with a phenomenon in which it is quite probable that no change will occur, there would seem to be little reason to use the multivariate T^2 to demonstrate that significant difference does occur over time. There is little value in demonstrating significant change over time in psychiatric symptomatology of patients treated with a particular drug because one might well expect change under almost any conditions. The above example is clinically important only if one really doubts that psychiatric patients evidence symptom reduction during short-term hospitalization. On the other hand, it might be of more importance to know that asymptotic performance levels on complex psychomotor tasks changed significantly after administration of a drug. This is simply to say that complexity of statistical tests is no substitute for adequate experimental design.

11.2.3 TESTING HYPOTHESES CONCERNING EQUALITY OF ELEMENTS IN MEAN VECTOR

Use of the T^2 statistic in testing the equality of elements in a mean vector from a single sample appears to provide a statistical basis for conclusions that are frequently advanced but seldom supported by statistical tests in the clinical literature. For example, an investigator frequently wants to make some statement about the pattern or relative magnitudes of effects from a particular treatment on different symptom factors. He concludes that drug D has greatest effect on symptom variables x_1, x_2, and x_3 and less effect on symptoms x_4 and x_5. Such conclusions have almost never been put to a test of statistical significance.

The use of the T^2 statistic for testing equality of values for different elements in a mean vector has been discussed by Rao (1952) and Anderson (1958) in terms of *symmetry*. The hypothesis is that all the elements in the mean vector are equal. That is, the hypothesis of symmetry is the hypothesis that the mean profile is flat without *pattern* or *variability*. If one were to subtract out the *elevation* or *intraprofile mean* from each individual profile, the hypothesis states that the mean of deviation scores across individual should be zero.

The fact that the hypothesis concerns deviations of individual mean scores about the total profile mean poses a slight problem for calculation of the T^2 statistic. By subtracting out the profile mean from each score in the profile, a degree of freedom is lost or, to say the same thing, a linear dependency is introduced among the residual measurements. This problem can be overcome by defining a new set of $p - 1$ arbitrary linear components and calculating the T^2 statistic using them. The coefficients for the transformations can be chosen in any manner so long as the elements in each transformation vector sum to zero and no set of coefficients can be defined as a linear combination of the other sets. For example,

$$y_1 = -3x_1 - 1x_2 + 1x_3 + 3x_4$$
$$y_2 = 3x_1 - 3x_2 - 3x_3 + 3x_4$$
$$y_3 = 2x_1 - 2x_2 + 2x_3 - 2x_4$$

where x_1, x_2, x_3, and x_4 are the original p variables and y_1, y_2, and y_3 are $p - 1$ transformed variables computed using transformation vectors containing elements that sum to zero. It is also necessary that no vector of transformation coefficients be linearly dependent on the others. As long as one defines $p - 1$ transformed variates, the solution is independent of how the variates are chosen.

Given a set of $p - 1$ transformed variates, the T^2 statistic can be calculated to test the hypothesis that the elements of the original mean vector $\bar{\mathbf{x}}' = [x_1 \quad x_2 \quad x_3 \quad x_4]$ are all equal

$$T^2 = n\bar{\mathbf{y}}'\mathbf{\Sigma}^{-1}\bar{\mathbf{y}} \tag{11.4}$$

where $\bar{\mathbf{y}}' = [\bar{y}_1 \quad \bar{y}_2 \quad \bar{y}_3]$ contains the means for the transformed variates and $\mathbf{\Sigma}^{-1}$ is the inverse of the matrix of covariance among the $p - 1$ *transformed variates*.

The resulting statistic is a T^2 with $n - 1$ degrees of freedom for $p - 1$ variates. It can be referred to the F distribution for evaluation of statistical significance of the differences in mean scores of the p original profile elements

$$F = \frac{T^2(n - p + 1)}{(n - 1)(p - 1)} \tag{11.5}$$

where F has $p - 1$ and $n - p + 1$ degrees of freedom. It will be appreciated that computation and evaluation of the T^2 statistic based on the $p - 1$ derived variates is identical to that described in Sec. 9.2.2, once the transformed variates have been obtained.

To illustrate the computations involved in testing the significance of a pattern effect, i.e., inequality of means for different variables, consider the 670 psychiatric patients on whom preratings and 4-week follow-up symptom ratings were obtained. It was hypothesized that different kinds of symptoms show different degrees of response to hospital treatment. In an effort to remove the artifact of pretreatment level, the 4-week follow-up factor scores for each patient were first expressed as percentages of that patient's initial scores. The statistical hypothesis then tested was that the percentage change in four different symptom factors is equal in the psychiatric hospital population from which the sample was drawn.

Let t represent percentage change for thinking disturbance; w, withdrawal-retardation; h, hostility-suspiciousness; and d, anxious depression. Three arbitrary transformed variates were defined as follows:

$$y_1 = t + w - h - d$$

$$y_2 = t - w - h + d$$

$$y_3 = t - w + h - d$$

Table 11.2 Statistics Needed to Test Significance of a Pattern Effect

MEAN PERCENTAGE-CHANGE SCORES FOR THE FOUR ORIGINAL FACTORS

Thinking disturbance	*Withdrawal-retardation*	*Hostile-suspiciousness*	*Anxious depression*
.6915	.9777	.8961	.6955

MEAN SCORES OF THE THREE TRANSFORMED VARIATES

y_1	y_2	y_3
.0776	−.4868	−.0855

VARIANCE-COVARIANCE MATRIX $\mathbf{\Sigma}$

$$\begin{bmatrix} 4.53767 & -.98554 & -1.92981 \\ -.98554 & 3.65672 & 1.20478 \\ -1.92981 & 1.20478 & 4.00031 \end{bmatrix}$$

INVERSE OF VARIANCE-COVARIANCE MATRIX $\mathbf{\Sigma}^{-1}$

$$\begin{bmatrix} .28113 & .03450 & .12522 \\ .03450 & .30783 & -.07606 \\ .12522 & -.07606 & .33330 \end{bmatrix}$$

The variates y_1, y_2, and y_3 were computed from the percentage-change scores for each of the 670 patients. Mean percentage-change scores for the four original factors and mean scores on the three transformed variates, together with the 3×3 variance-covariance matrix and its inverse, are presented in Table 11.2. The value $T^2 = 44.53$ was obtained by pre- and postmultiplying the matrix inverse by the vector of means for the transformed variates as in Eq. (11.1). The results were referred to an F distribution

$$F = \frac{T^2(670 - 3)}{669(3)} = 14.80$$

where F has 3 and 667 degrees of freedom. On the basis of these results, one can conclude that the percentage change in different symptom factors is indeed different during a brief period of hospitalization.

11.2.4 USE OF T^2 FOR TESTING HYPOTHESES CONCERNED WITH TRENDS

The repeated-measurement situation in which each subject is measured on one variable at several different points in time or under several different conditions appears to constitute an important potential application of the T^2 test for symmetry. Repeated-measurement data are often subjected to univariate analysis of variance of the treatments by subject type even when the assumption that correlations are equal across all trials cannot be met. Differences in correlations between measurements made at different times introduce no problem for the multivariate T^2 test because there is no requirement that they be equal.

Several types of problems involving multiple measurements of the same variable currently handled most frequently by univariate methods would appear to be more appropriately solved by the multivariate T^2. The question of significance of trends over time is one important example.

Suppose that n subjects are measured at p different times $t_1, t_2, \ldots, t_p$. The investigator wants to know whether there is any difference between the mean scores for the p time periods. The logic of the experiment is such that he can expect *in advance* that if there is a treatment effect, it will result in a general upward shift in performance over time and that the complexity of the curve will not exceed that of a third-degree polynomial. He defines three artificial transformed variates using orthogonal polynomial coefficients. With $p = 16$, the artificial variates were defined as follows:

$$\begin{aligned}
y_1 &= -15t_1 - 13t_2 - 11t_3 - 9t_4 - 7t_5 - 5t_6 - 3t_7 - 1t_8 + 1t_9 + 3t_{10} \\
&\qquad + 5t_{11} + 7t_{12} + 9t_{13} + 11t_{14} + 13t_{15} + 15t_{16} \\
y_2 &= -35t_1 - 21t_2 - 9t_3 + 1t_4 + 9t_5 + 15t_6 + 19t_7 + 21t_8 - 21t_9 - 19t_{10} \\
&\qquad - 15t_{11} - 9t_{12} - 1t_{13} + 9t_{14} + 21t_{15} + 35t_{16} \\
y_3 &= -455t_1 - 91t_2 + 143t_3 + 267t_4 + 301t_5 + 265t_6 + 179t_7 + 63t_8 \\
&\qquad - 63t_9 - 179t_{10} - 265t_{11} - 301t_{12} - 267t_{13} - 143t_{14} + 91t_{15} + 455t_{16}
\end{aligned}$$

The value for y_1, y_2, and y_3 were computed from the 16 scores for each individual, and the remainder of the analysis involved standard T^2 calculations for $p = 3$ variables

$$T^2 = n\bar{\mathbf{y}}'\mathbf{\Sigma}^{-1}\bar{\mathbf{y}}$$

where $\mathbf{\Sigma}^{-1}$ is the 3×3 covariance matrix computed from the scores on the three artificial trend variates. Then

$$F = \frac{T^2(n-3)}{(n-1)3}$$

was referred to an F distribution with 3 and $n - 3$ degrees of freedom.

Wherever serial correlations are present in repeated-measurement data and the general trend can be assumed to be reasonably simple, great advantage will accrue from reduction of dimensionality of the problem through use of rationally chosen artificial variates. The advantage of the multivariate T^2 approach over the univariate approach is that no assumptions are required concerning the consistency or equality of correlations between measurements made at different times. A possible disadvantage in some applications is that the number of subjects n should be at least 3 or 4 times the square of the number of artificial trend variates used in the analysis.

11.2.5 THE T^2 TEST OF DIFFERENCES BETWEEN MEAN VECTORS FOR TWO INDEPENDENT SAMPLES

The Hotelling T^2 statistic can be used to test the hypothesis that two groups can be considered to be random samples from populations having identical multivariate mean vectors. This is the null hypothesis of no difference in multivariate means. Rather than testing the significance of mean difference on each variable separately, an overall multivariate test of significance may be useful. The researcher frequently finds that one or more simple t tests will yield "significant" results, but he still may worry that one or two such results might obtain by chance because of the number of hypotheses tested. This is not the kind of situation where one can employ a range test or specify easily how many significant results might be expected by chance because of the correlations among the variables. In this situation the multivariate hypothesis of no difference in mean vectors is an appropriate hypothesis to test.

In the case of two independent groups, the T^2 statistic can be computed as follows:

$$T^2 = \frac{n_1 n_2}{n_1 + n_2} (\bar{\mathbf{x}}^{(1)} - \bar{\mathbf{x}}^{(2)})'\mathbf{\Sigma}^{-1}(\bar{\mathbf{x}}^{(1)} - \bar{\mathbf{x}}^{(2)}) \tag{11.6}$$

where $\bar{\mathbf{x}}^{(1)}$ and $\bar{\mathbf{x}}^{(2)}$ are the mean vectors for the two independent samples and $\mathbf{\Sigma}^{-1}$ is the inverse of the pooled within-groups covariance matrix (Sec. 2.7).

The statistic

$$F = T^2 \frac{n_1 + n_2 - p - 1}{(n_1 + n_2 - 2)p} \tag{11.7}$$

will be distributed as F with p and $n_1 + n_2 - p - 1$ degrees of freedom. It will be recalled that this is exactly the same test that was discussed with regard to simple discriminant-function analysis.

As an example, consider the question of whether patients who receive electroconvulsive therapy (ECT) differ symptomatically from those who do not. While convulsive therapy is reputed to have special implications for treatment of severe depressions, psychiatrists in the particular setting seemed to use it as frequently with schizophrenic patients. The question arose whether selection of this treatment was in any way systematically related to symptom-profile characteristics of the patients.

Pretreatment BPRS factor scores were computed for the sample of 670 inpatients previously described. The mean vectors for 427 patients who did receive ECT and 243 patients who did not, plus the within-groups covariance matrix and its inverse, are presented in Table 11.3. The value of $T^2 = 154.87$ was computed by pre- and postmultiplying the inverse of the within-groups covariance matrix by the vector of mean differences, as shown in Eq. (11.6). The T^2 was transformed to an F statistic using Eq. (11.7)

$$F = \frac{154.87(427 + 243 - 4 - 1)}{(427 + 243 - 2)4} = 38.54$$

Table 11.3 Statistics Needed to Test Significance of Difference between Two Groups

MEAN SCORE VECTORS

	Thinking disturbance	*Withdrawal-retardation*	*Hostile-suspiciousness*	*Anxious depression*
ECT Group	4.166276	4.007026	6.594848	9.044496
No ECT	3.020576	2.967078	6.502058	7.600823

VARIANCE-COVARIANCE MATRIX Σ

$$\begin{bmatrix} 15.189836 & 2.315935 & 6.767122 & -1.340093 \\ 2.315935 & 12.453140 & -.001157 & .623306 \\ 6.767122 & -.001157 & 14.642225 & -1.988207 \\ -1.340093 & .623306 & -1.988207 & 10.834659 \end{bmatrix}$$

INVERSE OF VARIANCE-COVARIANCE MATRIX Σ^{-1}

$$\begin{bmatrix} .086177 & -.016250 & -.039232 & .004394 \\ -.016250 & .083602 & .006759 & -.005579 \\ -.039232 & .006759 & .087906 & .010889 \\ .004394 & -.005579 & .010889 & .095159 \end{bmatrix}$$

The significance of the difference in multivariate mean vectors was evaluated by reference to the F distribution with 4 and 665 degrees of freedom. On the basis of this analysis, it was concluded that there was reliable evidence that patients who received ECT were symptomatically different from those who did not.

11.3 The Likelihood-ratio Criterion as a Multivariate Generalization of Analysis-of-variance F Test

11.3.1 DEFINITION

Just as Hotelling's T^2 can be used to test hypotheses representing multivariate generalizations of hypotheses that can be tested for a single variable using the simple t test, multivariate hypotheses that are similar to those usually tested by analysis-of-variance methods can be tested using the likelihood-ratio criterion. The multivariate generalization of analysis of variance, referred to by Rao (1952) as the *analysis of dispersion*, involves partitioning the total-sums-of-squares-and-cross-products matrix SP(T) in a manner that is identical to the partitioning of the sums of squares in univariate analysis of variance. The test statistic is then computed as the ratio of two determinants, or generalized variances

$$\Lambda = \frac{|W|}{|W + B|} \tag{11.8}$$

where W is the pooled within-groups SP matrix and B is the between-groups SP matrix (Sec. 2.7).

If the value of a determinant is conceived of as a scalar index of multivariate generalized variance, the Λ statistic is recognized to be the ratio of the within-groups error variance to the total variance. The larger the true group differences, the larger the denominator will be and the smaller the value of Λ. If Λ is sufficiently small, the hypothesis that multivariate means are equal for all groups can be rejected.

While calculation of the Λ statistic is straightforward and the rationale is appealingly clear, the distribution of the statistic is not a simple and familiar one. Numerous statisticians including Wilks (1932), Nair (1939), Bartlett (1938), Wald and Brookner (1941), and Rao (1948) have worked on the problem. An approximate test is available which Anderson (1958) concludes to be accurate to the third decimal place whenever n is greater than 3 times the square of the number of groups plus the square of the number of variables, $n > 3(k^2 + p^2)$. While this is greater *accuracy* than seems necessary in view of ever-present departures from assumptions and other practical problems, the multivariate analysis of dispersion should probably not be used except in the case of $n > 3(k^2 + p^2)$ because of uncertainty concerning *power* of the test when used with smaller samples.

When the total number of subjects is reasonably large, the evaluation of significance of Λ can be achieved simply by computing the statistic

$$\chi^2 = -m \log_e \Lambda \tag{11.9}$$

where $m = n - 1 - \frac{1}{2}(p + k)$. Since Λ will always be a fractional value less than unity, $\log_e \Lambda$ will be a negative number and the product $-m \log_e \Lambda$ will be a positive number. The chi-square statistic of Eq. (11.9) can be evaluated for significance by reference to the chi-square distribution (Table A.3) for $(k - 1)p$ degrees of freedom, where k is the number of groups and p is the number of variables.

11.3.2 MULTIVARIATE GENERALIZATION OF SIMPLE ONE-WAY CLASSIFICATION ANALYSIS OF VARIANCE

Suppose that $n_1, n_2, \ldots, n_k = n$ individuals are observed under k different treatment conditions and that each is measured on p characteristics. Assume that the p variables have independent multivariate normal distributions with equal covariance matrices within the populations represented by the k groups. Let the pooled within-groups corrected sums of squares-and-products matrix be designated W and the total SP matrix designated $T = W + B$. The Λ statistic is the ratio of the determinant of W relative to the determinant of $W + B$

$$\Lambda = \frac{|W|}{|T|} \tag{11.10}$$

If k is the number of groups and p is the number of variables, then the statistic

$$\chi^2 = -m \log_e \Lambda$$

will have (approximately) a chi-square distribution with $(k - 1)p$ degrees of freedom, where $m = n - 1 - \frac{1}{2}(k + p)$.

To illustrate the one-way classification multivariate analysis of variance, the BPRS profiles of the 670 patients in the previous example will be used to determine whether psychiatric patients assigned to different major types of drug treatments differ significantly in symptom characteristics. The mean factor-score vectors for patients who were treated with antidepressants, major tranquilizers, minor tranquilizers, and no psychiatric drug treatment are presented in Table 10.8, as are the within- and between-groups SP matrices. The determinants were evaluated using the square-root method (Sec. 2.6), and Λ was computed

$$\Lambda = \frac{|W|}{|T|} = \frac{443 \times 10^{13}}{513 \times 10^{13}} = .864$$

The $\chi^2 = -m \log_e \Lambda = 97.210$, where $m = 669 - .5(4 + 4) = 665$. The chi square with $(k - 1)p = 12$ degrees of freedom has probability less than .001

under the null hypothesis. The result leads us to conclude that the mean profiles for the four groups of patients treated with different drugs do differ significantly.

11.3.3 OTHER MULTIVARIATE TESTS OF SIGNIFICANCE FOR ONE-WAY CLASSIFICATION WITH k GROUPS

An alternative distribution function that can be used in evaluating the significance of Λ is discussed by Rao (1952). Let

$$z = \text{antilog}_e \frac{\log_e \Lambda}{s} \tag{11.11}$$

where

$$s = \sqrt{\frac{p^2(k-1)^2 - 4}{p^2 + (k-1)^2 - 5}} \qquad \text{and} \qquad m = n - 1 - \tfrac{1}{2}(k+p)$$

Then

$$F = \frac{1-z}{z} \frac{ms - [p(k-1) - 2]/2}{p(k-1)} \tag{11.12}$$

can be used as an F ratio with $p(k-1)$ and $ms - [p(k-1) - 2]/2$ degrees of freedom.

Applying this method to testing the significance of differences in mean symptom profiles for patients treated in four different ways, we recall the value of $\Lambda = .864$ that was obtained as the ratio of determinant values. $\log_e \Lambda = \log_e .864 = -.14618$, and

$$s = \frac{16.9 - 4}{16 + 9 - 5} = \frac{140}{20} = 2.6458$$

Then

$$z = \text{antilog}_e \frac{\log_e \Lambda}{s} = .946$$

and the statistic

$$F = \frac{1-z}{z} \frac{ms - [p(k-1) - 2]/2}{p(k-1)}$$

$$= \frac{.054}{.946} \frac{665(2.6458) - 5}{12} = 8.346$$

Reference to the distribution of an F ratio with 12 and 1,754 degrees of freedom leads to the conclusion that the treatment groups differ significantly at $<.001$ level. The consistency of this conclusion with that previously derived from the $\chi^2 = -m \log_e \Lambda$ test should be noted.

In the chapter concerned with multiple discriminant analysis yet another overall test of the significance of difference between multivariate mean vectors was described (Sec. 10.4). The trace, or sum of diagonal elements, of the matrix of the determinant $|B - \lambda W| = 0$ has an approximate chi-square distribution with $p(k - 1)$ degrees of freedom. We shall not repeat an extended discussion of the computational method, but it can be noted that the matrix from which diagonal elements are taken is the between-groups SP matrix pre- and post-multiplied by the triangular square-root inverse of the within-groups covariance matrix $C = V^{-1}BV^{-1}$ of Eq. (10.5). The sum of the diagonal elements in C is distributed approximately as chi square with $p(k - 1)$ degrees of freedom.

Several writers, including Rao (1952), have discouraged use of overall tests of significance, such as those based on the likelihood-ratio criterion, because of the realization that they are not sensitive to group differences which tend to fall along a particular dimension or continuum in the multivariate measurement space. It should be obvious that the likelihood ratio Λ is a function of all the roots of the determinantal equation $|B - \lambda W| = 0$. Similarly, the total chi square defined by the trace of the matrix $V^{-1}BV^{-1}$ can be partitioned into p or $k - 1$ (whichever is smaller) component chi-square values which are the separate roots of the determinantal equation (Sec. 10.4). Where differences between the several groups tend to fall along a single continuum, the test of the significance of the largest root will be more powerful than the test of significance of all roots combined. Jones (1966) has explored conditions under which power is greatest for each of several alternative multivariate tests of significance. It has been the experience of the present authors that the largest-root test frequently yields significant results even when the trace test (sum of diagonal elements) fails to do so.

11.4 Test of Departure from an a Priori Discriminant Function

A multivariate test of significance that has particular clinical implications is the test of an a priori discriminant function. On the basis of preexisting information, it is frequently possible to define a weighted combination of several variables which one would expect to represent the difference between two groups. For example, on the basis of diagnostic-stereotype profiles one may hypothesize that the only difference between schizophrenic and depressive patients is the contrast or difference in relative prominence of thinking-disturbance vs. anxious-depression symptoms. Given two groups of adequately diagnosed patients, one can test the hypothesis that the best possible linear discriminant function does not differ significantly from the simple a priori contrast.

Another similar type of application might arise from more theoretical considerations. For example, an investigator might postulate that the only basis for symptom-profile differences between single and married patients is the relatively higher incidence of schizophrenia in the single group and relatively higher incidence of depression in the married group. Given a set of discriminant-function coefficients defined on large n to represent the difference between schizophrenia and depression, it is possible to test the hypothesis that the schizo-depression contrast function accounts for all significant symptom-profile differences between single and married patients.

Rao (1952) describes a test which he attributes to R. A. Fisher. Let $y = a_1x_1 + a_2x_2 + \cdots + a_px_p$ be an a priori function and let $V(y) = \mathbf{a}'\mathbf{\Sigma}\mathbf{a}$ be the variance of the a priori linear function, as described in Sec. 2.7.8. In this case, $\mathbf{\Sigma}$ is the pooled within-groups covariance matrix, as described in Sec. 2.7.6. The squared distance between the two groups on the a priori assigned function can be computed as follows:

$$D_y^2 = \frac{(\bar{y}^{(1)} - \bar{y}^{(2)})^2}{V(y)} \tag{11.13}$$

where $\bar{y}^{(1)}$ and $\bar{y}^{(2)}$ are the group means on the a priori function and $V(y)$ is the within-groups variance of the a priori function.

Let D_p^2 be the Mahalanobis D^2 between the two groups computed from the p original variables using Eq. (11.14). This squared distance is then the maximum possible distance in terms of the p measurements. An F ratio with $p - 1$ and $n - p + 1$ degrees of freedom can be computed as follows:

$$D_p^2 = \mathbf{d}'\mathbf{\Sigma}^{-1}\mathbf{d} \tag{11.14}$$

where $\mathbf{d}$ is the vector of mean differences for the two groups on the p variables and $\mathbf{\Sigma}^{-1}$ is the inverse of the pooled within-groups covariance matrix. Then

$$U = \frac{1 + n_1n_2D_p^2/(n_1 + n_2)n}{1 + n_1n_2D_y^2/(n_1 + n_2)n} - 1 \tag{11.15}$$

where n_1 and n_2 are the sample sizes used in calculating the mean vectors and n is the degrees of freedom for the pooled within-groups covariance matrix (which can be different from $n_1 + n_2 - 2$ if the covariance matrix is estimated from a different data source). Then

$$F = U\frac{n - p + 1}{p - 1} \tag{11.16}$$

is the F ratio with $p - 1$ and $n - p + 1$ degrees of freedom.

To illustrate the procedure for testing departure from an a priori discriminant function, let us test the hypothesis that all symptom-profile differences between black and white psychiatric patients can be accounted for by a function previously derived in the analysis of differences between clinically schizophrenic and clinically depressed patients. In Sec. 9.4, the discriminant function separating schizophrenic and depressive groups was derived

$$y = -.36X_{\text{think}} + .15X_{\text{withdr}} - .14X_{\text{host}} + .31X_{\text{dep}} \tag{11.17}$$

Mean vectors and pooled within-groups covariance matrix for Anglo and Negro patients were presented in Table 9.7. The D_p^2 between the two racial groups computed from the four factor scores was found to be

$$D_p^2 = \mathbf{d}'\mathbf{\Sigma}^{-1}\mathbf{d} = .169$$

The variance of the a priori schizo-depression contrast function within the racial groups was computed by pre- and postmultiplying the pooled within-groups covariance matrix (Table 9.7) by the a priori weighting coefficients

$$V(y) = \mathbf{a}'\mathbf{\Sigma a} = 3.75$$

The group means on the a priori function were calculated by applying the weighting coefficients in Eq. (11.17) to the group means in Table 9.7.

$$\bar{y}^{(1)} = -.36(5.30) + .15(5.71) - .14(6.10) + .31(7.52) = .43$$

$$\bar{y}^{(2)} = -.36(5.35) + .15(5.56) - .14(5.05) + .31(5.99) = .06$$

The squared distance $D_y{}^2$ calculated from the a priori function was

$$D_y{}^2 = \frac{(\bar{y}^{(1)} - \bar{y}^{(2)})^2}{V(y)} = .037$$

The statistic U, which can be transformed to an F ratio, can be calculated from $D_p{}^2$ and $D_y{}^2$:

$$U = \frac{1 + n_1 n_2 D_p{}^2/(n_1 + n_2)n}{1 + n_1 n_2 D_y{}^2/(n_1 + n_2)n} - 1 = .01956$$

The F ratio with $p - 1$ and $n - p + 1$ degrees of freedom is then

$$F = \frac{U(n - p + 1)}{p - 1} = 20.91$$

where n is the degrees of freedom on which the within-groups covariance matrix is estimated. The analysis indicates that racial-group differences cannot be fully explained in terms of the symptom contrast which describes the difference between schizophrenic and depressed patients.

11.5 Complex Multivariate Analysis of Variance

The likelihood ratio can be used to test hypotheses that are analogous to main-effect and interaction hypotheses in a cross-classification analysis of variance. As Anderson (1958) has described, each F test in the univariate analysis of variance has an analogous U test in the multivariate analysis of variance. Multivariate score *vectors* replace individual variables in the generalized analysis of variance. Corresponding to each sum of squares in a complex analysis of variance design, a sums-of-squares-and-cross-products matrix can be computed. That is, the SP(T) matrix can be partitioned in exactly the same manner that the total sum of squares is partitioned in the univariate analysis of variance. Given the component SP matrices, appropriate test statistics of the type

$$U = \frac{|W|}{|W + B|}$$

can be constructed to test any conventional analysis-of-variance hypothesis.

A detailed discussion of all types of analysis-of-variance designs for which multivariate tests can be constructed is clearly beyond the scope of this book. The purpose of this section is to develop insight into the direct relationship between multivariate and univariate analyses of variance. To illustrate the partitioning of sums of squares and cross products, we shall consider specifically the two-way analysis of variance. It is hoped that the reader will find it possible to generalize the concepts to more complex designs as required.

The multivariate *vector* analog of the square of a single score is the *matrix product*, or *outer product*, of a score vector and its transpose.

$$[X^2] = \mathbf{x}\mathbf{x}'$$

where each element in the principal diagonal is the square of one element in score vector $\mathbf{x}$ and each off-diagonal element is a cross product.

The multivariate analog of a simple raw sum of squares is a sum of the matrix products

$$[\sum X^2] = \sum_{i}^{n} \mathbf{x}_i\mathbf{x}_i'$$

where each diagonal element is the raw sum of squares for one element in score vector $\mathbf{x}$ and each off-diagonal element is a raw sum of cross products.

Let $\mathbf{x}_{jki}$ be the score vector for the ith individual at the jth level of A and the kth level of B. The sum of score vectors for all individuals in the jkth cell of the cross-classification matrix will be represented by $\mathbf{x}_{jk.}$. The sum of score vectors for all individuals in the jth level of A will be represented by $\mathbf{x}_{j..}$, and the sum of score vectors for all individuals in the kth level of B will be represented by $\mathbf{x}_{.k.}$. The sum of score vectors for all individuals in all treatment combinations will be designated by $\mathbf{x}_{...}$. With this notation in mind, we now generalize some familiar formulas from the univariate analysis of variance.

Consider a two-way cross-classification design in which each individual is measured on p different characteristics. The total sum of squares in a two-way univariate analysis of variance can be represented as follows:

$$SS_T = \sum_{j}^{a}\sum_{k}^{b}\sum_{i}^{n} X_{jki}^2 - \frac{\left(\sum^{a}\sum^{b}\sum^{n} X\right)^2}{n}$$

The multivariate generalization is the total-sums-of-squares-and-cross-products matrix SP(T)

$$[\text{SP}(T)] = \sum_{j}^{a}\sum_{k}^{b}\sum_{i}^{n} \mathbf{x}_{jki}\mathbf{x}_{jki}' - \frac{\mathbf{x}_{...}\mathbf{x}_{...}'}{n} \tag{11.18}$$

where $\mathbf{x}_{...}$ is the sum vector obtained by summing score vectors for all individuals in all treatment combinations.

The total between-cells sum of squares, which is the sum of main effects and interaction sums of squares, can be calculated as follows in the univariate case:

$$SS_C = \frac{1}{n}\sum_{j}^{a}\sum_{k}^{b}\left(\sum_{i}^{n} X_{jki}\right)^2 - \frac{\left(\sum^{a}\sum^{b}\sum^{n} X_{jki}\right)^2}{n}$$

The multivariate analog is the total-between-cells-sums-of-squares-and-cross-products matrix SP(C):

$$[\mathrm{SP}(C)] = \frac{1}{n}\sum_{j}^{a}\sum_{k}^{b} \mathbf{x}_{jk\cdot}\mathbf{x}'_{jk\cdot} - \frac{\mathbf{x}_{\dots}\mathbf{x}'_{\dots}}{n} \tag{11.19}$$

where $\mathbf{x}_{jk\cdot}$ is the sum vector representing the sum of score vectors for n individuals in the jkth cell.

The sum of squares for main effect of classification A is computed as follows in the univariate analysis of variance:

$$SS_A = \frac{1}{bn}\sum_{j}^{a}\left(\sum_{k}^{b}\sum_{i}^{n} X_{jki}\right)^2 - \frac{(\sum\sum\sum X_{jki})^2}{n}$$

The multivariate analog is the between levels of A sums-of-squares-and-cross-products matrix SP(A):

$$[\mathrm{SP}(A)] = \frac{1}{bn}\sum_{j}^{a} \mathbf{x}_{j\cdot\cdot}\mathbf{x}'_{j\cdot\cdot} - \frac{\mathbf{x}_{\dots}\mathbf{x}'_{\dots}}{n} \tag{11.20}$$

where $\mathbf{x}_{j\cdot\cdot}$ is the sum vector obtained by summing score vectors for n individuals in each of b levels of factor B.

The sum of squares for main effect of classification B is computed as follows in the univariate analysis of variance:

$$SS_B = \frac{1}{an}\sum_{k}^{b}\left(\sum_{j}^{a}\sum_{i}^{n} X_{jki}\right)^2 - \frac{(\sum\sum\sum X_{jki})^2}{n}$$

The multivariate analog is the between levels of B sums-of-squares-and-cross products matrix SP(B):

$$[\mathrm{SP}(B)] = \frac{1}{an}\sum_{k}^{b} \mathbf{x}_{\cdot k\cdot}\mathbf{x}'_{\cdot k\cdot} - \frac{\mathbf{x}_{\dots}\mathbf{x}'_{\dots}}{n} \tag{11.21}$$

where $\mathbf{x}_{\cdot k\cdot}$ is the sum vector obtained by summing score vectors for n individuals in each of the a levels of factor A.

The within-cells-error-sums-of-squares-and-cross-products matrix can be computed as a residual:

$$[\mathrm{SP}(W)] = [\mathrm{SP}(T)] - [\mathrm{SP}(C)] \tag{11.22}$$

The multivariate analog of the univariate interaction sum of squares can also be computed as a residual:

$$[\mathrm{SP}(AB)] = [\mathrm{SP}(C)] - [\mathrm{SP}(A)] - [\mathrm{SP}(B)] \tag{11.23}$$

Tests of multivariate hypotheses concerning independent effects of A and B, and of the AB interaction, can be accomplished using chi-square tests derived from likelihood ratios. The main effect for A, or the average effect of A independent of B, can be tested as follows:

$$U_A = \frac{|\mathrm{SP}(W)|}{|\mathrm{SP}(W) + \mathrm{SP}(A)|} \tag{11.24}$$

where the numerator is the determinant of SP(W) and the denominator is the determinant of the sum SP(W) + SP(A).

The significance of U_A can be evaluated by transformation to a chi square with pq_1 degrees of freedom, where q_1 is equal to the number of levels of A minus 1

$$\chi^2 = -m \log_e U_A$$

where $m = n - \frac{1}{2}(p - q_1 + 1)$ and n is the number of degrees of freedom for error in the particular analysis of variance model, p is the number of variables, and q_1 is the number of levels of A minus 1. In a fixed-model two-way analysis-of-variance design with n_{jk} observations per cell, $n = n_t - ab$, where a and b are the number of levels of A and B and n_t is the total number of observations. In general, n is the number of degrees of freedom associated with the matrix of the determinant in the numerator of the U statistic.

The significance of the B effect can be tested using the likelihood ratio

$$U_B = \frac{|\mathrm{SP}(W)|}{|\mathrm{SP}(W) + \mathrm{SP}(B)|} \tag{11.25}$$

where the numerator is the determinant of SP(W) and the denominator is the determinant of SP(W) + SP(B). The significance of U_B can be evaluated by conversion to a chi square with pq_1 degrees of freedom, where q_1 is the number of levels for B minus 1

$$\chi^2 = -m \log_e U_B$$

where $m = n - \frac{1}{2}(p - q_1 + 1)$ and n is the number of degrees of freedom for error in the particular analysis-of-variance design. In a two-way fixed-model design with n_{jk} observations per cell, $n = n_t - ab$.

The interaction effect can be tested in an analogous fashion, by intuitive generalization of corresponding univariate tests, but at this point we should raise the problem of interpretation. The tests of main effects are, essentially, tests of the similarities of mean multivariate score vectors for various levels of one factor, say A, across all levels of the other factor. Because of artificial balancing of cell frequencies, the effect of A can be evaluated independent of effects of B. If a significant difference is found, one can examine the nature of differences in multivariate profiles between various categories of A. A significant multivariate interaction means that the nature of changes in profile *patterns* between various levels of A differs as a function of levels of B. The present authors offer the caution that the difficulties of interpretation may seriously limit the practical importance of tests of more complex multivariate hypotheses. There are occasions (to be illustrated) where tests of simple main effects are especially meaningful in clinical research.

A final word of caution is in order concerning sample sizes required for appropriate use of the asymptotic chi-square distribution test of significance in multivariate analysis of variance. Anderson (1958) has suggested that

$$\chi^2 = -m \log_e U$$

is distributed as chi square with pq_1 degrees of freedom if $m \geq 3(p^2 + q_1^2)$, where p is the number of variables and q_1 is the number of degrees of freedom for treatments (or interaction). Perhaps one can rely on the test as a reasonable approximation with somewhat smaller samples, but in general the multivariate analysis of variance as presented here is intended for use with large samples.

11.6 References

Anderson, T. W.: "Introduction to Multivariate Statistical Analysis," Wiley, New York, 1958.

Bartlett, M. S.: Further Aspects of the Theory of Multiple Regression, *Proc. Cambridge Phil. Soc.*, **34**:33–40 (1938).

Hotelling, H.: The Generalization of Student's Ratio, *Ann. Math. Statist.*, **2**:360–378 (1931).

Jones, L. V.: Analysis of Variance in Its Multivariate Developments, in R. B. Cattell (ed.), "Handbook of Multivariate Experimental Psychology," Rand McNally, Chicago, 1966.

Nair, U. S.: The Application of the Moment Function in the Study of Distribution Laws in Statistics, *Biometrika*, **30**:274–294 (1939).

Rao, C. R.: Tests of Significance in Multivariate Analysis, *Biometrika*, **35**:58–79 (1948).

———: "Advanced Statistical Methods in Biometric Research," Wiley, New York, 1952.

Wald, A., and R. J. Brookner: On the Distribution of Wilks' Statistic for Testing the Independence of Several Groups of Variates, *Ann. Math. Statist.*, **12**:137–152 (1941).

Wilks, S. S.: Certain Generalizations in the Analysis of Variance, *Biometrika*, **24**:471–494 (1932).

PART FOUR
CLASSIFICATION AMONG GROUPS

12

Decision Procedures for Assigning Individuals among Several Groups

12.1 Introduction

This is the first of four chapters concerned with quantitative methods for assignment of an individual to one of several groups on the basis of multiple measurements or other types of observations. These procedures are used where it is assumed that each individual belongs to one of several specified groups or populations but it is not known to which he belongs. Several relevant characteristics of the individual are observed, and on the basis of this information a classification decision is made. Assignment is based on estimates of the probabilities of occurrence of the patterns of characteristics (say symptom profile patterns) within each of the several groups.

The problems discussed in this and the following chapters should be clearly distinguished from the problem of developing classification concepts or classification groups where none are assumed beforehand. In Chap. 8, a variety of empirical methods for examination of natural groupings of individuals and for the development of classification typologies on the basis of empirical analyses were discussed. In the present chapter, it will be assumed that the classification groups are already defined and that the problem is one of assigning individuals to them with minimum probability of error.

The problem of assignment of individuals to a specified number of previously defined groups is well illustrated by the clinical diagnostic decision process. On the basis of certain preliminary information it can be assumed with (almost) certainty that the individual belongs to one of several recognized diagnostic groups. In the case of psychiatric disorders, the set of possible alternatives might include all official psychiatric diagnoses recognized by the professional organization or a subset of diagnostic alternatives dictated by the presence of a particular target symptom, such as hallucinations. A number of different observations, either qualitative or quantitative, are available to characterize the individual. The clinician must somehow combine the available observations to reach a diagnostic decision.

Statistical classification decisions, like clinical diagnostic decisions, are only probabilistically correct. The clinician realizes this when he lists a secondary diagnosis. The statistician recognizes it more explicitly when he is able to assign a probability estimate to each classification alternative. In its simplest form, statistical decision making involves assigning each individual to the most probable alternative with recognized and estimated probabilities of error. An advantage of the quantitative approach to classification of individuals is inherent in the recognition that some decisions are virtually certain to be correct, while the degree of uncertainty associated with others can be explicitly estimated.

In the research setting, quantitative statistical decision procedures can be employed for a variety of purposes. The first and perhaps most important basis for their use is that they provide an objective and operationally specified way of describing how individual patients get into particular groups. From a purely research point of view this has substantial advantage over the use of subjective classification methods that are not likely to be understood and used in the same way by different clinicians. Research results contributed by investigators who are professedly skeptical that anyone else can "correctly" classify patients in the same manner that they do are obviously of dubious general value. Sadly enough this is a common problem among many would-be clinical researchers.

In another context, objective quantitative classification methods can be used to assess the relevance of specified information for describing differences among groups. The extent to which clinical classification decisions actually depend upon certain specified information can be tested by examining the accuracy of classification decisions based on that information alone. To the extent that the statistical or quantitative decisions do not agree perfectly with clinical decisions one must assume that (1) other information is entering into the clinical decisions, (2) the specified information is being combined in a manner different from that specified by the quantitative model, or (3) that the clinical use of the specified information is not consistent and reliable.

In view of what has been said about quantitative approaches to assignment of individuals among several groups, several requirements for the development of such procedures should be evident. First, it is necessary that all the alternative groups or populations to which an individual might belong be specified in advance and included in the decision model. This is much more of a problem in some fields than in others. In most areas of psychiatry and psychology, at the present time, we are concerned more with "classification" than with "diagnosis." Thus, we are relatively free to define the number and nature of categories into which it seems useful that individuals should be classified.

Having defined the classification alternatives, the next problem is one of identifying and specifying the information that should be considered in reaching a classification decision. In order for the operational decision procedures to prove effective, it is necessary that the most relevant measurements or other types of observations be chosen. The relevant observations are those which tend best to distinguish members of one group from those of other groups. A large portion of the research currently under way in the psychological and psychiatric field is concerned with identification of the most relevant and reliably measurable discriminanda. This research includes investigations using factor analysis as well as discriminant analysis and regression techniques.

The third problem that must be faced in development of an operational quantitative decision procedure is to estimate on the basis of large samples of individuals of *known group membership* the probabilities of occurrence of various patterns of relevant characteristics. This presupposes that *initially* one is able to obtain groups of individuals that can be considered to be random samples from the populations into which the decision procedure to be developed is intended to assign individuals. This requirement has been one of the big deterrents for some investigators in psychiatric research. How does one collect preliminary samples which can be considered with confidence to belong to specified populations? For the present writers, this does not seem to be as big a problem as it does for some other investigators. First of all, there are few established diagnostic entities in this area of research. Diagnoses are really phenomenological classifications; and if firmer bases for differentiation are ever demonstrated, then they can be used to define groups. Even in areas where *true* diagnostic populations exist, a reasonably high proportion of misclassifications can exist in the normative samples without substantially affecting the derived statistical classification functions. Finally, in many applications, the initial classification groups are themselves empirically defined, using methods such as those described in Chap. 8, so that the problem becomes nonexistent.

12.2 Maximum-likelihood and Bayesian Models

This section presents the classification models which are the basis for procedures for the assignment of individuals discussed in the next three chapters. These models are given initially in quite abstract form, without concern for *how* the required probability estimates are derived. This is an organization of discourse with which the mathematically untrained reader tends to experience discomfort, having been accustomed for many years to an order of reasoning that goes from the specific to the general. Nevertheless, for several of the topics discussed in this book, it is advantageous first to acquire an overview or a general model to which specific procedures can later be related.

All the methods for classification of individuals discussed in the next few chapters require an estimate of the probability of occurrence of each complex multivariate observation pattern (profile) within each classification group. Given that samples of individuals have been obtained from the various populations, a number of different methods are available for estimating these probabilities of occurrence for the various patterns, and these differences in methods of estimating pattern probabilities constitute the subject matter of succeeding chapters. For the present, let us assume that the probability of (symptom) pattern $\mathbf{x}_i$ in group j is represented by the notation $P(\mathbf{x}_i \mid j)$. A *relative-frequency* definition of the probability of pattern $\mathbf{x}_i$ in group j can be stated in terms of the frequency of occurrence of the particular pattern relative to the frequencies of occurrence for all possible patterns in the jth group

$$P(\mathbf{x}_i \mid j) = \frac{f(\mathbf{x}_i \mid j)}{f(\mathbf{x}_1 \mid j) + f(\mathbf{x}_2 \mid j) + \cdots + f(\mathbf{x}_i \mid j)} \tag{12.1}$$

As we shall see, there are several less direct but more feasible ways of estimating the probability of a complex pattern in each group than by actually counting the frequencies of occurrence for the various patterns. The relative-frequency definition is presented only to elaborate the notion of the probability of occurrence of *patterns* of characteristics among individuals *within specified groups.*

If it can be assumed that the numbers of individuals in the several populations (alternative classification groups) are approximately equal, or if the *base rates* are not known, then a reasonable decision rule is to assign each individual to the group in which his particular symptom pattern has the greatest probability of occurrence. This is known as the *maximum-likelihood rule.* The probability of an individual's belonging to group j is considered equal to the ratio of the probability of his score pattern in group j relative to the sum of probabilities associated with his score pattern in all k groups

$$P(j \mid \mathbf{x}_i) = \frac{P(\mathbf{x}_i \mid j)}{P(\mathbf{x}_i \mid 1) + P(\mathbf{x}_i \mid 2) + \cdots + P(\mathbf{x}_i \mid j) + \cdots + P(\mathbf{x}_i \mid k)} \tag{12.2}$$

where $P(j \mid \mathbf{x}_i)$ is the probability that an individual belongs to group j given that he has score pattern $\mathbf{x}_i$. Obviously, the adequacy of such a decision procedure depends upon the accuracy of estimates of pattern probabilities $P(\mathbf{x}_i \mid j)$ within the various groups, and that accuracy in turn depends upon the size and representativeness of original samples from which the estimates were developed.

If the numbers of individuals in the different populations to which an individual might belong are substantially different, then assignment of each individual to the group in which his symptom or score pattern has the highest probability will not result in minimizing total errors of misclassification. The

Bayesian conditional-probability model takes base rates, or a priori probabilities, into account. The a priori probabilities are probabilities that a randomly selected individual will belong to each group disregarding his particular symptom or score classification. If 90 percent of all individuals in a mixed sample composed of individuals from several populations actually belong to one specific population, then the probability is $\pi_j = .90$ that a randomly chosen individual will belong to group j without even considering his particular characteristics. Let π_1 be the relative proportion in population 1, π_2 be the relative proportion in population 2, and so on. These relative proportions are taken as the a priori probabilities $\pi_1 + \pi_2 + \cdots + \pi_k = 1.0$.

A decision rule which will minimize total errors of misclassification in the face of known unequal population proportions is derived from estimating the probability of an individual's belonging to each group and then assigning him to the most probable group. This will be referred to as the *Bayesian model.*

$$P(j \mid \mathbf{x}_i) = \frac{\pi_j P(\mathbf{x}_i \mid j)}{\pi_i P(\mathbf{x}_i \mid 1) + \pi_2 P(\mathbf{x}_i \mid 2) + \cdots + \pi_j P(\mathbf{x}_i \mid k)} \tag{12.3}$$

where $\pi_j P(\mathbf{x}_i \mid j)$ is the joint probability of a randomly chosen individual's belonging to population j and at the same time having score pattern $\mathbf{x}_i$. It is the product of the probability of belonging to population j and the probability of symptom pattern $\mathbf{x}_i$ within population j. As mentioned at the outset, there are several ways to estimate the probabilities of various patterns within each population, but for the present we shall specify only that the pattern probability $P(\mathbf{x}_i \mid j)$ can be estimated.

A final elaboration on the basic model permits one to minimize the total loss or total seriousness of errors of misclassification. The *cost* of failure to classify an individual from population j correctly can be represented by the coefficient C_j, where $C_1 + C_2 + \cdots + C_j = 1$. The resulting classification rule involves evaluation of Eq. (12.4) with regard to each group and assigning the individual to the group for which $CP(j \mid \mathbf{x}_i)$ is greatest

$$CP(j \mid \mathbf{x}_i) = \frac{C_j \pi_j P(\mathbf{x}_i \mid j)}{C_1 \pi_1 P(\mathbf{x}_i \mid 1) + \cdots + C_j \pi_j P(\mathbf{x}_i \mid j) + \cdots + C_k \pi_k P(\mathbf{x}_i \mid k)} \tag{12.4}$$

where $C_j \pi_j P(\mathbf{x}_i \mid j)$ is the joint probability of belonging to population j and at the same time having pattern $\mathbf{x}_i$ multiplied by the relative cost of misclassifying an individual from population j.

To use any one of these three variations in assigning individuals to groups it is necessary to obtain estimates of the probabilities of particular patterns of observed scores within each population. This means that to develop a quantitative procedure for assignment of individuals one must have, to start with, a sample of individuals from each of the various populations into which individuals are later to be classified. The intention here has been to present the theoretical model in most general form and to emphasize that classification of individuals among k groups reduces to the problem of evaluating the probability of his particular pattern of scores in each of the several groups. The individual can be classified by the maximum-likelihood rule, which may be modified in the Bayesian manner by use of a priori probabilities or cost coefficients or both. The estimation of probabilities of score patterns within groups can be handled in several different ways, depending upon whether the multiple variables are quantitative or qualitative, independent or mutually correlated. The comparative utility of various methods depends to some extent upon the number of cases available in the samples from which probability estimates are to be derived.

12.3 A Computational Example

Consider first what is perhaps the simplest of all ways to estimate the probabilities $P(\mathbf{x}_i \mid j)$ of score patterns within each of several groups. It involves simply counting the frequency of occurrence of each possible pattern of the several characteristics in each group separately. Quantitative scores can be dealt with in this manner by artificially dichotomizing or trichotomizing them. Of course, qualitative variables, like signs and symptoms which have been recorded only as present or absent, are especially suited for this approach and cannot be dealt with adequately by methods used with continuous measurements having a multivariate normal distribution. The procedure of actually counting frequencies of occurrence for all possible patterns avoids tenuous assumptions regarding distributions and relationships among the several variables in the patterns. This method of estimating pattern probabilities can be recommended, however, only where very large samples are available and only where a relatively small number of variables are involved in the patterns. The approach is generally not practical because most classification problems require consideration of several variables and the number of possible patterns increases exponentially with the number of variables. For simple dichotomous present-absent items, the number of possible patterns to be considered is NPAT $= 2^p$, where p is the number of variables. For example, the number of possible patterns for 10 dichotomous variables is $2^{10} = 1{,}024$. In spite of the rather severe limitation on number of variables that can reasonably be considered, the procedure is precise and simple and avoids numerous assumptions.

For a sample of 670 hospitalized psychiatric patients, doctor's choice of drug treatment was recorded as antidepressant, major tranquilizer, minor tranquilizer, or no psychoactive drug. One of the three clinical psychologists interviewed and rated the symptomatology of each patient using the BPRS soon after admission and without knowing the drug assignment. It is of interest to examine whether patients can be accurately assigned among these four major treatment groups on the basis of presence or absence of three symptoms of anxiety, depressive mood, and thinking disturbance. The number of possible patterns of these three symptoms represented in dichotomous form is $2^3 = 8$.

The first step in the analysis was to count the frequency with which each possible pattern $+{+}+$, $+{+}0$, $+0+$, $0{+}+$, $+00$, $0{+}0$, $00+$, and 000

Table 12.1 Frequencies of Eight Possible Patterns of Anxiety, Depressive Mood, and Unusual Thought Content in Four Treatment Groups

	Treatment group			
Pattern	*Antidepressant*	*Major tranquilizer*	*Minor tranquilizer*	*No drug*
+ + +	8	20	2	9
+ + 0	83	46	52	75
+ 0 +	3	19	0	11
0 + +	0	1	0	0
+ 0 0	21	34	21	20
0 + 0	28	18	8	28
0 0 +	3	15	1	13
0 0 0	34	42	11	44
Total	180	195	95	200

occurred in each of the four treatment groups. The raw frequencies are shown in Table 12.1 for each of the treatment groups. Next, the raw frequencies were converted to probability estimates by dividing the frequency of each pattern by the total n in the particular drug group. The resulting values, shown in Table 12.2, are the estimates of probability of occurrence of pattern $\mathbf{x}_i$ given drug group j. These values of $P(\mathbf{x}_i \mid j)$ obtained using Eq. (12.1) are the within-group pattern-probability estimates required for Eqs. (12.2) to (12.4).

This is probably not the kind of application in which the use of differential base rates or a priori probabilities is most appropriate. There are no natural populations here, and the relative frequencies of use of different kinds of drug treatment depend solely on clinical practice and could easily change as the result of new information. Nevertheless, for purposes of this illustration,

Table 12.2 Probabilities of Eight Possible Patterns of Anxiety, Depressive Mood, and Unusual Thought Content in Each of Four Treatment Groups Separately, $P(\mathbf{x}_i \mid j)$

	Treatment group			
Pattern	*Antidepressant*	*Major tranquilizer*	*Minor tranquilizer*	*No drug*
+ + +	.044	.103	.021	.045
+ + 0	.461	.236	.547	.375
+ 0 +	.017	.097	.000	.056
0 + +	.000	.005	.000	.000
+ 0 0	.117	.174	.221	.100
0 + 0	.156	.092	.084	.140
0 0 +	.017	.077	.011	.065
0 0 0	.189	.215	.116	.220

assume that there is a true "phenothiazine-type patient," a true "antidepressant-type patient," and so on. Also assume that the relative frequencies of these different types are proportional to the sample frequencies of use of the four types of treatment. In the sample of 670 approximately equal numbers of patients received antidepressant, major tranquilizer, and no drug, while the number receiving minor tranquilizer was less than one-half the numbers in other treatment groups. Arbitrarily, the base rates $\pi_1 = .30$, $\pi_2 = .30$, $\pi_3 = .10$, and $\pi_4 = .30$ can be assigned such that $\pi_1 + \pi_2 + \pi_3 + \pi_4 = 1.00$.

The question of cost or seriousness of errors of misclassification is perhaps more relevant. It is widely thought that certain types of patients who are normally treated with major tranquilizers tend to manifest exacerbation of symptomatology when given antidepressant drugs. Similarly, patients who are appropriately treated with major tranquilizers tend to evidence little or no response to minor tranquilizers or no drug. The reverse is not generally true. With some exceptions, patients who might appropriately be treated with either antidepressant or minor tranquilizer tend to do well when given major tranquilizer. Whether these statements are entirely correct is not a point of major concern here. They have been advanced as the basis for illustrating the role of cost coefficients in the decision model. If it is considered to be relatively more serious to misclassify a phenothiazine patient and less serious to misclassify a no-drug patient into a drug group, the following cost coefficients could be adopted: $C_1 = .20$, $C_2 = .50$, $C_3 = .20$, and $C_4 = .10$.

Disregarding base rates and differential costs associated with misclassification, the maximum-likelihood classification rule results in the assignment of each individual to the treatment group in which the probability of occurrence of his particular symptom pattern is maximum. These within-group pattern probabilities can be read directly from Table 12.2. Since there are only eight possible patterns for the three symptoms, it is easy to specify which patients will be assigned to each group according to the classification rule. Any patient having a symptom pattern in which unusual thought content is present should be assigned to the major tranquilizer group. The four patterns involving presence of unusual thought content are $+++$, $+0+$, $0++$, and $00+$. Any patient having only depressive mood without anxiety or unusual thought content, i.e., pattern $0+0$, should be assigned to the antidepressant treatment group. Patients having both anxiety and depression ($++0$) or anxiety alone ($+00$) should be assigned to the minor tranquilizer group. Finally, the decision rule results in patients who do not have at least moderate elevation of symptomatology in any of these three areas being assigned to the no-drug group. These conclusions derived from an examination of the pattern probabilities appear satisfyingly in accord with clinical expectations and, in fact, represent the most frequent practice in our hospital setting.

Although there is risk in applying an empirically derived decision procedure to the same data from which it was derived, such an exercise can serve the purpose of a preliminary appraisal of the validity of the model or the extent of unsystematic behavior on the part of the original clinical decision makers. In problems such as the present example, where there are no definitive naturally occurring groups, discrepancies between the statistical decision and the clinical decision can be attributed either to imperfection of the model or to imperfection of the clinical decision makers. The total number of errors of misclassification resulting from use of the simple maximum-likelihood rule [Eq. (12.2)] can be easily evaluated from Table 12.1. A total of 50 patients who had symptom patterns involving presence of at least moderate unusual thought content (+++, +0+, 0++, 00+) were not assigned to major tranquilizer group; thus there was misclassification of 50 patients who would be assigned by the decision rule to major tranquilizer but who were not actually so assigned. A total of 54 patients whose symptom pattern included only elevated depressive mood (0+0) were not actually given antidepressant treatment. Since the decision rule results in all such patients being classified into the antidepressant group, another 54 errors can be counted. A total of 279 patients with either anxiety alone (+00) or anxiety plus depression (++0) were not treated with minor tranquilizer. Since the maximum-likelihood decision rule results in assignment of all such patients to the minor-tranquilizer group, an additional 279 errors must be credited to the model. Finally, 87 asymptomatic patients (000) were improperly assigned to one of the active drug groups. This results in a grand total of $50 + 54 + 279 + 87 = 470$ errors out of a possible 670, or a staggering 70 percent. In all fairness to the model, it must be recognized that clinical assignment may have been in error at least some of the time and that for purposes of this illustration a number of additional relevant variables were omitted from consideration that might have improved agreement with clinical assignment considerably. The proportion of misclassifications merely reflects the variability in individual clinical practice about the mode based on all clinicians. In spite of the variability, the decision rules resulting from examination of the pattern-probability estimates still seem logical and clinically sound in terms of what is known about the drugs.

Next, the effect of including base rates or a priori probabilities into the decision model will be examined. Assuming $\pi_1 = .30$, $\pi_2 = .30$, $\pi_3 = .10$, and $\pi_4 = .30$, the probability of an individual patient with each particular symptom pattern belonging to each of the four treatment groups can be calculated using Eq. (12.3). The probability with which a patient having symptom pattern +++ belongs to each of the four groups is estimated as follows:

Pattern +++:
$$P(1 \mid \mathbf{x}_1) = \frac{\pi_1 P(\mathbf{x}_1 \mid 1)}{\pi_1 P(\mathbf{x}_1 \mid 1) + \pi_2 P(\mathbf{x}_1 \mid 2) + \cdots + \pi_4 P(\mathbf{x}_1 \mid 4)}$$
$$= \frac{.3(.084)}{.3(.084) + .3(.103) + .1(.021) + .3(.045)}$$
$$= .351$$

Similarly, for each of the other three treatment groups, the probability of pattern $x_1 = +++$ can be calculated:

$$P(2 \mid \mathbf{x}_1) = \frac{.3(.103)}{.3(.084) + .3(1.03) + .1(.021) + .3(.045)} = .431$$

$$P(3 \mid \mathbf{x}_1) = \frac{.1(.021)}{.3(.084) + \cdots + .3(.045)} = .029$$

$$P(4 \mid \mathbf{x}_1) = \frac{.3(.045)}{.3(.084) + \cdots + .3(.045)} = .188$$

Even after taking a priori probabilities into account, patients with elevated levels of all three symptoms should be assigned to the major-tranquilizer group according to the Bayesian model.

Pattern $+0+$: $P(1 \mid \mathbf{x}_3) = .100$ $P(2 \mid \mathbf{x}_3) = .571$
$P(3 \mid \mathbf{x}_3) = .000$ $P(4 \mid \mathbf{x}_3) = .329$

Decision rule: Assign patient to major-tranquilizer group.

Pattern $0++$: $P(1 \mid \mathbf{x}_4) = .000$ $P(2 \mid \mathbf{x}_4) = 1.000$
$P(3 \mid \mathbf{x}_4) = .000$ $P(4 \mid \mathbf{x}_4) = .000$

Decision rule: Assign patient to major-tranquilizer group.

Pattern $+00$: $P(1 \mid \mathbf{x}_5) = .252$ $P(2 \mid \mathbf{x}_5) = .374$
$P(3 \mid \mathbf{x}_5) = .159$ $P(4 \mid \mathbf{x}_5) = .215$

Decision rule: Assign patient to major-tranquilizer group.

Pattern $0+0$: $P(1 \mid \mathbf{x}_6) = .375$ $P(2 \mid \mathbf{x}_6) = .221$
$P(3 \mid \mathbf{x}_6) = .067$ $P(4 \mid \mathbf{x}_6) = .337$

Decision rule: Assign patient to antidepressant group.

Pattern $00+$: $P(1 \mid \mathbf{x}_7) = .105$ $P(2 \mid \mathbf{x}_7) = .473$
$P(3 \mid \mathbf{x}_7) = .022$ $P(4 \mid \mathbf{x}_7) = .400$

Decision rule: Assign patient to major-tranquilizer group.

Pattern 000: $P(1 \mid \mathbf{x}_8) = .285$ $P(2 \mid \mathbf{x}_8) = .324$
$P(3 \mid \mathbf{x}_8) = .058$ $P(4 \mid \mathbf{x}_8) = .332$

Decision rule: Assign patient to no-drug group.

In summary, the Bayesian conditional-probability model resulted in the following decision rules. Classify any patient having at least moderate unusual thought content into the major-tranquilizer group, plus any patient with anxiety only. Classify into the antidepressant group any patient with depressive mood only or with anxiety plus depressive mood. Classify into the no-drug group any patient with no symptom elevated. It will be noted that consideration of a priori probabilities resulted in change of the decisions involving patterns of anxious-depression and anxiety alone.

When the total errors of misclassification are examined, a modest improvement can be noted. The total number of patients in the original sample who would be classified differently by the model than in actual clinical assignment is 426 out of the total of 670, or 64 percent. Although the gain was modest, the Bayesian model resulted in a reduction in total errors from 70 to 64 percent. It is interesting from a clinical point of view that the Bayesian model resulted in recommendation of either antidepressant or major tranquilizer for all patterns manifesting some presence of any of the three symptoms.

Inclusion of cost coefficients is not intended to minimize the total number of classification errors but instead to minimize the total seriousness of errors that are made. Classification rules based upon this model are derived from Eq. (12.4). At this point, it should no longer be necessary to illustrate the actual calculations involved in securing each $P(\mathbf{x}_i \mid j)$ estimate. The calculation involved in estimating the cost-weighted probability for classification into group 1 of a patient with pattern $\mathbf{x}_1 = +++$ is

$$P(1 \mid \mathbf{x}_1) = \frac{C_1\pi_1 P(\mathbf{x}_1 \mid 1)}{C_1\pi_1 P(\mathbf{x}_1 \mid 1) + C_2\pi_2 P(\mathbf{x}_1 \mid 2) + \cdots + C_4\pi_4 P(\mathbf{x}_1 \mid 4)}$$

$$= \frac{.2(.3)(.084)}{.2(.3)(.084) + \cdots + .1(.3)(.045)} = .226$$

$$P(2 \mid \mathbf{x}_1) = \frac{.5(.3)(.103)}{.2(.3)(.084) + \cdots + .1(.3)(.045)} = .694$$

$$P(3 \mid \mathbf{x}_1) = \frac{.2(.1)(.021)}{.2(.3)(.084) + \cdots + .1(.3)(.045)} = .019$$

$$P(4 \mid \mathbf{x}_1) = \frac{.1(.3)(.045)}{.2(.3)(.084) + \cdots + .1(.3)(.045)} = .061$$

Decision rule: Assign patient to major-tranquilizer group.

Similarly, applying Eq. (12.4) for patterns 2, 3, . . . , 8:

Pattern $++0$: $P(1 \mid \mathbf{x}_2) = .325$ $P(2 \mid \mathbf{x}_2) = .415$
$P(3 \mid \mathbf{x}_2) = .128$ $P(4 \mid \mathbf{x}_2) = .132$

Decision rule: Assign patient to major-tranquilizer group.

Pattern $+0+$: $P(1 \mid \mathbf{x}_3) = .059$ $P(2 \mid \mathbf{x}_3) = .843$
$P(3 \mid \mathbf{x}_3) = .000$ $P(4 \mid \mathbf{x}_3) = .097$

Decision rule: Assign patient to major-tranquilizer group.

Pattern $0++$: $P(1 \mid \mathbf{x}_4) = .000$ $P(2 \mid \mathbf{x}_4) = 1.000$
$P(3 \mid \mathbf{x}_4) = .000$ $P(4 \mid \mathbf{x}_4) = .000$

Decision rule: Assign patient to major-tranquilizer group.

Pattern +00: $P(1 \mid \mathbf{x}_5) = .173$ $P(2 \mid \mathbf{x}_5) = .644$
$P(3 \mid \mathbf{x}_5) = .109$ $P(4 \mid \mathbf{x}_5) = .074$

Decision rule: Assign patient to major-tranquilizer group.

Pattern 0+0: $P(1 \mid \mathbf{x}_6) = .322$ $P(2 \mid \mathbf{x}_6) = .475$
$P(3 \mid \mathbf{x}_6) = .058$ $P(4 \mid \mathbf{x}_6) = .145$

Decision rule: Assign patient to major-tranquilizer group.

Pattern 00+: $P(1 \mid \mathbf{x}_7) = .069$ $P(2 \mid \mathbf{x}_7) = .784$
$P(3 \mid \mathbf{x}_7) = .015$ $P(4 \mid \mathbf{x}_7) = .132$

Decision rule: Assign patient to major-tranquilizer group.

Pattern 000: $P(1 \mid \mathbf{x}_8) = .216$ $P(2 \mid \mathbf{x}_8) = .614$
$P(3 \mid \mathbf{x}_8) = .044$ $P(4 \mid \mathbf{x}_8) = .126$

Decision rule: Assign patient to major-tranquilizer group.

When one applies the cost coefficients derived from consideration of what is known about the liabilities and dangers of treating patients inappropriately with each of the major types of treatments, the interesting result is that the most complex model uniformly recommends use of the major-tranquilizer treatment across all patterns of the three symptoms of anxiety, depression, and thinking disturbance. It is difficult not to consider, at this point, the recent research literature, which suggests that major tranquilizers are superior therapy across a wide variety of depressed and schizophrenic patients (Overall and Hollister, 1964; Overall et al., 1966; Hollister et al., 1967; Hollister and Overall, 1968; Overall and Tupin, 1969; and Henry et al., 1969). It is intriguing to the present writers that the complete cost-weighted model, using estimates of a priori probabilities derived from the sample and using cost coefficients which seemed justified on quite a different empirical basis, resulted in decision rules consistent with this body of empirical data.

12.4 Diagnostic-classification Example

The problem of diagnostic classification is a more familiar and interesting one to many clinicians. As a second example of the direct method of counting relative frequencies of occurrence in order to estimate probabilities of complex patterns within groups, we shall consider briefly the diagnostic-stereotype data collected from experts in four European countries to represent hypothetical typical patients in various diagnostic categories. By omitting certain types, it was possible to group the prototype ratings into three broad categories: 1,419 profiles represented various types of nonparanoid schizophrenia, 649 represented hypothetical typical depressions, and 888 represented paranoid conditions including paranoid schizophrenia.

For the problem of discriminating among these three major diagnostic groups in terms of pattern probabilities, four BPRS symptom constructs of depressive mood, hostility, unusual thought content, and blunted affect were selected. Dichotomizing each rating variable into present-absent yielded $2^4 = 16$ patterns displayed in Table 12.3. The four symptoms entering into these patterns are in the order specified above. The frequency of each pattern among profiles purported to represent typical patients in the three major diagnostic classes is also shown in Table 12.3.

Table 12.3 Frequencies of 16 Possible Patterns of Depressive Mood, Hostility, Unusual Thought Content, and Blunted Affect in Three Diagnostic Groups

	Diagnostic group		
Pattern	*Schizophrenia*	*Depressed*	*Paranoid*
1111	154	52	83
0111	376	3	188
1011	75	104	19
1101	8	16	3
1110	43	17	86
0011	316	1	64
0101	53	0	20
0110	84	2	240
1001	29	97	2
1010	32	96	28
1100	9	19	5
1000	13	239	4
0100	16	0	36
0010	55	1	85
0001	119	1	3
0000	37	1	22
Total	1,419	649	888

The probability of a particular symptom pattern's occurring with each diagnostic group can be estimated by dividing the frequency for the specific pattern by the total frequency. For example, the probability of pattern 1011 (depressive mood, no hostility, unusual thought content, and blunted affect) in the schizophrenic group is $P(\mathbf{x}_3 \mid 1) = 75/1{,}419 = .0529$, where $\mathbf{x}_3$ indicates the third pattern, namely, 1011. The estimated probabilities of occurrence for the 16 symptom patterns *within* each diagnostic group are presented in Table 12.4.

Using the maximum-likelihood rule, each profile is classified into the category in which that particular pattern has greatest likelihood of occurrence. Scanning the first column of Table 12.4 reveals that a total of 6 out of the 16 possible patterns were judged to be relatively more representative of the schizophrenia class. A total of 6 patterns was judged to be relatively more representative of

Table 12.4 Relative Frequency-probability Estimates for Patterns of Depressive Mood, Hostility, Unusual Thought Content, and Blunted Affect in Three Diagnostic Groups

	Diagnostic group		
Pattern	*Schizophrenic*	*Depressed*	*Paranoid*
1111	.1085	.0801	.0935
0111	.2650	.0046	.2117
1011	.0529	.1602	.0214
1101	.0056	.0247	.0034
1110	.0303	.0262	.0968
0011	.2227	.0015	.0721
0101	.0374	.0000	.0225
0110	.0592	.0031	.2703
1001	.0204	.1495	.0023
1010	.0226	.1479	.0315
1100	.0063	.0293	.0056
1000	.0092	.3683	.0045
0100	.0113	.0000	.0405
0010	.0388	.0015	.0957
0001	.0839	.0015	.0034
0000	.0261	.0015	.0248
Total	1.0000	1.0000	1.0000

the depression class, and 4 patterns to be more representative of the paranoid class. All symptom patterns in which depressive mood (element 1) was positive were found to be indicative of the depression class except for the two patterns in which *both* hostility and unusual thought content were also present. All patterns in which blunted affect (element 4) was positive without positive depression were found to be relatively most probable in the schizophrenia category. Of the four patterns that were found to be relatively most probable in the paranoid class, none contained positive blunted affect (element 4), and only one contained positive depressive mood (element 1). These are some of the kinds of general rules of thumb that can be derived from examination of the pattern probabilities shown in Table 12.4.

To obtain estimates of the probability with which a particular pattern belongs in each diagnostic category, disregarding possible differences in base rates, one uses Eq. (12.2). These values can be obtained by dividing the entries in each row of Table 12.4 by the row total. The probabilities of group membership derived in this manner are shown in Table 12.5. It is obvious from inspection of Table 12.5 that some patterns can be classified with little probability of error while others are almost equally likely to belong in two or more categories.

Following the rule that each profile should be assigned to the diagnostic category in which it is most likely to occur, according to Eq. (12.2), a total of 2,074, or 70 percent, of the 2,956 stereotype profiles were assigned to the category that the expert judge intended to be represented by the profile. A cross

Table 12.5 Probabilities of 16 Patterns in Three Diagnostic Groups

	Diagnostic group		
Pattern	*Schizophrenic*	*Depressed*	*Paranoid*
1111	.3846	.2839	.3314
0111	.5506	.0096	.4399
1011	.2256	.6832	.0913
1101	.1662	.7329	.1009
1110	.1977	.1709	.6314
0011	.7516	.0051	.2433
0101	.6244	.0000	.3756
0110	.1780	.0093	.8127
1001	.1185	.8682	.0134
1010	.1119	.7322	.1559
1100	.1529	.7112	.1359
1000	.0241	.9641	.0118
0100	.2181	.0000	.7819
0010	.2853	.0110	.7037
0001	.9448	.0169	.0383
0000	.4981	.0286	.4733

tabulation of the maximum-likelihood pattern-probability classification against the diagnostic classification purported to be represented by each profile is shown in Table 12.6. The accuracy with which profiles purported to represent each of the three classes were correctly assigned by the maximum-likelihood rule varied from 88 percent correct for depressions and 74 percent correct for schizophrenia to only 50 percent correct for the paranoid profiles. The locus of greatest misclassification was between paranoid and schizophrenic categories.

One potential way the accuracy of classification might be improved is to add more symptom variables, but this is not really practical where the direct pattern-frequency-count method is used to estimate pattern probabilities. Another way to increase accuracy somewhat in this particular case would be to vary the cutting point at which symptom ratings are dichotomized. Finally, one can

Table 12.6 Classification into Diagnostic Groups Using Profile Patterns of Four BPRS Symptoms: Maximum-likelihood Rule

Probability-model classification	*Expert-specified diagnostic type*			
	Schizophrenic	*Depressed*	*Paranoid*	*Total*
Schizophrenic	1,055	58	380	1,493
Depressed	166	572	61	799
Paranoid	198	19	447	664
Total	1,419	649	888	2,956

improve the proportion of correct classifications somewhat by including consideration of base rates in the model [Eq. (12.3)]. Variation in cost coefficients [Eq. (12.4)] can also be used to effect a change in the frequency with which paranoid profiles are classified in the schizophrenic group. We shall leave these possibilities as exercises for the interested student, who can examine the effects of varying base rates and cost coefficients by starting with the $P(\mathbf{x}_i \mid j)$ pattern-probability estimates given in Table 12.4.

12.5 References

Henry, B. W., J. R. Markette, R. L. Emken, and J. E. Overall: Drug Treatment of Anxious Depression in Psychiatric Outpatients, *Dis. Nerv. Syst.*, **30**:675–679 (1969).

Hollister, L. E., and J. E. Overall: Phenothiazine Derivatives as Antidepressants, Second International Symposium on Action Mechanism and Metabolism of Psychoactive Drugs Derived from Phenothiazine and Structurally Related Compounds, *Aggressologie*, **9**:289–292 (1968).

———, ———, J. Shelton, V. Pennington, I. Kimbell, Jr., and M. Johnson: Amitryptilene, Perphenazine and Amitryptilene-Perphenazine Combination in Different Depressive Syndromes, *Arch. Gen. Psychiat.*, **17**:486–493 (1967).

Overall, J. E., and L. E. Hollister: Imipramine and Thioridazine in Depressed and Schizophrenic Patients, *J. Amer. Med. Assoc.*, **189**:605–608 (1964).

———, ———, M. Johnson, and V. Pennington: Nosology of Depression and Differential Response to Drugs, *J. Amer. Med. Assoc.*, **195**:946–948 (1966).

———, and J. P. Tupin: Investigation of Clinical Outcome in a Doctor's Choice Treatment, *Dis. Nerv. Syst.*, **30**:305–313 (1969).

13

Normal-probability-density Model for Classification among Several Groups

13.1 Introduction

In this chapter, a particular method for classification of individuals among several groups will be considered. The general model is the same as that described in the preceding chapter, special attention being directed toward the problem of estimating the probability of the particular multivariate score pattern within each group. Given that an adequate method is available for estimating the probability of a particular score pattern in each group, the general methods of arriving at classification decisions described in Chap. 12 can be utilized. In the present chapter, we shall be concerned with utilizing the multivariate normal-probability-density function in arriving at estimates of probability of occurrence for multivariate profile patterns.

Methods for classification of individuals among multiple groups on the basis of p normally distributed quantitative variables are considered in this chapter. As before, it will be assumed that each individual belongs to one of several specified populations. Characteristics of each individual are represented in a quantitative score vector or profile. The several variables in the score vector or profile are assumed to have a multivariate normal distribution with equal variances and covariances within the different groups; i.e., the variance-covariance matrix is assumed to be the same for each group.

In Chap. 9, the problem of assigning individuals between *two* groups on the basis of p measurement characteristics which could be assumed to have a multivariate normal distribution was considered. The simple discriminant-function analysis is a way of reducing several variables to a single composite variable that can be considered to have a univariate normal distribution. Probabilities of belonging to each group can be estimated from unit-normal probability densities, as previously described in Sec. 9.5.

The probability density in a *multivariate* normal distribution is a concept concerned with the relative number of individuals in a total population having a particular score pattern. In a *univariate* normal distribution, density at a particular point in the distribution is proportional to the likelihood of a particular score value. The univariate normal density is thus identified with the height of the normal curve at a particular point. The ordinate value or height of a normal-probability-distribution function has previously been defined (Sec. 9.5) to be the value at point x of the function

$$f(x)^i = \frac{1}{\sqrt{2\pi\sigma^2}} \exp\left[\frac{-\frac{1}{2}(x - \mu_i)^2}{\sigma^2}\right] \tag{13.1}$$

By direct analogy, the density of a multivariate normal distribution at a point identified with score vector $\mathbf{x}$ is

$$f(\mathbf{x})^{(i)} = \frac{1}{\sqrt{2\pi p\,|\mathbf{\Sigma}|}} \exp\left[-\tfrac{1}{2}(\mathbf{x} - \mathbf{u}^{(i)})'\mathbf{\Sigma}^{-1}(\mathbf{x} - \mathbf{u}^{(i)})\right] \tag{13.2}$$

The variance-covariance matrix $\mathbf{\Sigma}$ is the multivariate generalization of the variance σ^2 in the univariate case. Similarly, the multivariate mean vector $\boldsymbol{\mu}^{(i)}$ is the multivariate generalization of the univariate mean μ. In all respects, the multivariate normal-distribution function is a direct generalization of the univariate normal-distribution function.

Numerous texts have attempted to increase intuitive understanding of the concept of a multivariate normal-density function by technical and nontechnical analogies and illustrations. The concept is an important one, but for the present purpose it is sufficient simply to recognize that the multivariate normal case is a direct generalization or extension of the concept of a univariate normal-density function. The *variances* and *covariances* in the dispersion matrix $\mathbf{\Sigma}$ replace the variance of the single variable, and the vector of means $\boldsymbol{\mu}$ replaces the single mean of the univariate case. It will be appreciated that the multivariate normal-distribution function depends upon only the two parameters $\boldsymbol{\mu}$ and $\mathbf{\Sigma}$, just as the univariate normal-distribution function depends only on parameters μ and σ^2. The multivariate normal *density* can be conceived as a measure of the relative number of p-variate profiles that will fall within a particular *very small* region in the multivariate space. As in the case of the univariate normal distribution, the region is taken as infinitesimally small, so that one can conceive of the multivariate probability density as being an index that is proportional to the relative frequency in the population of a particular profile pattern.

If there are several groups or populations with known or estimated multivariate means and covariance matrices, one can estimate the relative probability of any particular p-variate score vector or profile pattern in each of the populations by using the multivariate normal-probability function given in Eq. (13.2). This is accomplished by inserting the population-parameter values $\boldsymbol{\mu}^{(i)}$ and $\boldsymbol{\Sigma}^{(i)}$ (or their estimates) into the equation and solving for $f(\mathbf{x})^{(1)}, f(\mathbf{x})^{(2)}, \ldots, f(\mathbf{x})^{(k)}$, where $f(\mathbf{x})^{(i)}$ is the multivariate normal probability density associated with score vector $\mathbf{x}$ in population $i = 1, 2, \ldots, k$.

13.2 Classification Using Normal Probability Densities

Classification of an individual among k groups proceeds exactly as described for the general case in Chap. 12. The probability of an individual with a particular score pattern $\mathbf{x}$ belonging to each of the k groups is evaluated using the multivariate normal-density function of Eq. (13.2). Disregarding a priori probabilities, or assuming that they are equal, the maximum-likelihood decision is achieved by calculating the probability ratio $P(i \mid \mathbf{x})$ for each group $i = 1, 2, \ldots, j, \ldots, k$ and assigning the individual to the group for which this value is greatest

$$P(i \mid \mathbf{x}) = \frac{f(\mathbf{x})^{(i)}}{f(\mathbf{x})^{(1)} + f(\mathbf{x})^{(2)} + \cdots + f(\mathbf{x})^{(k)}} \tag{13.3}$$

where $f(\mathbf{x})^{(i)}$ is the multivariate normal probability density associated with pattern $\mathbf{x}$ in the ith group as defined in Eq. (13.2).

It will be noted that the factor

$$\frac{1}{\sqrt{2\pi^p \, |\boldsymbol{\Sigma}|}}$$

is common to all terms in both the numerator and denominator and can thus be factored out and disregarded. The classification of individuals among k groups using the multivariate normal model reduces to evaluating the exponent

$$\exp\left[-\tfrac{1}{2}(\mathbf{x} - \boldsymbol{\mu}^{(i)})'\boldsymbol{\Sigma}^{-1}(\mathbf{x} - \boldsymbol{\mu}^{(i)})\right]$$

for each group $i = 1, 2, \ldots, k$. Remembering that the *natural* logarithm of a number is equal to the power to which the base e would have to be raised to equal the number, the desired densities can be obtained as the antilogarithms of the quantity $-\tfrac{1}{2}(\mathbf{x} - \boldsymbol{\mu}^{(i)})'\boldsymbol{\Sigma}^{-1}(\mathbf{x} - \boldsymbol{\mu}^{(i)})$

$$P(i \mid \mathbf{x}) = \frac{\text{antilog}_e\left[-\tfrac{1}{2}(\mathbf{x} - \boldsymbol{\mu}^{(i)})'\boldsymbol{\Sigma}^{-1}(\mathbf{x} - \boldsymbol{\mu}^{(i)})\right]}{\sum_j^k \text{antilog}_e\left[-\tfrac{1}{2}(\mathbf{x} - \boldsymbol{\mu}^{(j)})'\boldsymbol{\Sigma}^{-1}(\mathbf{x} - \boldsymbol{\mu}^{(j)})\right]} \tag{13.4}$$

If base rates or a priori probabilities are to be considered, an equation in the form of Eq. (12.3) can be written:

$$P(i \mid \mathbf{x}) = \frac{\pi_i f(\mathbf{x})^{(i)}}{\pi_1 f(\mathbf{x})^{(1)} + \pi_2 f(\mathbf{x})^{(2)} + \cdots} \tag{13.5}$$

or

$$P(i \mid \mathbf{x}) = \frac{\pi_i \text{ antilog}_e \, [-\frac{1}{2}(\mathbf{x} - \boldsymbol{\mu}^{(i)})'\boldsymbol{\Sigma}^{-1}(\mathbf{x} - \boldsymbol{\mu}^{(i)})]}{\sum_j^k \pi_j \text{ antilog}_e \, [-\frac{1}{2}(\mathbf{x} - \boldsymbol{\mu}^{(j)})'\boldsymbol{\Sigma}^{-1}(\mathbf{x} - \boldsymbol{\mu}^{(j)})]} \tag{13.6}$$

Similarly, cost coefficients in addition to base rates can be considered using an equation in the form of Eq. (12.4):

$$P(i \mid \mathbf{x}) = \frac{C_i \pi_i f(x)^{(i)}}{C_1 \pi_1 f(\mathbf{x})^{(1)} + C_2 \pi_2 f(\mathbf{x})^{(2)} + \cdots} \tag{13.7}$$

$$P(i \mid \mathbf{x}) = \frac{C_i \pi_i \text{ antilog}_e \, [-\frac{1}{2}(\mathbf{x} - \boldsymbol{\mu}^{(i)})'\boldsymbol{\Sigma}^{-1}(\mathbf{x} - \boldsymbol{\mu}^{(i)})]}{\sum_j^k C_j \pi_j \text{ antilog}_e \, [-\frac{1}{2}(\mathbf{x} - \boldsymbol{\mu}^{(j)})'\boldsymbol{\Sigma}^{-1}(\mathbf{x} - \boldsymbol{\mu}^{(j)})]} \tag{13.8}$$

It is worthwhile, from the point of view of understanding the nature of the multivariate probability-density function and the classification rules derived from it, to consider carefully the term $(\mathbf{x} - \boldsymbol{\mu}^{(i)})'\boldsymbol{\Sigma}^{-1}(\mathbf{x} - \boldsymbol{\mu}^{(i)})$. Reference to Sec. 11.4 reveals that it is simply the squared distance D_i^2 between the individual profile $\mathbf{x}$ and the mean profile for the ith group. Thus, classification of individuals among multiple groups using the multivariate normal model depends upon evaluation of the *distance* of the individual's score vector from the mean score vector for each group. In fact, assuming base rates to be equal, the probability of a particular score vector's belonging to a given multivariate normal distribution depends only upon the distance of the individual score vector from the mean vector in the multivariate distribution. In more clinical terminology, the probability of an individual's belonging to a particular group depends upon the degree of similarity between his profile or score vector and the mean vector for the group. It will not be surprising then to note that this entire section on multivariate normal classification methods is concerned simply with quantitative methods for assessing and defining the degree of similarity of an individual patient profile to group means or prototype profiles.

With regard to the estimation of probabilities of group membership, one should recall that the antilog of a negative number is a fraction less than unity. The larger the negative logarithm, the smaller the fraction that is its antilog. Thus, the greater the distance of the individual profile $\mathbf{x}$ from group mean $\boldsymbol{\mu}^{(i)}$, the smaller the quantity

$$\text{antilog}_e \, [-\tfrac{1}{2}(\mathbf{x} - \boldsymbol{\mu}^{(i)})'\boldsymbol{\Sigma}^{-1}(\mathbf{x} - \boldsymbol{\mu}^{(i)})]$$

and the smaller the estimated probability of membership in the ith group.

13.3 A Computational Example

13.3.1 BASIC DATA

As an illustration of the computational procedure, the drug-treatment assignment problem considered in Chap. 12 will again be used so that results can be compared with those obtained by other methods. In order to illustrate the actual calculations, we shall first be concerned with only the four higher-order factor scores. Later, to obtain results that can be compared with those obtained by other methods, classification based on all 16 symptom ratings will be examined. BPRS factor scores for the sample of 670 newly hospitalized patients were used for this illustrative analysis with the goal of developing a procedure for assigning patients to antidepressant, major-tranquilizer, minor-tranquilizer, or no-drug treatment groups.

The pooled within-groups SP matrix, computed as described in Sec. 2.7, has previously been presented in Table 10.8. The pooled within-groups covariance matrix can be obtained by dividing the sums of products by appropriate degrees of freedom. It will be assumed that variances and covariances for the four factor scores are equal in the four treatment populations, so that a single pooled within-groups covariance matrix can be taken as the estimate of $\mathbf{\Sigma}$ in Eqs. (13.3) to (13.8). The inverse of the pooled within-groups variance-covariance matrix, which is accepted as an estimate of $\mathbf{\Sigma}^{-1}$, is presented in Table 13.1. Mean factor-score vectors $\boldsymbol{\mu}^{(1)}$, $\boldsymbol{\mu}^{(2)}$, $\boldsymbol{\mu}^{(3)}$, and $\boldsymbol{\mu}^{(4)}$ for the four treatment groups taken from Table 10.8 can be used as estimates of those population parameters in Eqs. (13.3) to (13.8).

Table 13.1 Inverse of Within-groups Covariance Matrix Computed from Sums of Products Shown in Table 10.8

.089894	−.018197	−.036837	−.001318
−.018197	.082535	.005838	−.007892
−.036837	.005838	.086748	.007654
−.001318	−.007892	.007654	.089007

The factor-score vectors for three patients were as follows:

Patient	*Thinking disturbance*	*Withdrawal-retardation*	*Hostile-suspiciousness*	*Anxious depression*
A	4	4	6	8
B	6	4	8	9
C	3	3	6	9

The classification procedure based on the multivariate normal model will be illustrated with reference to these three patients and then applied to the complete sample to permit evaluation of results.

13.3.2 PATIENT A

The difference between the factor-score profile for patient A and the mean profile for the antidepressant treatment group was computed to be

$$\mathbf{d}_1' = (\mathbf{x} - \boldsymbol{\mu}^{(1)})' = [1.5 \quad .5 \quad .2 \quad -1.2]$$

The quadratic form in the matrix inverse was evaluated next:

$$(\mathbf{x} - \boldsymbol{\mu}^{(1)})'\boldsymbol{\Sigma}^{-1}(\mathbf{x} - \boldsymbol{\mu}^{(1)}) = \mathbf{d}_1'\boldsymbol{\Sigma}^{-1}\mathbf{d}_1 = .3168$$

The antilog of $-\frac{1}{2}\mathbf{d}_1'\boldsymbol{\Sigma}^{-1}\mathbf{d}_1$ was found to be .853:

$$f(\mathbf{x})^{(1)} = \text{antilog}_e -\tfrac{1}{2}(.3168) = \text{antilog}_e - .1584 = .853$$

Similar calculations resulted in evaluation of the distance of patient A from mean vectors for each of the other three treatment groups.

$$\mathbf{d}_2' = (\mathbf{x} - \boldsymbol{\mu}^{(2)})' = [-1.5 \quad .4 \quad -1.9 \quad .2]$$

$$\mathbf{d}_2'\boldsymbol{\Sigma}^{-1}\mathbf{d}_2 = .3289$$

and

$$f(\mathbf{x})^{(2)} = \text{antilog}_e -\tfrac{1}{2}\mathbf{d}_2'\boldsymbol{\Sigma}^{-1}\mathbf{d}_2 = \text{antilog}_e - .1644 = .848$$

$$\mathbf{d}_3' = (\mathbf{x} - \boldsymbol{\mu}^{(3)})' = [1.7 \quad .3 \quad .6 \quad -1.2]$$

$$\mathbf{d}_3'\boldsymbol{\Sigma}^{-1}\mathbf{d}_3 = .3351$$

and

$$f(\mathbf{x})^{(3)} = \text{antilog}_e -\tfrac{1}{2}\mathbf{d}_3'\boldsymbol{\Sigma}^{-1}\mathbf{d}_3 = \text{antilog}_e - .1675 = .846$$

$$\mathbf{d}_4' = (\mathbf{x} - \boldsymbol{\mu}^{(4)})' = [.1 \quad .2 \quad -.5 \quad -.3]$$

$$\mathbf{d}_4'\boldsymbol{\Sigma}^{-1}\mathbf{d}_4 = .0390$$

and

$$f(\mathbf{x})^{(4)} = \text{antilog}_e -\tfrac{1}{2}\mathbf{d}_4'\boldsymbol{\Sigma}^{-1}\mathbf{d}_4 = \text{antilog}_e - .0195 = .981$$

Disregarding base rates, or assuming them to be equal, the probability that patient A properly belongs to each of the four treatment groups can be estimated using Eq. (13.3)

$$P(1 \mid \mathbf{x}) = \frac{.853}{.853 + .848 + .846 + .981} = .241$$

$$P(2 \mid \mathbf{x}) = \frac{.848}{.853 + .848 + .846 + .981} = .240$$

$$P(3 \mid \mathbf{x}) = \frac{.846}{.853 + .848 + .846 + .981} = .240$$

$$P(4 \mid \mathbf{x}) = \frac{.981}{.853 + .848 + .846 + .981} = .278$$

The probability that a patient having a factor-score profile like patient A belongs to the no-drug group (group 4) is higher than the probabilities for the other three treatment groups. The maximum-likelihood rule would result in patient A's being assigned to the no-drug group. As the probability values of the other three groups are of about the same magnitude, the likelihood of error in the assignment of patient A is recognized to be substantial.

13.3.3 PATIENT B

For classification of patient B using the normal-probability-density model, the essential calculations are as follows.

Antidepressant: $\mathbf{d}_1' = (\mathbf{x} - \boldsymbol{\mu}^{(1)})' = [3.5 \quad .5 \quad 2.2 \quad -1.2]$

$$\mathbf{d}_1'\boldsymbol{\Sigma}^{-1}\mathbf{d}_1 = 1.0319$$

$$f(\mathbf{x})^{(1)} = \text{antilog}_e -\tfrac{1}{2}\mathbf{d}_1'\boldsymbol{\Sigma}^{-1}\mathbf{d}_1 = \text{antilog}_e - .5159 = .597$$

$$P(1 \mid \mathbf{x}) = \frac{f(\mathbf{x})^{(1)}}{\sum_j^4 f(\mathbf{x})^{(j)}} = \frac{.597}{2.974} = .201$$

Major tranquilizer: $\mathbf{d}_2' = (\mathbf{x} - \boldsymbol{\mu}^{(2)})' = [.5 \quad .4 \quad .1 \quad .2]$

$$\mathbf{d}_2'\boldsymbol{\Sigma}^{-1}\mathbf{d}_2 = .0284$$

$$f(\mathbf{x})^{(2)} = \text{antilog}_e -\tfrac{1}{2}\mathbf{d}_2'\boldsymbol{\Sigma}^{-1}\mathbf{d}_2 = \text{antilog}_e - .0142 = .986$$

$$P(2 \mid \mathbf{x}) = \frac{f(\mathbf{x})^{(2)}}{\Sigma f(\mathbf{x})^{(j)}} = \frac{.986}{2.974} = .332$$

Minor tranquilizer: $\mathbf{d}_3' = (\mathbf{x} - \boldsymbol{\mu}^{(3)})' = [3.7 \quad .3 \quad 2.6 \quad -1.2]$

$$\mathbf{d}_3'\boldsymbol{\Sigma}^{-1}\mathbf{d}_3 = 1.1823$$

$$f(\mathbf{x})^{(3)} = \text{antilog}_e -\tfrac{1}{2}\mathbf{d}_3'\boldsymbol{\Sigma}^{-1}\mathbf{d}_3 = \text{antilog}_e - .5911 = .554$$

$$P(3 \mid \mathbf{x}) = \frac{f(\mathbf{x})^{(3)}}{\Sigma f(\mathbf{x})^{(j)}} = \frac{.554}{2.974} = .186$$

No drug: $\mathbf{d}_4' = (\mathbf{x} - \boldsymbol{\mu}^{(4)})' = [2.1 \quad .2 \quad 1.5 \quad -.3]$

$$\mathbf{d}_4'\boldsymbol{\Sigma}^{-1}\mathbf{d}_4 = .3548$$

$$(\mathbf{x})^{(4)} = \text{antilog}_e -\tfrac{1}{2}\mathbf{d}_4'\boldsymbol{\Sigma}^{-1}\mathbf{d}_4 = \text{antilog}_e - .1774 = .837$$

$$P(4 \mid \mathbf{x}) = \frac{f(\mathbf{x})^{(4)}}{\Sigma f(\mathbf{x})^{(j)}} = \frac{.837}{2.974} = .281$$

The probability of a factor-score profile like that of patient B is judged to be greatest for the major-tranquilizer group. The maximum-likelihood rule would result in assignment of patient B to that group.

13.3.4 PATIENT C

Calculations required for assignment of patient C to one of the four treatment groups using the normal-probability model are as follows.

Antidepressant: $\mathbf{d}_1' = (\mathbf{x} - \boldsymbol{\mu}^{(1)})' = [.5 \quad -.5 \quad .2 \quad -.2]$

$$\mathbf{d}_1'\boldsymbol{\Sigma}^{-1}\mathbf{d}_1 = .04877$$

$$f(\mathbf{x})^{(1)} = \text{antilog}_e -\tfrac{1}{2}\mathbf{d}_1'\boldsymbol{\Sigma}^{-1}\mathbf{d}_1 = \text{antilog}_e - .0244 = .976$$

$$P(1 \mid \mathbf{x}) = \frac{f(\mathbf{x})^{(1)}}{\Sigma f(\mathbf{x})^{(j)}} = \frac{.976}{3.591} = .272$$

Major tranquilizer: $\mathbf{d}_2' = (\mathbf{x} - \boldsymbol{\mu}^{(2)})'[-2.5 \quad -.6 \quad -1.9 \quad 1.2]$

$$\mathbf{d}_2'\boldsymbol{\Sigma}^{-1}\mathbf{d}_2 = .6260$$

$$f(\mathbf{x})^{(2)} = \text{antilog}_e -\tfrac{1}{2}\mathbf{d}_2'\boldsymbol{\Sigma}^{-1}\mathbf{d}_2 = \text{antilog}_e - .3130 = .731$$

$$P(2 \mid \mathbf{x}) = \frac{f(\mathbf{x})^{(2)}}{\Sigma f(\mathbf{x})^{(j)}} = \frac{.731}{3.591} = .204$$

Minor tranquilizer: $\mathbf{d}_3' = (\mathbf{x} - \boldsymbol{\mu}^{(3)})' = [.7 \quad -.7 \quad .6 \quad -.2]$

$$\mathbf{d}_3'\boldsymbol{\Sigma}^{-1}\mathbf{d}_3 = .0976$$

$$f(\mathbf{x})^{(3)} = \text{antilog}_e -\tfrac{1}{2}\mathbf{d}_3'\boldsymbol{\Sigma}^{-1}\mathbf{d}_3 = \text{antilog}_e - .0488 = .952$$

$$P(3 \mid \mathbf{x}) = \frac{f(\mathbf{x})^{(3)}}{\Sigma f(\mathbf{x})^{(j)}} = \frac{.952}{3.591} = .265$$

No drug: $\mathbf{d}_4' = (\mathbf{x} - \boldsymbol{\mu}^{(4)})' = [-.9 \quad -.8 \quad -.5 \quad .7]$

$$\mathbf{d}_4'\boldsymbol{\Sigma}^{-1}\mathbf{d}_4 = .1414$$

$$f(\mathbf{x})^{(4)} = \text{antilog}_e -\tfrac{1}{2}\mathbf{d}_4'\boldsymbol{\Sigma}^{-1}\mathbf{d}_4 = \text{antilog}_e - .0707 = .932$$

$$P(4 \mid \mathbf{x}) = \frac{f(\mathbf{x})^{(4)}}{\Sigma f(\mathbf{x})^{(j)}} = \frac{.932}{3.591} = .260$$

The probability of a score profile like that of patient C is greatest in the antidepressant treatment group. The maximum-likelihood rule would result in assignment of the patient to that group. Again, the apparent probability of error is substantial since the probability of the score pattern in the minor-tranquilizer or the no-drug group is almost as great as it is in the antidepressant group.

To evaluate the adequacy of the model for assignment of patients to treatment groups on the basis of symptom ratings alone, a computer program was written to accomplish the calculations illustrated above for each patient. The 16-variable BPRS symptom-rating profiles were used as basic data rather than the four higher-order factor scores, and the group of patients treated with no drug was eliminated, to leave a total sample of 470 patients who received antidepressant, major-tranquilizer, or minor-tranquilizer drugs. The BPRS profile for each patient was classified by calculating the ratio of densities as specified in Eq. (13.4), with the patient being assigned to the drug-treatment group for which the resulting $P(i \mid \mathbf{x})$ was found to be greatest. A cross tabulation of the results obtained by profile analysis with the actual drug assignment for each patient is presented in Table 13.2.

Table 13.2 Assignment of Patients to Treatment Groups Using Normal-probability-density Model and Maximum-likelihood Rule

	Actual treatment group			
Profile assignment	*Antidepressant*	*Major tranquilizer*	*Minor tranquilizer*	*Total*
Antidepressant	87	47	26	160
Major tranquilizer	28	115	12	155
Minor tranquilizer	65	33	57	155
Total	180	195	95	470

Obviously, the problem of assignment to drug treatments was not selected in an effort to convince the reader of the power of the method. Among the three major drug-treatment groups, errors of misclassification reached 45 percent. The practical importance of the problem of correct treatment assignment was the major basis for choice of this example. As the results clearly attest, there is at present no set of clear-cut rules for assignment of patients to drug treatments. The groups are substantially overlapping, especially the antidepressant and minor-tranquilizer groups. The results do suggest that profile-analysis method can be used to identify groups of patients that are most typical of those treated in practice with each major class of drugs. One is stimulated to wonder whether the treatment responses of the "misclassified" groups would not differ from the treatment responses of patients who are more typical within each treatment group.

The neglect of base rates in the maximum-likelihood profile classification should be noted in passing. In actuality only about one-half as many patients received minor tranquilizers as received major tranquilizers or antidepressants. Because no base-rate coefficients were included in the model, profile classification based on the multivariate normal-density function resulted in approximately an equal number of patients being assigned to each type of treatment (160, 155, and 155). The extra number assigned to the minor-tranquilizer group tended to be patients who actually received antidepressant drugs. This situation could be remedied to some extent by taking into account the different frequencies with which the types of drugs were actually used. For example, one might let $\pi_1 = .35$, $\pi_2 = .45$, and $\pi_3 = .20$. For each patient, the probability of belonging to each treatment group could then be calculated using Eq. (13.5)

$$P(i \mid \mathbf{x}) = \frac{\pi_i f(\mathbf{x})^{(i)}}{\pi_1 f(\mathbf{x})^{(1)} + \pi_2 f(\mathbf{x})^{(2)} + \cdots}$$

The patient would then be assigned to the treatment group for which his calculated $P(i \mid \mathbf{x})$ is highest.

Although the use of differential base rates would, in fact, tend to reduce "errors" of misclassification in dealing with these particular data, one might well hesitate to accept the concept of base rates or a priori probabilities in connection with this type of assignment problem because we are not dealing with real or naturally defined populations. The drug-treatment populations exist only in terms of the clinical practice of a particular group of psychiatrists. That practice can easily change, even as the result of feedback from this type of analysis. The base-rate coefficients tend to become fudge factors that can be adjusted to bring results into better alignment with clinical practice and do not have the same theoretical significance as in problems concerned with more permanent, natural clinical populations. In spite of such doubts, the use of base-rate coefficients in connection with this assignment problem will be considered in the next chapter, primarily to illustrate how base rates can be used with the simpler profile-analysis methods.

In the second example of the previous chapter, typical (hypothetical) patient profiles were classified as schizophrenic, depressed, or paranoid on the basis of the frequency of profile patterns formed by dichotomizing four BPRS symptom-construct ratings. The effectiveness of this approach was shown in Table 12.3. These same patients were also classified by the normal-probability-density model using profile vectors of the four higher-order scores of the BPRS. The mean profiles representing the three diagnostic groups were presented in Table 10.12, which also showed the within-groups SP matrix. The within-groups covariance matrix can be obtained by dividing elements in the SP matrix by appropriate degrees of freedom. The required inverse of the within-groups covariance matrix appears in Table 13.3, and resulting classification in Table 13.4. It

Table 13.3 Inverse of Within-groups Covariance Matrix Computed from SP Matrix Shown in Table 10.12

$$\begin{bmatrix} .0787 & -.0160 & -.0282 & -.0096 \\ -.0160 & .0748 & -.0163 & .0032 \\ -.0282 & -.0163 & .0926 & -.0124 \\ -.0096 & .0032 & -.0124 & .0954 \end{bmatrix}$$

will be seen that 2,432 of the 2,956 profiles, or 82 percent, were correctly classified by using the normal-probability-density model. To extend this example somewhat further, all 16 BPRS symptom constructs were used to classify the prototype profiles. The mean vectors for the three groups are shown in Table 13.5, the dispersion matrix in Table 13.6, and the resulting classifications in

Table 13.4 Classification into Diagnostic Group Using BPRS Higher-order Factor Scores and the Normal-probability-density Model

	Diagnostic group			
Classification	*Schizophrenic*	*Depressive*	*Paranoid*	*Total*
Schizophrenic	1,033	22	99	1,154
Depressive	53	623	13	689
Paranoid	333	4	776	1,113
Total	1,419	649	888	2,956

Table 13.7. Classification was improved modestly using the 16 symptom constructs: 2,579, or 87 percent, were correctly classified. The mean vectors and covariance matrix are presented to provide the necessary basis for use of this procedure in clinical research and to make them available for use in a variety of other types of analyses for which such intermediate results are necessary.

13.4 Other Diagnostic Examples and Exercises

To illustrate further the use of the multivariate normal model for diagnostic decision making, diagnostic-stereotype data will be used from a study conducted in Czechoslovakia in which the World Health Organization International Classification System was employed. These data, like the other sets of psychiatric diagnostic-stereotype data described in Sec. 1.4.5, represent BPRS

Table 13.5 Mean BPRS Symptom-construct Scores for Three Diagnostic-stereotype Groups

	Schizophrenic	*Depressed*	*Paranoid*
Somatic concern	1.98	4.11	2.05
Anxiety	2.12	5.00	3.05
Emotional withdrawal	4.16	2.79	.85
Conceptual disorganization	3.84	1.60	3.15
Guilt feelings	1.02	4.56	.11
Tension	2.60	3.40	3.03
Mannerisms and posturing	3.65	1.45	2.36
Grandiosity	1.59	0.21	3.14
Depressive mood	1.53	5.40	1.60
Hostility	2.50	1.06	3.53
Suspiciousness	2.48	1.84	4.58
Hallucinatory behavior	2.56	1.18	4.18
Motor retardation	2.76	4.17	0.96
Uncooperativeness	3.39	2.15	2.81
Unusual thought content	3.79	2.10	4.54
Blunted affect	3.95	2.10	2.23

Table 13.6 Within-groups Covariance Matrix for Three Diagnostic-stereotype Groups Using 16 BPRS Symptom Constructs

	1	*2*	*3*	*4*	*5*	*6*	*7*	*8*	*9*	*10*	*11*	*12*	*13*	*14*	*15*	*16*
1	2.51	1.07	.03	.04	.61	.21	−.08	−.02	.68	.19	.42	.45	.23	−.03	.30	−.04
2	1.07	2.36	.05	.07	.84	.93	−.02	−.09	.91	.37	.63	.58	.12	.11	.25	−.49
3	.03	.05	2.62	1.05	.11	.14	.97	−.19	.02	.50	.41	.34	.92	1.30	.69	1.41
4	.04	.07	1.05	2.89	.28	.51	1.48	.40	.04	.45	.35	1.04	.47	.99	1.30	.93
5	.61	.84	.11	.28	1.80	.53	.12	.15	.82	.13	.34	.41	.38	.18	.28	−.11
6	.21	.93	.14	.51	.53	3.05	.76	.36	.39	.66	.65	.57	−.35	.52	.30	−.56
7	−.08	−.02	.97	1.48	.12	.76	2.96	.44	−.03	.62	.38	.82	.43	1.17	1.17	.85
8	−.02	−.09	−.19	.40	.15	.36	.44	2.50	−.12	.69	.50	.41	−.60	.20	.59	−.08
9	.68	.91	.02	.04	.82	.39	−.03	−.12	1.66	.15	.32	.30	.49	.07	.10	−.03
10	.18	.37	.50	.45	.13	.66	.62	.69	.15	2.42	1.30	.47	−.13	1.01	.55	.34
11	.42	.63	.41	.35	.34	.65	.38	.50	.32	1.30	2.38	.89	−.08	.78	.64	.17
12	.45	.58	.34	1.04	.41	.57	.82	.41	.30	.47	.89	3.32	.10	.74	1.24	.17
13	.23	.12	.92	.47	.38	−.35	.43	−.60	.49	−.13	−.08	.10	2.86	.72	.25	1.00
14	−.02	.11	1.30	.99	.18	.52	1.17	.20	.07	1.01	.78	.74	.72	2.70	.95	1.00
15	.30	.25	.69	1.30	.28	.30	1.17	.59	.10	.55	.64	1.24	.25	.95	2.71	.72
16	−.04	−.49	1.41	.93	−.11	−.56	.85	−.08	−.03	.34	.17	.17	1.00	1.00	.72	3.15

Table 13.7 Classification into Diagnostic Groups Using 16 BPRS Symptom Constructs in Normal-probability-density Model

	Diagnostic group			
Classification	*Schizophrenic*	*Depressed*	*Paranoid*	*Total*
Schizophrenic	1,162	18	87	1,267
Depressed	43	628	12	683
Paranoid	214	3	789	1,006
Total	1,419	649	888	2,956

rating profiles provided by expert judges for hypothetical "typical" cases belonging to each of 12 diagnostic categories.

The pooled within-groups covariance matrix for the 16 BPRS rating variables is shown in Table 13.8. The inverse of the pooled within-groups covariance matrix, which will be accepted as the estimate of $\mathbf{\Sigma}^{-1}$ to be used in classification computations, is presented in Table 13.9. Mean BPRS score vectors for the 12 diagnostic types are presented in Table 13.10, with severity for each symptom scored on a 0 to 6 scale.

For this problem, neither base rates nor cost coefficients are included in the model because neither were considered adequately known. The general normal-probability-density-model computer program presented at the end of this chapter was used to calculate the probability of membership in each of the 12 diagnostic groups for each of 1,047 profiles. This was done by first computing the probability-density index

$$f(\mathbf{x})^{(i)} = \text{antilog}_e \; -\tfrac{1}{2}\mathbf{d}_i'\mathbf{\Sigma}^{-1}\mathbf{d}_i$$

where $\mathbf{d}_i$ is the difference between the individual score vector $\mathbf{x}$ and the mean vector for the ith group ($i = 1, 2, \ldots, 12$). The probability that the profile belongs to each of the 12 groups was then calculated using Eq. (13.3) for $i = 1, 2, \ldots, 12$, and the profile was classified as belonging to the particular group for which the calculated $P(i \mid \mathbf{x})$ was greatest. A comparison of the assignments made by the computer with the actual designation for the profiles by the psychiatrists who provided them is presented in Fig. 13.1, where each dot represents five profiles, rounded up or down to the nearest multiple of five in each cell. The entries in the principal diagonal represent profiles that were "correctly" assigned, while the entries in the off-diagonal cells represent profiles that were assigned to a group other than the one originally designated by the psychiatrist. Once again, it should be pointed out that some of the "errors" may be due to atypical conceptions of the specified disorders by the psychiatrists who provided the profiles. The analysis suggests a method for studying the consensus among experts and for identifying judges whose conceptions are atypical.

Table 13.8 Within-groups Covariance Matrix of BPRS Symptom Ratings Pooled over 12 Diagnostic-stereotype Categories

	1	*2*	*3*	*4*	*5*	*6*	*7*	*8*	*9*	*10*	*11*	*12*	*13*	*14*	*15*	*16*
1	1.47	.56	.08	.07	.36	.26	.08	.16	.40	.18	.28	.20	.18	.01	.09	.01
2	.56	1.59	.06	.10	.38	.55	.15	.26	.52	.37	.52	.43	.14	.14	.33	−.01
3	.08	.06	1.74	.44	−.07	.07	.38	.10	.09	.21	.07	.24	.22	.44	.31	.56
4	.07	.10	.44	1.91	−.02	.26	.65	.24	.04	.23	.10	.48	.27	.41	.59	.47
5	.36	.38	−.07	−.02	1.10	.26	.04	.16	.43	.07	.23	.17	.14	−.04	.06	−.06
6	.26	.55	.07	.26	.26	2.08	.37	.13	.35	.36	.30	.26	.11	.23	.14	−.10
7	.08	.15	.38	.65	.04	.37	2.01	.19	.09	.35	.13	.43	.27	.32	.55	.57
8	.16	.26	.10	.24	.16	.13	.19	1.66	.14	.28	.42	.53	.01	.19	.27	.05
9	.40	.52	.09	.04	.43	.35	.09	.14	1.15	.19	.30	.30	.24	.07	.20	−.02
10	.18	.37	.21	.23	.07	.36	.35	.28	.19	1.78	.68	.44	−.03	.59	.35	.21
11	.28	.52	.07	.10	.23	.30	.13	.42	.30	.68	1.42	.70	.04	.39	.37	.04
12	.20	.43	.24	.48	.17	.26	.43	.53	.30	.44	.70	2.16	.20	.44	.82	.30
13	.18	.14	.22	.27	.14	.11	.27	.01	.24	−.03	.04	.20	1.83	.32	.22	.49
14	.01	.14	.44	.41	−.04	.23	.32	.19	.07	.59	.39	.44	.32	2.01	.51	.45
15	.09	.33	.31	.59	.06	.14	.55	.27	.20	.35	.37	.82	.22	.51	2.12	.62
16	.01	−.01	.56	.47	−.06	−.10	.57	.05	−.02	.21	.04	.30	.49	.45	.62	2.09

Table 13.9 Inverse of Within-groups Covariance Matrix Shown in Table 13.8

	1	*2*	*3*	*4*	*5*	*6*	*7*	*8*	*9*	*10*	*11*	*12*	*13*	*14*	*15*	*16*
1	.84	−.20	−.03	−.01	−.14	−.00	.00	−.01	−.12	−.01	−.04	.00	−.04	.03	.02	−.00
2	−.20	.92	.00	.01	−.09	−.13	.00	−.02	−.19	−.04	−.15	−.03	−.00	.02	−.07	.01
3	−.03	.00	.67	−.08	.06	.00	−.04	−.00	−.05	−.01	.02	−.01	−.00	−.08	.00	−.12
4	−.01	.01	−.08	.66	.02	−.05	−.12	−.04	.03	−.00	.05	−.06	−.03	−.04	−.09	−.03
5	−.14	−.09	.06	.02	1.14	−.04	−.00	−.05	−.30	.04	−.06	−.00	−.04	.03	.01	.00
6	−.00	−.13	.00	−.05	−.04	.57	−.08	.01	−.07	−.05	−.01	.00	−.00	−.03	.02	.06
7	.00	.00	−.04	−.12	−.00	−.08	.63	−.01	.01	−.07	.04	−.04	−.03	.01	−.06	−.09
8	−.01	−.02	−.00	−.04	−.05	.01	−.01	.68	.01	−.01	−.11	−.10	.02	−.00	−.00	.01
9	−.12	−.19	−.05	.03	−.30	−.07	.01	.01	1.20	−.02	−.05	−.05	−.10	.02	−.04	.04
10	−.01	−.04	−.01	−.00	.04	−.05	−.07	−.01	−.02	.76	−.28	.00	.07	−.14	−.00	−.03
11	−.04	−.15	.02	.05	−.06	−.01	.04	−.11	−.05	−.28	1.09	−.21	.01	−.08	−.03	.02
12	.00	−.03	−.01	−.06	−.00	.00	−.04	−.10	−.05	.00	−.21	.66	−.01	−.02	−.15	−.00
13	−.04	−.00	−.00	−.03	−.04	−.00	−.03	.02	−.10	.07	.01	−.01	.62	−.08	.01	−.12
14	.03	.02	−.08	−.04	.03	−.03	.01	−.00	.02	−.14	−.08	−.02	−.08	.63	−.06	−.04
15	.02	−.07	.00	−.09	.01	.02	−.06	−.00	−.04	−.00	−.03	−.15	.01	−.06	.64	−.11
16	−.00	.01	−.12	−.03	.00	.06	−.09	.01	.04	−.03	.02	−.00	−.12	−.04	−.11	.62

Table 13.10 Prototypes for 12 Diagnostic Groups from the International Nomenclature as Provided by Czechoslovakian Psychiatrists†

	1	*2*	*3*	*4*	*5*	*6*	*7*	*8*	*9*	*10*	*11*	*12*	*13*	*14*	*15*	*16*
1. Schizophrenia simplex	1.52	1.31	4.61	3.73	.59	1.24	2.58	.83	1.46	2.29	1.72	1.50	2.85	3.16	3.38	4.75
2. Schizophrenia paranoidní	1.79	3.20	3.62	3.37	.86	2.89	2.35	3.19	1.63	4.61	5.63	5.01	1.31	3.52	4.92	3.30
3. Schizophrenia hebefrenická	1.02	.88	3.51	3.92	.62	1.90	3.95	2.95	.91	3.14	2.38	2.08	1.46	3.50	4.42	4.03
4. Schizophrenia katatonická stuporozní	.34	.98	5.08	5.23	.37	2.85	4.93	.56	.82	1.82	1.27	2.24	5.24	4.76	3.37	4.70
5. Schizophrenia katatonická agitovaná	.62	2.14	4.44	4.87	.78	3.94	5.33	.78	1.00	3.14	2.18	2.91	1.72	4.48	4.05	3.87
6. Paranoidní stavy	2.18	3.54	2.87	2.12	1.34	3.12	1.43	2.83	2.20	4.40	5.47	3.20	1.03	2.92	4.16	1.90
7. Deprese psychotická	4.00	5.13	2.81	3.31	5.09	4.35	2.31	.70	5.79	1.31	2.45	2.89	4.73	2.72	3.39	2.60
8. Manie	.30	.39	.89	3.97	.24	3.49	2.87	5.23	.15	2.43	1.68	1.20	.14	1.91	3.29	1.36
9. Schizoafektivní psychóza	2.07	3.22	2.98	3.56	2.27	3.46	2.96	2.45	2.82	2.96	3.24	3.35	2.08	2.74	3.50	2.83
10. Chronická schizophrenia s floridními příznaky	1.94	1.76	3.80	4.48	1.07	2.27	3.54	2.42	1.58	3.05	3.39	3.82	2.31	3.12	4.24	4.29
11. Chronická schizophrenia bez výraznějšich příznaky	1.12	.73	3.79	3.51	.56	1.02	2.46	1.19	1.04	1.86	2.08	2.43	2.48	2.53	3.27	4.18
12. Jiné funckční psychotické stavy	2.53	3.13	2.46	3.06	2.03	3.17	2.34	1.63	2.89	2.45	2.84	3.16	2.15	2.37	2.82	2.10

† Because schizophrenia katatonická is omitted, the number assigned to a group does not always agree with the number in Fig. 3.5.

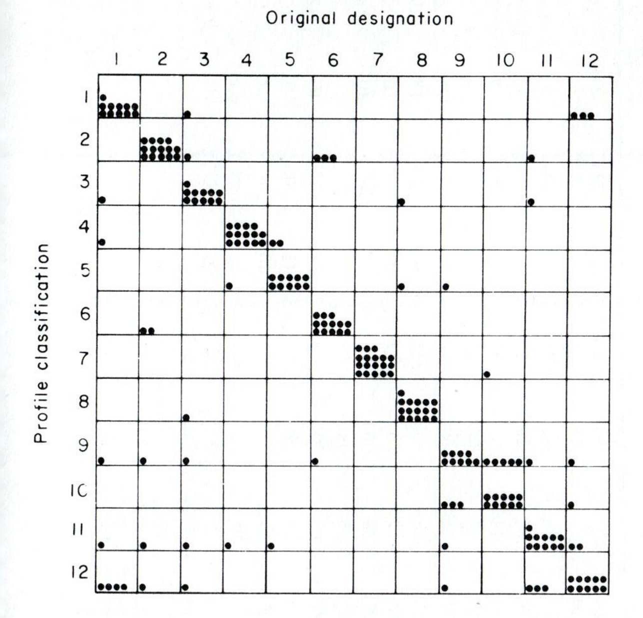

Figure 13.1 Tabulation of profile classification against original psychiatrists' designation for Czechoslovakian data.

As an exercise in use of the multivariate normal-density-function method for assignment of psychiatric patients among the 12 classes defined in the International Classification System (as interpreted by a large group of Czechoslovakian psychiatrists), the reader is encouraged to consider the 13 mean profiles for American psychiatric diagnostic types as if they were actual patient profiles. The American diagnostic prototypes are presented in Table 8.1. Using the mean vectors $\boldsymbol{\mu}^{(i)}$ from Table 13.10 and the inverse of the within-groups covariance matrix from Table 13.9, determine the probability of each American diagnostic prototype's belonging to each of the 12 classes taken from the International Classification. In which International category is each American prototype most likely to occur? Do the results appear valid from a clinical point of view? Do the results suggest any differences between the two diagnostic systems which might not be expected from simple translation of terminology?

The question of shrinkage or lack of reproducibility is an important one where parameters $\boldsymbol{\mu}^{(i)}$ and $\boldsymbol{\Sigma}^{-1}$ are estimated from the same sample to which the resulting procedure is applied to evaluate its validity. Because the parameter estimates are derived from the particular sample, it is likely that the resulting classification procedure will be more effective in assigning profiles from that sample than in assigning profiles from another independently drawn sample. To examine this problem empirically, the French diagnostic stereotype data were used. Approximately 120 BPRS rating profiles were obtained from as many different expert

Table 13.11 Within-groups Covariance Matrix for French Stereotype Ratings of 12 Diagnostic Groups, Sample A

	1	*2*	*3*	*4*	*5*	*6*	*7*	*8*	*9*	*10*	*11*	*12*	*13*	*14*	*15*	*16*
1	2.01	.74	.14	.16	.51	.21	.18	.08	.54	.19	.35	.26	.27	.08	.24	.08
2	.74	1.70	.19	.13	.50	.55	.12	−.02	.60	.23	.34	.21	.08	.14	.13	−.16
3	.14	.19	2.47	.78	.22	.09	.81	−.04	.15	.44	.60	.21	.45	.99	.73	1.22
4	.16	.13	.78	1.99	.21	.19	.84	.06	.02	.24	.26	.40	.23	.33	.72	.52
5	.51	.50	.22	.21	1.23	.27	.06	.04	.50	.11	.18	.13	.24	.15	.17	.08
6	.21	.55	.09	.19	.27	2.04	.48	.21	.18	.53	.47	.21	−.18	.27	.12	−.10
7	.18	.12	.81	.84	.06	.48	2.50	.16	−.04	.45	.41	.39	.27	.53	.75	.68
8	.08	−.02	−.04	.06	.04	.21	.16	1.43	.01	.38	.28	.02	−.06	.16	.18	.05
9	.54	.60	.15	.02	.50	.18	−.04	.01	1.22	.19	.23	.00	.35	.19	.10	.08
10	.19	.23	.44	.24	.11	.53	.45	.38	.19	1.88	1.06	.15	.10	.83	.30	.25
11	.35	.34	.60	.26	.18	.47	.41	.28	.23	1.06	1.98	.50	.14	.75	.42	.46
12	.26	.21	.21	.40	.13	.21	.39	.02	.00	.15	.50	1.80	.11	.33	.40	.22
13	.27	.08	.45	.23	.24	−.18	.27	−.06	.35	.10	.14	.11	1.41	.40	.18	.46
14	.08	.14	.99	.33	.15	.27	.53	.16	.19	.83	.75	.33	.40	2.36	.68	.89
15	.24	.13	.73	.72	.17	.12	.75	.18	.10	.30	.42	.40	.18	.68	2.43	.70
16	.08	−.16	1.22	.52	.08	−.10	.68	.05	.08	.25	.46	.22	.46	.89	.70	2.62

judges to represent each of 12 French diagnostic types. For purposes of the present analyses, the total samples were divided randomly into two subsets including approximately 60 profiles per diagnostic group, or a total of $60 \times 12 = 720$ in each subsample. These two independent random samples of French diagnostic-stereotype profiles will be called sample A and sample B.

The pooled within-groups covariance matrix derived from sample A is presented in Table 13.11, and the inverse of the sample A dispersion matrix is presented in Table 13.12. Mean BPRS profiles for the 12 French diagnostic groups, as derived from sample A, are shown in Table 13.13. These are the parameter estimates required for maximum-likelihood classification of individuals using the normal-probability-density-function method. When all 720 profiles in sample A were classified using the parameter estimates derived from the sample, 586, or 81 percent, were correctly assigned. The amount of shrinkage in percent of correct classification is illustrated in Fig. 13.2, which shows the classification of the sample B patients using the sample A parameters. The number of correct classifications declined to 551, or 76 percent, of the 720 sample B profiles. The shrinkage from 81 percent to 76 percent in this example is relatively small, but the two samples were randomly constructed and were not expected to be appreciably different. In practical applications, the sample used to obtain estimates of the parameters for classification purposes may differ systematically from the sample to be classified, and the increase in errors of classification can be expected to be correspondingly greater.

The stability of the multivariate profile-classification procedure was examined in still another way. The mean profile vectors and within-groups covariance matrix of sample B provided a second set of parameter estimates that could be used for maximum-likelihood classification based on the normal-probability-density model. The two sets of parameter estimates would be expected to be reasonably similar, as they were based on the two random subsamples, but it is important to know the extent to which they lead to the same classification of individuals. As a test, 6,000 BPRS profiles representing ratings of real American psychiatric patients were classified among the 12 French diagnostic groups, first using the parameter estimates derived from the French sample A and then again using the parameter estimates from sample B. A cross tabulation of these two classifications of the same patients into French diagnostic categories is shown in Table 13.14. As can be seen there is substantial agreement; yet 21.7 percent of the 6,000 profiles were classified differently using the parameter estimates derived from the two different samples.

For additional problems and exercises, the reader is encouraged to consider classifying the French diagnostic profiles as if they were actual patient profiles, using the means and covariances derived from the Czechoslovakian data. Into which Czechoslovak category should each French diagnostic prototype be assigned using the maximum-likelihood model? Assuming as base rates for the Czechoslovakian classes $\pi_1 = .05$, $\pi_2 = .20$, $\pi_3 = .05$, $\pi_4 = .05$, $\pi_5 = .05$, $\pi_6 = .10$, $\pi_7 = .20$, $\pi_8 = .10$, $\pi_9 = .05$, $\pi_{10} = .05$, $\pi_{11} = .05$, and $\pi_{12} = .05$,

Table 13.12 Inverse of Within-groups Covariance Matrix Shown in Table 13.11

	1	*2*	*3*	*4*	*5*	*6*	*7*	*8*	*9*	*10*	*11*	*12*	*13*	*14*	*15*	*16*
1	.66	−.20	.03	.00	−.13	.03	−.03	−.03	−.13	−.00	−.05	−.05	−.06	.05	−.03	−.01
2	−.20	.89	−.07	−.01	−.13	−.16	−.00	.06	−.28	.01	−.04	−.04	.06	−.00	−.00	.10
3	.03	−.07	.66	−.13	−.04	.03	−.07	.07	.02	−.00	−.08	.05	−.06	−.14	−.04	−.19
4	.00	−.01	−.13	.68	−.08	−.01	−.14	−.00	.05	−.03	.03	−.09	−.03	.05	−.11	−.01
5	−.13	−.13	−.04	−.08	1.09	−.09	.05	−.03	−.29	.03	.01	−.02	−.08	−.00	−.02	.01
6	.03	−.16	.03	−.01	−.09	.63	−.11	−.04	−.01	−.10	−.05	−.02	.12	−.03	.03	.04
7	−.03	−.00	−.07	−.14	.05	−.11	.55	−.02	.07	−.06	.02	−.04	−.06	.00	−.07	−.05
8	−.03	.06	.07	−.00	−.03	−.04	−.02	.76	.01	−.12	−.05	.03	.04	−.01	−.04	−.01
9	−.13	−.28	.02	.05	−.29	−.01	.07	.01	1.21	−.05	−.03	.08	−.22	−.02	−.02	−.02
10	−.00	.01	−.00	−.03	.03	−.10	−.06	−.12	−.05	.89	−.38	.08	.01	−.19	.02	.05
11	−.05	−.04	−.08	.03	.01	−.05	.02	−.05	−.03	−.38	.84	−.17	.02	−.04	−.02	−.05
12	−.05	−.04	.05	−.09	−.02	−.02	−.04	.03	.08	.08	−.17	.65	−.02	−.05	−.05	−.00
13	−.06	.06	−.06	−.03	−.08	.12	−.06	.04	−.22	.01	.02	−.02	.88	−.09	.03	−.07
14	.05	−.00	−.14	.05	−.00	−.03	.00	−.01	−.02	−.19	−.04	−.05	−.09	.64	−.08	−.10
15	−.03	−.00	−.04	−.11	−.02	.03	−.07	−.04	−.02	.02	−.02	−.05	.03	−.08	.53	−.05
16	−.01	.10	−.19	−.01	.01	.04	−.05	−.01	−.02	.05	−.05	−.00	−.07	−.10	−.05	.56

Table 13.13 Mean Profiles for 12 French Diagnostic Stereotypes, Sample A

	1	*2*	*3*	*4*	*5*	*6*	*7*	*8*	*9*	*10*	*11*	*12*	*13*	*14*	*15*	*16*
1. Schizophrénie paranoïde	2.73	2.51	5.01	5.11	1.06	2.23	4.51	2.66	1.58	3.31	3.75	4.78	2.20	4.18	5.60	4.85
2. Manie	.51	.53	1.46	3.10	.20	4.86	4.20	5.25	.18	2.96	1.33	.58	.00	2.11	2.93	1.25
3. État confusionnel	1.31	3.51	4.31	5.81	1.18	3.50	3.18	.28	1.80	1.16	1.13	4.40	3.60	3.06	3.33	2.75
4. Psychose paranoïaque	2.41	2.40	2.40	.81	.35	3.16	.90	4.78	1.25	5.41	5.68	1.06	.26	3.76	2.18	2.10
5. Excitation atypique	1.30	2.05	2.94	3.50	.84	4.38	4.00	3.35	.91	3.13	2.66	1.98	.15	3.20	3.67	2.83
6. Bouffée délirante	2.27	3.91	3.13	4.32	2.22	4.04	3.50	2.31	2.29	2.62	3.34	4.78	1.19	2.96	4.60	1.93
7. Psychose hallucinatoire chronique	2.45	2.98	2.25	2.15	.95	2.86	2.75	2.58	1.93	3.61	4.90	5.83	.76	2.90	4.13	2.21
8. Dépression réactionnelle	4.03	5.13	1.05	.61	3.66	3.88	.81	.10	5.33	.78	1.31	.15	3.11	1.11	.68	.81
9. Dépression atypique	3.86	4.20	3.59	2.44	3.30	3.18	2.42	.33	4.55	1.86	2.49	1.61	3.67	2.77	3.08	3.27
10. Mélancolie	4.50	5.68	3.29	1.70	5.80	3.45	1.80	.03	5.91	.73	2.13	1.40	5.13	2.85	2.26	2.11
11. Paraphrénie	1.40	1.33	2.33	3.57	.57	2.05	2.62	4.72	.86	2.16	2.72	3.98	.64	1.67	5.05	2.44
12. Hébéphrénie	2.55	2.21	5.44	4.68	1.27	2.78	4.90	1.06	1.88	2.67	3.00	2.75	4.42	4.68	4.86	5.40

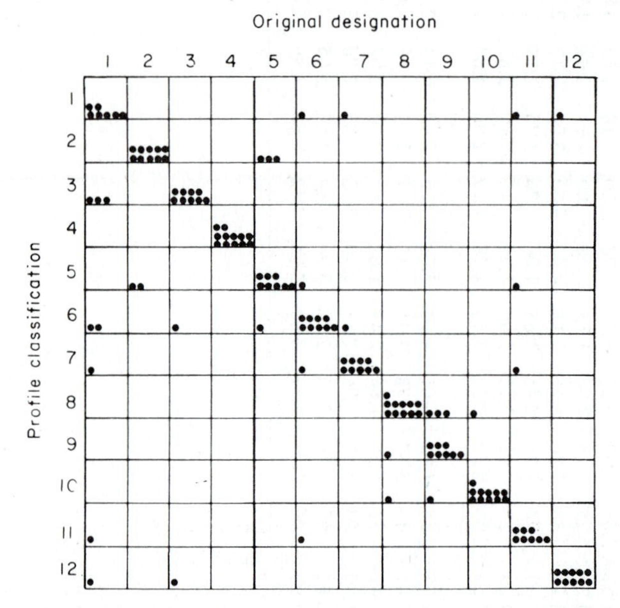

Figure 13.2 Tabulation of profile classification against original psychiatrists' designation for cross-validation sample in French data.

into which Czechoslovakian category should each French diagnostic prototype be placed using the Bayesian conditional-probability model and the normal-probability-density estimates? Various other permutations and combinations of parameter estimates and profiles to be classified can be obtained from Tables 13.5 to 13.13 to provide additional exercises.

Considering the four profiles listed below, into which American,

Table 13.14 Cross-Tabulation of Classification of 6,000 Patients Using Two Different Sets of Normal-density-function-parameter Estimates

	1	*2*	*3*	*4*	*5*	*6*	*7*	*8*	*9*	*10*	*11*	*12*
1	105	0	5	0	0	1	10	0	3	0	9	2
2	0	64	0	1	2	0	0	0	2	0	4	0
3	11	0	212	0	19	4	10	5	14	2	0	28
4	0	0	0	272	26	1	3	4	35	0	29	0
5	5	7	7	26	1,099	2	7	18	149	0	91	4
6	26	0	12	0	11	114	24	3	26	5	8	0
7	16	0	1	3	4	4	191	2	3	0	17	0
8	0	0	1	2	3	0	0	789	105	8	0	0
9	1	0	16	9	47	9	8	157	1,190	10	2	7
10	0	0	0	0	0	0	0	0	4	28	0	0
11	9	4	1	0	4	6	3	1	31	0	402	0
12	15	2	33	2	59	0	3	0	16	0	11	284

Czechoslovakian, and French categories should each be placed (using the Czechoslovakian dispersion matrix in lieu of a separate American dispersion matrix)?

Patient A:	2	4	1	0	5	3	0	0	5	3	1	0	2	1	0	1
Patient B:	2	4	4	4	1	1	3	2	3	0	0	5	4	4	4	5
Patient C:	2	1	2	5	0	2	2	4	1	5	4	0	1	4	5	1
Patient D:	1	1	4	1	0	0	3	0	0	0	0	3	3	0	4	

Before using the maximum-likelihood statistical model, inspect the profiles and arrive at a clinical judgment concerning whether each should be classed as depressed, paranoid, or schizophrenic. Does the statistical model confirm your subjective judgment?

13.5 Computer Program for Classifying Individuals among Several Groups

A computer program capable of classifying individuals among several specified groups on the basis of multiple-measurement profiles will be presented here. This program provides options of using complete multivariate normal-density function, simple profile-analysis d^2 approximation to normal density function, transformation to canonical variates, or the normalized vector-product method. Prior probabilities or base rates are considered by the program; however, if such are not known, or if they are to be disregarded, entry of equal values for all groups nullifies this option.

The input required by the program varies depending upon which option is to be used. The program will be presented here in spite of the fact that several of the alternative methods provided as options represent the subject matter of the next two chapters. In all cases, two control cards containing problem parameters and base rates, respectively, are required. In all cases, prototypes or mean vectors for the various groups into which individuals are to be classified are required. If the complete multivariate normal-density function (Chap. 13) is to be used, the within-groups covariance matrix must be read in. If an orthogonal transformation is to be used, the transformation vectors must be provided. The setup required for use of the program under each option will be described briefly.

1 Normal-density function

a One control card containing the following parameters in successive four-column fields:
NGPS = number of groups
NVAR = number of variables
NOBS = number of individuals to be classified
METH = method to be used; code 0001 for normal-density function.

b One card containing a priori probabilities or base rates in successive four-column fields. If base rates are not to be considered, enter 0001 in each field.

c Within-groups covariance matrix. The elements of the covariance matrix are read under format 2. This format statement should be modified to conform to the manner in which the covariance matrix is punched.

d Mean vectors for NGPS groups. The mean vectors are read under format 6, which specifies up to 16 successive four-column fields on one card for each mean vector.

e Data cards for NOBS individuals to be assigned among the groups. The individual profiles are read under format 3. This format should be modified according to manner in which data cards are punched.

2 Simple d^2 profile analysis

a One control card containing the following parameters in successive four-column fields:
NGPS = number of groups
NVAR = number of variables
NOBS = number of individuals to be classified
METH = method to be used; code 0002 for simple d^2.

b One card containing a priori probabilities or base rates in successive four-column fields. If base rates are not to be considered, enter 0001 in each field.

c Mean vectors for NGPS groups. The mean vectors are read under format 6, which specifies up to 16 successive four-column fields on one card for each mean vector.

d Data cards for NOBS individuals to be assigned among the groups. The individual profiles are read under format 3, which can be modified according to manner in which data appear in cards.

e *Important note.* To use this general multivariate classification program for simple d^2 profile analysis, a rescaling of the input data may be required if profile elements are large. This can be accomplished by altering statement 112 in the mainline program as follows:

112 A(J,J) = 1.0 for range 0–10
112 A(J,J) = 0.01 for range 10–100

and

112 A(J,J) = 0.001 for larger scores

3 Transformation to orthogonal variates

a One control card containing the following parameters in successive four-column fields:
NGPS = number of groups
NVAR = number of original untransformed variables
NOBS = number of individuals to be classified
METH = method to be used; code 0003 for transformation to orthogonal variates
NTRAN = number of transformed variates to be used in classification.

b One card containing a priori probabilities or base rates in successive four-column fields. If base rates are not to be considered, enter 0001 in each field.

c Mean vectors for NGPS groups on the original untransformed variables. The mean vectors are read under format 6, which specifies up to 16 successive four-column fields on one card for each mean vector.

d Weighting coefficients defining NTRAN transformed variates. The vectors of weighting coefficients are read under format 107, which specifies up to 16 coefficients representing one transformed variate on one card.

e Data cards for NOBS individuals to be assigned among the groups. The individual profiles are read under format 3, which can be modified according to how data appear in cards.

4 Normalized vector product

a One control card containing the following parameters in successive four-column fields:
NGPS = number of groups
NVAR = number of variables
NOBS = number of individuals to be classified
METH = method to be used; code 0004 for normalized vector product.

b One card containing a priori probabilities or base rates in successive four-column fields. If base rates are not to be considered, enter 0001 in each field.

c Mean vectors for NGPS groups. The mean vectors are read under format 6, which specifies up to 16 successive four-column fields on one card for each mean vector.

d Data cards for NOBS individuals to be assigned among the groups. The individual profiles are read under format 3, which can be modified according to manner in which data appear in cards.

The program, as listed here, is set up to classify each individual using the method indicated and then to *punch* the classification code into the last column of the patient's data card. The statement resulting in the punching of code into the data card is statement 1189, and the format which determines the column into which the classification code will be punched is format 1190. It is always dangerous to use a program that punches results back into the data cards because one can by carelessness punch into a column that is used for some other purpose. Care should be exercised each time this program is used to ensure that format 1190 indicates the proper column into which results should be punched. The classification code that is punched $(1, 2, 3, \ldots, k)$ will be the number of the group corresponding to the order in which the group mean vectors were entered. Thus, care should be exercised to ensure that mean vectors are entered in proper order.

Computer Program for Classifying Individuals on the Basis of Quantitative Score Profiles

```
      DIMENSION A(30,30),XBAR(20,30),X(30),D(30),Y(30), B(30,30),YB
     1AR(20,30),SSQ(20),BRATE(20)
    1 FORMAT(20I4)
    2 FORMAT(4F8.4)
    3 FORMAT(16F1.0)
    6 FORMAT(16F5.2)
    7 FORMAT(4F16.8)
  107 FORMAT(16F5.2)
      L1=1
      READ(2,1)  NGPS,NVAR,NOBS,METH,NTRAN
      READ(2,91)(BRATE(I),I=1,NGPS)
   91 FORMAT(20F4.0)
      SUM=0.0
      DO 101 I=1,NGPS
  101 SUM=SUM+BRATE(I)
      DO 102 I=1,NGPS
  102 BRATE(I)=BRATE(I)/SUM
      IF(METH-1)110,111,110
  110 DO 112 J=1,NVAR
      DO 113 K=1,NVAR
  113 A(J,K)=0.0
  112 A(J,J)=1.0
      GO TO 114
C     READ WITHIN GROUPS COVARIANCE MATRIX
  111 DO 10 J=1,NVAR
   10 READ(2,2)(A(J,K),K=1,NVAR)
  114 IF(METH-3)116,115,116
C     READ ORIGINAL GROUP MEANS THAT ARE TO BE TRANSFORMED
  115 READ(2,6)((YBAR(J,K),K=1,NVAR),J=1,NGPS)
C     READ ORTHOGONAL TRANSFORMATION VECTORS
      READ(2,107)((B(J,K),K=1,NVAR),J=1,NTRAN)
      GO TO 117
C     READ ORIGINAL GROUP MEANS WHICH ARE NOT TO BE TRANSFORMED
  116 DO 11 J=1,NGPS
   11 READ(2, 6)(XBAR(J,K),K=1,NVAR)
      FNVAR=NVAR
      IF(METH-5) 308,309,308
  309 DO 311 J=1,NGPS
      SSS=0.0
      DO 310 K=1,NVAR
  310 SSS=SSS+XBAR(J,K)
      DO 311 K=1,NVAR
  311 XBAR(J,K)=XBAR(J,K)-SSS/FNVAR
  308 DO 210 J=1,NVAR
  210 SSQ(J)=0.0
      GO TO (117,217,217,216,216),METH
  216 DO 215 J=1,NGPS
      DO 215K=1,NVAR
  215 SSQ(J)=SSQ(J)+XBAR(J,K)*XBAR(J,K)
      DO 214 J=1,NGPS
      DO 214K=1,NVAR
  214 XBAR(J,K)=XBAR(J,K)/SQRT(SSQ(J))
      GO TO 217
C     COMPUTE INVERSE OF COVARIANCE MATRIX
  117 CALL UMINV(NVAR,A)
      WRITE(3,12)
   12 FORMAT(1X,'INVERSE')
      DO   13 J=1,NVAR
   13 WRITE(3,  7)(A(J,K),K=1,NVAR)
C     READ INDIVIDUAL DATA VECTORS
  217 DO 25 I=1,NOBS
      IF(METH-3)120,121,120
  121 READ(2,3)(Y(J),J=1,NVAR)
      DO 119 K=1,NTRAN
      X(K)=0.0
      DO 119 J=1,NVAR
  119 X(K)=X(K)+Y(J)*B(K,J)
      DO 125 J=1,NGPS
      DO 125 K=1,NTRAN
      XBAR(J,K)=0.0
      DO 125 II=1,NVAR
  125 XBAR(J,K)=XBAR(J,K)+YBAR(J,II)*B(K,II)
```

```
      NVAR=NTRAN
      GO TO 122
  120 READ(2,3) (X(J),J=1,NVAR)
      SQ=0.0
      IF(METH-5) 306,307,306
  307 SSS=0.0
      DO 411 J=1,NVAR
  411 SSS=SSS+X(J)
      SSS=SSS/FNVAR
      DO 412 J=1,NVAR
  412 X(J)=X(J)-SSS
  306 IF(METH-4) 122,222,222
  222 DO 223 J=1,NVAR
  223 SQ=SQ+X(J)*X(J)
      DO 224 J=1,NVAR
  224 X(J)=X(J)/SQRT(SQ)
  122 CONTINUE
C     COMPUTE DEVIATION VECTOR FROM GROUP MEANS
      SUM=0.0
      DO 16 K=1,NGPS
      DO 14 J=1,NVAR
   14 D(J)=X(J)-XBAR(K,J)
C     COMPUTE NORMAL DENSITY WITHIN GROUP K
      DEN=0.0
      DO 15 J=1,NVAR
      DO 15 J2=1,NVAR
   15 DEN=DEN + A(J,J2)*D(J)*D(J2)
      Z=-.5*DEN
      Y(K)=EXP(Z)*BRATE(K)
   16 SUM=SUM+Y(K)
C     COMPUTE GROUP MEMBERSHIP PROBABILITY
      DO 17 K=1,NGPS
   17 Y(K)=Y(K)/SUM
C     FIND LARGEST GROUP MEMBERSHIP PROBABILITY
      MAX=1
      PROB=Y(1)
      DO 20 K=2,NGPS
      IF(Y(K)-PROB) 20,20,19
   19 PROB=Y(K)
      MAX=K
   20 CONTINUE
C     MAX IS THE NUMBER OF THE GROUP TO WHICH INDIVIDUAL LIKELY BELONGS
C     PUNCH MAX INTO SPECIFIED COLLUMN OF DATA CARD
 1189 WRITE(2,1190) MAX
 1190 FORMAT(79X,I1)
   25 CONTINUE
      CALL EXIT
      END
```

Subroutine for Matrix Inversion Used by Classification Program

```
      SUBROUTINE UMINV(N,A)
      DIMENSION IPIVO(30),INDEX(30,2),PIVOT(30),A(30,30)
  111 FORMAT (5F18.8)
      DET=1.0
      DO 20 J=1,N
   20 IPIVO (J)=0
      DO 550 I=1,N
      AMAX=0.0
      DO 105 J=1,N
      IF (IPIVO (J)-1) 60,105,60
   60 DO 100 K=1,N
      IF (IPIVO (K)-1) 80,100,740
   80 IF(ABS(AMAX)-ABS(A(J,K)))85,100,100
   85 IROW=J
      ICOLU =K
      AMAX=A(J,K)
  100 CONTINUE
  105 CONTINUE
      IF(ABS (AMAX)-2.0E-7) 800,800,801
  800 WRITE (3,666)
      WRITE(1,666)
  666 FORMAT(20H DETERMINANT = ZERO )
      DET =0.0
      PAUSE
      RETURN
  801 CONTINUE
      IPIVO (ICOLU )=IPIVO (ICOLU )+1
      IF (IROW-ICOLU ) 140,260,140
  140 DET=-DET
      DO 200 L=1,N
      SWAP=A(IROW,L)
      A(IROW,L)=A(ICOLU ,L)
  200 A(ICOLU ,L)=SWAP
  260 INDEX(I,1)=IROW
      INDEX(I,2)=ICOLU
      PIVOT(I)=A(ICOLU ,ICOLU )
      DET=DET*PIVOT(I)
      A(ICOLU ,ICOLU )=1.0
      DO 350 L=1,N
  350 A(ICOLU ,L)=A(ICOLU ,L)/PIVOT(I)
      DO 550 L1=1,N
      IF(L1-ICOLU ) 400,550,400
  400 T=A(L1,ICOLU )
      A(L1,ICOLU )=0.0
      DO 450 L=1,N
  450 A(L1,L)=A(L1,L)-A(ICOLU ,L)*T
  550 CONTINUE
      DO 710 I=1,N
      L=N+1-I
      IF (INDEX(L,1)-INDEX(L,2)) 630,710,630
  630 IROW=INDEX(L,1)
      ICOLU =INDEX(L,2)
      DO 705 K=1,N
      SWAP=A(K,IROW)
      A(K,IROW)=A(K,ICOLU )
      A(K,ICOLU )=SWAP
  705 CONTINUE
  710 CONTINUE
  740 RETURN
      END
```

14

Use of Canonical Variates for Classification

14.1 Introduction

In Chap. 10, the method of multiple discriminant analysis was discussed primarily from the point of view of studying configural relationships among k groups in a reduced multivariate measurement space. It can now be appreciated that multiple discriminant analysis also provides the basis for a powerful and simple approach to classification of individuals among the several groups. The result of multiple discriminant analysis is transformation of p original correlated variables to a smaller set of statistically independent *discriminant functions* or *canonical variates.* The canonical variates are defined in such manner that they are uncorrelated within groups and are also scaled to unit variance within groups. This means that the pooled within-groups covariance matrix $\mathbf{\Sigma}$ and its inverse $\mathbf{\Sigma}^{-1}$ are identity matrices.

The use of statistically independent canonical variates with unit variance has substantial computational advantage for classification of individuals using the multivariate normal-probability-density function because $\mathbf{\Sigma} = \mathbf{\Sigma}^{-1} = I$. When the original scores for each individual have been transformed to r canonical variates by applying the discriminant-function weighting coefficients, the r canonical variates for the individual can be represented as the vector $\mathbf{y}$ and the group mean scores on the r canonical variates by the mean vectors $\mathbf{u}^{(1)}, \mathbf{u}^{(2)}, \ldots, \mathbf{u}^{(k)}$. The multivariate normal-density function for the ith group is

$$F(y)^{(i)} = \frac{1}{\sqrt{2\pi^r}\,|\mathbf{\Sigma}|} \exp\,[-\tfrac{1}{2}(\mathbf{y} - \mathbf{u}^{(i)})'\,\mathbf{\Sigma}^{-1}\,(\mathbf{y} - \mathbf{u}^{(i)})]$$

however, since $\mathbf{\Sigma} = \mathbf{\Sigma}^{-1} = I$, it reduces to

$$F(y)^{(i)} = \frac{1}{\sqrt{2\pi^r}} \exp\,[-\tfrac{1}{2}(\mathbf{y} - \mathbf{u}^{(i)})'(\mathbf{y} - \mathbf{u}^{(i)})]$$

It should be recognized that the vector $\mathbf{y} - \mathbf{u}^{(i)}$ is the vector of differences between *canonical-variate scores* for a particular individual and the mean scores for the canonical variates in the ith group. The vector product

$$D_i^2 = (\mathbf{y} - \mathbf{u}^{(i)})'(\mathbf{y} - \mathbf{u}^{(i)})$$

is simply the sum of squared deviations of the individual's canonical-variate scores about the mean canonical-variate scores for the ith group

$$F(y)^{(i)} = \frac{1}{\sqrt{2\pi^r}} \exp\,(-\tfrac{1}{2}D_i^2) \tag{14.1}$$

Equation (13.4) can be rewritten in terms of canonical-variate scores and using *natural* logs as follows:

$$P(j\,|\,\mathbf{y}) = \frac{\text{antilog}_e -\tfrac{1}{2}D_j^2}{\text{antilog}_e -\tfrac{1}{2}D_1^2 + \text{antilog}_e -\tfrac{1}{2}D_2^2 + \cdots + \text{antilog}_e -\tfrac{1}{2}D_k^2} \tag{14.2}$$

Simple maximum-likelihood classification, which does not take into account differential base rates or cost coefficients, can be achieved by calculating the D^2 distance of an individual from each group and assigning the individual to the group for which this distance is *smallest*. It will be recalled that a larger negative logarithm corresponds to a smaller positive fraction; thus the antilog$_e$ $-\tfrac{1}{2}D_i^2$ increases as D_i^2 decreases in size.

Because the denominator of Eq. (14.2) is constant with regard to all groups $j = 1, 2, \ldots, k$, the same classification decision will result from considering only the relative magnitudes of the antilog$_e$ $-\tfrac{1}{2}D_j^2$ with regard to the several groups, disregarding the denominator. Because antilogs have monotonic relationship to their associated logs, one need not even bother with taking the antilog if the sole purpose is maximum-likelihood classification. The value of D_j^2 can be computed to indicate the distance of the individual's canonical-variate profile from each of the several group means $j = 1, 2, \ldots, k$, and the individual can be assigned to the group for which this value is smallest. If, however, one is interested in an estimate of the exact probability of membership in each group, Eq. (14.2) should be used as written.

It should be obvious how a priori probabilities and cost coefficients can be included in the classification model in the now familiar manner. Rewriting

Eqs. (13.6) and (13.8) in terms of the simple-to-compute D_j^2 values yields

$$P(j \mid y) = \frac{\pi_j \text{ antilog}_e -\frac{1}{2}D_j^2}{\sum^k \pi_i \text{ antilog}_e -\frac{1}{2}D_i^2} \tag{14.3}$$

and

$$P(j \mid y) = \frac{c_j\pi_j \text{ antilog}_e -\frac{1}{2}D_j^2}{\sum^k c_i\pi_i \text{ antilog}_e -\frac{1}{2}D_i^2} \tag{14.4}$$

It does not seem feasible to eliminate the antilog transformation in Eqs. (14.3) and (14.4) when base rates or cost coefficients are to be considered. In a computer, of course, this makes no difference, but the transformation to antilog values is admittedly tedious for hand calculation. Many readers will find it necessary to go back to a high school mathematics text to refresh their memories on the use of natural log tables. The present writers have had to from time to time!

14.2 A Computational Example

Consider again the problem of classification of prototypical patient profiles as paranoid, schizophrenic, or depressed on the basis of four BPRS scores representing levels of severity for factors of thinking disturbance, withdrawal-retardation, hostile-suspiciousness, and anxious depression. As before, the problem in this example is to define procedures which will result in classification of the individual profile according to the relative probabilities of the particular profile pattern within the population of *stereotype* profiles for paranoid, schizophrenic, and depressive groups.

The within- and between-groups SP matrices and the vectors of mean factor scores for the three major diagnostic types were presented in Table 10.12 for use as an exercise in multiple discriminant analysis. For the present purpose, the discriminant-function analysis was completed, and two highly significant functions were found to account for differences among the three diagnostic types. The two discriminant functions were computed to be

$$y^{(1)} = -.1188x_1 + .0668x_2 - .1000x_3 + .2705x_4$$

and

$$y^{(2)} = .0432x_1 - .2385x_2 + .1237x_3 + .1035x_4$$

where x_1 = thinking disturbance, x_2 = withdrawal-retardation, x_3 = hostile-suspiciousness and x_4 = anxious depression.

Canonical-variate means for the three diagnostic groups were computed by applying the discriminant-function weighting coefficients to the means of the three groups on the four original symptom factors. The canonical-variate group means were found to be:

Schizophrenic: $\boldsymbol{\mu}^{(1)\prime} = [.2976 \quad -.5348]$

Depressed: $\boldsymbol{\mu}^{(2)\prime} = [3.9234 \quad .3222]$

Paranoid: $\boldsymbol{\mu}^{(3)\prime} = [-.1778 \quad 1.1171]$

(Because these canonical-variate means were actually calculated by a computer carrying more decimal places than are provided in previous tables, hand calculations will be found to agree to only two significant places.) The variance of each canonical variate is unity, and the covariance between the two canonical variates is zero within the groups; hence, the within-groups covariance matrix $\boldsymbol{\Sigma}$ can be taken as an identity matrix. Thus, all the basis for normal-probability-density classification is available from the discriminant-function analysis.

Let us examine the method of classifying three patients whose raw-score profiles are as follows:

Patient A: $\mathbf{x}^{(1)\prime} = [8 \quad 7 \quad 2 \quad 3]$

Patient B: $\mathbf{x}^{(2)\prime} = [3 \quad 9 \quad 1 \quad 1]$

Patient C: $\mathbf{x}^{(3)\prime} = [1 \quad 2 \quad 5 \quad 9]$

The first step is to transform the original raw-score profiles to canonical form. This is accomplished by applying the discriminant-function weights to the raw scores for each patient.

Patient A: $\mathbf{y}^{(1)\prime} = [.1287 \quad -.7660]$

where

$$y^{(1)} = -.1188\,(8) + .0668\,(7) + -.1000\,(2) + .2705\,(3)$$
$$= .1287$$

$$y^{(2)} = .0432\,(8) + -.2385\,(7) + .1237\,(2) + .1035\,(3)$$
$$= -.7660$$

Patient B: $\mathbf{y}^{(2)\prime} = [.4153 \quad -1.7897]$

Patient C: $\mathbf{y}^{(3)\prime} = [1.9493 \quad 1.1162]$

To classify patient A among the three groups, it is necessary to calculate the sum of squared deviations of his canonical-variate scores about the mean canonical variates for each of the three groups

$$D_1^2 = (\mathbf{y}^{(1)} - \boldsymbol{\mu}^{(1)})'(\mathbf{y}^{(1)} - \boldsymbol{\mu}^{(1)}) = \mathbf{d}_1'\mathbf{d}_1$$

where

$$\mathbf{d}_1' = [(.1287 - .2976) \quad (-.7660 + .5348)]$$

so that

$$D_1^2 = -.1689^2 + -.2312^2 = .0820$$

$$D_2^2 = (\mathbf{y}^{(1)} - \boldsymbol{\mu}^{(2)})'(\mathbf{y}^{(1)} - \boldsymbol{\mu}^{(2)}) = \mathbf{d}_2'\mathbf{d}_2$$
$$= (.1287 - 3.9234)^2 + (-.7660 - .3222)^2 = 15.5839$$

$$D_3^2 = (\mathbf{y}^{(1)} - \boldsymbol{\mu}^{(3)})'(\mathbf{y}^{(1)} - \boldsymbol{\mu}^{(3)}) = \mathbf{d}_3'\mathbf{d}_3$$
$$= (.1287 + .1778)^2 + (-.7660 - 1.1171)^2 = 3.6400$$

Having calculated the D_j^2 distance of the canonical-variate profile for patient A from the mean canonical-variate profile for each group, the probability can be calculated that a patient who has a profile like patient A will belong to each of the three groups, assuming equal base rates. Equation (14.2) is employed for this purpose.

Schizophrenic:

$$P(1 \mid \mathbf{y}) = \frac{\text{antilog}_e -\tfrac{1}{2}(.082)}{\text{antilog}_e -\tfrac{1}{2}(.082) + \text{antilog}_e -\tfrac{1}{2}(15.5839) + \text{antilog}_e -\tfrac{1}{2}(3.64)}$$
$$= \frac{.96}{1.122} = .856$$

Depressive:

$$P(2 \mid \mathbf{y}) = \frac{\text{antilog}_e -\tfrac{1}{2}(15.5839)}{\text{antilog}_e -\tfrac{1}{2}(.082) + \text{antilog}_e -\tfrac{1}{2}(15.5839) + \text{antilog}_e -\tfrac{1}{2}(3.64)}$$
$$= \frac{.000}{1.122} = .000$$

Paranoid:

$$P(3 \mid \mathbf{y}) = \frac{\text{antilog}_e -\tfrac{1}{2}(3.64)}{\text{antilog}_e -\tfrac{1}{2}(.082) + \text{antilog}_e -\tfrac{1}{2}(15.5839) + \text{antilog}_e -\tfrac{1}{2}(3.64)}$$
$$= \frac{.162}{1.122} = .144$$

It will be noted that the denominators are identical in all three ratios, so that the calculation is required only once. The probability of a randomly selected profile, like that of patient A, being a stereotype profile in the schizophrenic diagnostic category is greatest: $P(1 \mid \mathbf{y}) = .856$. The probability is essentially zero that patient A is typical of the depressive group.

The same procedure was applied in classification of profiles for patients B and C. For patient B, the sum of squared deviations of canonical scores about the group means were calculated to be

$$D_1^2 = 1.5886 \qquad D_2^2 = 13.5431 \qquad D_3^2 = 8.8013$$

The probabilities associated with membership in three stereotype populations

were calculated as follows, assuming equal base rates:
Schizophrenic:

$$P(1 \mid \mathbf{y}) = \frac{\text{antilog}_e -\frac{1}{2}(1.5886)}{\text{antilog}_e -\frac{1}{2}(1.5886) + \text{antilog}_e -\frac{1}{2}(13.5431) + \text{antilog}_e -\frac{1}{2}(8.8013)}$$
$$= .971$$

Depressive:

$$P(2 \mid \mathbf{y}) = \frac{\text{antilog}_e -\frac{1}{2}(13.5431)}{\text{antilog}_e -\frac{1}{2}(1.5886) + \text{antilog}_e -\frac{1}{2}(13.5431) + \text{antilog}_e -\frac{1}{2}(8.8013)}$$
$$= .002$$

Paranoid:

$$P(3 \mid \mathbf{y}) = \frac{\text{antilog}_e -\frac{1}{2}(8.8013)}{\text{antilog}_e -\frac{1}{2}(1.5886) + \text{antilog}_e -\frac{1}{2}(13.5431) + \text{antilog}_e -\frac{1}{2}(8.8013)}$$
$$= .027$$

The results obtained for patient C were $D_1^2 = 5.4539$, $D_2^2 = 4.5275$, and $D_3^2 = 4.5246$. The probabilities of diagnostic-group membership were estimated to be $P(1 \mid \mathbf{y}) = .239$, $P(2 \mid \mathbf{y}) = .380$, and $P(3 \mid \mathbf{y}) = .381$. The reader planning to use these methods is encouraged to verify the results for patients B and C to be sure that he understands the calculations. The computer program provided in the final section of Chap. 13 can be employed in practical application to classify patients using canonical variates derived from a previous multiple discriminant analysis.

To demonstrate the effectiveness of this procedure in classifying individual patients, canonical scores for all 2,956 hypothetical typical patients were calculated, and they were assigned as just described. Table 14.1 shows that 2,364 out of 2,956, or 80 percent, were correctly classified.

Table 14.1 Classification into Diagnostic Groups Using Canonical Variates: Maximum-likelihood Rule

	Diagnostic group			
Classification	*Schizophrenic*	*Depressed*	*Paranoid*	*Total*
Schizophrenic	1,009	24	137	1,170
Depressed	61	621	17	699
Paranoid	349	4	734	1,087
Total	1,419	649	888	2,956

14.3 Classification Using Arbitrary Orthogonal Variates

14.3.1 INTRODUCTION

Most of what was said in Sec. 14.1 with regard to the advantages of using canonical variates for evaluation of normal probability densities can also be said for any arbitrary orthogonal transformation of the within-groups covariance matrix. A number of different methods are available, such as the method of principal components described in Chap. 3, for computing mutually uncorrelated artificial variates by operating on the within-groups covariance matrix. Such variates can be scaled to unit variance within groups, and then the normal probability density can be evaluated by computing the sum of squared deviations of the individual's scores about the group means on the orthogonal variates.

If we let $\mathbf{y}$ represent the vector of scores on r orthogonal artificial variates for one individual and $\mathbf{u}^{(1)}, \mathbf{u}^{(2)}, \mathbf{u}^{(3)}, \ldots, \mathbf{u}^{(k)}$ the vectors of group means on the same orthogonal variates, the probability density for score vector $\mathbf{y}$ in group j is simply

$$F(\mathbf{y})^{(j)} = \frac{1}{\sqrt{2\pi^r}} \exp\left[-\tfrac{1}{2}(\mathbf{y} - \mathbf{u}^{(j)})'(\mathbf{y} - \mathbf{u}^{(j)})\right]$$

The probability that an individual who does have score vector $\mathbf{y}$ will belong to group j, disregarding possible differences in base rates, is simply

$$P(j \mid \mathbf{y}) = \frac{\text{antilog}_e -\tfrac{1}{2}D_j^2}{\sum^k \text{antilog}_e -\tfrac{1}{2}D_i^2}$$

where D_i^2 is the sum of squared deviations of the individual's scores about the mean scores for the ith group.

The Bayesian model which includes base rates $\pi_1, \pi_2, \ldots, \pi_k$ for the k groups is written

$$P(j \mid \mathbf{y}) = \frac{\pi_j \,\text{antilog}_e -\tfrac{1}{2}D_j^2}{\sum^k \pi_i \,\text{antilog}_e -\tfrac{1}{2}D_i^2}$$

Cost coefficients can be introduced in the obvious manner by associating with each group a cost-of-misclassification factor, $c_1\pi_1, c_2\pi_2, \ldots, c_k\pi_k$.

There are some significant differences between the use of canonical variates derived from multiple discriminant analysis and the use of principal components as orthogonal variates. The canonical variates are derived with specific reference to group differences and thus tend to be a more parsimonious and powerful basis for classification of individuals in situations where multiple discriminant analysis is feasible. There are problems in which it is not appropriate to undertake a multiple discriminant analysis but highly appropriate to reduce the number of variates and achieve a convenient orthogonalization at the same time.

It will be recalled that completion of the multiple discriminant analysis requires computation of the triangular square-root inverse of the within-groups covariance matrix (Sec. 10.3). Where variables are numerous and highly correlated, it is quite possible that one variable may be approximated very closely as a linear combination of the others. This means that the within-groups covariance matrix will be singular to all intents and purposes and that any inverse which does exist for that matrix will be due to measurement error only. Such matrix inverses are obviously highly unstable and difficult to replicate in different samples. In the presence of numerous, highly correlated variables, it is frequently an advantage to accomplish a preliminary orthogonal reduction using principal components or some related method.

Where principal-components analysis is employed to derive a reduced set of statistically independent linear functions, it should be remembered that the usual principal-components scaling defines variables which have variance λ_i

$$\mathbf{a}^{(i)\prime} W \mathbf{a}^{(i)} = \lambda_i$$

It is necessary to divide each element in the principal-component vector by the square root of the associated λ in order to scale the orthogonal components to unit variance

$$\lambda^{-\frac{1}{2}} \mathbf{a}^{(i)\prime} W \mathbf{a}^{(i)} \lambda^{-\frac{1}{2}} = \lambda^{-\frac{1}{2}} \lambda \lambda^{-\frac{1}{2}} = 1$$

14.3.2 COMPUTATIONAL EXAMPLE

The pooled within-groups covariance matrix of the 16 BPRS symptoms derived from the psychiatric diagnostic-stereotype data was used in an example in Chap. 13 and has been presented in Table 13.6. It will be recalled that the data were grouped into three major classes representing schizophrenic, paranoid, and depressive diagnostic types. The within-groups covariance matrix was estimated on 2,958 degrees of freedom by computing the SP matrix for each diagnostic group separately, summing the three matrices, and dividing by degrees of freedom, as described in Sec. 2.7.

Principal components of the pooled within-groups covariance matrix were computed using the iterative method described in Chap. 3. Four principal components were defined with variance $\lambda_1 + \lambda_2 + \lambda_3 + \lambda_4 = 23.8844$ out of a total of 41.9037, or 57 percent of the total within-groups variation represented in the principal diagonal of the within-groups covariance matrix.

Rather than leaving the principal-component vectors in normalized form $\mathbf{a}'\mathbf{a} = 1$, from which the variance $\lambda = \mathbf{a}' W \mathbf{a}$ results, *each vector was rescaled by dividing each element in it by the square root of the associated λ. This resulted in the orthogonal linear components having unit variance*, consistent with the desire for a dispersion matrix that is an identity matrix, $\mathbf{\Sigma} = I$. The rescaled principal-component vectors are presented in Table 14.2. By converting original BPRS profiles to the four orthogonal composite scores, one can accomplish maximum-likelihood classification by computing probability estimates as shown in Eq. (14.2). One simply applies the four sets of weights shown in Table 14.2 and then works with the four resulting scores as if they were the raw data.

Mean BPRS diagnostic-prototype vectors for paranoid, schizophrenic, and depressive types were presented in Table 13.5. Applying the weights in Table

Table 14.2 Scaled Principal-component Coefficients Derived from Within-diagnostic-group Covariance Matrix

	I	*II*	*III*	*IV*
Somatic concern	.0280	.1122	.1755	−.0452
Anxiety	.0360	.1653	.1523	.0164
Emotional withdrawal	.0945	−.1093	.0747	.1174
Conceptual disorganization	.1141	−.0472	−.0459	−.1735
Guilt feelings	.0351	.0908	.1233	−.0326
Tension	.0592	.1552	−.0543	.0950
Mannerisms and posturing	.1137	−.0468	−.0852	−.0715
Grandiosity	.0359	.0595	−.1907	.0078
Depressive mood	.0245	.0843	.1584	.0003
Hostility	.0763	.0585	−.0834	.2822
Suspiciousness	.0749	.0960	−.0315	.1939
Hallucinatory behavior	.0994	.0856	−.0332	−.2770
Motor retardation	.0494	−.1033	.2425	−.0043
Uncooperativeness	.1121	−.0493	−.0040	.1596
Unusual thought content	.1071	.0001	−.0548	−.1776
Blunted affect	.0803	−.1808	.0702	.0695

14.2 to the mean vectors, one obtains four new values to replace the 16 original values in each mean profile. The transformed group means are presented in Table 14.3 for paranoid, schizophrenic, and depressive types to emphasize that one can compute group mean vectors involving several weighted functions by applying the weights to the group means on the original variables. To classify individual patients, the same four sets of weights are applied to transform the original BPRS ratings to four new variables. Sums of squared deviations D_1^2, D_2^2, and D_3^2 are computed from scores on the four orthogonal functions, and classification probabilities are computed using Eq. (14.2) or (14.3). Table 14.4 presents the resulting classification. It is particularly appropriate to compare this table with Table 13.7, as it is precisely the same problem except for the transformation of the 16 BPRS symptoms to four orthogonal variates by means of principal-components analysis. There is an obvious loss of precision in classification (from 87 to 79 percent), but this is to be expected as the four

Table 14.3 Transformed Group Means of Diagnostic Groups

Schizophrenic	*Depressed*	*Paranoid*
4.67	3.72	4.63
.29	2.04	1.79
1.18	3.95	.28
.50	.60	.53

Table 14.4 Classification into Diagnostic Groups Using Transformed Scores in Normal-probability-density Model

	Diagnostic group			
Classification	*Schizophrenic*	*Depressed*	*Paranoid*	*Total*
Schizophrenic	961	22	113	1,096
Depressed	92	621	14	727
Paranoid	366	6	761	1,133
Total	1,419	649	888	2,956

principal components accounted for only 57 percent of the variance represented in the within-groups covariance matrix and some of this loss of information would have been useful for classification purposes. In fact if all 16 possible principal components had been used, the resulting classification would have been identical to that shown in Table 13.7. In a practical application where a classification system is being developed to apply to other independent sets of data, it might be desirable to transform by the use of principal components not simply to orthogonalize the variates but to reduce the rank, the expectation being that the use of the first few principal components would account for most of the variance and result in less shrinkage when applied to new data. It is also interesting to note, in comparing Tables 13.7 and 14.4, that the increase in misclassifications appears to be disproportionately greater in the schizophrenic group rather than random, which suggests that there was some systematic loss of information in the reduction of rank.

14.4 Use of Computer Program for Classification of Individuals on the Basis of Orthogonal Transformations

Given that one has a set of weighting coefficients which produce an orthogonal transformation of the original multiple measurements, the computer program listed in Sec. 13.5 can be used to effect the transformation and to classify individuals among several groups on the basis of the transformed variates. Multiple discriminant analysis and principal-components analysis of the within-groups covariance matrix have been discussed as possible orthogonal-transformation procedures. Various forms of factor analysis and other linear components analyses can also be used to obtain the transformation vectors. Whereas the transformation should, strictly speaking, be orthogonal *within* groups, it is sometimes useful to use an orthogonal transformation derived from principal-components analysis or from factor analysis of a total covariance or correlation matrix. If a correlation matrix is the basis for the analysis leading to an orthogonal transformation, the coefficients of the transformation vectors should be divided by the standard deviations of the respective profile elements for use with data in raw-score form. In any event, the orthogonal transformations should be scaled so that the resulting composite scores have equal (or approximately equal) variances. The scaling problem is handled nicely in multiple discriminant analysis. Where principal-components analysis has been used, as described in Sec. 14.3.2, vectors should be multiplied by the reciprocal of the square root of the associated λ_i.

To use the computer program for assigning individuals among several groups with an orthogonal transformation, option 3 should be employed. The setup and control cards required are described under option 3 in Sec. 13.5. In summary, the input sequence includes (1) a control card specifying number of groups, number of variables in the original profiles, total number of individuals to be classified, method to be used (option 3), and the number of transformation vectors to be used, (2) base rates or a priori probabilities, (3) mean vectors on the original untransformed variables, (4) vectors of weighting coefficients defining the new transformed variates as functions of the original variables, and (5) data cards.

14.5 Summary

In this chapter the possibility of using a variety of types of orthogonal transformations to simplify the problem of evaluating multivariate normal-density functions and related probabilities was discussed. Canonical variates obtained from multiple discriminant analysis and principal components from the within-groups covariance matrix were discussed as specific examples of orthogonal transformations useful for classification purposes. The advantage of using a properly scaled orthogonal transformation of the original correlated profile elements is that the covariance matrix becomes an identity matrix and need not be considered further. An additional possible advantage is that only the major "true" underlying factors or linear components need be considered, while the residual small "error factors" can be disregarded. This appears particularly important where the original profiles contain many highly correlated variables. Use of the computer program for classification of individuals among several groups with orthogonal transformation was described.

The use of an orthogonal transformation produces profile components that are statistically independent within groups and thus simplifies evaluation of probability densities. Following an orthogonal transformation, the probability of a profile pattern's belonging in a particular group is simply a function of the sum of squares of differences between the (transformed) individual's scores and the (transformed) group means. This suggests the next step in approximation of the complete multivariate normal probability-density function. Rather than actually transforming the original profile elements, it can simply be *assumed* that they are orthogonal and have equal variance. The next chapter discusses simple profile-analysis methods that, at least implicitly, involve these assumptions.

15

Simple Profile-analysis Methods for Classification

15.1 Introduction

A number of methods for classification of individuals on the basis of multivariate measurement profiles have been discussed. Although the methods based on evaluation of the normal-probability-density function are effective and mathematically elegant, it must be admitted that computational difficulties render them impractical except where a computer can be used. Fortunately with the widespread availability of computers today, the computations need not constitute an insurmountable practical problem, and these methods are recommended as being efficient where measurements can reasonably be assumed to have a multivariate normal distribution. It should nevertheless be mentioned that there are a number of simpler profile-analysis methods that have been used effectively by psychologists for some years. It is the purpose of this chapter to discuss the two most frequently used simple profile-analysis methods and to relate them to more complex models previously discussed.

There is a danger that an extended discussion of relationships to other models may mask the great simplicity of the basic profile-analysis methods. For this reason, each method will first be presented in its simplest and most operational form, without any attempt to explicate its theoretical basis. As practical methods, the *simple profile-distance function* and *normalized vector product* are exceptionally useful in certain circumstances. The computations are simple and do not actually depend on any theoretical considerations. However, to understand the relationships of the simple profile-analysis methods to more complex mathematical models, an examination of the profile methods within the context of a multivariate normal-probability-density model is useful.

15.2 Simple Distance-function Classification

The difference between two multivariate measurement profiles has frequently been defined in terms of the sum of squares of differences in scores on the several component variables. Geometric considerations have led to the conception that this simple index represents the "distance" between the profiles in a multidimensional euclidean space. If the observed scores within each profile are conceived as coordinate values defining the relationship of the profile vector to rectangular coordinate axes, a generalization of the pythagorean theorem can be invoked to define the distance between two profiles as the sum of squares of differences in projections on the coordinate axes. In the simple two-dimensional case, the square on the hypotenuse is equal to the sum of squares on the other two sides. Generalized to p dimensions, the square of the distance between two points is equal to the sum of squares of differences in component scores

$$d^2 = d_1^2 + d_2^2 + \cdots + d_p^2 \tag{15.1}$$

where $d_1^2, d_2^2, \ldots, d_p^2$ are squares of differences in scores on the p different measures. In vector notation, the sum of squares of difference scores can be represented as the scalar product of a vector of difference scores and its transpose, $d^2 = \mathbf{d}'\mathbf{d}$.

The simple d^2 index of profile distance can be used to evaluate the similarity of an individual's profile to the mean profiles for several groups. The individual can then be assigned to the group to which his profile is most similar. Under certain assumptions (discussed in later sections of this chapter) the simple d^2 indices can be used to provide estimates of the probabilities of group membership or as a basis for Bayesian conditional-probability classification which takes into account a priori probabilities or cost coefficients [Eq. (15.4) and Sec. 15.5.2].

To illustrate the simple d^2 distance-function approach to classification of individuals among several groups, before encumbering the method with any theoretical considerations, the mean factor-score profiles for three major diagnostic classes from Table 10.12 will be used:

Schizophrenic: $\boldsymbol{\mu}^{(1)\prime} = [13.2 \quad 13.9 \quad 11.4 \quad 7.7]$

Depressive: $\boldsymbol{\mu}^{(2)\prime} = [7.9 \quad 12.1 \quad 8.1 \quad 18.0]$

Paranoid: $\boldsymbol{\mu}^{(3)\prime} = [14.9 \quad 9.1 \quad 13.9 \quad 8.8]$

Suppose that a patient has the profile $\mathbf{x}' = [9 \quad 2 \quad 8 \quad 3]$.

The simple d^2 index relating the patient profile to each of the three groups, in turn, is computed as follows:

$$d^2_{(1)} = (9 - 13.2)^2 + (2 - 13.9)^2 + (8 - 11.4)^2 + (3 - 7.7)^2$$
$$= 192.90$$

$$d^2_{(2)} = (9 - 7.9)^2 + (2 - 12.1)^2 + (8 - 8.1)^2 + (3 - 18.0)^2$$
$$= 328.23$$

$$d^2_{(3)} = (9 - 14.9)^2 + (2 - 9.1)^2 + (8 - 13.9)^2 + (3 - 8.8)^2$$
$$= 153.67$$

The distance of the patient's profile from the mean profile for the paranoid group is found to be less than the distances from the other two mean profiles. The patient should therefore be assigned to the paranoid group.

15.3 Relationship of the Simple d^2 to the Normal-probability-density Model

The simple d^2 distance function can be used to classify patients among several groups by calculating the distance of the individual's profile from each of the group means and then assigning the individual to the group from which he has least distance. Under certain conditions (discussed in this section) this simple method of assignment to groups is recognized as constituting a maximum-likelihood classification procedure.

The differences between corresponding profile elements can be represented as a difference vector

$$\mathbf{d}^{(i)} = \mathbf{x} - \boldsymbol{\mu}^{(i)} = [d_1^{(i)} \quad d_2^{(i)} \quad \cdots \quad d_p^{(i)}]$$

where $d_1^{(i)}, d_2^{(i)}, \ldots, d_p^{(i)}$ are differences between the individual's scores and the mean scores for the ith group on the p variables. The simple d^2 index can then be represented in vector notation as the inner product of the difference vector and its transpose

$$d^2_{(i)} = \mathbf{d}^{(i)\prime}\mathbf{d}^{(i)} = (\mathbf{x} - \boldsymbol{\mu}^{(i)})'(\mathbf{x} - \boldsymbol{\mu}^{(i)})$$

This can be related to the multivariate normal-density function as used in evaluating probabilities of group membership. In Chap. 13, the multivariate density was shown to depend upon the vector of differences between individual scores and group means and on the within-groups covariance matrix

$$f(\mathbf{x}) = \frac{1}{\sqrt{2\pi^p\,|\boldsymbol{\Sigma}|}} \exp\,[-\tfrac{1}{2}(\mathbf{x} - \boldsymbol{\mu})'\boldsymbol{\Sigma}^{-1}(\mathbf{x} - \boldsymbol{\mu})]$$

or, in terms of the difference vectors,

$$f(\mathbf{x}) = \frac{1}{\sqrt{2\pi^p\,|\boldsymbol{\Sigma}|}} \exp\,(-\tfrac{1}{2}\mathbf{d}^{(i)\prime}\,\boldsymbol{\Sigma}^{-1}\,\mathbf{d}^{(i)})$$

If the within-groups covariance matrix is an identity matrix, it is obvious that the multivariate normal-density function becomes

$$f(\mathbf{x}) = \frac{1}{\sqrt{2\pi^p}} \exp\left(-\tfrac{1}{2}\mathbf{d}^{(i)\prime}\mathbf{d}^{(i)}\right) \tag{15.2}$$

where $\mathbf{d}^{(i)\prime}\mathbf{d}^{(i)} = d^2_{(i)}$ is the simple d^2 distance relating the individual's profile to the mean profile for ith group. Therefore, under the assumption that the within-groups covariance matrix is an identity matrix or a diagonal matrix containing elements that are all equal to some constant, the simple $d^2 = d_1{}^2 + d_2{}^2 + \cdots + d_p{}^2$ replaces the more complex $D^2 = \mathbf{d}^{(i)\prime}\,\boldsymbol{\Sigma}^{-1}\,\mathbf{d}^{(i)}$ in the multivariate normal-density function.

The model used in previous chapters for classification of individuals based on multivariate normal probability densities can be used under the simplifying assumption that the within-groups covariance matrix is an identity matrix or a matrix with all diagonal elements equal. Disregarding a priori probabilities, the probability that an individual with profile $\mathbf{x}$ belongs to group j is

$$P(j\,|\,\mathbf{x}) = \frac{\text{antilog}_e\ -\tfrac{1}{2}\mathbf{d}^{(j)\prime}\mathbf{d}^{(j)}}{\sum_i^p \text{antilog}_e\ -\tfrac{1}{2}\mathbf{d}^{(i)\prime}\mathbf{d}^{(i)}} \tag{15.3}$$

A priori probabilities, or base rates, can be taken into account in a Bayesian model

$$P(j\,|\,\mathbf{x}) = \frac{\pi_j\ \text{antilog}_e\ -\tfrac{1}{2}\mathbf{d}^{(j)\prime}\mathbf{d}^{(j)}}{\sum_i^p \pi_i\ \text{antilog}_e\ -\tfrac{1}{2}\mathbf{d}^{(i)\prime}\mathbf{d}^{(i)}} \tag{15.4}$$

Taking both base rates and cost coefficients into account, we have the familiar form previously presented:

$$P(j\,|\,\mathbf{x}) = \frac{c_j\pi_j\ \text{antilog}_e\ -\tfrac{1}{2}\mathbf{d}^{(j)\prime}\mathbf{d}^{(j)}}{\sum_i^p c_i\pi_i\ \text{antilog}_e\ -\tfrac{1}{2}\mathbf{d}^{(i)\prime}\mathbf{d}^{(i)}} \tag{15.5}$$

If differential base rates need not be considered, it is not necessary to go through the antilog transformation to accomplish maximum-likelihood classification. The denominator of Eq. (15.3) is constant for all groups, and the classification decision depends only on differences in the values of the numerator for different groups. The antilog of a negative number increases toward unity as the negative number goes toward zero. Thus, the nearer to zero the quantity $-\tfrac{1}{2}\mathbf{d}^{(j)\prime}\mathbf{d}^{(j)}$ is, the greater the estimated probability of group membership as defined in Eq. (15.3) will be. This leads precisely to the simple d^2 classification rule stated earlier: assuming that the p measures have a multivariate normal distribution and that they are uncorrelated and have the same variance, maximum-likelihood classification can be accomplished by computing the sum of squared deviations of the individual's scores about the group mean values and assigning the individual to the group for which this sum of squares is smallest. To facilitate evaluation of antilog values, it is useful to rescale profile elements by shifting the decimal point so that scores fall into the range 0 to 20. The shift in decimal point does not affect the value of ratios in Eqs. (15.3) to (15.5).

15.4 Unequal Variances among Profile Elements

If it can be assumed that variables within the profile are reasonably independent, the problem posed by unequal variances can easily be overcome by transforming each variable to standard or z-score form. This can be accomplished by dividing the deviation of the individual's score about the group mean by the standard deviation for each particular variable. In this manner, all deviation scores are expressed in a common unit, and the simple d^2 index is not affected by the original difference in variance. Where variances of the p measures cannot be assumed to be equal, the vector of standard difference scores can be defined as

$$\hat{\mathbf{d}}^{(i)} = \left[\frac{x_1 - \mu_1^{(i)}}{\sigma_1} \quad \frac{x_2 - \mu_2^{(i)}}{\sigma_2} \quad \cdots \quad \frac{x_p - \mu_p^{(i)}}{\sigma_p}\right] \tag{15.6}$$

The simple d^2 index of profile similarity is then

$$d^2_{(i)} = \hat{\mathbf{d}}^{(i)\prime}\hat{\mathbf{d}}^{(i)}$$

If it is assumed that the several measures are uncorrelated but have different variances, the within-groups covariance matrix should be a *diagonal* matrix containing the (unequal) variances in the principal diagonal. Let $\boldsymbol{\Sigma} = V$ be a diagonal matrix containing within-groups variances for the p variables. Then V^{-1} is a diagonal matrix containing reciprocals of the within-groups variances, and $V^{-\frac{1}{2}}$ is a diagonal matrix containing reciprocals of the standard deviations (square root of variances). It will be recognized that $V^{-1} = V^{-\frac{1}{2}}V^{-\frac{1}{2}}$ and that $\hat{\mathbf{d}}^{(2)}$ of Eq. (15.6) is a vector defined by the product of the simple difference vector $\mathbf{d}^{(i)}$ and the matrix $V^{-\frac{1}{2}}$.

The normal probability density can be written

$$f(\mathbf{x}) = \frac{1}{\sqrt{2\pi^p\,|V|}} \exp\left(-\tfrac{1}{2}\mathbf{d}^{(i)\prime}V^{-1}\mathbf{d}^{(i)}\right) \tag{15.7}$$

It will be recalled that the constant $1/\sqrt{2\pi^p\,|V|}$ can be disregarded because it appears in the same form in both numerator and denominator of the probability ratio.

The pre- and postmultiplication of a diagonal matrix by a vector is tantamount to multiplying each element in the vector by the square root of the corresponding element in the diagonal matrix and then computing the simple vector product

$$\hat{\mathbf{d}}^2_{(i)} = \mathbf{d}^{(i)\prime}V^{-1}\mathbf{d}^{(i)} = \mathbf{d}^{(i)\prime}V^{-\frac{1}{2}}V^{-\frac{1}{2}}\mathbf{d}^{(i)} \tag{15.8}$$

where $\mathbf{d}^{(i)\prime}V^{-\frac{1}{2}}$ is a vector containing as elements the deviations of individual scores about group mean scores multiplied by the square root of associated element V^{-1}

$$\hat{\mathbf{d}}^{(i)} = \mathbf{d}^{(i)\prime} V^{-\frac{1}{2}} \tag{15.9}$$

where the elements in $V^{-\frac{1}{2}}$ are reciprocals of the standard deviations for the p variables. The reader should recognize that $\hat{\mathbf{d}}^{(i)}$ defined in Eq. (15.9) is precisely the same as that defined in Eq. (15.6).

The multivariate normal probability density can thus be written directly in terms of the simple d^2 derived from standardized deviation scores, that is, $d^2_{(i)} = \hat{\mathbf{d}}^{(i)\prime}\hat{\mathbf{d}}^{(i)}$,

$$f(\mathbf{x}) = \frac{1}{\sqrt{2\pi^p\,|V|}} \exp\left(-\tfrac{1}{2}\hat{\mathbf{d}}^{(i)\prime}\hat{\mathbf{d}}^{(i)}\right) \tag{15.10}$$

Given the multivariate normal probability density, the probability that an individual with a particular profile pattern will belong to each group can be calculated as previously shown. Disregarding a priori probabilities and cost coefficients, maximum-likelihood classification can be achieved by computing the ratio of each probability density to the sum of densities for all groups and assigning the individual to the group for which $P(j \mid \mathbf{x})$ is greatest

$$P(j \mid \mathbf{x}) = \frac{\text{antilog}_e -\tfrac{1}{2}\hat{\mathbf{d}}^{(j)\prime}\hat{\mathbf{d}}^{(j)}}{\sum_i^p \text{antilog}_e -\tfrac{1}{2}\hat{\mathbf{d}}^{(i)\prime}\hat{\mathbf{d}}^{(i)}} \tag{15.11}$$

where $\hat{\mathbf{d}}^{(j)\prime}\hat{\mathbf{d}}^{(j)}$ is a scalar quantity equal to the sum of squares of elements in the *standard-score* difference vector [Eq. (15.6)].

Since the antilog of a negative number is a fraction that increases toward unity as the negative number approaches zero, larger probabilities of group membership are associated with smaller values of $d^2_{(i)} = \hat{\mathbf{d}}^{(i)\prime}\hat{\mathbf{d}}^{(i)}$. Maximum-likelihood classification can be achieved, under the assumptions previously stated, by calculating the sum of squares for standard-deviation scores with respect to each group and then assigning the individual to the group for which this index is smallest.

15.5 Computational Examples Using Simple Distance Function

15.5.1 DIAGNOSTIC CLASSIFICATION

To provide an example yielding results that can be compared with those from the other classification methods, psychiatric diagnostic-stereotype data will again be used. Since the simple d^2 index is easy to compute and useful in applications involving profiles having several elements, the complete 16-element BPRS profiles will be examined. This provides an interesting exercise because it is recognized that the 16 ratings are not uncorrelated and that the 7-point scales do not yield a multivariate normal distribution. On the other hand, it must be admitted that the four higher-order factors have approximately equal representation in the profiles, so that the problem of independence of individual variables is not as serious as it might be if some major factors were over represented.

Prototype profiles computed as means of the psychiatrists' stereotype ratings for schizophrenic, paranoid, and depressive diagnostic types have been presented in Table 13.5. The results of simple d^2 classification are tabulated against actual psychiatrists' classification in Table 15.1 in a format that lends

Table 15.1 Classification into Diagnostic Groups Using 16 BPRS Symptom Constructs and the d^2 Criterion

Simple d^2 classification	*Diagnosis* Schizophrenic	Depressed	Paranoid	Total
Schizophrenic	1,091	26	85	1,202
Depressed	65	620	12	697
Paranoid	263	3	791	1,057
Total	1,419	649	888	2,956

itself to comparison with previous results. In making such comparisons, one should keep in mind that the detailed 16-element BPRS profiles were used for simple d^2 classification, whereas more condensed factor-score profiles consisting of only four elements were used in some of the previous analyses. The ease of calculating simple d^2 indices of profile similarity facilitates applications in which numerous variables are considered; however, one should exercise some care with regard to problems of correlation and heterogeneous variances. If the several underlying primary factors in the measurement domain are approximately equally represented, the effect is much like choosing a smaller number of variables to represent each factor uniquely.

15.5.2 ASSIGNMENT TO DRUG-TREATMENT GROUPS

Another problem examined previously with regard to classification of individuals involves different major drug-treatment groups. Mean BPRS profiles for patients treated with antidepressant, major-tranquilizer, and minor-tranquilizer drugs are presented in Table 15.2. The 470 patients from whose BPRS ratings the drug-treatment prototypes were derived were classified according to simple d^2 similarity to the three major drug-treatment prototypes. It will be obvious from the paucity of pathology in some symptom areas that the ratings do not have a multivariate normal distribution. In such cases, the simple profile-analysis methods may actually be preferred over more complex methods for which normality assumptions are likely to be more critical.

For this problem it was considered relevant to take into account the base rates of drug use because major tranquilizers were used almost twice as frequently as minor tranquilizers. Accordingly, $\pi_1 = .33$, $\pi_2 = .42$, and $\pi_3 = .25$ were employed as a priori probabilities for antidepressant, major tranquilizer, and

Table 15.2 Mean BPRS Profiles for Patients Treated with Three Types of Drugs

	Antidepressants	*Major tranquilizers*	*Minor tranquilizers*
Somatic concern	2.06	1.68	2.64
Anxiety	3.76	3.59	3.98
Emotional withdrawal	1.40	1.80	1.24
Conceptual disorganization	1.43	2.56	1.62
Guilt feelings	1.70	1.22	1.46
Tension	1.64	1.96	1.63
Mannerisms and posturing	0.49	0.73	0.49
Grandiosity	0.36	0.89	0.37
Depressive mood	3.68	2.97	3.72
Hostility	2.57	3.25	2.43
Suspiciousness	1.85	3.01	1.88
Hallucinatory behavior	0.22	0.98	0.14
Motor retardation	0.95	0.68	1.17
Uncooperativeness	1.32	1.65	1.08
Unusual thought content	0.82	1.95	0.60
Blunted affect	1.18	1.11	1.27

minor tranquilizer, respectively. Because base rates were to be considered, Eq. (15.4) was used to estimate probabilities of group membership after the simple $d^2 = \mathbf{d}^{(i)\prime}\mathbf{d}^{(i)}$ had been computed to relate the individual profile to each treatment-group prototype. That is, each of the antilog$_e$ $-\frac{1}{2}\mathbf{d}^{(i)\prime}\mathbf{d}^{(i)}$ values was multiplied by the associated π_i coefficient before forming the probability ratio of Eq. (15.4). Each patient was then assigned to the treatment group in which an individual with his profile pattern was judged most likely to belong. A cross tabulation of profile pattern assignment with the actual drug-treatment assignment for each patient is presented in Table 15.3. While the results do not indicate that one can match the clinical decisions concerning treatment assignment using symptom profile patterns alone, a reasonably strong central tendency is present. It is possible using this method to assign each patient to the treatment group in which his particular profile occurs with greatest frequency.

Table 15.3 Drug-group Assignment by Simple d^2 with Differential Base Rates Considered

	Actual drug assignment			
Profile classification	*Antidepressant*	*Major tranquilizer*	*Minor tranquilizer*	*Total*
Antidepressant	87	53	33	173
Major tranquilizer	39	117	17	173
Minor tranquilizer	54	25	45	124
Total	180	195	95	470

Of the patients actually treated with each type of drug, the most probable assignment by profile analysis was to that type of drug. It would be interesting, from a research point of view, to compare the responses of patients whose profile patterns were typical of the profile patterns for patients receiving a certain type of treatment with the responses of patients whose profiles were not typical for the treatment received.

15.5.3 ASSIGNMENT TO INTERNATIONAL DIAGNOSTIC-CLASSIFICATION GROUPS

To provide an additional example for comparison of the simple d^2 approach with the normal-probability-density-function method (Chap. 13), the diagnostic-prototype profiles provided by Czechoslovakian psychiatrists to represent 12 diagnostic types from the World Health Organization International Diagnostic Classification were used. Mean BPRS rating profiles representing the 12 diagnostic classes were previously presented in Table 13.10. For simple d^2 classification of individuals among the 12 groups, the sum of squared deviations for each individual profile about the group mean vectors was computed, and the profile was assigned to the group for which the sum of squared deviations of profile scores was smallest. A cross tabulation of results from the simple d^2 classification against the psychiatrists' original designation is presented in graphic form in Fig. 15.1, just as results from the complete normal-probability-density method were presented in Fig. 13.1. Although there was some slight

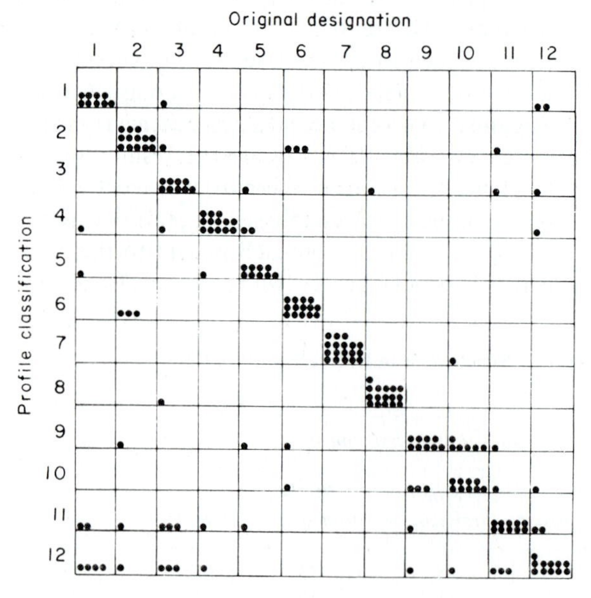

Figure 15.1 Classification of Czechoslovakian diagnostic profiles based on simple d^2 index.

loss in accuracy of classifying the diagnostic profiles when the simple d^2 index was relied upon, the results obtained in this particular case are in fact quite comparable to those previously obtained for the same data using the more complicated normal-probability-density method. One can expect that in general the similarities of results from the two methods will depend upon the degree of independence and comparability of variances among the profile components.

15.6 Normalized-vector-product-classification Model

A slightly different geometric conception is the basis for normalized-vector-product classification. Whereas the simple d^2 analysis of profile similarity is associated with the conception of multivariate profiles in terms of components of *elevation*, *variability*, and *pattern*, the vector conception of multivariate profiles suggests the two vector parameters of length and direction. Under appropriate assumptions, the *length* of a profile vector is defined as the sum of squares of scores within the profile. The *direction* of the profile vector is associated with the remaining information in the profile that is independent of vector length. Thus, the multivariate profile is reduced to the two parameters that define any vector, and relationships between profiles can be defined in terms of the vector parameters. The difference from ordinary profile-analysis concepts is more apparent than real since the normalized-vector-product model can be used meaningfully only when there is a true zero point or a rational origin for each measurement in the profile. This in effect eliminates the elevation factor by equating profiles with regard to a common origin.

The normalized-vector-product index of profile similarity is defined as the sum of cross products of elements in profiles that have been previously normalized to unit sum of squares. Normalizing can be accomplished by dividing each element within a profile by the square root of the sum of squares for all elements. After the normalizing, the profile *vectors* have unit length; thus they can differ only in direction. The angular cosine between two normalized profile vectors is defined as the vector product, or sum of cross products of elements within the profiles. The normalized-vector-product index of profile similarity can be conceived as a measure of the angle between the profile vectors under the assumption that the profile elements are uncorrelated and of equal variance. Apart from any geometric considerations, the product of normalized profile vectors is a useful, practical index of profile similarity

$$V^{(i)} = \mathbf{x}'\boldsymbol{\mu}^{(i)} = \sum_{j}^{p} x_j\mu_j \tag{15.12}$$

where $\mathbf{x}'$ is the normalized score profile for an individual and $\boldsymbol{\mu}^{(i)}$ is the normalized mean profile for the ith group.

The normalized vector product can also be defined in terms of raw profile elements.

$$V^{(i)} = \frac{\mathbf{x}'\boldsymbol{\mu}^{(i)}}{\sqrt{\mathbf{x}'\mathbf{x}}\sqrt{\boldsymbol{\mu}^{(i)\prime}\boldsymbol{\mu}^{(i)}}} = \frac{\sum x\mu}{\sqrt{\sum x^2}\sqrt{\sum \mu^2}} \tag{15.13}$$

where $\mathbf{x}$ and $\boldsymbol{\mu}^{(i)}$ are raw-score profiles. Obviously, Eqs. (15.12) and (15.13) are equivalent except that normalizing is accomplished after computing the profile product in Eq. (15.13).

Although the normalized-vector-product method has been presented as based on a unique geometric model, it can be related to the general multivariate normal-probability-density function. As with the other methods that have been discussed, the normalized vector product can be conceived as a special case of the more general model. The multivariate normal-density function has been represented as

$$\frac{1}{2\pi^p\,|\boldsymbol{\Sigma}|}\exp\left[-\tfrac{1}{2}(\mathbf{x}-\boldsymbol{\mu}^{(i)})'\,\boldsymbol{\Sigma}^{-1}\,(\mathbf{x}-\boldsymbol{\mu}^{(i)})\right]$$

Since classification decisions have been shown to depend entirely on the quadratic form in the exponent, attention will be concentrated there:

$$\begin{aligned} D^2 &= (\mathbf{x}-\boldsymbol{\mu}^{(i)})'\,\boldsymbol{\Sigma}^{-1}\,(\mathbf{x}-\boldsymbol{\mu}^{(i)}) \\ &= \mathbf{x}'\,\boldsymbol{\Sigma}^{-1}\,\mathbf{x} + \boldsymbol{\mu}^{(i)\prime}\,\boldsymbol{\Sigma}^{-1}\,\boldsymbol{\mu}^{(i)} - 2\mathbf{x}'\,\boldsymbol{\Sigma}^{-1}\,\boldsymbol{\mu}^{(i)} \end{aligned}$$

Thus, the probability of a particular profile $\mathbf{x}$ occurring in the ith group is a function of the two vector lengths $\mathbf{x}'\,\boldsymbol{\Sigma}^{-1}\,\mathbf{x}$ and $\boldsymbol{\mu}^{(i)\prime}\,\boldsymbol{\Sigma}^{-1}\,\boldsymbol{\mu}^{(i)}$ and the cosine of the angle between the vectors. If it is assumed that the within-groups covariance matrix is an identity matrix, $\boldsymbol{\Sigma}^{-1} = I$, the probability of group membership becomes a function of the sums of squares and cross products only

$$D^2 = \mathbf{x}'\mathbf{x} + \boldsymbol{\mu}^{(i)\prime}\boldsymbol{\mu}^{(i)} - 2\mathbf{x}'\boldsymbol{\mu}^{(i)} \tag{15.14}$$

where it is assumed that $\boldsymbol{\Sigma}^{-1} = I$.

If in addition the vectors have been *normalized* so that the sum of squares is unity for each profile, the probability of group membership is a function of the cross-product term only

$$D^2 = 2(1 - \mathbf{x}'\boldsymbol{\mu}^{(i)}) \tag{15.15}$$

$$V^{(i)} = \mathbf{x}'\boldsymbol{\mu}^{(i)} = 1 - \frac{D^2}{2} \tag{15.16}$$

where $V^{(i)}$ is the normalized vector product [Eq. (15.10)] and D^2 is the simple squared distance function as defined in Sec. (15.2) but computed using normalized profiles. Thus, exactly the same classification decisions will result from evaluating normalized vector products and choosing the *largest* or from computing d^2 distances between the normalized profiles and identifying the *smallest* one. This becomes important in that it enables us to use a single general computer program to classify patients according to all methods discussed so far. The differences in the methods are implicit in the transformations and assumptions employed.

15.7 Applications of the Normalized-vector-product Method

In this section, results from several applications of the normalized-vector-product method for assignment of individuals among several groups will be presented. At the risk of appearing repetitious, we shall use for these examples data that have previously been used to illustrate other methods.

15.7.1 DIAGNOSTIC-CLASSIFICATION PROBLEM

The BPRS profiles provided by Czechoslovakian psychiatrists to represent 12 diagnostic classes from the World Health Organization International Diagnostic Classification are of some general interest insofar as they represent a system for classification of psychiatric disorders that is supposed to have utility across national boundaries. The results obtained from the multivariate normal-probability-density model (Fig. 13.1) and from the simple profile-analysis d^2 distance function (Fig. 15.1) have already been considered. With regard to classifying profiles representing hypothetical typical patients, the simple d^2 method proved to be approximately 95 percent as efficient as the method based on the complete multivariate normal-probability-density model.

In order to classify BPRS among the 12 diagnostic classes representing the International Classification System using the normalized-vector-product method, the mean diagnostic-prototype profiles previously presented in Table 13.10 were first normalized. That is, the elements in each profile were divided by the square root of the sum of squares for all elements in the profile. In order to conserve space, the normalized profiles will not be presented here since they can be obtained, if desired, from the original prototype profiles shown in Table 13.10.

Once the diagnostic-prototype profiles were normalized, assignment of individual patients among the 12 classes was accomplished by computing the vector product (sum of cross products) of the individual profile with each of the 12 normalized prototype profiles and assigning the individual to the class for which this vector product was largest. It will be recognized that there is no necessity for normalizing the individual profile that is to be assigned to one of the groups because the normalizing coefficient simply enters as a constant in all vector products. A constant multiplier leaves relative magnitudes unchanged. Of course, it is necessary to normalize the group means or diagnostic prototypes because the normalizing coefficients will almost certainly be different for different prototype vectors.

Results from application of the normalized-vector-product method in assignment of individual profiles among the 12 classification groups are presented in Fig. 15.2. Each dot represents five profiles, rounded to the nearest multiple. The entries in each row represent a single class as designated by the experts who provided the profiles. The entries in each column represent a single class as assigned by a computer program based on the normalized-vector-product method. Entries in the principal diagonal represent (multiple of five) cases for which normalized-vector-product classification agreed with the original designation for the profiles. In this particular application, the normalized-vector-product method did not appear quite as effective as either the complete multivariate normal-probability-density function or the simple d^2 method.

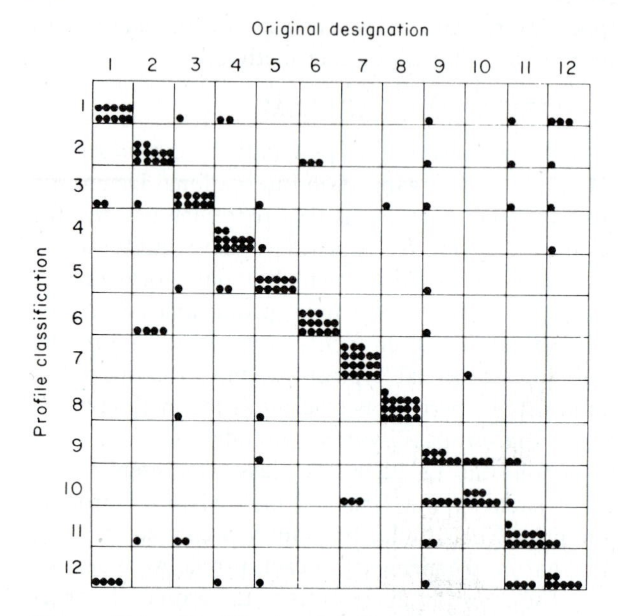

Figure 15.2 Classification of Czechoslovakian diagnostic profiles based on normalized-vector-product index.

The number of assignments that agreed with the original designation by the psychiatrist who provided the profile was approximately 90 percent as great as the number of correct assignments obtained from the method based on the multivariate normal-density function. In this application, the normalized-vector-product method was approximately 95 percent as efficient as the simple d^2 method. While one should not place too much emphasis on a single comparison, the results do suggest superiority of the more complex methods, a fact which should not be considered too disturbing in view of the widespread availability of computers. The results also suggest the importance of undertaking other comparisons of efficiencies of the various methods in an effort to arrive at a general recommendation.

15.7.2 TREATMENT ASSIGNMENT

Patients treated with different types of psychiatric drugs tend to have somewhat different symptom characteristics, although the distinctions are not sharp as one might suspect a priori. In a sense, we can conceive of choice of drug treatments as defining different patient populations. Mean BPRS profiles for patients treated with major tranquilizers, minor tranquilizers, and antidepressant drugs in a sample of inpatients studied at the University of Texas Medical Branch were computed. The sample of $n = 470$ was obtained by eliminating from a larger sample all patients who did not receive one of the three

Table 15.4 Normalized Profiles Representing Three Types of Drugs

	Antidepressants	*Major tranquilizer*	*Minor tranquilizer*
Somatic concern	.2736	.2011	.3410
Anxiety	.4995	.4298	.5141
Emotional withdrawal	.1860	.2155	.1602
Conceptual disorganization	.1900	.3065	.2092
Guilt feelings	.2258	.1461	.1886
Tension	.2179	.2346	.2105
Mannerisms and posturing	.0651	.0874	.0633
Grandiosity	.0478	.1065	.0478
Dépressive mood	.4889	.3556	.4805
Hostility	.3414	.3891	.3139
Suspiciousness	.2458	.3603	.2428
Hallucinatory behavior	.0292	.1173	.0181
Motor retardation	.1262	.0814	.1511
Uncooperativeness	.1754	.1975	.1395
Unusual thought content	.1089	.2334	.0775
Blunted affect	.1568	.1329	.1640

major classes of drug treatment. Each mean profile was normalized to eliminate the "severity of illness" factor. Normalized prototype profiles representing major-tranquilizer, minor-tranquilizer, and antidepressant patterns are presented in Table 15.4. In order to assign any particular patient to one of the drug-treatment groups, one simply computes the product of his profile vector with each of the normalized drug-treatment-group prototypes. The patient is then assigned to the treatment group for which this product is greatest.

To examine the validity of the method in classification of patients from the original sample, profiles for the 470 patients were assigned among the three treatment groups using the normalized-vector-product index of profile similarity. Results, cross-tabulated against actual drug assignment, are presented in Table 15.5. As previously noted in examining results from applications of other

Table 15.5 Classification into Three Drug-treatment Groups Using 16 BPRS Symptom Constructs and the Normalized-vector-product Index of Profile Similarity

	Drug			
Classification	*Antidepressant*	*Major tranquilizer*	*Minor tranquilizer*	*Total*
Antidepressant	87	45	25	157
Major tranquilizer	33	118	14	165
Minor tranquilizer	60	32	56	148
Total	180	195	95	470

classification methods, these data to not lend themselves to good discrimination among treatment groups. Clear trends are obvious, and the results obtained from the simple normalized-vector-product method appear slightly better than those obtained using the simple d^2 index of profile similarity (Table 15.3).

In order to examine further the validity of the normalized-vector-product method for assigning patients to drug-treatment groups, we undertook an

Table 15.6 Diagnostic Profiles Assigned to Each Major Type of Drug Treatment

Antidepressants

American:
- Psychotic depressive
- Manic-depressive, depressive

Czech:
- Deprese psychotická

French:
- Depression atypique
- Melancolie

Italian:
- Melancolia
- Depressive atipica
- Melancolia involutiva

Major tranquilizers

American:
- Paranoia
- Paranoid state
- Paranoid schizophrenic
- Acute undifferentiated schizophrenic
- Catatonic schizophrenic
- Hebephrenic schizophrenic
- Simple schizophrenic
- Chronic undifferentiated schizophrenic
- Residual schizophrenic
- Schizo-affective
- Manic-depressive, manic

Czech:
- Schizophrenia simplex
- Schizophrenia paranoidní
- Schizophrenia hebefrenická
- Schizophrenia katatonická
- Schizophrenia katatonická, stuporozní
- Schizophrenia katatonická agitovaná
- Paranoidní stavy
- Manie
- Schizoafektivní psychóza
- Chronickà schizophrenia s floridními příznaky
- Chronická schizophrenia bez výraznějších příznaky
- Jiné funckcní psychotické stavy

Major tranquilizers (Continued)

French:
- Schizophrénie paranoïde
- Manie
- État confusionnel
- Psychose paranoïaque
- Excitation atypique
- Bouffée délirante
- Psychose hallucinatoire chronique
- Paraphrénie
- Hébephrénie

German:
- Schubförmige katatone Schizophrenie
- Schubförmige paranoide Schizophrenie
- Schubförmige paranoide-halluzinatorische Schizophrenie
- Paranoia
- Caënasthetische Schizophrenie
- Hebephrene Schizophrenie
- Schizophrenie simplex
- Schizophrener Personlichkeitswandel
- Manie
- Mischpsychose

Italian:
- Schizophrenia ebefrenica
- Schizophrenia catatonica
- Schizophrenia paranoide
- Schizophrenia simplex
- Paranoia
- Paraphrenia
- Mania
- Stato confusionale o amenza

Minor tranquilizers

French:
- Dépression réactionnelle

German:
- Endogene Depression
- Endoreaktive Dysthymie

exercise involving the psychiatric diagnostic prototypes derived from psychiatrists' stereotypes in five different countries. Since the diagnostic stereotypes were provided to represent only major psychiatric disorders (functional psychoses), it should be expected a priori that few, if any, of the diagnostic-prototype profiles will correspond to patterns that are properly treated with minor tranquilizers. A total of 61 diagnostic prototypes representing various forms of schizophrenia, paranoid states, depression, and mania were available. The 12 Czechoslovakian prototypes shown in Table 13.10 and the 13 American prototypes shown in Table 8.1 were among the 61.

Each of the 61 diagnostic prototypes was assigned to major tranquilizer, antidepressant, or minor tranquilizer by computing the vector product with each of the three normalized drug-group profile vectors (Table 15.4) and assigning the diagnostic prototype to the treatment yielding the largest vector product. The results are listed in Table 15.6. The American, Czechoslovakian, French, and Italian depression prototypes were assigned by the computer program to treatment with antidepressant drug. Only the French dépression réactionnelle and the two German depression prototypes were assigned to treatment with minor tranquilizer, and it can be noted that two of the three prototypes assigned to minor tranquilizer were simple anxious-reactive depressions. All the other diagnostic prototypes representing all forms of schizophrenia, paranoid states, and manic states were assigned to treatment with major tranquilizer. In view of the natures of the disorders represented by the collection of diagnostic prototypes, and in view of the particular choices of treatment available, these results seem to the present authors to confirm the general validity of the method and to provide encouragement for further applications of profile-analysis methods in treatment assignment.

16

Pattern-probability Model for Classification Using Qualitative Data

16.1 Introduction

In the preceding chapters, several different methods for classification of patients on the basis of quantitative measurement profiles have been discussed. Methods based on the multivariate normal-distribution assumptions appear generally more powerful than alternative methods whenever distribution assumptions can be reasonably met. There are many practical problems, however, in which the information to be used in making classification decisions is categorical or qualitative in nature. Sex, race, religious preference, occupation, and marital status are examples of categorical data that may be relevant for classification decisions. The multivariate normal classification model and related quantitative profile-analysis techniques are not appropriate when classification decisions are to be based on such data. In the presence of qualitative data, the patterns upon which classification decisions are based may involve simple presence vs. absence of characteristics. Some qualitative attributes may have several alternative states, e.g., marital status.

In this chapter, methods for assigning individuals among multiple groups on the basis of patterns of qualitative variables will be considered. The status of an individual on one multicategory qualitative variable, such as sex or marital status, will be called an *attribute state.* A pattern of p qualitative variables consists of the coded attribute states for an individual on the p qualitative variables. Some qualitative variables, such as sex and sick vs. well, have only two attribute states. Other qualitative variables, such as religious preference and marital status, have several. Qualitative score patterns may contain variables having different numbers of categories or attribute states. If each qualitative variable is coded $1, 2, \ldots, m$, a pattern of coded attribute states may appear as a score vector $[1 \quad 1 \quad 2 \quad 1 \quad 5 \quad 1 \quad 2]$. Each attribute state, say married or male, has a certain probability of occurrence in any particular population of individuals. It is with the probabilities of occurrence of attribute states and patterns of attribute states that this chapter is concerned. The direct evaluation of relative frequencies of various patterns, as illustrated in Sec. 12.3, is feasible when the number of characteristics is small and the sample sizes very large. It is not generally feasible, however, to use such an approach when the number of elements in a pattern exceeds three or four because the number of possible patterns increases exponentially. With only 10 qualitative variables, each scored only as present or absent, the number of possible patterns is $2^{10} = 1{,}024$. Obviously, excessively large samples would be required to evaluate the relative frequency of occurrence of each of the possible patterns *within each group.*

Simple probability theory provides a rule for combining in a logical fashion information concerning qualitative characteristics to derive an estimate of the probability of occurrence of complex patterns. The *joint probability* of p independent events is equal to the product of the simple probabilities associated with occurrence of the separate events. This is known as the *multiplicative law* defining the probability of simultaneous occurrence of several independent events. The validity of this combinatorial probability rule can be verified by simple experiment. For example, the probability of occurrence of a pattern of heads-tails-heads can be calculated as the product of the probability of heads, the probability of tails, and the probability of heads again on three successive tosses of an unbiased coin: $P(\mathbf{x}) = P(H)P(T)P(H) = .5(.5)(.5) = .125$. If one were to toss in succession three coins on a large number of occasions, the long-run proportion of head-tail-head patterns would be found to approximate 12.5 percent. Since it is sometimes difficult to make the transition from "independent tosses of an unbiased coin" to probabilities of occurrence of patterns of clinical characteristics, the reader should take the time to consider other analogies to the coin-tossing situation. For example, consider four independent rolls of a six-sided die. What is the probability of the pattern 1166? It is $P(\mathbf{x}) = .167(.167)(.167)(.167)$.

If psychiatric symptoms or other characteristics can be considered to be independent events *within* classification groups or populations, the frequency of joint occurrence of any particular pattern can be estimated using the multiplicative rule. This means that it is not necessary to evaluate directly the relative frequencies of occurrence of entire patterns but instead only to obtain estimates of the relative frequency of occurrence for each specific characteristic. *Assuming independence of the several characteristics within groups*, the joint probability of a particular pattern **x** (say male, Negro, age <30) can be estimated as a multiplicative function of the separate probabilities within each group. Let $P(1_2 \mid k)$ be the probability of attribute state 2 (say female) on variable 1 within group k. In general, let $P(ij \mid k)$ be the probability of attribute state j on variable i within group k

$$P(\mathbf{x} \mid k) = P(1j \mid k)P(2j \mid k)P(3j \mid k) \cdots P(pj \mid k) \tag{16.1}$$

where $P(\mathbf{x} \mid k)$ is the estimated probability of pattern **x** within group k and $P(1j \mid k), P(2j \mid k), \ldots, P(pj \mid k)$ are probabilities of particular attribute states on p qualitative variables.

The simple within-group probabilities are usually estimated as relative frequencies of occurrence of attribute states

$$P(\mathbf{x} \mid k) = \frac{n(1j \mid k)}{n(1\cdot \mid k)} \frac{n(2j \mid k)}{n(2\cdot \mid k)} \cdots \frac{n(pj \mid k)}{n(p\cdot \mid k)} \tag{16.2}$$

where $n(ij \mid k)$ is the number of individuals in group k who have attribute state j on the ith variable and $n(i\cdot \mid k)$ is the total number of individuals who have been observed for status on variable i regardless of their particular attribute state on the variable.

Once an estimate of the probability of occurrence of pattern **x** in each group $(i = 1, 2, \ldots, k)$ has been obtained, the general maximum-likelihood model [Eq. (12.2)] or the Bayesian model [Eq. (12.3) or (12.4)] can be used to estimate the probability of group membership for an individual having any particular pattern **x**

$$P(j \mid \mathbf{x}) = \frac{P(\mathbf{x} \mid j)}{P(\mathbf{x} \mid 1) + P(\mathbf{x} \mid 2) + \cdots + P(\mathbf{x} \mid j) + \cdots + P(\mathbf{x} \mid k)} \tag{16.3}$$

Taking a priori probabilities into account,

$$P(j \mid k) = \frac{\pi_j P(\mathbf{x} \mid j)}{\pi_1 P(\mathbf{x} \mid 1) + \pi_2 P(\mathbf{x} \mid 2) + \cdots + \pi_j P(\mathbf{x} \mid j) + \cdots + \pi_k P(\mathbf{x} \mid k)} \tag{16.4}$$

and taking both cost coefficients and a priori probabilities into account,

$$P(j \mid k) = \frac{C_j \pi_j P(\mathbf{x} \mid j)}{C_1 \pi_1 P(\mathbf{x} \mid 1) + C_2 \pi_2 P(\mathbf{x} \mid 2) + \cdots + C_k \pi_k P(\mathbf{x} \mid k)} \tag{16.5}$$

Before considering a computational example, it is worthwhile to examine in more detail the assumption of *independence* required for validity of the joint probability estimates defined in Eqs. (16.1) and (16.2). The assumption of independence concerns relationships among characteristics *within* classification groups. If two or more characteristics are to be useful for classification, it is almost certain that they will be related across individuals in all groups. The utility of specified characteristics for classification depends upon attribute states on those characteristics having different probabilities of occurrence in the different groups, and this almost certainly implies correlation *across* groups. However, it does not mean that the characteristics will necessarily be correlated *within* groups.

Consider, for example, the symptoms of anxiety, guilt feelings, and depressive mood. Across a heterogeneous sample of patients including both depressive and nondepressive diagnostic types, the occurrences of these symptoms will be correlated. On the other hand, if one selects a relatively homogeneous group of depressive types, it is less likely that these same variables will correlate substantially; i.e., one cannot predict the level of anxiety from knowledge of the level of depressive mood *among depressed patients.* In the general psychiatric population, on the other hand, one can predict with reasonable accuracy that patients evidencing higher levels of depressive mood will also evidence substantially more anxiety and vice versa. The authors have used this model in a variety of situations where independence among variates within groups cannot be assumed to be perfect, and, in general, the results have been quite good. With regard to this model, as with other methods, we recommend that it be tried where reasonably appropriate and that replication samples be used to examine reliability and validity.

An advantage of the pattern-probability model is that it can be used with qualitative or categorical variables having different numbers of categories. For example, sex can be scored in two categories, marital status in four, age in three, and occupation in six. As for age, quantitative variables can be reduced to category form for consideration with other inherently qualitative variables. For each variable, the relative frequency of various attribute states can be evaluated within each group and the calculations of within-group pattern probabilities undertaken according to Eq. (16.1) or (16.2).

16.2 Computational Example

To illustrate the calculations required for pattern-probability classification, we shall consider the assignment of individuals (diagnostic-stereotype profiles) among schizophrenic, paranoid, and depressive groups using symptom ratings for depressive mood, hostility, unusual thought content, and blunted affect. This problem was selected to permit comparison of results with those obtained previously from direct evaluation of pattern frequencies (Table 12.4).

Simple frequencies of different levels of severity for the four symptoms were tabulated within each major diagnostic class. This was accomplished as follows: the entries in Table 16.1 represent actual counts of the numbers of individuals (stereotype profiles) having ratings of not present, very mild, mild, moderately severe, and extremely severe for each symptom variable considered separately. The entries in Table 16.2 were obtained by dividing the frequencies in various categories of severity by the total n in each diagnostic group. That is, each entry is equal to the corresponding entry in Table 16.1 divided by the row total. The

Table 16.1 Frequencies at Seven Levels of Severity for Four Symptoms among Schizophrenic, Paranoid, and Depressive Diagnostic Stereotypes

	Level of severity						
	1	*2*	*3*	*4*	*5*	*6*	*7*
Schizophrenic:							
Depressive mood	440	322	294	233	89	36	5
Hostility	195	230	251	324	268	123	28
Unusual thought content	60	85	139	286	316	301	232
Blunted affect	84	90	115	199	266	363	302
Paranoid:							
Depressive mood	223	229	202	143	71	12	8
Hostility	54	68	108	156	213	206	83
Unusual thought content	25	32	38	76	161	274	282
Blunted affect	161	165	177	174	118	76	17
Depressive:							
Depressive mood	5	1	3	14	56	184	386
Hostility	311	151	78	61	40	7	1
Unusual thought content	186	107	80	102	96	56	22
Blunted affect	226	77	72	87	82	72	33

relative frequencies in Table 16.2 will be accepted as estimates of the probabilities of occurrence for the particular symptom ratings within each diagnostic group.

To obtain an estimate of the probability of occurrence of any symptom pattern within each group, one multiplies together the appropriate simple

Table 16.2 Relative Frequencies at Seven Levels of Severity of Four Symptoms in Three Diagnostic Groups

	Level of severity						
	1	*2*	*3*	*4*	*5*	*6*	*7*
Schizophrenic:							
Depressive mood	.31	.22	.20	.16	.06	.02	.00
Hostility	.13	.16	.17	.22	.18	.08	.01
Unusual thought content	.04	.05	.09	.20	.22	.21	.16
Blunted affect	.05	.06	.08	.14	.18	.25	.21
Paranoid:							
Depressive mood	.25	.25	.22	.16	.07	.01	.00
Hostility	.06	.07	.12	.17	.23	.23	.09
Unusual thought content	.02	.03	.04	.08	.18	.30	.31
Blunted affect	.18	.18	.19	.19	.13	.08	.01
Depressive:							
Depressive mood	.00	.00	.00	.02	.08	.28	.59
Hostility	.47	.23	.12	.09	.06	.01	.00
Unusual thought content	.28	.16	.12	.15	.14	.08	.03
Blunted affect	.34	.11	.11	.13	.12	.11	.05

probability estimates. For example, the probability of the pattern $\mathbf{x}' = [2 \quad 4 \quad 2 \quad 1]$ in the schizophrenic group (group 1) is

$$P(\mathbf{x} \mid 1) = .22(.22)(.05)(.05) = .0001210$$

The probability of this same pattern in the paranoid group (group 2) is

$$P(\mathbf{x} \mid 2) = .25(.17)(.03)(.18) = .0002295$$

The probability of the symptom pattern in the depressive group (group 3) is

$$P(\mathbf{x} \mid 3) = .01(.09)(.16)(.34) = .00004896$$

Given these estimates of the probability of occurrence of the specific pattern in each group, it is an easy matter to arrive at an estimate of the probabilities of group membership for an individual who has the symptom pattern

$$P(1 \mid \mathbf{x}) = \frac{P(\mathbf{x} \mid 1)}{P(\mathbf{x} \mid 1) + P(\mathbf{x} \mid 2) + P(\mathbf{x} \mid 3)} = \frac{.0001210}{.0001210 + .0002295 + .00004896} = .303$$

$$P(2 \mid \mathbf{x}) = \frac{.0002295}{.0001210 + .0002295 + .00004896} = .575$$

$$P(3 \mid \mathbf{x}) = \frac{.00004896}{.0001210 + .0002295 + .00004896} = .123$$

Given a randomly selected individual having symptom pattern [2 4 2 1] on the four symptoms, the probability of belonging to the paranoid group is substantially greater than the probability of belonging to either of the other two groups (assuming roughly equivalent base rates).

Each of the diagnostic profiles provided by European psychiatrists was classified using estimates of pattern probability obtained from multiplication of simple probabilities of the four symptom ratings, as illustrated above. A cross tabulation of the results against the psychiatrists' original designation for each profile is presented in Table 16.3. The multiplicative pattern-probability model resulted in a classification of the prototype profiles that agreed 80 percent of the time with the psychiatrists' designation.

Table 16.3 Results from Pattern-probability Classification of Diagnostic-stereotype Profiles Using Four BPRS Symptoms

Pattern classification	*Diagnostic group*			
	Schizophrenic	*Depressed*	*Paranoid*	*Total*
Schizophrenic	1,072	34	182	1,288
Depressed	64	607	28	699
Paranoid	283	8	678	969
Total	1,419	649	888	2,956

It is interesting to note that the multiplicative pattern-probability model yielded a higher proportion of correct classification than the direct evaluation of symptom patterns (Table 12.4). The reason for this is that the four symptom variates were considered as present-absent dichotomies for direct count of pattern frequencies. The dichotomous scoring resulted in $2^4 = 16$ possible patterns, a feasible number for direct-frequency count. The number of possible patterns of four symptoms, each recorded in seven levels, is $7^4 = 2{,}401$. It would obviously have been impossible to evaluate directly the relative frequency of occurrence for each of 2,401 patterns within each diagnostic group because none of the groups contained that many profiles. On the other hand, it is quite possible to *estimate* the probability of each pattern using the multiplicative rule. In the present application, the advantage accruing from consideration of finer pattern detail (seven attribute states for each symptom variable) outweighed the possible disadvantages resulting from failure to satisfy fully the assumption of independence. The multiplicative pattern-probability model yielded better results than the direct pattern frequency-count approach.

16.3 Estimation of Base Rates in Specified Subpopulations

The pattern-probability model has particular utility for classification of individuals on the basis of qualitative or categorical data. A problem of special concern is the dependence of psychopathology on historical and sociocultural background data. It is generally recognized that incidences of various major psychiatric disorders vary as a function of age, race, sex, marital status, and social class. Although symptom profile patterns are important for diagnostic classification, the frequency of occurrence of psychiatric disorders differs as a function of the particular combination of background characteristics represented in patients from the specific treatment setting.

Background characteristics can be considered in the diagnostic decision process to define specific subpopulations with distinct a priori probabilities for each disorder. Subsequently symptom patterns can be examined and their probabilities of occurrence in each diagnostic population estimated. The Bayesian conditional-probability model, as defined in Eq. (12.3), can be used to combine the a priori probabilities derived from qualitative background information with the symptom-pattern information in reaching a final diagnostic decision.

The analyses described in this section were undertaken in an effort to develop procedures for utilizing relevant background information in classifying individuals among schizophrenic, paranoid, and depressive diagnostic groups. Although the probability estimates derived fróm the qualitative background data can be combined with those derived from quantitative profile analyses as described above, our purpose here is to examine the utility of the pattern-probability model and to specify procedures for using background data to estimate probabilities of major psychiatric disorders in carefully defined subpopulations.

The 3,498 case records with BPRS symptom ratings plus background data were first classified into schizophrenic, paranoid, and depressive types on the basis of their BPRS symptom profile patterns. The normal-probability-density model was used for this purpose, and, in fact, the computer program used to obtain the results shown in Fig. 13.1 was used to classify the patients among the three major diagnostic groups. Next, the frequency in each category of age,

Table 16.4 Frequencies of Background Characteristics in Three Profile Types†

	Schizophrenic	*Depressive*	*Paranoid*
Age, years:			
<30	308	209	234
30–39	309	248	255
40–49	361	317	252
50+	454	306	227
Race:			
Anglo	934	896	797
Negro	364	99	122
Latin	111	60	36
Sex:			
Male	817	570	519
Female	622	509	448
Education, years:			
8	319	166	160
8–11	374	301	259
12	267	290	228
13+	262	249	281
Occupation:			
Never worked	117	14	65
Housewife	86	53	45
Unskilled	680	417	361
Skilled	279	261	171
Clerical	103	147	114
Self-employed	4	15	21
Professional-managerial	86	87	108
Student	37	40	43
Marital status:			
Single	365	170	214
Married	568	526	440
Divorced or separated	392	274	258
Widowed	111	108	56
Treatment setting:			
Outpatient (acute)	239	177	33
Outpatient (chronic)	205	46	48
Veterans hospital	248	313	165
Private inpatient	386	459	454
State hospital	577	89	270

† Sum of frequencies is not constant across all variables because of missing data.

race, sex, marital status, educational achievement, and occupation was tabulated separately for the patients classified by symptom profile as schizophrenic, paranoid, or depressive types. The frequencies of various background characteristics within the diagnostic groups are presented in Table 16.4. The frequencies were converted to relative-frequency probability estimates by dividing the number in each category by the total n for diagnostic group. These are taken as the simple probabilities associated with occurrence of specific background characteristics within each diagnostic group and are presented in Table 16.5.

Table 16.5 Relative Frequencies in Various Categories of Background Variables for Three Profile Types

	Schizophrenic	*Depressive*	*Paranoid*
Age, years:			
<30	.22	.19	.24
30–39	.22	.23	.26
40–49	.25	.29	.26
50+	.32	.28	.23
Race:			
Anglo	.66	.85	.83
Negro	.26	.09	.13
Latin	.08	.06	.04
Sex:			
Male	.57	.53	.54
Female	.43	.47	.46
Education, years:			
8	.26	.17	.17
8–11	.31	.30	.28
12	.22	.29	.25
13+	.21	.25	.30
Occupation:			
Never worked	.08	.01	.07
Housewife	.06	.05	.05
Unskilled	.49	.38	.39
Skilled	.20	.24	.18
Clerical	.07	.14	.12
Self-employed	.00	.01	.02
Professional-managerial	.06	.08	.12
Student	.03	.04	.05
Marital status:			
Single	.25	.16	.22
Married	.40	.49	.45
Divorced or separated	.27	.25	.27
Widowed	.08	.10	.06
Treatment setting:			
Outpatient (acute)	.14	.16	.03
Outpatient (chronic)	.12	.04	.05
Veterans hospital	.15	.29	.17
Private inpatient	.23	.42	.47
State hospital	.35	.08	.28

Given the simple within-groups probability estimates for each background characteristic, the joint probability of any complex pattern of the six background variables within each diagnostic group can be estimated as a multiplicative function of the appropriate simple probabilities. For example, the pattern *male*, *Anglo*, *age twenty-five*, *single*, with *tenth-grade education* and *unskilled* occupation has the following probability within the schizophrenic group.

$$P(\mathbf{x} \mid 1) = .57(.66)(.22)(.25)(.31)(.49) = .00314$$

The probability of the pattern in the depressive group is

$$P(\mathbf{x} \mid 2) = .53(.85)(.19)(.16)(.30)(.38) = .00156$$

and the probability of the pattern in the paranoid group is

$$P(\mathbf{x} \mid 3) = .54(.83)(.24)(.22)(.28)(.39) = .00258$$

Given the probability of the pattern in each diagnostic group, the probability that a patient who has the particular pattern of background characteristics belongs to each diagnostic group can be estimated, disregarding base rates, as follows:

$$P(1 \mid \mathbf{x}) = \frac{.00314}{.00314 + .00156 + .00258} = .431$$

$$P(2 \mid \mathbf{x}) = \frac{.00156}{.00314 + .00156 + .00258} = .214$$

$$P(3 \mid \mathbf{x}) = \frac{.00258}{.00314 + .00156 + .00258} = .354$$

From this analysis, we conclude that a white male, age twenty-five, who is single, has tenth-grade education, and works as an unskilled laborer is most likely to belong in the schizophrenic diagnostic group if he becomes mentally ill. Similar calculations can be accomplished for any other particular pattern of the six background characteristics.

Each of the 3,498 patients was classified into schizophrenic, paranoid, or

Table 16.6 Pattern-probability Classification Derived from Demographic Data

	Symptom profile class			
Demographic classification	*Schizophrenic*	*Depressed*	*Paranoid*	*Total*
Schizophrenic	919	269	274	1,462
Depressed	228	513	271	1,012
Paranoid	297	302	425	1,024
Total	1,444	1,084	970	3,498

depressive group using pattern-probability estimates derived from his background characteristics. A cross tabulation of results from the background classification with the symptom-profile-pattern classification is presented in Table 16.6. This comparison is of interest with regard to establishing the validity of *both* methods of classification. Essentially, what has been demonstrated is a strong relationship between symptom-rating patterns and complex patterns of background characteristics.

16.4 Computer Programs for Classification Assignment Using Categorical Data

Two computer programs are included in this section. The first is useful in calculating the simple probability matrices which contain the probabilities of various attribute states within each classification group. This program is useful in compiling the basic probability matrices from sets of raw data for groups of previously classified or diagnosed individuals. The simple probability tables are then used with the second program for assignment of new individuals among the classification groups. Of course, if the simple p values are available from some outside source, there is no need to use the probability-generating program. For example, in diagnostic classification of psychiatric patients, the simple p values within schizophrenic, paranoid, and depressive groups are given in Table 16.2. These are the types of basic probability statistics generated by the within-groups probability program WGP.

To generate within-groups probability tables similar to those in Tables 16.2 and 16.5, the program WGP can be used as follows. First sort data cards containing up to 20 categorical variables, each with no more than 10 levels or attribute states coded 0, 1, 2, . . . , 9, according to criterion group. For example, sort cards containing symptom-rating profiles into diagnostic groups. Count the number in each group. One control card is used to specify the number of groups and the number of variables. Another control card precedes data for each group of S's containing the number in the group, in the first four-column field, and the name of the group in the remainder of card. Data cards to be read in format 3.

Input for Program WGP

1 Card 1 NGPS NVAR (in successive four-column fields)
2 Group card NOBS GROUP NAME (in first four-column field followed by the name of group 1)
3 Data cards for group 1
4 Group card NOBS GROUP NAME (in first four-column field followed by the name of group 2)
5 Data cards for group 2
6 Group card NOBS GROUP NAME (in first four-column field followed by name of group 3)
7 Data cards for group 3

Program to Generate Within-Groups Frequency Counts and Probability Estimates

```
      DIMENSION NA(10,100),NX(100),NDX(100),NAME(20)
C     GENERATES WITHIN GROUPS PROBABILITY (WGP)
    1 FORMAT(2I4)
    2 FORMAT( I3,4X,10I5)
    3 FORMAT(16I4)
    4 FORMAT(' FREQUENCIES IN EACH CATEGORY FOR THE',I4,' VARIABLES')
    5 FORMAT(//,' PROPORTIONS IN EACH CATEGORY FOR THE',I4,' VARIABLES')
    8 FORMAT(I3,4X,10I6)
   14 FORMAT(I4,20A2)
   15 FORMAT(//20A2)
      READ(2,1) NGPS,NVAR
      DO 100 JK=1,NGPS
      READ(2,14) NOBS,(NAME(I),I=1,20)
      WRITE(3,15)(NAME(I),I=1,20)
      FNOBS=NOBS
      DO 20 J=1,NVAR
      DO 20 J2=1,10
   20 NA(J2,J)=0
      DO 29 I=1,NOBS
      READ(2,3)(NX(J),J=1,NVAR)
      DO 29 J=1,NVAR
      NDX(J)=NX(J)+1
      MDX = NX(J) + 1
   29 NA(MDX,J) = NA(MDX,J) + 1
      NSUM=0.0
      DO 101 J=25,34
  101 NSUM=NSUM+NDX(J)
      IF(NSUM-10)102,102,103
  102 NA(10,25)=NA(10,25)+1
      DO 104 J=25,34
  104 NA(1,J)=NA(1,J)-1
  103 CONTINUE
      NSUM=0.0
      DO 105 J=35,43
  105 NSUM=NSUM+NDX(J)
      IF(NSUM-9) 106,106,107
  106 NA(10,35)=NA(10,35)+1
      DO 108 J=35,43
  108 NA(1,J)=NA(1,J)-1
  107 CONTINUE
      NSUM=0.0
      DO 110 J=44,53
  110 NSUM=NSUM+NDX(J)
      IF(NSUM-10) 111,111,112
  111 NA(10,44)=NA(10,44)+1
      DO 113 J=44,53
  113 NA(1,J)=NA(1,J)-1
  112 CONTINUE
      NSUM=0.0
      DO 120 J=54,70
  120 NSUM=NSUM+NDX(J)
      IF(NSUM-17) 121,121,122
  121 NA(10,54)=NA(10,54)+1
      DO 123 J=54,70
  123 NA(1,J)=NA(1,J)-1
  122 CONTINUE
      NSUM=0.0
      DO 130 J=71,87
  130 NSUM=NSUM+NDX(J)
      IF(NSUM-17) 131,131,132
  131 NA(10,71)=NA(10,71)+1
      DO 133 J=71,87
  133 NA(1,J)=NA(1,J)-1
  132 CONTINUE
      WRITE(3,4) NVAR
      DO 40 J=1,NVAR
   40 WRITE(3,2)J,(NA(J2,J),J2=1,10)
      WRITE(3,5) NVAR
      DO 50 J=1,NVAR
      DO 49 J2=1,10
      FNA=NA(J2,J)
   49 NA(J2,J)=FNA*100.0/FNOBS
   50 WRITE(3,8)J,(NA(J2,J),J2=1,10)
  100 CONTINUE
      CALL EXIT
      END
```

The pattern-probability classification program BAYCP uses simple probability tables to calculate probabilities of multiattribute patterns and classifies each individual into the group in which his score pattern is most probable. The program is written to handle up to 12 classification groups and up to 20 single-digit qualitative variables. Differential base rates are considered by the program, but if they are unknown, they can be considered to be equal for all groups. The input sequence contains one control card specifying the number of classification alternatives among which individuals are to be assigned (maximum 12), the number of qualitative variables entering into the classification decision (maximum 20), and the total number of individuals to be classified in the particular computer run.

Input Sequence for Program BAYCP

1 One control card NGPS NVAR NOBS
2 Simple probability tables organized by group and variable within group, with simple probabilities of attribute states 0, 1, 2, . . . , 9 punched into 10 successive four-column fields
3 One card containing NGPS base rates punched into successive four-column fields
4 Individual data cards containing NVAR single-digit qualitative variables read in format 2; the code for the group to which individual is to be assigned is punched into a column of data card (format 5)

Program to Assign Individuals among *K* Groups Using Qualitative Data

```
C     BAYESIAN CONDITIONAL PROBABILITY PROGRAM
      DIMENSION D(12,10,20),BASE(12),PROB(12),A(20)
    1 FORMAT(10F4.0)
    2 FORMAT(8X,2F1.0,4X,2F1.0)
    3 FORMAT(20I4)
    5 FORMAT(44X,I1)
      READ(2,3) NGPS,NVAR,NOBS
      DO 20 J=1,NGPS
      DO 20 I=1,NVAR
      READ(2,1)(D(J,K,I),K=1,10)
   20 CONTINUE
      READ(2,1)(BASE(I),I=1,NGPS)
      DO 30 KK=1,NOBS
      READ(2,2)(A(K),K=1,NVAR)
      DO 31 J=1,NGPS
   31 PROB(J)=BASE(J)
      SUM=0.0
      DO 100 J=1,NGPS
      DO 99 K=1,NVAR
      I=A(K)+1.0
   99 PROB(J)=PROB(J)*D(J,I,K)
  100 SUM=SUM+PROB(J)
      DO 41 J=1,NGPS
      IF(SUM)39,39,40
   39 PROB(J)=0.0
      GO TO 41
   40 PROB(J)=PROB(J)/SUM
   41 CONTINUE
      K=1
      DO 60 J=2,NGPS
      IF(PROB(J)-PROB(K))60,60,59
   59 K=J
   60 CONTINUE
   30 WRITE(2,5)K
      CALL EXIT
      END
```

PART FIVE
MULTIPLE REGRESSION AND ANALYSIS OF VARIANCE

17 Multiple Linear Regression and Correlation

17.1 Introduction

Multiple regression and multiple correlation are methods used to study the relationship of a single quantitative dependent variable (criterion measure) to several independent variables (predictors). The independent variables are frequently called *predictor variables* because of the usefulness of multiple-regression methods for developing equations to predict the value of a future event, say college-grade point average, from knowledge of the scores for an individual on several independent variables such as grades in high school courses. As we shall see, the multiple-regression model can serve as the basis for testing a variety of hypotheses concerning the nature and strengths of effects of several independent variables on a quantitative dependent variable. The analysis of variance and analysis of covariance are special cases of general linear regression analysis and, in clinical and social science research, regression analysis has what are perhaps its most important applications in evaluation of the effects of a variety of variables which cannot be subjected to experimental control but which are nevertheless of primary importance. The principal distinction between clinical and social research vs. laboratory research is the impossibility or unfeasibility of subjecting the phenomena of interest to rigid experimental control. For example, in dealing with human lives and welfare, it is encumbent upon us to develop fully our ability to utilize available information without jeopardizing maximum therapeutic benefit to the patients involved. In many situations, general regression methods provide a realistic alternative to experimental intervention.

17.2 Definitions and Derivations

At this point in the book, many readers will anticipate how the statistician approaches the study of the relationship of a single dependent variable to multiple independent variables. The several independent variables are combined into a single optimally weighted linear composite, and the problem reduces to that of studying the relationship of one quantitative variable to another. The several optimal weighting coefficients are called *partial-regression coefficients* or simply multiple-regression coefficients. The correlation coefficient relating a quantitative dependent variable to the optimal weighted composite is called a *multiple-correlation coefficient* and is usually represented by the symbol R, but since we have used this notation frequently in representing a complete correlation matrix, the notation Ψ will be adopted for the multiple-correlation coefficient in this chapter.

It is well to distinguish at the outset between *standard-score* multiple-regression equations and *raw-score* multiple-regression equations. A standard-score regression equation involves a set of *standard partial-regression coefficients* that can be applied to z-score values of the independent variables to estimate the z-score value of the dependent variable

$$\tilde{y} = \beta_1 z_1 + \beta_2 z_2 + \cdots + \beta_p z_p \tag{17.1}$$

where $z_1, z_2, \ldots, z_p$ are standard scores with zero means and unit variances and y is also a unit-variance standard score. The estimated value for the dependent variable $\tilde{y}$ is thus defined as a weighted combination of the independent, or predictor, variables. For purposes of the tests of significance to be discussed in later sections, it is assumed that the z_i are measured or defined without error, while the dependent variable y is a quantitative measure having a gaussian normal error distribution. In practice, multiple-regression methods have been found by applied researchers to be quite useful in prediction problems where the independent predictor variables are recognized to contain a certain degree of error.

The problem in multiple regression is to define a set of weighting coefficients $\beta_1, \beta_2, \ldots, \beta_p$ such that the average discrepancy of the estimated value $\tilde{y}$ about the actual standard-score value y is a minimum. The *least-squares* solution defines a set of weighting coefficients such that the sum of squared deviations of estimated scores about the actual observed scores for the dependent variable is a minimum. The criterion function to be minimized is

$$f(\beta)_i = \frac{1}{n}\sum (y - \tilde{y})^2 \tag{17.2}$$

and

$$f(\beta_i) = \frac{1}{n} \sum (y - \beta_1 z_1 - \beta_2 z_2 - \cdots - \beta_p z_p)^2 \tag{17.3}$$

Any student facile in elementary calculus can expand the squared term and take partial derivatives with respect to the β_i to obtain a system of normal equations that can be solved simultaneously to obtain optimal values for the weighting coefficients. The resulting equations are of the form

$$\begin{aligned} \beta_1 1.0 + \beta_2 r_{12} + \cdots + \beta_p r_{1p} &= r_{1y} \\ \beta_1 r_{21} + \beta_2 1.0 + \cdots + \beta_p r_{2p} &= r_{2y} \\ \cdots\cdots\cdots\cdots\cdots\cdots\cdots\cdots \\ \beta_1 r_{p1} + \beta_2 r_{p2} + \cdots + \beta_p 1.0 &= r_{py} \end{aligned} \tag{17.4}$$

in which the r_{ij} are product-moment correlation coefficients between the ith and jth independent variables and the r_{iy} on the right are correlations between the independent and dependent variables. This same system of equations can be written in matrix form as

$$R\boldsymbol{\beta} = \mathbf{v} \tag{17.5}$$

where R is the matrix of product-moment correlations among the independent variables, $\mathbf{v}$ is the vector of correlations with the criterion or dependent variable, and $\boldsymbol{\beta}$ is the vector of unknown weighting coefficients. Multiplying on both sides by the inverse of R, we obtain the solution equation in which elements of R^{-1} and $\mathbf{v}$ can be estimated from the data

$$\boldsymbol{\beta} = R^{-1}\mathbf{v} \tag{17.6}$$

Although it is important from a strict statistical point of view to distinguish between true population parameters and sample estimates, we propose to use a consistent notation throughout this chapter. The basic model, such as defined in Eq. (17.5), can be considered to involve true population parameters. In practical problems, one does not actually have population means, standard deviations, and correlations with which to work. Thus, the equations are solved by substituting best available unbiased estimates for the true population parameters. For example, in the solution of a practical problem, the matrix R^{-1} of Eq. (17.6) can be taken as the inverse of the matrix of intercorrelations among the independent variables as calculated from sample data, and the vector $\mathbf{v}$ will contain the product-moment correlations of the dependent variable with the independent variables. Where sample means and correlations have been substituted for population parameters, the regression equations that result are recognized to be only "best estimates" of the true population values. In the remainder of this chapter, we shall assume that sample statistics can be substituted for the population parameters they estimate without employing different notation to emphasize the fact.

In expanded form, the solution for optimal standard-score regression equations can be written as a set of linear equations involving elements of the matrix inverse R^{-1} and correlation coefficients relating the criterion variable to the several independent variables. Let r^{ij} represent the element in the ith row and jth column of R^{-1}

$$\begin{aligned} r_{1y}r^{11} + r_{2y}r^{12} + \cdots + r_{py}r^{1p} &= \beta_1 \\ r_{1y}r^{21} + r_{2y}r^{22} + \cdots + r_{py}r^{2p} &= \beta_2 \\ \cdots\cdots\cdots\cdots\cdots\cdots\cdots \\ r_{1y}r^{p1} + r_{2y}r^{p2} + \cdots + r_{py}r^{pp} &= \beta_p \end{aligned} \tag{17.7}$$

The *multiple-correlation coefficient* Ψ is the correlation between the best linear composite of the independent variables and the single quantitative dependent variable. It is an index, ranging from 0 to +1, of the degree of relationship between the dependent variable and the set of independent variables. The square of the multiple-correlation coefficient Ψ^2 represents the proportion of the total variation in the dependent variable that can be accounted for by linear regression on the several independent variables. The general matrix formula for the square of the multiple-correlation coefficient is

$$\mathbf{v}'R^{-1}\mathbf{v} = \Psi^2 \tag{17.8}$$

Recognizing that the standard partial-regression coefficients represent the product $\boldsymbol{\beta} = R^{-1}\mathbf{v}$, the proportion of total variance of the dependent variable that can be accounted for by linear regression on the set of p independent variables can be obtained as a simple vector product

$$\Psi^2 = \mathbf{v}'\boldsymbol{\beta} = r_{1y}\beta_1 + r_{2y}\beta_2 + \cdots + r_{py}\beta_p \tag{17.9}$$

where $r_{1y}, r_{2y}, \ldots, r_{py}$ are correlations of the dependent variable with the p independent variables.

The coefficient Ψ^2 can be used to partition the total sum of squares for an analysis-of-variance F test of the significance of regression, as will be developed in the following section.

17.3 Raw-score Multiple-regression Equations and Multiple Correlation

In practice, the estimation or prediction of raw-score values for the criterion variable is more often desired than the estimation of standard-deviate z-score values. While it is possible to compute raw-score partial-regression coefficients directly using variances and covariances, rather than correlations, it is also a simple matter to transform standard-score equations to raw-score form if the means and variances for the several variables are known or can be estimated from sample data. The z_i values in Eq. (17.1) are the deviations of raw scores x_i about their respective means $\bar{x}_i$ divided by their respective standard deviations σ_i

$$z_i = \frac{x_i - \bar{x}_i}{\sigma_i} \tag{17.10}$$

Similarly, the standard-score dependent variable y is calculated using the mean and standard deviation for that variable.

Substituting raw-score values into Eq. (17.1) gives the scalar equation

$$\frac{\tilde{y}-\bar{y}}{\sigma_y} = \beta_1 \frac{x_1 - \bar{x}_1}{\sigma_1} + \beta_2 \frac{x_2 - \bar{x}_2}{\sigma_2} + \cdots + \beta_p \frac{x_p - \bar{x}_p}{\sigma_p} \tag{17.11}$$

Thence,

$$\tilde{y} - \bar{y} = \beta_1 \frac{\sigma_y}{\sigma_1}(x_1 - \bar{x}_1) + \cdots + \beta_p \frac{\sigma_y}{\sigma_p}(x_p - \bar{x}_p)$$

and

$$\tilde{y} = b_1 x_1 + b_2 x_2 + \cdots + b_p x_p + k_0 \tag{17.12}$$

where

$$b_i = \beta_i \frac{\sigma_y}{\sigma_i} \quad \text{and} \quad k_0 = \bar{y} - (b_1\bar{x}_1 + b_2\bar{x}_2 + \cdots + b_p\bar{x}_p)$$

It will be recognized that the regression constant k_0 is the difference between the mean value of the weighted function as calculated from the $\bar{x}$ means for the independent variables and the mean $\bar{y}$ for the dependent variable. The raw-score partial-regression coefficients in Eq. (17.12) are equal to the standard partial-regression coefficients of Eq. (17.1) multiplied by the ratio of standard deviations for the dependent and independent variables. For most of the calculations described in this and the succeeding chapter, we shall use the standard-score formulas. Because the conversion to raw-score equations is so simple and self-evident, there seems to be little value in carrying through a complete set of essentially duplicate derivations and computations.

It should be noted in this section, however, that raw-score regression equations can be calculated directly by working with *covariances* rather than correlations. The raw-score partial-regression coefficients of Eq. (17.12) can be computed directly using the inverse of the matrix of covariances between the independent variables and the vector of covariances with the criterion variable

$$\mathbf{b} = C^{-1}\mathbf{u} \tag{17.13}$$

where C^{-1} is the inverse of the matrix of covariances among independent variables and $\mathbf{u}$ is the vector of covariances with the criterion or dependent variable.

The proportion of the total sum of squares for the dependent variable that can be accounted for by regression on the several independent variables can be computed from the same inverse covariance matrix or from the raw-score partial-regression coefficients

$$\Psi'^2 = \frac{\mathbf{u}'C^{-1}\mathbf{u}}{\sigma_y^{\,2}} = \frac{\mathbf{u}'\mathbf{b}}{\sigma_y^{\,2}} \tag{17.14}$$

where $\sigma_y^{\,2}$ is the variance of the dependent variable.

17.4 Example of Multiple-regression Calculations

Suppose that an investigator wants to know whether the level of total psychiatric pathology after 4 weeks of treatment can be predicted with meaningful degree of accuracy from knowledge of pretreatment symptom ratings. Pre- and posttreatment BPRS symptom ratings for the sample of 670 psychiatric inpatients were used to examine this question. The matrix of correlations among the four higher-order factor scores derived from pretreatment ratings and the vector of correlations of posttreatment total pathology scores with the pretreatment factor scores are presented in Table 17.1. The inverse of the 4×4 correlation matrix among the independent variables is also presented, as are means and standard deviations for all variables.

Optimum *standard* partial-regression coefficients [Eq. (17.1)] for use in predicting posttreatment total pathology from the pretreatment factor scores were computed using the elements of the matrix inverse and the correlations

Table 17.1 Data Used in Example of Multiple-regression Calculations

INTERCORRELATIONS AMONG PRETREATMENT SCORES ON FOUR HIGHER-ORDER SYMPTOM FACTORS AND VECTOR OF CORRELATIONS WITH POSTTREATMENT TOTAL PATHOLOGY

TD	*W-R*	*H-S*	*AD*	*Posttreat total pathology*
1.0000	.1892	.4551	−.0572	.2967
.1892	1.0000	.0115	.0938	.1479
.4551	.0115	1.0000	−.1231	.3473
−.0572	.0938	−.1231	1.0000	.0652

MEANS AND STANDARD DEVIATIONS

	Means	*Standard deviations*
Thinking disturbance	3.7507	3.9434
Withdrawal-retardation	3.6299	3.5720
Hostile-suspiciousness	6.5612	3.8575
Anxious depression	8.5209	3.4236
Posttreatment total pathology	20.0194	10.3160

INVERSE OF CORRELATION MATRIX AMONG FOUR HIGHER-ORDER SYMPTOM FACTORS

$$\begin{bmatrix} 1.3181 & -.2450 & -.5940 & .0253 \\ -.2450 & 1.0549 & .0867 & -.1023 \\ -.5940 & .0867 & 1.2836 & .1159 \\ .0253 & -.1023 & .1159 & 1.0253 \end{bmatrix}$$

with the criterion [Eq. (17.6) or (17.7)]

$$\beta_1 = .2967(1.3181) + .1479(-.2450) + \cdots + .0652(.0253) = .1502$$

$$\beta_2 = .2967(-.2450) + .1479(1.0549) + \cdots + .0652(-.1023) = .1068$$

$$\beta_3 = .2967(-.5940) + .1479(.0867) + \cdots + .0652(.1159) = .2899$$

$$\beta_4 = .2967(.0253) + .1479(-.1023) + \cdots + .0652(1.0253) = .0995$$

The raw-score multiple-regression equation was defined as follows using Eq. (17.12):

$$\tilde{y} = .3929x_1 + .3084x_2 + .7753x_3 + .2998x_4 - 9.7848$$

The square of the multiple correlation between posttreatment total pathology and pretreatment factor scores can be computed by pre- and postmultiplying the inverse of the correlation matrix by the vector of correlation with the dependent variable. Since the optimum regression coefficients were obtained by postmultiplying R^{-1} by the vector of correlations with the criterion, one can obtain Ψ^2 simply as the weighted sum of correlations with the dependent variable

$$\Psi^2 = \mathbf{v}'R^{-1}\mathbf{v} = \mathbf{v}'\mathbf{a} \qquad (17.15)$$

$$\Psi^2 = .1502(.2967) + .1068(.1479) + .2899(.3473) + .0995(.0652) = .1675$$

and thus the multiple correlation $\Psi = \sqrt{.1675} = .4093$.

The square of the multiple-correlation coefficient Ψ^2 is the percentage of variance of the dependent variable that can be accounted for by linear regression on the composite of several independent variables. It is often referred to as the *coefficient of determination.* The quantity $1 - \Psi^2$ (coefficient of nondetermination) is the percentage of variance of the dependent variable that cannot be accounted for by regression on the several independent variables. This means that Ψ^2 and $1 - \Psi^2$ can be used to partition the total sum of squares SS_T into

Source	*SS*	*df*
Regression	$\Psi^2 SS_T$	p
Deviations from regression	$(1 - \Psi^2)SS_T$	$n - p - 1$
Total	SS_T	$n - 1$

regression and residual error components. Assuming that the residual errors (deviations of observed scores about the predicted values) are independent, normally distributed, and equal in variance for all combinations of the independent variables, the analysis-of-variance F ratio can be used to test the null hypothesis that there is no linear relationship between the dependent variable and the several independent variables

$$F = \frac{\Psi^2 SS_T/p}{[(1 - \Psi^2)SS_T]/n - p - 1} \tag{17.16}$$

Since the total sum of squares appears in both numerator and denominator, it can be eliminated from the calculation entirely

$$F = \frac{\Psi^2(n - p - 1)}{(1 - \Psi^2)p} \tag{17.17}$$

where n is the total number of subjects, p is the number of independent variables, and F is an F ratio with p and $n - p - 1$ degrees of freedom. If the F ratio is statistically significant, one can conclude that the dependent variable is related to some extent to one or more of the independent variables.

17.5 Tests of Significance for Independent Contributions of Individual Predictor Variables

Given a standard multiple-regression equation of the type shown in Eq. (17.1), it is frequently of interest to know whether each independent variable contributes significantly toward prediction of the criterion. The variance of an individual standard partial-regression coefficient β_j can be expressed as the product of the residual variance $(1 - \Psi^2)/(n - p - 1)$ and the diagonal element r^{jj} of the inverse of the correlation matrix R^{-1}. The standard error of an individual standard-score regression coefficient can be computed as follows:

$$\mathrm{SE}_{\beta_j} = \sqrt{\frac{r^{jj}(1 - \Psi^2)}{n - p - 1}} \tag{17.18}$$

The statistic

$$t = \frac{\beta_j}{\mathrm{SE}_{\beta_j}} \tag{17.19}$$

is distributed as Student's t with $n - p - 1$ degrees of freedom.

If the regression equation is in raw-score form [Eq. (17.12) or (17.13)], the partial-regression coefficients can be tested in a similar manner. The standard error of the raw-score coefficient b_j can be expressed as a product of the residual variance $(1 - \Psi^2)\mathrm{SS}_T/(n - p - 1)$ and the corresponding diagonal element σ^{jj} of the inverse of the *covariance* matrix C^{-1}

$$\mathrm{SE}_{b_j} = \sqrt{\frac{\sigma^{jj}(1 - \Psi^2)\mathrm{SS}_T}{n - p - 1}} \tag{17.20}$$

It will be noted that the total corrected sum of squares SS_T for the dependent variable has to be inserted because the variables have not been standardized.

The statistic

$$t = \frac{b_j}{\mathrm{SE}_{b_j}} \tag{17.21}$$

is distributed as Student's t with $n - p - 1$ degrees of freedom.

An alternative test of significance which makes use of the F-ratio statistic can be calculated using the diagonal elements of the inverse of the matrix of covariances among the predictor variables. The sum of squares for variance accounted for by each predictor variable, over and above that accounted for by the others, can be calculated by dividing the *square* of the raw-score partial-regression coefficient by the associated diagonal element from the matrix inverse

$$\mathrm{SS}_{b_1} = \frac{b_1^{\,2}}{\sigma^{11}}$$

$$\mathrm{SS}_{b_2} = \frac{b_2^{\,2}}{\sigma^{22}} \tag{17.22}$$

. . . .

$$\mathrm{SS}_{b_p} = \frac{b_p^{\,2}}{\sigma^{pp}}$$

Since each partial-regression coefficient represents only one degree of freedom, the appropriate F ratio for testing the independent contribution of each variable is obtained as follows:

$$F_{b_1} = \frac{\mathrm{SS}_{b_1}(n-p-1)}{(1-\Psi^2)\mathrm{SS}_T}$$

$$F_{b_2} = \frac{\mathrm{SS}_{b_2}(n-p-1)}{(1-\Psi^2)\mathrm{SS}_T} \tag{17.23}$$

. . . .

$$F_{b_p} = \frac{\mathrm{SS}_{b_p}(n-p-1)}{(1-\Psi^2)\mathrm{SS}_T}$$

Finally, one can test the significance of the additional contribution of each independent variable by calculating Ψ^2 with and without the particular variable included. Let $\Psi_p^{\,2}$ be the square of the multiple correlation calculated from Eq. (17.9) or (17.14) with all p variables included. Let Ψ^2_{p-j} be the square of the multiple correlation calculated with all variables *except* the jth one included. Then

$$\mathrm{SS}_{b_j} = (\Psi_p^{\,2} - \Psi^2_{p-j})\mathrm{SS}_T \tag{17.24}$$

where SS_T is the total corrected sum of squares for the dependent variable and SS_{b_j} is the sum of squares for the independent contribution of the jth variable over and above all others. The appropriate F ratio for testing the independent effect of each variable is then

$$F_{b_j} = \frac{(\Psi_p^2 - \Psi_{p-j}^2)SS_T(n - p - 1)}{(1 - \Psi_p^2)SS_T} \tag{17.25}$$

Since the total sum of squares appears in both numerator and denominator, it can be eliminated by cancellation

$$F_{b_j} = \frac{(\Psi_p^2 - \Psi_{p-j}^2)(n - p - 1)}{1 - \Psi_p^2} \tag{17.26}$$

17.6 Testing the Significance of an a Priori Regression Equation

There is frequently a need in clinical research to test the hypothesis that the best predictive equation differs significantly from some prespecified a priori function. For example, it may be of interest to know whether the best predictive function differs significantly from a simple sum of the independent variables. Do different pretreatment symptoms differ in value for predicting treatment outcome? This type of question can be answered by testing the significance of the difference in predictive efficacy of the best linear equation relative to the a priori equal weighting.

Let $\tilde{y}_\alpha = \boldsymbol{\alpha}'\mathbf{z}$ be the a priori function and let $\tilde{y}_a = \mathbf{a}'\mathbf{z}$ be the best-fit least-squares regression equation calculated using the same p predictor variables. The square of the correlation between the a priori function and the criterion is

$$\Psi_\alpha^2 = \frac{(\boldsymbol{\alpha}'\mathbf{v})^2}{\boldsymbol{\alpha}'R\boldsymbol{\alpha}} \tag{17.27}$$

where $\mathbf{v}$ is a vector containing correlations between the dependent variable and the several independent variables and R is the matrix of intercorrelations among the p independent variables.

The square of the multiple correlation between the dependent variable and the best linear combination of the p independent variables is

$$\Psi_a^2 = \mathbf{v}'R^{-1}\mathbf{v} = \mathbf{a}'R\mathbf{a} \tag{17.28}$$

The significance of the difference between the a priori function and the best linear combination can be tested using an F ratio with $p - 1$ and $n - p - 1$ degrees of freedom

$$F = \frac{(\Psi_a^2 - \Psi_\alpha^2)(n - p - 1)}{(1 - \Psi_a^2)(p - 1)} \tag{17.29}$$

To illustrate the application of this important test, the reader is encouraged to compute Ψ_α^2 from the data given in Table 17.1 and test the hypothesis that the BPRS posttreatment total pathology scores can be predicted best as a simple unweighted sum of the four pretreatment factor scores, using Eq. (17.29). Rejection of this hypothesis would suggest that different symptom factors have different prognostic value.

17.7 General Linear Regression with Qualitative Independent Variables

There is rather widespread misunderstanding concerning the appropriateness of general multiple-regression methods for use where the independent variables are qualitative or categorical in nature. In multiple-regression analyses, it is assumed that the *dependent* variable is a quantitative measure with normal distribution and equal variance for all combinations of the independent variables. It is not assumed that the independent variables are normally distributed or even that they are quantitative measurements. In fact, from a strict theoretical point of view, many of the tests of significance in regression analysis are based on the assumption that the independent variables assume only fixed values and are measured without error. It is perfectly acceptable to use qualitative or categorical variables as independent variates.

Some precautions should be mentioned, however. If a multicategory variable involves no logical or empirically derived ordinal scale of measurement, it would be meaningless to assign arbitrary (nominal) numbers to the various categories and to treat the resulting scores as if they represented a single quantitative variable. For example, six categories of marital status are *different;* yet the mere assigning of numbers 1, 2, . . . , 6 to the various categories does not make a meaningful quantitative variable that can be used in a linear prediction model. There are actually five degrees of freedom available, so that the solution involves expanding the "single" multicategory variable into several dummy variables, say, married (yes or no), single (yes or no), and so on. No information is lost by leaving out the sixth category because its value is known, given knowledge of status on the first five. More will be said of this problem in the discussion of the general linear regression approach to the analysis of variance. The purpose here is merely to emphasize that categorical data can be used quite appropriately in the development of multiple linear regression functions.

Dichotomous variables occupy a somewhat special place in the continuum between quantitative and categorical predictors. Since there are only two possible categories, any arbitrary assignment of score values, say 1 and 2, will result in an ordinal scale. There is no problem of differences in score intervals since only one interval is involved. As we shall see in a later section, the assignment of score values $+1$ and -1 has certain advantages from point of view of interpretation of regression coefficients, but that is not a matter of concern here. It is quite legitimate to include in a single multiple-regression equation a combination of quantitative and dichotomous independent variables.

17.8 The Simple Analysis of Variance as a Linear Regression Problem

It will be assumed here that the reader has a general familiarity with traditional methods of the analysis of variance for partitioning total variation into within-groups and between-groups sums of squares in the calculation of F ratios for testing the null hypothesis that means for several groups are equal. In this section, it will be demonstrated that exactly this same type of analysis can be accomplished by general multiple-regression and correlation techniques. The key to use of regression methods for an analysis of variance concerned with group differences is the development of a *design matrix* using dummy variates to represent group membership. The total sum of squared deviations for the dependent variable can then be partitioned into a sum of squares due to group differences and a residual sum of squares due to variation within groups. These sums of squares are precisely proportional to the within-groups and between-groups sums of squares that could be calculated for the same data by traditional hand-calculation methods.

For the simple one-way-classification randomized design, there are $k - 1$ degrees of freedom associated with the between-treatments sum of squares. A vector containing $k - 1$ dummy variates scored 1, 0, and -1 is generated for each individual with the dependent variable y entered at the right. If the individual belongs to category 1, his score vector contains 1 in the first position and zeros in positions $2, 3, \ldots, k - 1$. The last element is his score on the dependent variable, say the posttreatment anxious-depression factor score. In general, if the individual belongs in one of the first $k - 1$ categories, he is assigned a score of 1 for the element representing the category to which he belongs and zero for the others. If the individual belongs in the last category, i.e., the kth treatment group, he is assigned a score of -1 for all $k - 1$ dummy variates. The design matrix thus consists of zero-one dummy variates with a quantitative dependent variable entered at the right.

Having constructed a design matrix as indicated, with the dependent variable entered on the right, the next step is to compute the regression of the dependent variable on the $k - 1$ dummy variates. Let Ψ_b^2 represent the squared multiple correlation obtained from regression of the criterion measure on the $k - 1$ dummy variates derived from group membership

$$\Psi_b^2 = \frac{\mathbf{u}'C^{-1}\mathbf{u}}{\sigma_y^2} \tag{17.30}$$

where C^{-1} is the inverse of the matrix of covariances among the $k - 1$ dummy variates and $\mathbf{u}$ is the vector of covariances between the dependent variate and the $k - 1$ dummy variates. The sum of squares due to mean differences among the k treatment groups is

$$\mathrm{SS}_B = \Psi_b^2 \mathrm{SS}_T \tag{17.31}$$

where SS_T is the total mean-corrected sum of squares for the dependent variable. The sum of squares for residual within-groups error variation is

$$\mathrm{SS}_W = (1 - \Psi_b^2)\mathrm{SS}_T \tag{17.32}$$

The summary of the analysis of variance for the simple one-way randomized design is as follows:

Source	*SS*	*df*
Between groups	$\Psi_b^2 SS_T$	$k - 1$
Within groups	$(1 - \Psi_b^2)SS_T$	$n - k$
Total	SS_T	$n - 1$

For interpretation of the analysis of variance, it is necessary to examine the nature of differences in group means. Using the general linear regression approach, the *analysis-of-variance parameters* are estimated by the *partial-regression coefficients* in the multiple-regression equation. Each of the $k - 1$ partial-regression coefficients (raw-score beta weights) represents the deviation of a group mean about the grand mean. Since the analysis-of-variance parameters are scaled according to the restriction that they sum to zero, the parameter for the kth group is equal to *minus* the algebraic sum of the other $k - 1$ parameters. Addition of the grand-mean to the group-mean deviation scores results in the estimates of treatment means.

To illustrate the calculation of simple analysis of variance by regression methods, consider the data in Table 17.2 representing beta IQ test performance scores for randomly selected patients with clinical diagnoses of depression, schizophrenia, and organic brain syndromes. The 30 cases in each group were randomly selected from among a total of 460 cases studied by Dudley, Williams, and Overall (1970) at the Rusk State Hospital. The mean scores for the group are shown at the bottom of the columns. The problem for analysis of variance is whether these means differ significantly relative to the variability within groups. Application of the regression method will also permit the examination of how the regression coefficients can be used to estimate group means.

The design matrix constructed from the data in Table 17.2 is illustrated for the first four subjects in each group in Table 17.3. The reader should confirm that he could construct the complete design matrix for the set of data. The intercorrelation matrix between independent and dependent variables is shown at the bottom of Table 17.3.

The multiple correlation representing the regression of the dependent variable on the two dummy variates was calculated to be $\Psi = .45508$, and the total corrected sum of squares for the dependent variable was $SS_T = 27{,}307.82$. Ψ^2 could have been computed in general matrix form

$$\Psi_b^2 = \mathbf{v}'R^{-1}\mathbf{v}$$

where R^{-1} is the inverse of the matrix of intercorrelations among the $k - 1$ independent variates. In this particular problem, with only two independent variates, the multiple correlation Ψ can be calculated from the normal equations

Table 17.2 Age, Sex, Education, and Raw Beta IQ Total Score for Patients in Three Diagnostic Groups

Depression				*Schizophrenia*				*Organic brain syndrome*			
Age	*Sex*	*Education*	*IQ*	*Age*	*Sex*	*Education*	*IQ*	*Age*	*Sex*	*Education*	*IQ*
4	1	2	73	4	2	2	52	1	1	2	26
3	2	2	56	4	2	1	41	5	1	2	59
6	2	2	44	4	1	1	10	2	2	2	64
4	2	2	67	3	2	2	62	6	1	1	35
4	2	2	55	3	2	2	64	3	2	2	06
5	2	2	44	5	2	2	24	3	2	1	19
2	2	2	51	3	1	2	27	2	2	1	67
2	2	2	73	6	1	3	41	3	2	3	59
2	2	2	64	5	2	1	16	2	2	2	35
2	2	2	51	4	2	3	47	4	2	1	35
3	2	2	62	3	2	2	36	5	1	2	72
2	2	1	49	3	2	2	63	2	2	1	44
3	1	3	63	3	1	2	59	6	2	1	40
4	2	2	54	3	2	2	31	6	1	1	07
3	2	3	66	3	1	2	43	6	1	2	56
3	1	2	73	2	1	2	49	3	2	2	32
4	2	2	73	2	1	1	12	5	1	2	24
2	2	3	64	3	2	2	44	3	2	2	63
4	2	2	45	5	2	2	39	5	1	1	43
2	2	2	65	3	2	2	47	5	1	1	44
4	2	1	55	3	2	3	46	6	1	1	54
2	2	2	62	4	1	2	59	6	1	1	48
5	2	2	61	2	2	3	60	3	1	3	58
5	1	1	37	2	1	2	28	3	1	1	57
3	2	1	44	3	2	1	19	1	2	1	12
4	1	2	57	2	1	2	61	5	2	2	22
5	2	2	56	4	2	2	14	1	2	3	54
4	2	2	50	5	2	2	16	2	2	1	33
5	2	3	44	4	1	2	49	1	2	2	37
5	2	2	49	5	2	1	10	5	1	1	47
Means			56.9				39.0				41.7

by simple elimination and substitution.

$$1.0000\beta_1 + .5000\beta_2 = .3575$$
$$.5000\beta_1 + 1.0000\beta_2 = -.0652$$

$$1.0000\beta_1 + .5000\beta_2 = .3575$$
$$-1.0000\beta_1 - 2.0000\beta_2 = .1304$$

and

$$\beta_2 = -.3252$$
$$\beta_1 = .5201$$

Table 17.3 Construction of Design Matrix for Simple One-way Analysis of Variance

DESIGN MATRIX

Depressed	*Schizophrenic*	*Raw IQ*
1	0	73
1	0	56
1	0	44
1	0	67
0	1	52
0	1	41
0	1	10
0	1	62
−1	−1	26
−1	−1	59
−1	−1	64
−1	−1	35
−1	−1	37
−1	−1	47

INTERCORRELATIONS AMONG GROUP MEMBERSHIP VARIATES AND DEPENDENT VARIATE

$$\begin{bmatrix} 1.0000 & .5000 & .3575 \\ & 1.0000 & -.0652 \\ & & 1.0000 \end{bmatrix}$$

Using Eq. (17.9), we obtain

$$\Psi_b^2 = .3575(.5201) + -.3252(-.0652) = .2071$$

The resulting analysis of variance for differences in mean beta IQ total score performance for patients from the three diagnostic groups is summarized as follows:

Source	*SS*	*df*	*Mean square*	*F*
Diagnoses	.2071(27,307.82)	2	2,827.72	11.36
Within groups	.7929(27,307.82)	87	248.88	
Total	27,307.82	89	306.83	

The standard-score regression equation relating the independent variable to the two group-membership variates is

$$\tilde{y} = .5201z_1 - .3252z_2$$

The variances for the two dummy variables were found to be $\sigma_{a_1}{}^2 = .674157$ and $\sigma_{a_2}{}^2 = .674157$. The total variance for the dependent variable was $\sigma_y{}^2 = 313.42$. Using Eq. (17.12), the standard regression equation can be converted to raw-score form

$$\tilde{y} = 11.0333a_1 - 6.9000a_2 + 45.8667$$

The raw-score partial-regression coefficients provide estimates of the deviation of group means for the first $k - 1$ groups about the grand mean for all groups. The deviation of the kth group mean about the grand mean is equal to *minus* the sum of the other $k - 1$ mean deviations. Thus, the regression equation provides the following estimates of group means in connection with the analysis of variance:

$$\bar{x}_{\text{dep}} = 11.0333 + 45.8667$$

$$\bar{x}_{\text{schizo}} = -6.9000 + 45.8667$$

$$\bar{x}_{\text{organic}} = -4.1333 + 45.8667$$

The reader should verify that the group means and the analysis of variance obtained by the regression method are equivalent to those obtained by other methods. The practical importance of the regression solution, of course, is not obvious in such a simple example. The means can be obtained more easily by adding up the data columns, and the analysis of variance can be accomplished more easily *in this instance* by adding up the sums of squares. In more complex cases, the regression approach has very great practical advantage. The intention here has been merely to illustrate that analysis of variance can be conceived as a multiple-regression problem. The reader should be warned that in order for the partial-regression coefficients to provide direct estimates of the group means, the dummy variates should be generated as illustrated. Assign the value of 1 in the column representing the group to which the individual belongs and zeros elsewhere, unless the individual belongs to the kth group. If the individual belongs to the kth group, not represented by a separate column in the matrix of dummy variates, assign scores of -1 for all dummy variates representing the other $k - 1$ classes.

17.9 Simple Analysis of Covariance by General Regression Method

There are situations where it is known that some extrinsic variable, such as the age of the patient, is likely to influence treatment effects. One way of handling this type of problem is to create a carefully stratified cross-classification analysis of variance design, classifying patients into different age levels to control for this variable. Since age is a quantitative variable, another way the problem can be handled is to use regression methods to partial out the effects of age. This approach has generally been called the *analysis of covariance.*

Before discussing how the simple analysis of covariance can be accomplished using general multiple-regression methods, one further parallel with the two-way cross-classification analysis of variance should be mentioned. A reason that is frequently given for preferring a *treatments-by-levels* type of analysis of variance over the analysis of covariance is that it provides a ready test of the significance of *interaction* of treatments with levels of the control variable. In the next chapter it will be seen how a complex cross-classification analysis of variance can be accomplished using a design matrix constructed to represent treatment classifications with interactions entered as cross products of the treatment variates. At that point, the analogy between the test of interaction effects in the analysis of variance and the test for *heterogeneity of regression* in the analysis of covariance will become obvious.

The test for equality of regression slopes (say regression of the dependent variable on age) within the several treatment groups is a test of the hypothesis that the influence of age on the dependent variable is the same in all treatment groups. This is the regression analogy of the interaction hypothesis in the analysis of variance. If heterogeneous regression is evident, one usually stops and examines the separate group regressions rather than proceeding directly to evaluation of treatment differences. In the corresponding fixed-effects analysis of variance, a significant interaction is grounds for examining simple effects rather than main effects.

In order to use the general multiple-regression method to accomplish simple one-way-classification analysis of covariance, it is necessary to construct a design matrix indicating group membership (1, 0, and −1), the scores on the covariate, *plus the cross product of covariate scores with each dummy variate*, and finally the score on the dependent variable. Using the data presented in Table 17.2, a design matrix was developed assigning scores of 1 for group membership if an individual belonged in one of the first $k - 1 = 2$ diagnostic groups and scores of −1 if he was in the third group. Next the age of the individual (scored in decades 1, 2, 3, . . . , 6) was entered as a quantitative variable, followed by cross products of age and group-membership variates. Finally, the beta IQ score was entered on the right as the dependent variable. The columns of the design matrix are shown for the first four individuals in each diagnostic group in Table 17.4. The student should be able to fill out the remainder of the design matrix and calculate the multiple-regression and correlation coefficients as indicated.

The analysis of covariance is accomplished as follows. First, the regression of the dependent variable on the total of $2(k - 1) + 1 = 2k - 1$ variates is computed. Let $\Psi'^2_{\alpha,\beta,\alpha\beta}$ represent the total regression on all dummy variates and covariates. Next, the columns headed a_1x and a_2x, representing products of group-membership coefficients with the covariate, are eliminated from the model. A new squared multiple-correlation coefficient $\Psi'^2_{\alpha,\beta}$ is calculated representing the regression of the dependent variable on the group-membership vectors and the covariate. Finally, the regressions on the group-membership vectors alone Ψ'^2_{α} and on the covariate alone Ψ'^2_{β} are calculated. From these several squared multiple-correlation coefficients and the total sum of squared deviations about

Table 17.4 Construction of Design Matrix for Simple One-way Analysis of Covariance with Test for Heterogeneity of Regression†

a_1	a_2	x	a_1x	a_2x	y
1	0	4	4	0	73
1	0	3	3	0	56
1	0	6	6	0	44
1	0	4	4	0	67
0	1	4	0	4	52
0	1	4	0	4	41
0	1	4	0	4	10
0	1	3	0	3	62
−1	−1	1	−1	−1	26
−1	−1	5	−5	−5	59
−1	−1	2	−2	−2	64
−1	−1	6	−6	−6	35
−1	−1	1	−1	−1	37
−1	−1	5	−5	−5	47

† a_1 = depressive.
a_2 = schizophrenic.
x = age.
a_1x = depressive by age.
a_2x = schizophrenic by age.
y = raw beta IQ.

the grand mean for the dependent variable the complete analysis of covariance can be accomplished.

Source	*SS*	*df*
Difference in group regression	$(\Psi'^2_{\alpha,\beta,\alpha\beta} - \Psi'^2_{\alpha,\beta})\text{SS}_T$	$k - 1$
Adjusted group means	$(\Psi'^2_{\alpha,\beta} - \Psi'^2_{\beta})\text{SS}_T$	$k - 1$
Covariance adjustment	$(\Psi'^2_{\alpha,\beta} - \Psi'^2_{\alpha})\text{SS}_T$	1
Residual error	$(1 - \Psi'^2_{\alpha,\beta,\alpha\beta})\text{SS}_T$	$n - 2k$
Total	SS_T	$n - 1$

Regression analyses were accomplished using various combinations of the columns of the complete design matrix (Table 17.4) as independent variates. The simple product-moment correlation coefficients used in computing the various Ψ'^2 values are shown in Table 17.5. The results were as follows:

Regression on all group variates and covariates:

$\Psi'^2_{\alpha,\beta,\alpha\beta} = .2666 \qquad \text{df} = 5$

Table 17.5 Intercorrelations among Group-membership Variates and Covariates for Analysis of Raw Beta IQ with Age as a Covariate†

	a_1	a_2	x	a_1x	a_2x	y
a_1	1.0000	.5000	−.0401	.9245	.4748	.3575
a_2		1.0000	−.0501	.4708	.9279	−.0652
x			1.0000	−.1639	−.1950	−.1436
a_1x				1.0000	.5473	.2898
a_2x					1.0000	−.1019
y						1.0000

† a_1 = depressive.
a_2 = schizophrenic.
x = age.
a_1x = depressive by age.
a_2x = schizophrenic by age.
y = raw beta IQ.

Regression on group membership variates and the covariate:

$\Psi'^2_{\alpha,\beta} = .2265 \qquad \text{df} = 3$

Regression on the two group membership variates only:

$\Psi'^2_{\alpha} = .2071 \qquad \text{df} = 2$

Total sum of squared deviations about grand mean for the dependent variable:

$SS_T = 27{,}004.40 \qquad \text{df} = 84$

The summary of the analysis of covariance is presented in Table 17.6. There is no evidence that the nature of the regression of IQ scores on age differs between schizophrenic, depressive, and organic groups. The covariance-adjusted mean IQ scores for the three diagnostic groups differed significantly ($p < .001$), indicating that one should not assume that test performance is unrelated to clinical diagnosis. The group-mean adjusted regression on the covariate (common within-groups regression) was not statistically significant in

Table 17.6 Summary of Analysis of Covariance for Differences in Beta IQ Performance among Three Diagnostic Groups with Age as Covariate

Source	*SS*	*df*	*Mean square*	*F*
Difference in group regression	(.2666 − .2265)(27,004.40)	2	541.44	2.30
Adjusted means	(.2265 − .0206)(27,004.40)	2	2,779.80	11.79
Covariance adjustment	(.2265 − .2071)(27,004.40)	1	523.87	2.22
Residual error	(1.0 − .2666)(27,004.40)	84	235.77	
Total	27,004.40	89		

this small-sample analysis. In general, however, there is ample evidence that performance does vary with age, and one would thus want to retain age as a covariate to feel confident that the observed group differences were not a function of age.

The analysis of covariance has been considered most often as a method for controlling for the effects of extrinsic sources of variance in applications where the primary concern is with differences in group means on the dependent variable. It should be recognized also, however, as a method for studying differences in the nature of regression lines or multiple-regression surfaces within two or more groups. Frequently the question of similarities and differences in regression lines (or multivariate regression surfaces) is of primary importance, and the question whether differences in group means exist is of secondary importance. The analysis of covariance in such cases might better be termed the analysis of differences in within-group regressions.

The important thing about the general multiple-regression approach to the analysis of covariance is that it provides for direct examination of the nature of the regression effects that are tested. Associated with each Ψ^2 is a regression equation. In the present example, the variables in these equations are group-membership scores (scored 1, 0, and -1), values on the covariate x, and the interaction or cross-product, terms a_1x and a_2x. The equations for regression of beta IQ on various combinations of the independent variables were calculated to be as follows:

Regression on all group variates and covariates: $\Psi^2 = .2666$

$$y = 13.307a_1 + 2.698a_2 - 2.910x - .620a_1x - 2.746a_2x + 56.067 \qquad (17.33)$$

This equation can be used to plot the specific group regression lines as a function of age. The slopes of the separate group regression lines will not be the same because the cross-product terms a_1x and a_2x are included in the model. It is the difference in these slopes that is tested in the test for heterogeneity of regression. The separate group regression lines can be plotted by constructing any two (arbitrary) dummy variates for two (hypothetical) members of each group having substantially different covariate scores. For example, two members of the depression group (scored 1, 0 for group membership) having covariate ages of twenty and sixty years (decades 2 and 6) would be represented as follows:

Depressions:

a_1	a_2	x	a_1x	a_2x
[1	0	2	2	0]
[1	0	6	6	0]

Similarly, two members of the organic-brain-syndrome group (scored -1, -1 for group membership) having ages of twenty and sixty years would be represented as follows:

Brain syndromes:

a_1	a_2	x	a_1x	a_2x
[-1	-1	2	-2	-2]
[-1	-1	6	-6	-6]

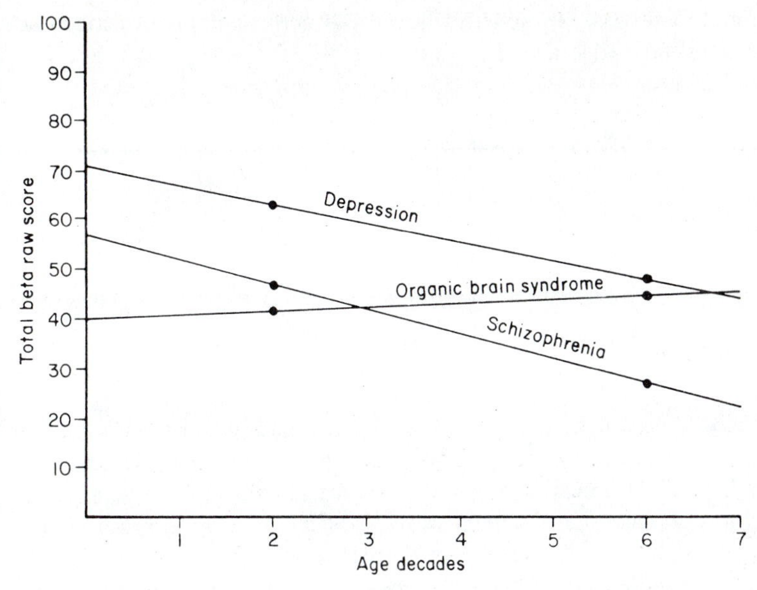

Figure 17.1 Separate group regression lines based on the full equation including interaction terms.

By applying the weights in the complete regression equation to these dummy score vectors, two points on each group regression line can be located and the regression line for each group can be drawn through the pairs of points. The within-groups regression lines for IQ of schizophrenic, organic, and depressive groups as a function of age are plotted in Fig. 17.1. Each of these lines was plotted using Eq. (17.33) to locate two points for (hypothetical) members of each group. The student should verify the correctness of the line for the schizophrenic group, in which the dummy variates would be of the following form for two patients having ages of twenty and sixty years:

Schizophrenics:
$$\begin{matrix} a_1 & a_2 & x & a_1x & a_2x \\ [0 & 1 & 2 & 0 & 2\,] \\ [0 & 1 & 6 & 0 & 6\,] \end{matrix}$$

In this particular example it was found that no evidence exists to suggest that the slopes of the regression lines differ significantly in the different diagnostic groups. The apparent differences in slopes of the regression lines in Fig. 17.1 can be considered to be due to sampling error. In such a case, it is meaningful to drop back to a simpler equation which does not involve the group-by-covariate product terms. The regression of IQ on group membership and age, disregarding the group-by-age product terms, was found to be as follows for $\Psi'^2_{\alpha,\beta}$:

$$\tilde{y} = 10.974a_1 - 7.019a_2 - 1.778x + 52.210 \qquad (17.34)$$

Constructing a vector for two hypothetical patients in each group generated the following variates:

	a_1	a_2	x
Depression	1	0	2
	1	0	6
Schizophrenia	0	1	2
	0	1	6
Organic syndrome	−1	−1	2
	−1	−1	6

Applying the weighting coefficients in Eq. (17.34) to these independent variates yielded two points on each of the common-slope regression line. The results are plotted in Fig. 17.2. The analysis of covariance (Table 17.6) revealed that the mean separation between the regression lines for the three groups was highly significant. This effect is clearly evident in Fig. 17.2. The analysis-of-covariance test for differences among group means is equivalent to difference in y-intercept values for several regression lines having the same slope. If the slopes cannot be assumed to be equal, then the test of covariance-adjusted mean differences is not meaningful.

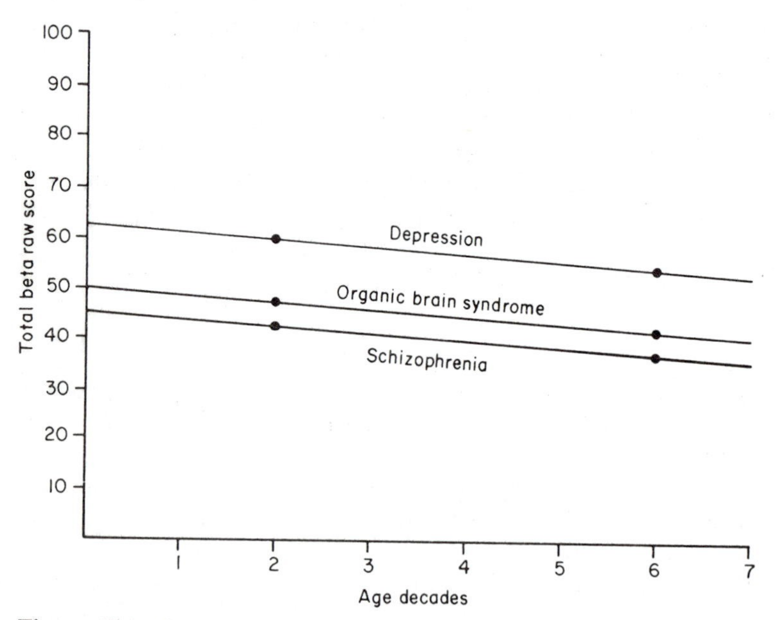

Figure 17.2 Separate group regression lines based on the reduced equation not including interaction terms.

17.10 Multiple Covariance Analysis and Multiple Within-groups Regression

The extension of linear regression methods to multiple covariance analysis is so direct and straightforward that the actual calculations will be left as an exercise for the student. Rather than including a single covariate x, as in Table 17.3, two or more quantitative covariates can be included in the model. The regression of the dependent variable on the group-membership variates plus the multiple covariates is first calculated. The covariates are then excluded from the model, and the regression on the group-membership variates is calculated. Finally, the group-membership variates are excluded from the model, and regression on the multiple covariates is calculated. The difference in Ψ'^2 with and without group-membership variates in the model represents the proportion of total sum of squares accounted for by differences in group means after effects of linear regression on the several covariates have been partialed out.

Since the partitioning of the total sum of squares involves calculations of various Ψ'^2 values, the complete analysis can be represented exactly as in the case of simple covariance and group regression. Let $\Psi'^2_{\alpha,\beta,\alpha\beta}$ represent Ψ'^2 obtained from regression of the dependent variable on $k - 1$ group variates, b covariates, and $(k - 1)b$ product terms. Let $\Psi'^2_{\alpha,\beta}$ represent regression of the dependent variable on the $k - 1$ group-membership variates and the b covariates, without the product terms included in the model. Let Ψ'^2_{α} represent the Ψ'^2 for regression on the group-membership variates alone, without covariates in the model, and let Ψ'^2_{β} represent regression of the dependent variable on the several covariates, without group-membership variates included in the model. Each of these Ψ'^2 values can be calculated using Eq. (17.8) or (17.9) with the appropriate independent variates.

Using these several squared multiple-correlation coefficients, the total sum of squares can be partitioned as follows:

Source	*SS*	*df*
Difference in group regression	$(\Psi'^2_{\alpha,\beta,\alpha\beta} - \Psi'^2_{\alpha,\beta})SS_T$	$b(k - 1)$
Adjusted group means	$(\Psi'^2_{\alpha,\beta} - \Psi'^2_{\beta})SS_T$	$k - 1$
Covariance adjustment	$(\Psi'^2_{\alpha,\beta} - \Psi'^2_{\alpha})SS_T$	b
Residual error	$(1 - \Psi'^2_{\alpha,\beta,\alpha\beta})SS_T$	$n - k(b + 1)$
Total	SS_T	$n - 1$

If the difference in between-group regression surfaces is significant when tested against the residual error, then one should concentrate on examining the individual group regression functions. If the difference between group regression surfaces is not significant, then the adjusted group means can be tested against the residual mean square to determine whether the group means are different after statistical adjustment for regression on the multiple covariates. This is the test of mean differences in the multiple covariance analysis. The significance of the covariance adjustment can be tested to determine whether inclusion of the multiple covariates in the model actually improved the precision of the analysis.

17.11 Exercises in Multiple-regression Prediction

Pre- and posttreatment BPRS ratings for a large sample of schizophrenic patients from several Veterans Administration hospitals were used in constructing the correlation matrix in Table 17.7. The first four variables in the correlation matrix represent pretreatment factor scores, and the final variable is *change* in total pathology scores. For convenience, the inverse of the matrix of intercorrelations among the four factor scores is also presented. Note that the problem is essentially the same as that presented in Table 17.1, except that the criterion measure is change in total pathology rather than the simple posttreatment score. Also, different raw data were used.

1 Calculate the standard partial-regression coefficients and Ψ^2. To provide a check for accuracy of calculations, the authors determined that $\beta_2 = .1883$ and $\Psi^2 = .28$.

2 Calculate the standard errors of the (standard-score) regression coefficients

Table 17.7 Data for Exercises in Multiple-regression Prediction

MATRIX OF INTERCORRELATIONS AMONG PRETREATMENT SCORES ON FOUR SYMPTOM FACTORS AND CORRELATIONS WITH CHANGE DURING TREATMENT

TD	*W-R*	*H-S*	*AD*	*Change*
1.0000	.1917	.2013	.0131	.3663
.1917	1.0000	.0746	.0250	.2643
.2013	.0746	1.0000	.2461	.3499
.0131	.0250	.2461	1.0000	.2612

MEANS AND STANDARD DEVIATIONS

	Means	*Standard deviations*
Thinking disturbance	8.1372	4.70
Withdrawal-retardation	6.2934	4.04
Hostile-suspiciousness	5.6814	4.11
Anxious depression	4.8076	4.13
Change	12.5710	12.60

INVERSE OF MATRIX OF INTERCORRELATIONS AMONG THE FOUR SYMPTOM FACTORS

1.079237	−.192039	−.213561	.043220
−.192039	1.039815	−.035271	−.014800
−.213561	−.035271	1.112068	−.270001
.043220	−.014800	−.270001	1.066251

and use the t statistic to test the hypothesis that each coefficient does not differ from zero. What are the values of the various t statistics? The authors found $t = 6.4969$ to be the value associated with β_3.

3 Convert the standard partial-regression coefficients to raw-score form. What is the regression constant k_0 that must be added to the raw-score regression equation? The authors found the partial-regression coefficient associated with the fourth variable to be $b_4 = .5971$.

The examples of multiple-regression analysis considered throughout this chapter have used the intercorrelations among predictor and criterion variables to obtain standard regression equations. In Sec. 17.3 the direct computation of raw-score regression equations using covariances among the original variables was described. Table 17.8 shows the covariances among four symptom-factor scores and change in total pathology. The covariances were computed from the same data as the correlations in Table 17.7.

Table 17.8 Data for Multiple-regression Exercise Using Covariances

MATRIX OF COVARIANCES AMONG PRETREATMENT SCORES ON FOUR SYMPTOM FACTORS AND VECTOR OF COVARIANCES WITH CHANGE DURING TREATMENT

$$
\begin{array}{cccc}
TD & W\text{-}R & H\text{-}S & AD \\
\end{array}
\qquad Change
$$

$$
\begin{bmatrix}
22.0964 & 3.6405 & 3.8952 & 0.2555 \\
3.6405 & 16.3213 & 1.2405 & 0.4183 \\
3.8952 & 1.2405 & 16.9409 & 4.1818 \\
0.2555 & 0.4183 & 4.1818 & 17.0371
\end{bmatrix}
\begin{bmatrix}
21.6545 \\
13.4293 \\
18.1111 \\
13.5618
\end{bmatrix}
$$

MEANS AND STANDARD DEVIATIONS

	Means	*Standard deviations*
Thinking disturbance	8.1372	4.70
Withdrawal-retardation	6.2934	4.04
Hostile-suspiciousness	5.6814	4.11
Anxious depression	4.8076	4.13
Change	12.5710	12.60

INVERSE OF MATRIX OF COVARIANCES AMONG THE FOUR SYMPTOM FACTORS

$$
\begin{bmatrix}
.048843 & -.010112 & -.011039 & .002225 \\
-.010112 & .063709 & -.002120 & -.000892 \\
-.011039 & -.002120 & .065646 & -.015895 \\
.002225 & -.000892 & -.015895 & .062586
\end{bmatrix}
$$

4 Calculate the raw-score regression equation for predicting change in total pathology as a function of pretreatment factor scores. How do the raw-score regression coefficients calculated in this manner correspond to those obtained by first solving for standard partial regression coefficients using

intercorrelations in Table 17.7? The authors found the raw-score partial regression coefficient associated with the fourth variable to be $b_4 = .5971$. What are the values of the others? What is the value of the regression constant k_0? *Hint:* See the definition of regression constant associated with Eq. (17.12).

5 Use the t statistic to test the significance of each raw-score regression coefficient. How do the t values compare with those obtained for testing the standard-score coefficients derived from same data (Table 17.7)?

6 Numerous correlation and covariance matrices have been presented throughout this book. Choose a variable of interest in one of these matrices and derive an equation for predicting it from other variables included in the matrix. What proportion of the total variance in the selected variable can be accounted for by the regression?

17.12 Reference

Dudley, H. K., J. D. Williams, and J. E. Overall: Relationships of Beta IQ Scores to Socio-cultural Factors in a Psychiatric Population, *J. Clin. Psychol*, **27**:68–74 (1971).

18

Complex Least-squares Analysis of Variance[1]

18.1 Introduction

In this chapter the general multiple-regression approach to the analysis of variance and analysis of covariance will be examined in more detail. Of particular concern are variations on the general method that can be employed to test different kinds of hypotheses in complex multifactor analyses involving interaction effects. In the final section, a computer program that can be used to accomplish multifactor analyses of variance-covariance in the presence of unequal and disproportionate cell frequencies is presented.

Most contemporary psychologists have learned analysis of variance in the experimental-design tradition. Only in recent years has it been widely recognized that the same types of analyses can be accomplished using more general linear regression methods (Bottenberg and Ward, 1960; Bradley, 1968; Cohen, 1968*a*; Jennings, 1967; Ward, 1962). While it is true that conventional analyses of variance can be accomplished by least-squares regression methods, it is important to emphasize that such methods can also lead to a variety of different solutions only superficially similar to the familiar analysis

[1] The content of this chapter is essentially that of an article previously published by Overall and Spiegel (1969) and is reproduced here by permission.

of variance. The purpose of the present chapter is to define a proper generalization of familiar analysis of variance and to discuss distinctive features of alternative solutions.

Since most researchers are already familiar with other computational procedures for analysis of data from balanced (equal cell frequency) experimental designs, interest in least-squares regression solutions has tended to center about applications involving unequal and disproportionate cell frequencies. Ordinary analysis-of-variance calculations cannot be employed in the presence of unequal and disproportionate cell frequencies because the unequal frequencies result in correlation among the classification variables. Where two or more factors are correlated, it is difficult to know which is influencing the criterion to what extent. Least-squares methods can be used to estimate the independent effect of each variable adjusted for relationships to other classification variables.

Using least-squares regression methods, analyses of variance can be accomplished on data from arbitrary experimental designs in which no attempt is made to control cell frequencies. This is especially important in clinical and social psychological research, where the number and nature of classification variables preclude selection of equal numbers of subjects for each combination of factors. In some research, differential frequencies of occurrence are an inherent feature of the problem under investigation, and in such situations artificial stratification can have undesirable effects. A least-squares analysis is appropriate in such circumstances.

At a time when psychologists and other applied researchers are beginning to employ least-squares regression methods as a generalized approach to the analysis of variance, it is not widely appreciated that only under very restricted circumstances are results equivalent to those of conventional analysis of variance. Jennings (1967) and Cohen (1968*a*) have emphasized the flexibility of least-squares multiple-regression analysis for testing hypotheses that are not ordinarily examined in conventional analyses of variance. Although this flexibility is important, researchers should be cautioned against placing conventional analysis-of-variance interpretations on tests of hypotheses that are not like those provided by the usual balanced experimental designs. The problem is a real and serious one because a number of "least-squares analysis-of-variance" computer programs are currently available which accomplish different types of analyses, and the method used is frequently not properly understood by the user.

Although it is true that the general least-squares method can be employed to accomplish simple analyses of variance which might otherwise be undertaken using more conventional calculations, a great deal of intuitive generalization is required for application to more complex problems. The justification for applying a particular least-squares regression technique to a complex disproportionate-cell-frequency problem is often that the method yields conventional results when applied to data from a conventional balanced experimental design and that all calculations can be accomplished in exactly the same manner in the presence of unequal cell frequencies. The fact that a method of analysis yields conventional results in the equal-cell frequency case is not adequate justification for use with unequal and disproportionate cell frequencies. Several different least-squares methods can be shown to yield identical results in the balanced-design situation but yield substantially different results when applied to data involving unequal cell frequencies.

Three least-squares methods for the analysis of experimental data are of particular interest. They yield identical results when applied to equal-cell-frequency problems but different results in other cases. The *complete least-squares*, or *general linear-model, analysis* is simply a conventional least-squares multiple-regression solution in which each effect, whether it is main effect or interaction, is adjusted for relationship to all other effects in the model. The second method, here termed the *experimental-design method*, takes into account the experimental-design hierarchy of main effects and interactions, adjusting each effect for all other effects at an equal or lower level and ignoring higher-order effects (Snedecor and Cochran, 1967; Winer, 1962). The third method depends on an a priori ordering of all terms in the model, and each effect is adjusted only for those preceding it in the ordering (Cohen, 1968*b*). All these methods are "correct," but each results in the testing of a different set of hypotheses, and the differences are not apparent in the usual analysis-of-variance summary tables.

Before discussing the alternative least-squares methods that appear to be promising extensions of the analysis of variance, a brief description of the general least-squares regression approach to analysis of experimental data will be presented. The utilization of standard multiple-regression calculations in analysis of experimental data requires definition of a set of independent variates representing main effects and interactions. There are many ways in which such (dummy) variates can be defined, although most of the alternatives do not incorporate the usual analysis-of-variance restrictions. The method of developing a design matrix described in the next section imposes the usual analysis-of-variance restrictions and results directly in conventional analysis-of-variance parameter estimates (adjusted means) and sums of squares. There is considerable conceptual and computational advantage in this approach.

18.2 Construction of the Design Matrix for Regression Analysis

To employ standard multiple-regression methods for the analysis of multiple-classification experimental data, it is necessary to construct a matrix of dummy variates specifying treatment classifications and interactions. If such a design matrix includes proper restrictions, the estimates of parameters for the analysis of variance can be obtained directly as partial-regression coefficients.

One reason the analysis of variance has not generally been conceived in regression terms is simply a matter of differences in the conventional notations. The analysis-of-variance model is usually written in very abbreviated notation such that several parameter values (say α_1, α_2, α_3, and α_4) are all represented by the symbol α_i. The familiar analysis-of-variance model for the two-way case with interaction is

$$X = \mu + \alpha_i + \beta_j + \alpha\beta_{ij} + \varepsilon_{ijk} \tag{18.1}$$

where μ is the grand mean, the α_i are deviations of row means about the grand mean, β_j are deviations of column means about the grand mean, the $\alpha\beta_{ij}$ are deviations of the cell means about row and column effects, and ε_{ijk} is random error: $n(0,\sigma^2)$. Once again, it is important to note that several different α_i, β_j, and $\alpha\beta_{ij}$ are implicit in this abbreviated notation. *In the regression model, these* α_i, β_i, *and* $\alpha\beta_{ij}$ *are the partial-regression coefficients.*

To arrive directly at a regression solution in which the coefficients are estimates of parameters for the analysis of variance, it is necessary to impose proper restrictions. Since only $a - 1$ degrees of freedom are available for A and only $b - 1$ degrees of freedom for B, a total of only $a - 1$ plus $b - 1$ parameters associated with these main effects can be estimated directly. A customary analysis-of-variance restriction is to require that the sum of the row effects equals zero, and the same for column effects. Similarly, it is customary to impose the restriction that the interaction parameters sum to zero across both rows and columns separately,

$$\sum_i \alpha_i = \sum_j \beta_j = \sum_i \alpha\beta_{ij} = \sum_j \alpha\beta_{ij} = 0 \tag{18.2}$$

Development of a design matrix incorporating these restrictions can best be described in terms of a simple example. Table 18.1 presents hypothetical data for a two-way analysis involving disproportionate cell frequencies. Each individual belongs to one level of A, one level of B, and one AB combination. The design matrix is formed to include $a - 1$ columns representing the first $a - 1$ levels of A, $b - 1$ columns representing the first $b - 1$ levels of B, and $(a - 1)(b - 1)$ columns representing the various combinations of the first $a - 1$ levels of A with the first $b - 1$ levels of B. It will be noted that one column is included to represent each degree of freedom for rows, columns, and interaction. The reader should be able to generalize these considerations to more complex designs.

For the present, the discussion disregards interaction effects. The columns associated with A and B main effects are filled. If an individual belongs in one of the first $a - 1$ categories of A, he is scored 1 in that column and 0 in any other columns associated with A. If the individual belongs in the last category of A, he is scored -1 in *all* $a - 1$ columns associated with the A main effect. In this way, the category membership of the individual is uniquely specified. In a similar manner, if an individual belongs in one of the first $b - 1$ categories of B, he receives an entry of 1 in the column associated with that particular category and 0 in other columns associated with B. If he belongs in the last category of B, he receives a score of -1 in *all* $b - 1$ columns representing the B main effect.

Table 18.1 Data for Two-way Analysis of Variance

	Effect		
Effect	A_1	A_2	A_3
B_1	61	42	96
	73	53	81
	52		92
B_2	79	37	45
	65	32	37
	81	50	
B_3	43	87	75
	35	81	59
		65	
B_4	56	72	98
	25	84	77
	19		91
	35		

The entries in columns of the design matrix associated with AB interaction are obtained as products of entries in the corresponding main-effect columns. For example, entries in the column headed by A_1B_1 are obtained by multiplying the corresponding entries in columns A_1 and B_1. There is an interaction column in the design matrix for each combination of an A column with a B column.

The complete design matrix including proper restrictions is presented in Table 18.2 for the two-way example. Although there are several different points at which the proper analysis-of-variance restrictions can be imposed, the simplest and safest procedure is to incorporate restrictions in the initial matrix of dummy variates, as shown here. Given a design matrix constructed in this manner, one can employ a standard least-squares multiple-regression solution to obtain directly estimates of parameters and sums of squares for the analysis of variance.

The effect of unequal and disproportionate cell frequencies is to introduce correlation between columns of the design matrix. In the orthogonal, or equal-cell-frequency, case, considerable simplification is achieved because correlations between columns in the design matrix are precisely zero. The reader not already familiar with this fact may gain insight by constructing a dummy-variate design matrix for a two- or three-way equal-cell-frequency problem. In the absence of correlations among independent variables, the squared multiple correlation can be obtained directly as the sum of simple squared correlations between the individual predictors and the criterion. Thus, the complex least-squares regression solution is not required in the case of equal cell frequencies.

Table 18.2 Design Matrix for Two-way Analysis of Variance

A_1	A_2	B_1	B_2	B_3	A_1B_1	A_1B_2	A_1B_3	A_2B_1	A_2B_2	A_2B_3	X
1	0	1	0	0	1	0	0	0	0	0	61
1	0	1	0	0	1	0	0	0	0	0	73
1	0	1	0	0	1	0	0	0	0	0	52
1	0	0	1	0	0	1	0	0	0	0	79
1	0	0	1	0	0	1	0	0	0	0	65
1	0	0	1	0	0	1	0	0	0	0	81
1	0	0	0	1	0	0	1	0	0	0	43
1	0	0	0	1	0	0	1	0	0	0	35
1	0	−1	−1	−1	−1	−1	−1	0	0	0	56
1	0	−1	−1	−1	−1	−1	−1	0	0	0	25
1	0	−1	−1	−1	−1	−1	−1	0	0	0	19
1	0	−1	−1	−1	−1	−1	−1	0	0	0	35
0	1	1	0	0	0	0	0	1	0	0	42
0	1	1	0	0	0	0	0	1	0	0	53
0	1	0	1	0	0	0	0	0	0	1	37
0	1	0	1	0	0	0	0	0	1	0	32
0	1	0	1	0	0	0	0	0	1	0	50
0	1	0	0	1	0	0	0	0	0	1	87
0	1	0	0	1	0	0	0	0	0	1	81
0	1	0	0	1	0	0	0	0	0	1	65
0	1	−1	−1	−1	0	0	0	−1	−1	−1	72
0	1	−1	−1	−1	0	0	0	−1	−1	−1	84
−1	−1	1	0	0	−1	0	0	−1	0	0	96
−1	−1	1	0	0	−1	0	0	−1	0	0	81
−1	−1	1	0	0	−1	0	0	−1	0	0	92
−1	−1	0	1	0	0	−1	0	0	−1	0	45
−1	−1	0	1	0	0	−1	0	0	−1	0	37
−1	−1	0	0	1	0	0	−1	0	0	−1	75
−1	−1	0	0	1	0	0	−1	0	0	−1	59
−1	−1	−1	−1	−1	1	1	1	1	1	1	98
−1	−1	−1	−1	−1	1	1	1	1	1	1	77
−1	−1	−1	−1	−1	1	1	1	1	1	1	91

The partial-regression coefficients associated with the first $a - 1$ categories of A are estimates of the deviations of those category means about the grand mean. Since $\sum \alpha_i = 0$ for all categories, the parameter estimate for the final category is minus the sum of the $a - 1$ values obtained in the regression solution. Similarly, since $\sum \beta_j = 0$ for all categories of B, the parameter estimate for the final category is minus the sum of the $b - 1$ values obtained for the other categories. The missing interaction parameters can be estimated in a similar manner from the restrictions that $\alpha\beta_{ij}$ sum to zero across both rows and columns.

The analysis-of-variance parameter estimates derived directly from regression of the criterion measure on the dummy variates of the design matrix are shown in the left-hand column of Table 18.3. The regression constant provides an estimate of the grand mean. Any interested reader can verify that the raw data

Table 18.3 Analysis-of-variance Parameters Derived from Multiple-regression Equation†

Parameter	*Coefficients*	*Analysis-of-variance parameters*
α_1	−9.1	−9.1
α_2	−.9	−.9
α_3		10.0
β_1	4.8	4.8
β_2	−9.7	−9.7
β_3	−.4	−.4
β_4		5.3
$\alpha_1\beta_1$	4.7	4.7
$\alpha_1\beta_2$	32.3	32.3
$\alpha_1\beta_3$	−13.1	−13.1
$\alpha_1\beta_4$		−23.9
$\alpha_2\beta_1$	−18.0	−18.0
$\alpha_2\beta_2$	−11.4	−11.4
$\alpha_2\beta_3$	17.3	17.3
$\alpha_2\beta_4$		12.1
$\alpha_3\beta_1$		13.3
$\alpha_3\beta_2$		−20.9
$\alpha_3\beta_3$		−4.2
$\alpha_3\beta_4$		11.8
C	61.6	61.6

† Regression equation:

$$Y = -9.1X_1 - .9X_2 + 4.8X_3 - 9.7X_4 - .4X_5 + 4.7X_6 + 32.3X_7 - 13.1X_8 - 18.0X_9 - 11.4X_{10} + 17.3X_{11} + 61.6.$$

$$\mu_{11} = -9.1 + 4.8 + 4.7 + 61.6 = 62.0$$

$$\mu_{23} = -.9 - .4 + 17.3 + 61.6 = 77.6$$

$$\mu_{34} = 10.0 + 5.3 + 11.8 + 61.6 = 88.7$$

(dummy variates and criterion) shown in Table 18.2 will yield these parameter estimates as raw-score beta weights in a multiple-regression analysis. The missing parameter estimates derived from the restrictions $\sum \alpha_i = \sum \beta_j = \sum \alpha\beta_{ij} = 0$ are entered in the right-hand column of Table 18.3. The **ij**th cell mean can be estimated by summing α_i, β_j, $\alpha\beta_{ij}$, and C for any **i** and **j**

$$\mu_{ij} \approx \alpha_i + \beta_j + \alpha\beta_{ij} + C \tag{18.3}$$

18.3 Partitioning Sums of Squares for the Analysis of Variance

Sums of squares representing variation in parameter values can be obtained in several different ways. The method to be described involves computing the square of the multiple correlation[1] R^2 for regression of the criterion measure on different subsets of the dummy variates. As is well known, the square of the

[1] Because there is no chance of confusion in the use of R as a symbol for multiple correlation in this chapter, we revert to the standard notation.

multiple-correlation coefficient relating a criterion measure to several independent variates represents the proportion of the total sum of squares for the criterion that can be accounted for by regression on the set of independent variates. The *independent* contribution of each subset of independent variates (say α_i, $i = 1, 2, \ldots, a - 1$) can be estimated by calculating the multiple R^2 with the α_i parameters in the model and then with the α_i parameters disregarded. Similarly for β_j and $\alpha\beta_{ij}$ effects.

For the two-way analysis with interaction, the R^2 associated with regression of the criterion measure on all row, column, and interaction components will be represented by $R^2(\alpha_i,\beta_j,\alpha\beta_{ij})$. The R^2 derived from regression on only the row and column variates will be represented by $R^2(\alpha_i,\beta_j)$. The sum of squares representing the independent contribution of interaction effects *over* and *above* row and column effects is

$$SS_{AB} = SS_T[R^2(\alpha_i,\beta_j,\alpha\beta_{ij}) - R^2(\alpha_i,\beta_j)] \tag{18.4}$$

where SS_T is the total mean-corrected sum of squares for the criterion variable. The SS_{AB} has associated with it degrees of freedom equal to the number of dummy variates representing $\alpha\beta_{ij}$ in the design matrix. The independent contributions of α_i and β_j can be calculated in a similar manner; however, this

Source	*SS*	*df*
	METHOD 1	
A	$SS_T[R^2(\alpha_i,\beta_j,\alpha\beta_{ij}) - R^2(\beta_j,\alpha\beta_{ij})]$	$a - 1$
B	$SS_T[R^2(\alpha_i,\beta_j,\alpha\beta_{ij}) - R^2(\alpha_i,\alpha\beta_{ij})]$	$b - 1$
AB	$SS_T[R^2(\alpha_i,\beta_j,\alpha\beta_{ij}) - R^2(\alpha_i,\beta_j)]$	$(a - 1)(b - 1)$
Error	$SS_T[1 - R^2(\alpha_i,\beta_j,\alpha\beta_{ij})]$	$n - ab$
Total	SS_T	$n - 1$
	METHOD 2	
A	$SS_T[R^2(\alpha_i,\beta_j) - R^2(\beta_j)]$	$a - 1$
B	$SS_T[R^2(\alpha_i,\beta_j) - R^2(\alpha_i)]$	$b - 1$
AB	$SS_T[R^2(\alpha_i,\beta_j,\alpha\beta_{ij}) - R^2(\alpha_i,\beta_j)]$	$(a - 1)(b - 1)$
Error	$SS_T[1 - R^2(\alpha_i,\beta_j,\alpha\beta_{ij})]$	$n - ab$
Total	SS_T	$n - 1$
	METHOD 3 (assuming the order to be $\alpha_i,\beta_j,\alpha\beta_{ij}$)	
A	$SS_T[R^2(\alpha_i)]$	$a - 1$
B	$SS_T[R^2(\alpha_i,\beta_j) - R^2(\alpha_i)]$	$b - 1$
AB	$SS_T[R^2(\alpha_i,\beta_j,\alpha\beta_{ij}) - R^2(\alpha_i,\beta_j)]$	$(a - 1)(b - 1)$
Error	$SS_T[1 - R^2(\alpha_i,\beta_j,\alpha\beta_{ij})]$	$n - ab$
Total	SS_T	$n - 1$

is where a distinction between three methods becomes apparent. Method 1, called here the *complete linear-model analysis*, involves estimation of independent effects of each factor adjusted for all others included in the model. Method 2, called the *experimental-design analysis*, involves estimation of main effects disregarding interactions and then estimation of interactions adjusted for main effects. Method 3, called the *step-down analysis*, involves an initial ordering of the effects and then estimating each effect adjusted for those preceding it in the ordering and ignoring those following it.

Where the design matrix has been constructed to incorporate the analysis-of-variance restrictions, all three methods will yield exactly the same (conventional) results in application to data involving equal cell frequencies. On the other hand, quite different results derive from the three methods in applications involving disproportionate cell frequencies. Summary tables for results obtained from application of the three methods to the data from Table 18.1 are presented in Table 18.4. Method 1 utilizes a single consistent model including interaction terms. Method 2 derives main-effect estimates from an additive (no-interaction)

Table 18.4 Summary of Least-squares Analyses of Variance for Two-way Unequal-cell-frequency Problems

Source	*SS*	*df*	*Mean square*	*F*
		METHOD 1		
A	1,904.0	2	952.0	8.4
B	1,130.3	3	376.8	3.3
AB	9,560.0	6	1,593.3	14.0
Error	2,269.9	20	113.5	
Sum	14,864.2			
Total	15,552.9	31		
		METHOD 2		
A	2,729.7	2	1,364.8	12.0
B	779.1	3	259.7	2.3
AB	9,560.0	6	1,593.3	14.0
Error	2,269.9	20	113.5	
Sum	15,338.7			
Total	15,552.9	31		
		METHOD 3		
A	2,943.9	2	1,472.0	13.0
B	779.1	3	259.7	2.3
AB	9,560.0	6	1,593.3	14.0
Error	2,269.9	20	113.5	
Sum	15,552.9			
Total	15,552.9	31		

model and includes interaction parameters to evaluate deviations from the additive model. Method 3 adds a term to the model for each effect tested. Superficially, method 1 would seem to be the most direct and consistent since all sums of squares included in the summary table are derived from the same linear model. The summary table obtained by method 2 includes sums of squares actually calculated from two different models, one with interactions and one without interactions, while the summary table for method 3 is based upon as many models as there are effects to be tested.

A survey of the statistical literature suggests that method 1 is consistent with the general linear-model analysis described in abstract terms by mathematical statisticians for the equal-cell-frequency case (Graybill, 1961). This is not particularly helpful with regard to the present problem because all three methods yield the same results in the equal-cell-frequency case. In this case

$$R^2(\alpha_i,\beta_j) = R^2(\alpha_i) + R^2(\beta_j) \tag{18.5}$$

and

$$R^2(\alpha_i,\beta_j,\alpha\beta_{ij}) = R^2(\alpha_i,\beta_j) + R^2(\alpha\beta_{ij}) \tag{18.6}$$

Although theoretical statisticians provide few specific recommendations for handling unequal cell frequencies, the brief discussions that are presented imply a similar treatment.

The more practical statistical texts, in which actual computational procedures are described for analyses of variance involving unequal cell frequencies, provide support for method 2. Snedecor and Cochran (1967) and Rao (1965) give examples in which main effects are estimated disregarding interactions but interactions are subsequently adjusted for main effects. Method 2 as described herein yields the solutions presented by those authors, and methods 1 and 3 produce different results.

18.4 The Proportionate-cell-frequency Case

There may seem to be little basis for saying that any one of these methods is a proper or conventional analysis of variance. The conventional analysis of variance was developed in the context of balanced experimental designs. All three methods yield identical results for such problems. Since conventional analysis-of-variance calculations cannot be employed in the presence of disproportionate cell frequencies, there is no way to compare and to say, for example, that method 1 is the proper generalization while methods 2 and 3 are not. There is a type of problem, however, that can be handled by conventional analysis of variance and for which method 1 yields different results from methods 2 and 3. This is the *proportional*-cell-frequency problem.

An example taken from Lindquist (1953) is presented in Table 18.5. The design matrix including main effects and all simple interactions (but not the triple interaction) is illustrated in Table 18.6 by inclusion of one observation from each treatment combination. As in the previous example, each individual

Table 18.5 Example from Lindquist

	B_1			B_2			B_3			B_4		
Effect	C_1	C_2	C_3	C_1	C_2	C_3	C_1	C_2	C_3	C_1	C_2	C_3
A_1	9	15	10	14	8	9	19	18	20	16	17	17
	7	10	10	7	6	13	16	13	17	9	16	11
	7	13	15	5	10	7	12	9	13	17	19	15
	14	16	18	15	13	13	16	13	16	9	10	12
				12	13	12	10	7	9	22	21	14
				11	14	8	9	13	14			
							11	7	14			
							12	9	12			
A_2	13	16	16	17	7	15	29	21	19	30	21	15
	12	10	9	22	16	23	25	18	24	21	15	16
	18	18	14	14	7	16	18	19	15	9	6	17
	18	20	19	17	13	15	25	21	22	13	16	23
				10	16	11	15	15	17	18	18	17
				12	9	13	21	15	17			
							18	12	11			
							30	18	24			

Source: Lindquist, 1953.

receives a score of either 0 or 1 on each dummy variate associated with a main effect if he belongs to one of the first $k - 1$ categories. If he belongs in the last category for a particular classification variable, he receives scores of -1 on all dummy variates associated with the main effect for that classification variable. Simple interaction variates are calculated as products of appropriate pairs of main-effect variates.

Application of methods 1, 2, and 3 to the Lindquist problem resulted in the sums of squares shown in Table 18.7. Except for omission of the triple interaction and consequent pooling of that sum of squares into the error term, methods 2 and 3 yielded results identical to those presented by Lindquist. The results obtained from method 1 were distinctly different with regard to those effects where orthogonality was not maintained in the design. It seems clear from these results that method 2 is the proper generalization of traditional experimental-design analysis of variance and that method 1 is something different. In most instances it would seem undesirable for the magnitude of a main effect to be adjusted for interaction effects, although certain logical applications of method 1 will be mentioned.

The fact that method 3 yielded results identical to those obtained from method 2 in this example is due to the particular a priori ordering that was specified. Whenever main effects all precede interaction terms and lower-order interactions precede higher-order interactions, the only differences between methods 2 and 3 will be found in the within-levels adjustments. In a proportionate-cell-frequency design, main effects are mutually orthogonal, even though the unequal frequencies result in correlations between interactions and main effects.

Table 18.6 Selected Rows from Design Matrix for Lindquist Three-way Problem†

Subject	A_1	B_1	B_2	B_3	C_1	C_2	A_1B_1	A_1B_2	A_1B_3	A_1C_1	A_1C_2	B_1C_1	B_1C_2	B_2C_1	B_2C_2	B_3C_1	B_3C_2	X
1	1	1			1		1			1		1						9
5	1	1				1	1			0	1		1					15
9	1	1			−1	−1	1			−1	−1	−1	−1					10
13	1		1		1			1		1				1				14
19	1		1			1		1			1				1			8
25	1		1		−1	−1		1		−1	−1			−1	−1			9
31	1			1	1				1	1						1		18
39	1			1		1			1		1						1	19
47	1			1	−1	−1			1	−1	−1					−1	−1	20
55	1	−1	−1	−1	1		−1	−1	−1	1		−1		−1		−1		16
60	1	−1	−1	−1		1	−1	−1	−1		1		−1		−1		−1	17
65	1	−1	−1	−1	−1	−1	−1	−1	−1	−1	−1	1	1	1	1	1	1	17
70	−1	1			1		−1			−1		1						13
74	−1	1				1	−1				−1		1					16
78	−1	1			−1	−1	−1			1	1	−1	−1					16
82	−1		1		1			−1		−1				1				17
88	−1		1			1		−1			−1				1			7
94	−1		1		−1	−1		−1		1	1			−1	−1			15
100	−1			1	1				−1	−1						1		29
108	−1			1		1			−1		−1						1	21
116	−1			1	−1	−1			−1	1	1					−1	−1	19
124	−1	−1	−1	−1	1		1	1	1	−1		−1		−1		−1		30
129	−1	−1	−1	−1		1	1	1	1		−1		−1		−1		−1	21
134	−1	−1	−1	−1	−1	−1	1	1	1	1	1	1	1	1	1	1	1	15

† The first subject in each treatment combination is represented.

Table 18.7 Sums of Squares for Lindquist Three-way Proportionate-cell-frequency Problem

		SS			
Source	*df*	*Lindquist*	*Method 1*	*Method 2*	*Method 3*
A	1	617.8	480.7	617.9	617.9
B	3	394.1	394.7	394.1	394.1
C	2	49.9	19.5	49.9	49.9
AB	3	119.1	119.1	119.1	119.1
AC	2	92.3	92.3	92.3	92.3
BC	6	110.1	109.2	109.2	109.2
ABC	6	70.5			
Error	114	2,003.4	2,074.8	2,074.8	2,074.8
Total	137	3,457.2	3,457.2	3,457.2	3,457.2
Sum of effects		3,457.2	3,290.3	3,457.3	3,457.3

It is especially important to note that component sums of squares always add up to the total sum of squares when method 3 is used. This is not generally true for methods 1 and 2 if disproportionate cell frequencies are present. The latter methods are concerned with the independent contribution of each classification variable over and above the others included in the analysis. It is possible for a criterion variable to be highly dependent on each of several classification factors and yet for no one classification factor to account for significant variation over and above the others. In using either method 1 or 2, care should be exercised not to include classification variables that are highly related. If two classification variables are so highly related that it would prove difficult to fill all cells in a cross-classification matrix, it would seem wise to select only one of them for consideration in an analysis. Method 3 can be used with less concern over dependencies among the classification variables. If a criterion variable is substantially dependent on the total set of classification variables, the dependence will be reflected in the partitioned sums of squares.

Selection of one of the three methods should depend upon conceptualization of the problem and the nature of questions to be asked. Method 2 should be employed whenever the problem is conceived as a multiclassification factorial design and where conventional analysis of variance might have been employed except for unequal cell frequencies or other computational difficulties. Method 2 is appropriate where one hopes to account for systematic variation in terms of simple additive main effects but wishes to test the significance of interactions as a safeguard against nonadditivity.

Method 1 should be used if conceptualization of the problem is in general linear-regression terms and not in experimental-design terms. Method 1 can be used to determine whether classification factors have pervasive main effects that cannot be accounted for entirely by more specific cell (interaction) effects. In using method 1, care should be exercised not to interpret significant "main effects" in the same way that one might interpret a significant main effect from a balanced factorial design. Powerful suppressor effects are not uncommon between classification variables and their interactions. It is perhaps best to interpret the parameter estimates as regression coefficients and not as estimated mean values when method 1 is employed. Method 3 can be used if a logical a priori ordering exists among hypotheses to be tested. For example, if certain classification variables have logical priority in a theoretical or causal sense, their effects can be tested disregarding secondary factors and then the effects of secondary factors tested after adjustment for the primary factors. One may wish to order effects to be tested in terms of a priori probabilities of significance, testing first the effects which are expected to be strongest and subsequently the weaker effects to determine whether they add anything. In dealing with highly interdependent classification variables, it may not be useful to test the significance of the independent effect of each factor adjusted for all others. Method 3 has a pronounced advantage in minimizing the possibility that significant effects will cancel one another. Again, one should be careful in the interpretation of results to take into account the different levels of adjustment for other effects.

In an effort to clarify the special areas of usefulness for the different methods, this chapter concludes with the discussion of a concrete example of research for each of the three methods.

18.5 Examples

18.5.1 APPLICATION OF METHOD 1 IN STUDY OF THE INDEPENDENT CONTRIBUTIONS OF BACKGROUND VARIABLES AND THEIR INTERACTIONS

Standard history and psychiatric symptom-rating profiles were obtained for a sample of 3,516 patients from a variety of state hospitals, Veterans Administration hospitals, and university hospital inpatient and outpatient services. The background data included age, race, sex, education, work achievement, and marital status. The severity of 16 psychiatric symptoms was recorded on the BPRS. The study was undertaken to investigate the dependence of manifest psychopathology on background factors. Simple interactions between background variables were included to test for significant pattern effects. A criterion measure of particular concern was the severity of anxious depression, scored as the sum of ratings for anxiety, guilt feelings, and depressive mood.

Method 1 was chosen because the problem was viewed essentially as a multiple-regression problem involving categorical or qualitative independent variates. The object was to determine the significance of the independent contribution of each factor and each interaction included in the original model and thence to arrive at a minimum adequate model by eliminating factors. Terms were eliminated from the model in several stages by discarding the least significant factor identified in the previous analysis. The final analysis based on a reduced

Table 18.8 Summary of Method 1 Analysis of Variance for Anxious-depression-factor Scores

Source	*SS*	*df*	*Mean square*	*F*
Age	12,418.75	1	12,418.75	7.34
Race	141,749.34	2	70,874.67	41.87
Sex	4,988.38	1	4,988.38	2.95
Education	43,172.38	1	43,172.38	25.51
Work level	61,055.88	1	61,055.88	36.07
Marital status	13,182.25	1	13,182.25	7.79
Age × education	9,151.50	1	9,151.50	5.41
Sex × age	28,225.63	1	28,225.63	16.68
Sex × race	13,702.00	2	6,851.00	4.05
Sex × work	95,987.89	1	95,987.89	56.71
Sex × marital status	10,671.50	1	10,671.50	6.30
Error	5,927,497.00	3,502	1,692.60	

model in which each factor and each interaction contributes significantly is shown in Table 18.8.

It should be noted that the approach led to a reduced model in which each of the original classification variables except sex contributed a highly significant additive effect. With one exception, each of the interactions involved the sex factor. Results from the analysis led to recognition that the level of manifest anxious depression relates to background variables differently among male and female patients. Subsequently, a separate model was fitted to the data for men and women independently to eliminate the interaction effects.

18.5.2 APPLICATION OF METHOD 2 IN CLINICAL DRUG STUDY

Method 2 is considered most appropriate for experimental research viewed in the context of traditional analysis of variance. In a double-blind clinical drug study, 86 male schizophrenic patients were randomly assigned to either chlorpromazine or to a new test drug. Each patient was interviewed and rated independently by two professional observers using the BPRS. A standard, coded psychiatric history was completed at the start of treatment. After 6 weeks of treatment, the same two clinical observers completed follow-up BPRS symptom ratings. Because the patients differed in age, course of illness, work achievement, marital status, and a variety of other characteristics, it was considered important to examine whether interactions between drug treatments and these other variables accounted for significant variance and, if not, to compare the drug treatments in an additive factorial analysis of variance. Patients were classified into paranoid and core-schizophrenic groups by computer analysis of pretreatment BPRS rating profiles. The interaction between drug treatments and patient profile types was considered to be of special interest in

Table 18.9 Summary of Method 2 Analysis of Variance for Post-treatment Thinking-disturbance-factor Scores

Source	*SS*	*df*	*Mean square*	*F*
Pretreat (covariate)	574.19	1	574.19	7.94
Hospitals	244.13	3	81.38	1.13
Age	5.53	1	5.53	.76
Course of illness	658.60	2	329.30	4.55
Work achievement	2.81	1	2.81	.04
Marital status	119.76	2	59.88	.83
Alcohol behavior	4.64	1	4.64	.06
Religious involvement	63.52	2	31.76	.44
Profile type	134.56	1	134.56	1.86
Drugs	176.82	1	176.82	2.45
Drug × age	3.84	1	3.84	.05
Drug × course	258.30	2	129.15	1.79
Drug × alcohol	.34	1	.34	.01
Drug × religious	271.23	2	135.62	1.88
Drug × profile type	295.85	1	295.85	4.09
Error	4,555.92	63	72.32	

evaluating whether the two drugs differed in specific indications for different types of patients.

Results of the least-squares (method 2) analysis of variance are presented in Table 18.9. It will be noted that only the interactions involving drug treatments were included in the model. Technically, it should be assumed that other interactions are nonexistent if they are left out of the model, otherwise the residual error will contain nonrandom components (Graybill, 1961). In practice, the bias resulting from pooling of one or more minor interactions into the error term should be negligible where the number of degrees for error is reasonably large. In addition, the bias should be a conservative one. Considering all the other systematic effects, distribution problems, and other biasing factors, the fact that one cannot reasonably include all possible effects in an analysis should not be a major concern. The present analysis revealed a significant drug-by-profile-type interaction. Examination of adjusted cell means revealed that chlorpromazine was substantially superior to the test drug in treatment of the core-schizophrenic profile type but that there was no appreciable difference between the two drugs in treatment of patients with paranoid hostile-suspiciousness profile patterns. The results from the analysis were interpreted as one would interpret results from an ordinary factorial-design analysis of variance.

18.5.3 APPLICATION OF METHOD 3 IN TESTING A LOGICAL HIERARCHY OF HYPOTHESES

A sample of 1,149 psychiatric patients was drawn in approximately equal proportions from university hospital inpatient and outpatient services for a study of influences of social-class factors on the nature of manifest psychopathology. Psychiatric symptomatology of each patient was recorded on the

BPRS. Background data including father's occupation, composite parental social-class rating, patient's education, and patient's work achievement were recorded.

It was felt that much previous research had failed to distinguish adequately between parental social class and patient social achievement. Whereas patient social achievement may be, at least partially, a result of the incipient psychopathology, it seems less likely that father's occupation and parental social class should be substantially affected by the patient's psychopathology. Accordingly, method 3 was utilized to test a logically ordered hierarchy of hypotheses. Father's work achievement (unskilled vs. other) was tested disregarding other factors. The influence of parental social class (low, low middle, middle, or above) was tested after adjustment for father's work achievement but disregarding patient factors. Patient education was entered into the model next because of temporal and logical priority over patient's work achievement. Finally, the relevance of patient's work achievement for manifest psychopathology was tested after adjustment for relationship to parental social-class variables and patient education.

The criterion measure of primary interest in this study was the schizo-depression contrast function which was computed as the difference between symptom ratings in the area of thinking disturbance (conceptual disorganization, hallucinatory behavior, and unusual thought content) and ratings in the area of anxious depression (anxiety, guilt feelings, and depressive mood). Results of the hierarchical method 3 analysis are presented in Table 18.10. Each of the factors, examined in the sequence specified, proved to be highly significant in relationship to the schizo-depression profile balance. A proneness toward relative prominence of thinking disturbance was apparent in patients whose fathers had lower occupational achievement. Low parental social-class rating after adjustment for father's work achievement contributed further to the thinking-disturbance imbalance. Even after adjustment for both parental factors, low educational achievement of the patient was consistent with still greater prominence of thinking disturbance. Finally, after adjustment for both parental factors plus patient's own educational achievement, low work achievement on the part of the patient was associated with increased prominence of thinking disturbance relative to anxious depression. Had the reverse ordering

Table 18.10 Summary of Method 3 Analysis of Variance for Hierarchy of Hypotheses

Source	*SS*	*df*	*Mean square*	*F*
Father work	14,478.06	1	14,478.06	8.61
Parental social class	10,284.06	2	5,142.03	3.06
Patient education	47,427.21	3	15,809.07	9.37
Patient work	29,684.33	1	29,684.33	17.60
Error	1,924,878.20	1140	1,687.01	

of hypotheses been employed, or had method 1 or 2 been used to test the independent contribution of each factor, neither of the parental variables would have been judged to contribute significantly to differences in relative prominence of thinking-disturbance vs. anxious-depressive symptoms.

18.6 Summary and Conclusions

Important differences exist among least-squares analyses that appear superficially similar to the familiar analysis of variance. Method 2 of this chapter appears to be the most direct generalization of conventional experimental-design analysis of variance. There are special circumstances in which the complete least-squares solution represented by method 1 or the step-down procedure of method 3 may be considered more appropriate. It is important in this era of canned computer programs for the investigator to be aware of differences between the apparently similar methods, to know which method has been employed, and to define the method in sufficient detail so that an informed scientific audience can appreciate what was done and know how to interpret the results.

18.7 Computer Program for General Linear Regression Analysis of Variance and Covariance

18.7.1 INTRODUCTION

Although any program for multiple-regression and multiple-correlation analysis can be used with a properly constructed matrix of dummy variables, the procedure can be awkward due to the necessity for several passes through the computer to calculate the various R^2 values. Also, the construction of the matrix of dummy variates is a laborious process. In this section, a computer program written in Fortran IV for the IBM 1130 disk-oriented computer is presented. This program accepts raw data, including qualitative classification variables which are to be "expanded" into dummy variates, quantitative covariates, and quantitative dependent variables. The classification variables, each with up to 10 categories coded in a single card column, are each expanded into $k - 1$ dummy variates as described in Sec. 8.2. The covariates are treated as quantitative independent variables as described in Chap. 17. One to five different dependent variables can be analyzed in a single pass through the computer.

The program LINAR accomplishes method 2 experimental-design analysis of variance. As a consequence, it is considered especially appropriate where the conception of the problem is that of an ordinary analysis of variance but where unequal and disproportionate cell frequencies render conventional calculations inappropriate. The program permits testing of two-way and three-way interactions, as desired, so long as the total nonerror degrees of freedom do not exceed the order of the matrix which can be stored in available core. (In a 16-K IBM 1130, the model can include something in excess of 50 degrees of freedom for hypotheses to be tested.)

The program LINAR consists of a mainline program and five subroutines.

The subroutines accomplish the following:

1 XPAND reads data cards, constructs the design matrix consisting of dummy variates and quantitative covariates (if any), and then computes the matrix of intercorrelations among all independent and dependent variables.
2 COUNT produces a frequency count for k categories of each independent variable and for cross-classification cells of any two-way interactions that are to be tested. These frequency counts are listed in order of main effects and then two-way interaction tables in sequence.
3 PART partitions the total correlation matrix into independent and dependent variables.
4 MINV computes the inverse of the matrix of intercorrelations among the independent variables (dummy variates).
5 AV2 computes the regression coefficients and sums of squares for the analysis of variance covariance.

18.7.2 INPUT FOR GENERAL LINEAR ANALYSIS (METHOD 2)

1 One card containing number of dependent variables (NFAC), number of independent variables (prior to expansion) (NVAR), number of observations on each dependent variable (NOBS), and number of two-way and three-way interactions to be tested (NINT). These parameters are read from successive four-column fields.
2 One card containing in successive four-column fields the ordinal positions of qualitative (categorical) independent variables that are to be expanded to form dummy variates of the design matrix. The four-column fields corresponding to quantitative covariates are punched zero; for example 12045008, where the independent variables occupying positions 1, 2, 4, 5, and 8 are to be expanded into zero-one dummy variates and variables 3, 6, and 7 are to be treated as quantitative covariates.
3 One card containing in successive four-column fields the number of degrees of freedom associated with each of the NVAR independent variables. Covariates have 1 degree of freedom each, and categorical classification variables have $k - 1$ degrees of freedom.
4 Identification cards: NVAR cards with the name of effects to be tested punched in alphabetic characters in the first 15 columns of each card.
5 Category-grouping cards: one card for each of the NVAR effects to be tested. Each card contains in the first 10 columns the new category designation for individuals coded 0, 1, 2, . . . , 9 on one of the categorical independent variables. For example, 1112220000, where individuals originally coded 0, 1, or 2 are to be placed into treatment group 1, individuals originally coded 3, 4, or 5 are to be placed in treatment group 2, and all other individuals are to be placed in treatment group 3. Note that there are $k - 1$ different category-identification codes on each of the category-grouping cards, with the kth category represented by the zero punches. The category-grouping card associated with a covariate is punched 0000000000.
6 Interaction cards: One card for each two-way and three-way interaction to be tested. All two-way interaction cards should precede any three-way

interaction cards. For two-way interactions, the factor identification numbers for the two factors should be punched into the first two four-column fields. For three-way interactions, the factor identification numbers for the three factors should be punched into the first three four-column fields. Each right-hand term should be larger than the left-hand term:

Two-way:	1	3	0	correct
	3	1	0	incorrect
Three-way:	2	3	4	correct
	2	4	3	incorrect

7 Data cards: Data for each individual read according to format 3 of SUBROUTINE EXPAND. Usually, the independent variables are entered first and the dependent variable or variables second. The independent variables are read as fixed point or integer variables [M(I), I = 1, NVAR] and the dependent variables are read as floating-point or real variables [DATA (I), I = 1, NFAC].

18.7.3 OUTPUT

The output from the general linear-analysis program includes the following:

1. Listing of effects to be tested plus interactions identified by ordinal position of factors, with column of design matrix in which first dummy variable associated with each factor appears.
2. Subclass frequencies for categories of each main effect and then two-way interaction cell frequencies.
3. Partial-regression coefficients for the full model with interactions included; these partial-regression coefficients are associated with the dummy variates and covariates of the design matrix.
4. Partial-regression coefficients associated with main effects after interaction variables are omitted.
5. Summary table for the analysis of variance.

Mainline Program for General Linear Regression Analysis of Variance and Covariance (Method 2)

```
C     GENERAL LINEAR ANALYSIS OF VARIANCE
      DIMENSION M(54),NDF(55),INT(30,3),CELL(160),VCRT(5),XCRT(5),VP(54)
     1,XP(54),E(54,5),N(54)
      COMMON L2,L3,L4,L5,L6,L7,L8
      DEFINE FILE 2(60,10,U,L2), 3(60,120,U,L3), 4(60,120,U,L4), 5(80,16
     10,U,L5),6(60,120,U,L6), 7(60,120,U,L7), 8(32,160,U,L8)
C     XPANDS- XPANDS VARIABLES AND BUILDS CORRELATION MATRIX
C     COUNT- COUNTS CELL FREQUENCIES
C     PART-PARTITIONS MATRIX
C     MINV- INVERTS  MATRIX
C     AV2-EXPERIMENTAL DESIGN MODEL
      CALL XPAND(MVAR,NOBS,NPRE,NCRIT,NINT,M,NDF,INT,NVAR,N,N3WAY,NJ2)
      CALL COUNT(CELL,NVAR,NJ2,NDF,NOBS,INT,M)
      CALL PART (  NPRE,NCRIT,VCRT,XCRT,VP,XP,E)
      CALL MINV(NPRE)
      CALL AV2(NPRE,NCRIT,NOBS,VCRT,XCRT,VP,XP,E,NVAR,N,NINT,NDF,INT,N3W
     1AY,NJ2)
      CALL EXIT
      END
```

Subroutine to Generate Dummy Variates and Compute Intercorrelation Matrix

```
      SUBROUTINE XPAND(MVAR,NOBS,MDM,NFAC,NINT,N,NDF,INT,NVAR,KJK,N3WAY,
     1NJ2)
      DIMENSION N(54),NAME(30,8), NN(54,10),DATA(18),M(54),ND(54),
     1NDF(55),INT(30,3),KJK(54),D(55),R(54,54),XM(54),V(54),XXM(54)
      COMMON L2,L3,L4,L5,L6,L7,L8
    1 FORMAT(20I4)
    2 FORMAT (8A2)
C     FORMAT NUMBER 3 MUST BE CHANGED TO AGREE WITH FORMAT OF INPUT DATA
    3 FORMAT(3I4,F4.0)
    4 FORMAT(1H0,(10F8.3))
    5 FORMAT(10I1)
    6 FORMAT(3I4)
      READ(2,1)NFAC,NVAR,NOBS,NINT
      READ(2,1)(N(J),J=1,NVAR)
      READ (2,1) (NDF(J),J=1,NVAR)
      READ(2,2)((NAME(J,K),K=1,8),J=1,NVAR)
      READ(2,5)((NN(J,K),K=1,10),J=1,NVAR)
      IF(NINT) 8,8,7
    7 READ(2,6)((INT(J,K),K=1,3),J=1,NINT)
    8 FOBS=NOBS
      NJ2=NINT+1
      DO 10 J=1,54
      XXM(J)=0.0
      XM(J)=0.0
      DO 9 K=1,54
    9 R(J,K)=0.0
   10 V(J)=0.0
      L8=1
      DO 41 K=1,NOBS
      READ(2,3)(M(I),I=1,NVAR),(DATA(I),I=1,NFAC)
      NVAR2=NVAR
      NCOL=1
      L=1
      I=1
      DO 23 J=1,NVAR
      IF(J-N(L)) 11,16,11
   11 ND(NCOL)=M(J)
      L=L+1
   12 IF(K-1)23,13,23
   14 FORMAT(4H INT,4X,I2,1H+,I2,1H+,I2,6X,I2)
   15 FORMAT(1H,8A2,5X,I2,5X,13I1)
   13 WRITE(3,15)(NAME(J,KD),KD=1,8),NCOL,(NN(J,JK),JK=1,10)
      WRITE(8'L8)(NAME(J,KD),KD=1,8),NCOL,(NN(J,JK),JK=1,10)
      KJK(I)=NCOL
      I = I + 1
      GO TO 23
   16 MM=M(J)+1
      MM=NN(L,MM)
      MD=NDF(J)
      DO 17 KK=1,MD
      MDM=NCOL+KK-1
   17 ND(MDM)=0
      IF(MM)20,18,20
   18 DO 19 KK=1,MD
      MDM=NCOL+KK-1
      XXM(MDM)=XXM(MDM)+1.0
   19 ND(MDM)=-1
      GO TO 21
   20 MDM=NCOL+MM-1
      ND(MDM)=1
   21 IF(L-NVAR)22,12,12
   22 L=L+1
      GO TO 12
   23 NCOL=NCOL+NDF(J)
      MDM=NCOL-1
      DO 24 J=1,MDM
   24 D(J)=ND(J)
      N3WAY=0
C     CONSTRUCT SIMPLE INTERACTION TERMS
      IF(NINT) 38,38,25
   25 DO 37 J2=1,NINT
      IF(K-1) 27,26,27
   26 WRITE(3,14)(INT(J2,KD),KD=1,3),NCOL
```

```
 27 NR1=INT(J2,1)
    NR2=INT(J2,2)
    NR3=INT(J2,3)
    IF(N3WAY)128,128,127
128 IF(NR3) 127,127,129
129 N3WAY=NCOL
    NJ2=J2
127 NR11=NR1-1
    N1=0
    IF(NR11)28,30,28
 28 DO 29 J3=1,NR11
 29 N1=N1+NDF(J3)
 30 N2=N1+NDF(NR1)
    N1=N1+1
    NR22=NR2-1
    N3=0
    DO 31 J3=1,NR22
 31 N3=N3+NDF(J3)
    N4=N3+NDF(NR2)
    N3=N3+1
    IF(NR3) 131,131,132
132 NR33=NR3-1
    N5=0
    DO 137 J3=1,NR33
137 N5=N5+NDF(J3)
    N6=N5+NDF(NR3)
    N5=N5+1
    GO TO 133
131 N5=55
    N6=55
    D(55)=1
133 JJJ=MDM
    DO 36 J=N1,N2
    DO 35 JJ=N3,N4
    DO 35 JJ3=N5,N6
    JJJ=JJJ+1
    D(JJJ)=D(J)*D(JJ)*D(JJ3)
    IF(D(JJJ)) 32,35,33
 32 XXM(JJJ)=XXM(JJJ)+1.0
    GO TO 35
 33 IF(D(J)) 34,35,35
 34 XXM(JJJ)=XXM(JJJ)-1.0
 35 CONTINUE
 36 CONTINUE
    NVAR2=NVAR2+1
    KJK(NVAR2) = MDM+1
    IF(NR3) 136,136,139
136 NR3=55
    NDF(NR3)=1
139 MDM=MDM+NDF(NR1)*NDF(NR2)*NDF(NR3)
    NCOL=MDM+1
 37 NDF(NVAR2)=NDF(NR1)*NDF(NR2)*NDF(NR3)
 38 DO 39 J=1,NFAC
    JJ=J+MDM
 39 D(JJ)=DATA(J)
    MVAR = MDM + NFAC
    DO 40 J=1,MVAR
    XM(J)=XM(J)+D(J)
    DO 40 JK=1,MVAR
 40 R(J,JK)=R(J,JK)+D(J)*D(JK)
 41 CONTINUE
    DO 42 J=1,MVAR
 42 V(J)=1.0/SQRT(R(J,J)-XM(J)*XM(J)/FOBS)
    DO 44 J=1,MVAR
    DO 43 K=1,MVAR
 43 R(J,K)=(R(J,K)-XM(J)*XM(K)/FOBS)*V(J)*V(K)
 44 WRITE(3'J)(R(J,K),K=1,MVAR)
    L2=1
    DO 45 J=1,MVAR
    D(J)=XM(J)/FOBS
    V(J)=1.0/(V(J)*V(J)*(FOBS-1.0))
    XM(J)=XM(J)+XXM(J)
    WRITE(2'J) D(J),V(J),XM(J)
 45 CONTINUE
    WRITE(3,46)
```

```
46 FORMAT(//,10X,'MEANS')
   WRITE(3,4)(D(J),J=1,MVAR)
   WRITE(3,47)
47 FORMAT(//,10X,'VARIANCES')
   WRITE(3,4)(V(J),J=1,MVAR)
   RETURN
   END
```

Subroutine to Produce Frequency Counts for Main-effect and Simple-interaction Categories

```
      SUBROUTINE COUNT(CELL,NVAR,NJ2,NDF,NOBS,INT,NM)
C     CALCULATE SUBCLASS NUMBERS FOR VARIABLES
      DIMENSION NDF(55),INT(30,3),NM(54),CELL(160)
      COMMON L2,L3,L4,L5,L6,L7,L8
    1 FORMAT(1H0,'SUBCLASS NUMBERS')
  300 FORMAT (1H ,10F5.0)
      WRITE(3,1)
      NJ1=NJ2-1
      FOBS=NOBS
      L2=1
      L=1
      DO 21 K=1,NVAR
      DO 70 NX=1,NVAR
      IF(K-NM(NX))70,71,70
   70 CONTINUE
      CELL(L)=FOBS
      WRITE(3,300) CELL(L)
      READ(2'L2) XM,D,SUM
      L2=L2-1
      WRITE(2'L2) XM,D,CELL(L)
      L=L-1
      GO TO 21
   71 M=NDF(K)+L-1
      CELL(M+1)=0.0
      DO 20 I=L,M
      READ(2'L2)XM,D,SUM
      CELL(I)=SUM
   20 CELL(M+1)=CELL(M+1)+SUM
      CELL(M+1)=FOBS-CELL(M+1)
      IC=M+1
      WRITE(3,300)(CELL(II),II=L,IC)
   21 L=L+NDF(K)+1
      IF(NJ1)80,80,800
  800 DO 30 I=1,NJ1
      IF (INT(I,3))80,2,80
    2 JJ=INT(I,1)
      JJ=INT(I,1)
      KP=INT(I,2)
      JJJ=NDF(JJ)
      KKK=NDF(KP)
      III=NVAR+I
      NDF(III)=JJJ*KKK
      M=L+KKK-1
      DO 31 J=1,JJJ
      CELL(M+1)=0.0
      DO 32 K=L,M
      READ(2'L2)XM,D,SUM
      CELL(K)=SUM
   32 CELL(M+1)=CELL(M+1)+SUM
      ME=0
      IF (JJ-1)80,35,36
   35 ME=J
      GO TO 37
   36 MJJ=JJ-1
      DO 50 MJ=1,MJJ
      ME=ME+NDF(MJ)+1
      DO 42 IX=1,NVAR
      IF(MJ-NM(IX))42,50,42
   42 CONTINUE
      ME=ME-1
   50 CONTINUE
      ME=ME+J
   37 CELL(M+1)=CELL(ME)-CELL(M+1)
```

```
   IC=M+1
   WRITE(3,300)(CELL(II),II=L,IC)
   L=M+2
31 M=M+1+KKK
   KKK=KKK+1
   M=M+1-KKK
   L=L-(JJJ*KKK)
   DO 40 K=1,KKK
   CELL(M+1)=0.0
   DO 33 J=L,M,KKK
33 CELL(M+1)=CELL(M+1)+CELL(J)
   ME=0
   MKK=KP-1
   DO 39 MK=1,MKK
   ME=ME+NDF(MK)+1
   DO 41  IX=1,NVAR
   IF(MK-NM(IX))41,39,41
41 CONTINUE
   ME=ME-1
39 CONTINUE
   ME=ME+K
   CELL(M+1)=CELL(ME)-CELL(M+1)
   M=M+1
40 L=L+1
   L=M-KKK+1
   WRITE(3,300)(CELL(II),II=L,M)
30 L=M+1
80 RETURN
   END
```

Subroutine to Partition Total Correlation Matrix

```
      SUBROUTINE PART(  NPRE,NCRIT,VCRT,XCRT,VP,XP,E)
C     SUB-PROGRAM TO FORM REDUCED CORRELATION MATRIX
      DIMENSION A(54)    ,E(54,5),XP(54),VP(54),XCRT(5),VCRT(5)
      COMMON L2,L3,L4,L5,L6,L7,L8
      NVAR2 = NPRE + NCRIT
      L2=1
      DO 10 J=1,NVAR2
      READ(2'J)XUM,P,SUM
      NX = J
      IF(NX-NPRE) 8,8,7
    8 XP(NX)=XUM
      VP(NX)=SQRT(P)
      GO TO 10
    7 NX=NX-NPRE
      XCRT(NX)= XUM
      VCRT(NX)= SQRT(P)
   10 CONTINUE
      L3=1
      NSTAR=NPRE+1
      DO 15 J=1,NVAR2
      READ(3'J)(A(K),K=1,NVAR2)
      DO 15 KK=NSTAR,NVAR2
      K=KK-NPRE
   15 E(J,K)=A(KK)
      RETURN
      END
```

Subroutine to Compute Inverse of Matrix of Intercorrelations among Independent Variables

```
   SUBROUTINE MINV(N)
   DIMENSION IPIVO(54),INDEX(54,2),PIVOT(54), A(54,54)
   COMMON L2,L3,L4,L5,L6,L7,L8
   DO 21 J=1,N
21 READ(3'J)(A(J,K),K=1,N)
   DET=1.0
   DO 20 J=1,N
20 IPIVO (J)=0
   DO 550 I=1,N
   AMAX=0.0
   DO 105 J=1,N
```

```
      IF (IPIVO (J)-1) 60,105,60
 60   DO 100 K=1,N
      IF (IPIVO (K)-1) 80,100,740
 80   IF(ABS(AMAX)-ABS(A(J,K)))85,100,100
 85   IROW=J
      ICOLU =K
      AMAX=A(J,K)
100   CONTINUE
105   CONTINUE
      IF(ABS (AMAX)-2.0E-7) 800,800,801
800   WRITE(1,666)
666   FORMAT(20H DETERMINANT = ZERO )
      DET =0.0
      PAUSE
801   CONTINUE
      IPIVO (ICOLU )=IPIVO (ICOLU )+1
      IF (IROW-ICOLU ) 140,260,140
140   DET=-DET
      DO 200 L=1,N
      SWAP=A(IROW,L)
      A(IROW,L)=A(ICOLU ,L)
200   A(ICOLU ,L)=SWAP
260   INDEX(I,1)=IROW
      INDEX(I,2)=ICOLU
      PIVOT(I)=A(ICOLU ,ICOLU )
      DET=DET*PIVOT(I)
      A(ICOLU ,ICOLU )=1.0
      DO 350 L=1,N
350   A(ICOLU ,L)=A(ICOLU ,L)/PIVOT(I)
      DO 550 L1=1,N
      IF(L1-ICOLU ) 400,550,400
400   T=A(L1,ICOLU )
      A(L1,ICOLU )=0.0
      DO 450 L=1,N
450   A(L1,L)=A(L1,L)-A(ICOLU ,L)*T
550   CONTINUE
      DO 710 I=1,N
      L=N+1-I
      IF (INDEX(L,1)-INDEX(L,2)) 630,710,630
630   IROW=INDEX(L,1)
      ICOLU =INDEX(L,2)
      DO 705 K=1,N
      SWAP=A(K,IROW)
      A(K,IROW)=A(K,ICOLU )
      A(K,ICOLU )=SWAP
705   CONTINUE
710   CONTINUE
      DO 10 I=1,N
 10   WRITE(4'I )(A(I,J),J=1,N)
740   RETURN
      END
```

Subroutine to Calculate Various Regression Functions and Related Sums of Squares

```
      SUBROUTINE AV2(   NPRE,NCRIT,NOBS,VCRT,XCRT,VP,XP,E,NVAR,N,NINT,NDF
     1,INT,N3WAY,NJ2)
      DIMENSION A(54,54),E(54,5),XP(54),VP(54),XCRT(5),VCRT(5),IX(80),
     1SUM(54),Q(54),D(55),N(54),F(54),SS(54),XS(54),NDF(55),INT(30,3),
     2 NAME(8),NN(10)
      COMMON L2,L3,L4,L5,L6,L7,L8
131   FORMAT(1H ,10HSS DELETE=,E18.8,2X,3HDF=,I4,2X,3HMS=,E18.8,2X,2HF=,
     1F10.4)
120   FORMAT(1H ,//23H ANALYSIS FOR CRITERION,I5)
135   FORMAT(1H ,10X,7HSTD. WT,5X,6HRAW WT)
907   FORMAT(1H ,I2,5X,E18.8,10X,E18.8,10X,F6.0)
133   FORMAT(1H ,9H CONSTANT, E18.8)
 16   FORMAT(1H ,12H RESIDUAL SS,E18.8,6X,3H MS,E18.8,6X,3H DF,F6.0)
 15   FORMAT(1H ,3H R=,F10.4,2X,5HRSQR=,F10.4,2X,3HSS=,E18.8,2X,3HDF=,F8
     1.0)
      L3=1
      L5 = 1
      NWAY = NVAR + NINT
```

```
      NVAR2 = NPRE + NCRIT
      JK = NWAY + 1
      N(JK) = NPRE + 1
      NSTAR = 1
      JJK=JK+1
      DO 318 K=1,JJK
      DO 317 I=1,80
  317 IX(I)= 0
  318 WRITE(5'L5)(IX(I),I=1,80)
      DO 319 M=1,80
  319 IX(M) = M
      L5 = 1
      I=2
      IF (NINT)648,648,647
  648 DO 651 J=1,NVAR
      NSTOP = N(I)-1
      WRITE(5'L5)(IX(K),K=NSTAR,NSTOP)
      NSTAR=N(I)
  651 I=I+1
      GO TO 341
  647 MINT=NVAR + 1
      MK=N(MINT)
      DO 320 J =1,NVAR
      NSTOP =N(I)-1
      WRITE(5'L5)(IX(K),K=NSTAR,NSTOP),(IX(L),L=MK,NPRE)
      NSTAR=N(I)
  320 I=I+1
      NJ1=NJ2-1
      IF(NJ1-NINT) 321,3321,321
 3321 DO 3340 J=1,NINT
      NSTOP=N(I)-1
      WRITE(5'L5)(IX(K),K=NSTAR,NSTOP)
      NSTAR=N(I)
 3340 I=I+1
      GO TO 421
  321 DO 340 J=1,NJ1
      NSTOP =N(I)-1
      WRITE(5'L5)(IX(K),K=NSTAR,NSTOP),(IX(K),K=N3WAY,NPRE)
      NSTAR=N(I)
  340 I = I+1
      IF(I-JJK  ) 420,421,420
  420 DO 400 J=NJ2,NINT
      NSTOP=N(I)-1
      WRITE(5'L5)(IX(K),K=NSTAR,NSTOP)
      NSTAR=N(I)
  400 I=I+1
  421 WRITE(5'L5)(IX(K),K=MK,NPRE)
      IF(NJ1-NINT) 423,341,423
  423 WRITE(5'L5)(IX(K),K=N3WAY,NPRE)
  341 IX(1)=0
      IX(1)= 0
      WRITE(5'L5)IX
      FNOBS=NOBS
      L4=1
      DO 741 J=1,NPRE
  741 READ(4'L4)(A(J,K),K=1,NPRE)
C     ANALYZE SUCCESSIVELY K CRITERION VARIABLES
      DO 950 KK=1,NCRIT
      L3=1
      L2=1
      L7 = 1
      L5=1
      M=0
      M2=0
      K2=NPRE+KK
      WRITE(3,120)KK
      WRITE(3,135)
      DO 905 I=1,NPRE
      D(I)=0.0
      DO 905 J=1,NPRE
  905 D(I)=D(I)+A(I,J)*E(J,KK)
      L6=1
      WRITE(6'L6) D
      DFR=NOBS-NPRE-1
      FNPRE=NPRE
```

```
      SST=VCRT(KK)*VCRT(KK)*(FNOBS-1.0)
 270  CONTINUE
      RSQR=0.0
      DO 906 I=1,NPRE
 906  RSQR=RSQR+D(I)*E(I,KK)
      SSE1=SST*(1.0-RSQR)
      IF(M)802,802,803
 802  ERR=SSE1/DFR
      ERROR = ERR
      SSE2 =SSE1
 803  M=M+1
      SUM(M)=SST*RSQR
      XYZ=0
      DO 1905 I=1,NPRE
      Q(I)=(VCRT(KK)/VP(I))*D(I)
      IF(M-1) 1906,1904,1906
1906  IF(JJK  -M)1905,1804,1905
1804  DDSUM=0
      XYZ=1
      GO TO 904
1904  WRITE(3'L3) Q(I)
      XYZ=1
      READ(2'L2) XM,DDD,DDSUM
 904  WRITE(3,907) I,D(I),Q(I),DDSUM
1905  CONTINUE
      CON=XCRT(KK)
      DO 925 I=1,NPRE
 925  CON=CON-Q(I)*XP(I)
      IF(XYZ-1)1889,1888,1889
1888  WRITE(3,133) CON
1889  R=SQRT(RSQR)
      IF(XYZ-1) 1779,1777,1779
1777  WRITE(3,15) R,RSQR,SUM(M),DFR
      WRITE(3,16)SSE1,ERR,DFR
1779  DFM2=M2
      DIFSS=SUM(1)-SUM(M)
      IF(M2)926,926,927
 927  VM2=DIFSS/DFM2
      RATIO=VM2/ERR
      IF(XYZ-1)1669,1666,1669
1666  WRITE(3,131) DIFSS,M2,VM2,RATIO
1669  WRITE(7'L7) DIFSS,M2,VM2,RATIO
 926  CONTINUE
      READ(5'L5)IX
      M2=0
      L4=1
      DO 777 J=1,NPRE
 777  READ(4'L4)(A(J,K),K=1,NPRE)
      L6=1
      READ(6'L6)D
      IF (IX(1))948,948,931
 931  M2=0
      DO 935 K6=1,80
      IF(IX(K6))270,270,236
 236  K=IX(K6)
      M2=M2+1
      DK= D(K)
      DO 243 J=1,NPRE
      IF(D(J))246,243,246
 246  D(J)= D(J)-DK*A(J,K)/A(K,K)
 243  CONTINUE
      D(K)=0.0
      DO 888 I=1,NPRE
 888  Q(I)= A(I,K)
      DO 889 I=1,NPRE
      DO 889 J=I,NPRE
      A(I,J)= A(I,J)- Q(I)*Q(J)/Q(K)
 889  A(J,I)=A(I,J)
 935  CONTINUE
 948  IF(NINT)950,950,949
 949  L7=1
      DO 10 I=1,JJK
      READ(7'L7) DIFSS,M2,VM2,RATIO
      SS(I)=DIFSS
      XS(I)=VM2
```

```
 10 F(I)=RATIO
    DO 12 I=1,NVAR
    SS(I)=SS(I)-SS(JK)
    XS(I)=SS(I)/NDF(I)
 12 F(I)=XS(I)/ERROR
    IF(NJ1-NINT) 251,250,251
251 DO 110 I=1,NJ1
    II=NVAR+I
    SS(II)=SS(II)-SS(JJK)
    XS(II)=SS(II)/NDF(II)
110 F(II)=XS(II)/ERROR
250 L8=1
    DO 14 I=1,NVAR
    READ(8'L8) NAME,NCOL,NN
 14 WRITE(3,112) I,SS(I),NDF(I),XS(I),F(I),NAME,NCOL,NN
    I = NVAR + 1
    DO 18 J=1,NJ1
    WRITE(3,114) INT(J,1),INT(J,2),SS(I),NDF(I),XS(I),F(I)
 18 I=I+1
    IF(NJ1-NINT) 213,219,213
213 DO 218 J=NJ2,NINT
    WRITE(3,214) INT(J,1),INT(J,2),INT(J,3),SS(I),NDF(I),XS(I),F(I)
214 FORMAT(I2,1X,I2,1X,I2,E15.8,I3,E18.8,F7.3)
218 I=I+1
219 WRITE(3,118) SSE2,DFR,ERROR
114 FORMAT(I2,1X,I2, E18.8,I3,E18.8,F7.3)
112 FORMAT(I5,E18.8,I3,E18.8,F7.3,8A2,5X,I2,5X,13I1)
118 FORMAT(6H ERROR, E17.8,F4.0,E16.8)
950 CONTINUE
    RETURN
    END
```

18.8 References

Bottenberg, R., and J. H. Ward: "Applied Multiple Linear Regression Analysis," Department of Commerce, Office of Technical Services, 1960.

Bradley, H. E.: Multiple Classification Analysis for Arbitrary Experimental Design, *Technometrics*, **10**:13–28 (1968).

Cohen, J.: Multiple Regression as a General Data-analytic System, *Psychol. Bull.*, **70**:426–443 (1968*a*).

———: Prognostic Factors in Functional Psychosis: A Study in Multivariate Methodology, *Trans. N.Y. Acad. Sci.*, **30**:833–840 (1968*b*).

Graybill, F. A.: "An Introduction to Linear Statistical Models," McGraw-Hill, New York, 1961.

Jennings, E.: Fixed Effects Analysis of Variance by Regression Analysis, *Multivar. Behav. Res.*, **2**:95–108 (1967).

Lindquist, E. F.: "Design and Analysis of Experiments in Psychology and Education," Houghton Mifflin, Boston, 1953.

Overall, J. E., and D. K. Spiegel: Concerning Least Squares Analysis of Experimental Data, *Psychol. Bull.*, **72**:311–322 (1969).

Rao, C. R.: "Linear Statistical Inference and Its Applications," Wiley, New York, 1965.

Snedecor, G. W., and W. G. Cochran: "Statistical Methods," 6th ed., Iowa State University Press, Ames, 1967.

Ward, J. H.: Multiple Linear Regression Models, in H. Borko (ed.), "Computer Applications in the Behavioral Sciences," Prentice-Hall, Englewood Cliffs, N.J., 1962.

Winer, B. J.: "Statistical Principles in Experimental Design," McGraw-Hill, New York, 1962.

Appendix

Table A.1 Areas and Ordinates of the Normal Curve in Terms of x/σ

(1) z Standard Score $\left(\frac{x}{\sigma}\right)$	(2) A Area from Mean to $\frac{x}{\sigma}$	(3) B Area in Larger Portion	(4) C Area in Smaller Portion	(5) y Ordinate at $\frac{x}{\sigma}$
0.00	.0000	.5000	.5000	.3989
0.01	.0040	.5040	.4960	.3989
0.02	.0080	.5080	.4920	.3989
0.03	.0120	.5120	.4880	.3988
0.04	.0160	.5160	.4840	.3986
0.05	.0199	.5199	.4801	.3984
0.06	.0239	.5239	.4761	.3982
0.07	.0279	.5279	.4721	.3980
0.08	.0319	.5319	.4681	.3977
0.09	.0359	.5359	.4641	.3973
0.10	.0398	.5398	.4602	.3970
0.11	.0438	.5438	.4562	.3965
0.12	.0478	.5478	.4522	.3961
0.13	.0517	.5517	.4483	.3956
0.14	.0557	.5557	.4443	.3951
0.15	.0596	.5596	.4404	.3945
0.16	.0636	.5636	.4364	.3939
0.17	.0675	.5675	.4325	.3932
0.18	.0714	.5714	.4286	.3925
0.19	.0753	.5753	.4247	.3918
0.20	.0793	.5793	.4207	.3910
0.21	.0832	.5832	.4168	.3902
0.22	.0871	.5871	.4129	.3894
0.23	.0910	.5910	.4090	.3885
0.24	.0948	.5948	.4052	.3876
0.25	.0987	.5987	.4013	.3867
0.26	.1026	.6026	.3974	.3857
0.27	.1064	.6064	.3936	.3847
0.28	.1103	.6103	.3897	.3836
0.29	.1141	.6141	.3859	.3825
0.30	.1179	.6179	.3821	.3814
0.31	.1217	.6217	.3783	.3802
0.32	.1255	.6255	.3745	.3790
0.33	.1293	.6293	.3707	.3778
0.34	.1331	.6331	.3669	.3765

Table A.1 Areas and Ordinates of the Normal Curve in Terms of x/σ *(Continued)*

(1) z Standard Score $\left(\frac{x}{\sigma}\right)$	(2) A Area from Mean to $\frac{x}{\sigma}$	(3) B Area in Larger Portion	(4) C Area in Smaller Portion	(5) y Ordinate at $\frac{x}{\sigma}$
0.35	.1368	.6368	.3632	.3752
0.36	.1406	.6406	.3594	.3739
0.37	.1443	.6443	.3557	.3725
0.38	.1480	.6480	.3520	.3712
0.39	.1517	.6517	.3483	.3697
0.40	.1554	.6554	.3446	.3683
0.41	.1591	.6591	.3409	.3668
0.42	.1628	.6628	.3372	.3653
0.43	.1664	.6664	.3336	.3637
0.44	.1700	.6700	.3300	.3621
0.45	.1736	.6736	.3264	.3605
0.46	.1772	.6772	.3228	.3589
0.47	.1808	.6808	.3192	.3572
0.48	.1844	.6844	.3156	.3555
0.49	.1879	.6879	.3121	.3538
0.50	.1915	.6915	.3085	.3521
0.51	.1950	.6950	.3050	.3503
0.52	.1985	.6985	.3015	.3485
0.53	.2019	.7019	.2981	.3467
0.54	.2054	.7054	.2946	.3448
0.55	.2088	.7088	.2912	.3429
0.56	.2123	.7123	.2877	.3410
0.57	.2157	.7157	.2843	.3391
0.58	.2190	.7190	.2810	.3372
0.59	.2224	.7224	.2776	.3352
0.60	.2257	.7257	.2743	.3332
0.61	.2291	.7291	.2709	.3312
0.62	.2324	.7324	.2676	.3292
0.63	.2357	.7357	.2643	.3271
0.64	.2389	.7389	.2611	.3251
0.65	.2422	.7422	.2578	.3230
0.66	.2454	.7454	.2546	.3209
0.67	.2486	.7486	.2514	.3187
0.68	.2517	.7517	.2483	.3166
0.69	.2549	.7549	.2451	.3144

Table A.1 Areas and Ordinates of the Normal Curve in Terms of x/σ (*Continued*)

(1) z Standard Score $\left(\frac{x}{\sigma}\right)$	(2) A Area from Mean to $\frac{x}{\sigma}$	(3) B Area in Larger Portion	(4) C Area in Smaller Portion	(5) y Ordinate at $\frac{x}{\sigma}$
0.70	.2580	.7580	.2420	.3123
0.71	.2611	.7611	.2389	.3101
0.72	.2642	.7642	.2358	.3079
0.73	.2673	.7673	.2327	.3056
0.74	.2704	.7704	.2296	.3034
0.75	.2734	.7734	.2266	.3011
0.76	.2764	.7764	.2236	.2989
0.77	.2794	.7794	.2206	.2966
0.78	.2823	.7823	.2177	.2943
0.79	.2852	.7852	.2148	.2920
0.80	.2881	.7881	.2119	.2897
0.81	.2910	.7910	.2090	.2874
0.82	.2939	.7939	.2061	.2850
0.83	.2967	.7967	.2033	.2827
0.84	.2995	.7995	.2005	.2803
0.85	.3023	.8023	.1977	.2780
0.86	.3051	.8051	.1949	.2756
0.87	.3078	.8078	.1922	.2732
0.88	.3106	.8106	.1894	.2709
0.89	.3133	.8133	.1867	.2685
0.90	.3159	.8159	.1841	.2661
0.91	.3186	.8186	.1814	.2637
0.92	.3212	.8212	.1788	.2613
0.93	.3238	.8238	.1762	.2589
0.94	.3264	.8264	.1736	.2565
0.95	.3289	.8289	.1711	.2541
0.96	.3315	.8315	.1685	.2516
0.97	.3340	.8340	.1660	.2492
0.98	.3365	.8365	.1635	.2468
0.99	.3389	.8389	.1611	.2444
1.00	.3413	.8413	.1587	.2420
1.01	.3438	.8438	.1562	.2396
1.02	.3461	.8461	.1539	.2371
1.03	.3485	.8485	.1515	.2347
1.04	.3508	.8508	.1492	.2323

Table A.1 Areas and Ordinates of the Normal Curve in Terms of x/σ (*Continued*)

(1) z Standard Score $\left(\frac{x}{\sigma}\right)$	(2) A Area from Mean to $\frac{x}{\sigma}$	(3) B Area in Larger Portion	(4) C Area in Smaller Portion	(5) y Ordinate at $\frac{x}{\sigma}$
1.05	.3531	.8531	.1469	.2299
1.06	.3554	.8554	.1446	.2275
1.07	.3577	.8577	.1423	.2251
1.08	.3599	.8599	.1401	.2227
1.09	.3621	.8621	.1379	.2203
1.10	.3643	.8643	.1357	.2179
1.11	.3665	.8665	.1335	.2155
1.12	.3686	.8686	.1314	.2131
1.13	.3708	.8708	.1292	.2107
1.14	.3729	.8729	.1271	.2083
1.15	.3749	.8749	.1251	.2059
1.16	.3770	.8770	.1230	.2036
1.17	.3790	.8790	.1210	.2012
1.18	.3810	.8810	.1190	.1989
1.19	.3830	.8830	.1170	.1965
1.20	.3849	.8849	.1151	.1942
1.21	.3869	.8869	.1131	.1919
1.22	.3888	.8888	.1112	.1895
1.23	.3907	.8907	.1093	.1872
1.24	.3925	.8925	.1075	.1849
1.25	.3944	.8944	.1056	.1826
1.26	.3962	.8962	.1038	.1804
1.27	.3980	.8980	.1020	.1781
1.28	.3997	.8997	.1003	.1758
1.29	.4015	.9015	.0985	.1736
1.30	.4032	.9032	.0968	.1714
1.31	.4049	.9049	.0951	.1691
1.32	.4066	.9066	.0934	.1669
1.33	.4082	.9082	.0918	.1647
1.34	.4099	.9099	.0901	.1626
1.35	.4115	.9115	.0885	.1604
1.36	.4131	.9131	.0869	.1582
1.37	.4147	.9147	.0853	.1561
1.38	.4162	.9162	.0838	.1539
1.39	.4177	.9177	.0823	.1518

Table A.1 Areas and Ordinates of the Normal Curve in Terms of x/σ (*Continued*)

(1) z Standard Score $\left(\frac{x}{\sigma}\right)$	(2) A Area from Mean to $\frac{x}{\sigma}$	(3) B Area in Larger Portion	(4) C Area in Smaller Portion	(5) y Ordinate at $\frac{x}{\sigma}$
1.40	.4192	.9192	.0808	.1497
1.41	.4207	.9207	.0793	.1476
1.42	.4222	.9222	.0778	.1456
1.43	.4236	.9236	.0764	.1435
1.44	.4251	.9251	.0749	.1415
1.45	.4265	.9265	.0735	.1394
1.46	.4279	.9279	.0721	.1374
1.47	.4292	.9292	.0708	.1354
1.48	.4306	.9306	.0694	.1334
1.49	.4319	.9319	.0681	.1315
1.50	.4332	.9332	.0668	.1295
1.51	.4345	.9345	.0655	.1276
1.52	.4357	.9357	.0643	.1257
1.53	.4370	.9370	.0630	.1238
1.54	.4382	.9382	.0618	.1219
1.55	.4394	.9394	.0606	.1200
1.56	.4406	.9406	.0594	.1182
1.57	.4418	.9418	.0582	.1163
1.58	.4429	.9429	.0571	.1145
1.59	.4441	.9441	.0559	.1127
1.60	.4452	.9452	.0548	.1109
1.61	.4463	.9463	.0537	.1092
1.62	.4474	.9474	.0526	.1074
1.63	.4484	.9484	.0516	.1057
1.64	.4495	.9495	.0505	.1040
1.65	.4505	.9505	.0495	.1023
1.66	.4515	.9515	.0485	.1006
1.67	.4525	.9525	.0475	.0989
1.68	.4535	.9535	.0465	.0973
1.69	.4545	.9545	.0455	.0957
1.70	.4554	.9554	.0446	.0940
1.71	.4564	.9564	.0436	.0925
1.72	.4573	.9573	.0427	.0909
1.73	.4582	.9582	.0418	.0893
1.74	.4591	.9591	.0409	.0878

Table A.1 Areas and Ordinates of the Normal Curve in Terms of x/σ (*Continued*)

(1) z Standard Score $\left(\frac{x}{\sigma}\right)$	(2) A Area from Mean to $\frac{x}{\sigma}$	(3) B Area in Larger Portion	(4) C Area in Smaller Portion	(5) y Ordinate at $\frac{x}{\sigma}$
1.75	.4599	.9599	.0401	.0863
1.76	.4608	.9608	.0392	.0848
1.77	.4616	.9616	.0384	.0833
1.78	.4625	.9625	.0375	.0818
1.79	.4633	.9633	.0367	.0804
1.80	.4641	.9641	.0359	.0790
1.81	.4649	.9649	.0351	.0775
1.82	.4656	.9656	.0344	.0761
1.83	.4664	.9664	.0336	.0748
1.84	.4671	.9671	.0329	.0734
1.85	.4678	.9678	.0322	.0721
1.86	.4686	.9686	.0314	.0707
1.87	.4693	.9693	.0307	.0694
1.88	.4699	.9699	.0301	.0681
1.89	.4706	.9706	.0294	.0669
1.90	.4713	.9713	.0287	.0656
1.91	.4719	.9719	.0281	.0644
1.92	.4726	.9726	.0274	.0632
1.93	.4732	.9732	.0268	.0620
1.94	.4738	.9738	.0262	.0608
1.95	.4744	.9744	.0256	.0596
1.96	.4750	.9750	.0250	.0584
1.97	.4756	.9756	.0244	.0573
1.98	.4761	.9761	.0239	.0562
1.99	.4767	.9767	.0233	.0551
2.00	.4772	.9772	.0228	.0540
2.01	.4778	.9778	.0222	.0529
2.02	.4783	.9783	.0217	.0519
2.03	.4788	.9788	.0212	.0508
2.04	.4793	.9793	.0207	.0498
2.05	.4798	.9798	.0202	.0488
2.06	.4803	.9803	.0197	.0478
2.07	.4808	.9808	.0192	.0468
2.08	.4812	.9812	.0188	.0459
2.09	.4817	.9817	.0183	.0449

Table A.1 Areas and Ordinates of the Normal Curve in Terms of x/σ (*Continued*)

(1) z Standard Score $\left(\frac{x}{\sigma}\right)$	(2) A Area from Mean to $\frac{x}{\sigma}$	(3) B Area in Larger Portion	(4) C Area in Smaller Portion	(5) y Ordinate at $\frac{x}{\sigma}$
2.10	.4821	.9821	.0179	.0440
2.11	.4826	.9826	.0174	.0431
2.12	.4830	.9830	.0170	.0422
2.13	.4834	.9834	.0166	.0413
2.14	.4838	.9838	.0162	.0404
2.15	.4842	.9842	.0158	.0396
2.16	.4846	.9846	.0154	.0387
2.17	.4850	.9850	.0150	.0379
2.18	.4854	.9854	.0146	.0371
2.19	.4857	.9857	.0143	.0363
2.20	.4861	.9861	.0139	.0355
2.21	.4864	.9864	.0136	.0347
2.22	.4868	.9868	.0132	.0339
2.23	.4871	.9871	.0129	.0332
2.24	.4875	.9875	.0125	.0325
2.25	.4878	.9878	.0122	.0317
2.26	.4881	.9881	.0119	.0310
2.27	.4884	.9884	.0116	.0303
2.28	.4887	.9887	.0113	.0297
2.29	.4890	.9890	.0110	.0290
2.30	.4893	.9893	.0107	.0283
2.31	.4896	.9896	.0104	.0277
2.32	.4898	.9898	.0102	.0270
2.33	.4901	.9901	.0099	.0264
2.34	.4904	.9904	.0096	.0258
2.35	.4906	.9906	.0094	.0252
2.36	.4909	.9909	.0091	.0246
2.37	.4911	.9911	.0089	.0241
2.38	.4913	.9913	.0087	.0235
2.39	.4916	.9916	.0084	.0229
2.40	.4918	.9918	.0082	.0224
2.41	.4920	.9920	.0080	.0219
2.42	.4922	.9922	.0078	.0213
2.43	.4925	.9925	.0075	.0208
2.44	.4927	.9927	.0073	.0203

Table A.1 Areas and Ordinates of the Normal Curve in Terms of x/σ (*Continued*)

(1) z Standard Score $\left(\frac{x}{\sigma}\right)$	(2) A Area from Mean to $\frac{x}{\sigma}$	(3) B Area in Larger Portion	(4) C Area in Smaller Portion	(5) y Ordinate at $\frac{x}{\sigma}$
2.45	.4929	.9929	.0071	.0198
2.46	.4931	.9931	.0069	.0194
2.47	.4932	.9932	.0068	.0189
2.48	.4934	.9934	.0066	.0184
2.49	.4936	.9936	.0064	.0180
2.50	.4938	.9938	.0062	.0175
2.51	.4940	.9940	.0060	.0171
2.52	.4941	.9941	.0059	.0167
2.53	.4943	.9943	.0057	.0163
2.54	.4945	.9945	.0055	.0158
2.55	.4946	.9946	.0054	.0154
2.56	.4948	.9948	.0052	.0151
2.57	.4949	.9949	.0051	.0147
2.58	.4951	.9951	.0049	.0143
2.59	.4952	.9952	.0048	.0139
2.60	.4953	.9953	.0047	.0136
2.61	.4955	.9955	.0045	.0132
2.62	.4956	.9956	.0044	.0129
2.63	.4957	.9957	.0043	.0126
2.64	.4959	.9959	.0041	.0122
2.65	.4960	.9960	.0040	.0119
2.66	.4961	.9961	.0039	.0116
2.67	.4962	.9962	.0038	.0113
2.68	.4963	.9963	.0037	.0110
2.69	.4964	.9964	.0036	.0107
2.70	.4965	.9965	.0035	.0104
2.71	.4966	.9966	.0034	.0101
2.72	.4967	.9967	.0033	.0099
2.73	.4968	.9968	.0032	.0096
2.74	.4969	.9969	.0031	.0093
2.75	.4970	.9970	.0030	.0091
2.76	.4971	.9971	.0029	.0088
2.77	.4972	.9972	.0028	.0086
2.78	.4973	.9973	.0027	.0084
2.79	.4974	.9974	.0026	.0081

Table A.1 **Areas and Ordinates of the Normal Curve in Terms of** x/σ (*Continued*)

(1) z Standard Score $\left(\frac{x}{\sigma}\right)$	(2) A Area from Mean to $\frac{x}{\sigma}$	(3) B Area in Larger Portion	(4) C Area in Smaller Portion	(5) y Ordinate at $\frac{x}{\sigma}$
2.80	.4974	.9974	.0026	.0079
2.81	.4975	.9975	.0025	.0077
2.82	.4976	.9976	.0024	.0075
2.83	.4977	.9977	.0023	.0073
2.84	.4977	.9977	.0023	.0071
2.85	.4978	.9978	.0022	.0069
2.86	.4979	.9979	.0021	.0067
2.87	.4979	.9979	.0021	.0065
2.88	.4980	.9980	.0020	.0063
2.89	.4981	.9981	.0019	.0061
2.90	.4981	.9981	.0019	.0060
2.91	.4982	.9982	.0018	.0058
2.92	.4982	.9982	.0018	.0056
2.93	.4983	.9983	.0017	.0055
2.94	.4984	.9984	.0016	.0053
2.95	.4984	.9984	.0016	.0051
2.96	.4985	.9985	.0015	.0050
2.97	.4985	.9985	.0015	.0048
2.98	.4986	.9986	.0014	.0047
2.99	.4986	.9986	.0014	.0046
3.00	.4987	.9987	.0013	.0044
3.01	.4987	.9987	.0013	.0043
3.02	.4987	.9987	.0013	.0042
3.03	.4988	.9988	.0012	.0040
3.04	.4988	.9988	.0012	.0039
3.05	.4989	.9989	.0011	.0038
3.06	.4989	.9989	.0011	.0037
3.07	.4989	.9989	.0011	.0036
3.08	.4990	.9990	.0010	.0035
3.09	.4990	.9990	.0010	.0034
3.10	.4990	.9990	.0010	.0033
3.11	.4991	.9991	.0009	.0032
3.12	.4991	.9991	.0009	.0031
3.13	.4991	.9991	.0009	.0030
3.14	.4992	.9992	.0008	.0029

Table A.1 Areas and Ordinates of the Normal Curve in Terms of x/σ (*Continued*)

(1) z Standard Score $\left(\frac{x}{\sigma}\right)$	(2) A Area from Mean to $\frac{x}{\sigma}$	(3) B Area in Larger Portion	(4) C Area in Smaller Portion	(5) y Ordinate at $\frac{x}{\sigma}$
3.15	.4992	.9992	.0008	.0028
3.16	.4992	.9992	.0008	.0027
3.17	.4992	.9992	.0008	.0026
3.18	.4993	.9993	.0007	.0025
3.19	.4993	.9993	.0007	.0025
3.20	.4993	.9993	.0007	.0024
3.21	.4993	.9993	.0007	.0023
3.22	.4994	.9994	.0006	.0022
3.23	.4994	.9994	.0006	.0022
3.24	.4994	.9994	.0006	.0021
3.30	.4995	.9995	.0005	.0017
3.40	.4997	.9997	.0003	.0012
3.50	.4998	.9998	.0002	.0009
3.60	.4998	.9998	.0002	.0006
3.70	.4999	.9999	.0001	.0004

Table A.2 The 5 (Roman Type) and 1 (Boldface Type) Percent Points for the Distribution of F

n_2	n_1 degrees of freedom (for greater mean square)																							
	1	2	3	4	5	6	7	8	9	10	11	12	14	16	20	24	30	40	50	75	100	200	500	∞
1	161	200	216	225	230	234	237	239	241	242	243	244	245	246	248	249	250	251	252	253	253	254	254	254
	4,052	**4,999**	**5,403**	**5,625**	**5,764**	**5,859**	**5,928**	**5,981**	**6,022**	**6,056**	**6,082**	**6,106**	**6,142**	**6,169**	**6,208**	**6,234**	**6,258**	**6,286**	**6,302**	**6,323**	**6,334**	**6,352**	**6,361**	**6,366**
2	18.51	19.00	19.16	19.25	19.30	19.33	19.36	19.37	19.38	19.39	19.40	19.41	19.42	19.43	19.44	19.45	19.46	19.47	19.47	19.48	19.49	19.49	19.50	19.50
	98.49	**99.00**	**99.17**	**99.25**	**99.30**	**99.33**	**99.34**	**99.36**	**99.38**	**99.40**	**99.41**	**99.42**	**99.43**	**99.44**	**99.45**	**99.46**	**99.47**	**99.48**	**99.48**	**99.49**	**99.49**	**99.49**	**99.50**	**99.50**
3	10.13	9.55	9.28	9.12	9.01	8.94	8.88	8.84	8.81	8.78	8.76	8.74	8.71	8.69	8.66	8.64	8.62	8.60	8.58	8.57	8.56	8.54	8.54	8.53
	34.12	**30.82**	**29.46**	**28.71**	**28.24**	**27.91**	**27.67**	**27.49**	**27.34**	**27.23**	**27.13**	**27.05**	**26.92**	**26.83**	**26.69**	**26.60**	**26.50**	**26.41**	**26.35**	**26.27**	**26.23**	**26.18**	**26.14**	**26.12**
4	7.71	6.94	6.59	6.39	6.26	6.16	6.09	6.04	6.00	5.96	5.93	5.91	5.87	5.84	5.80	5.77	5.74	5.71	5.70	5.68	5.66	5.65	5.64	5.63
	21.20	**18.00**	**16.69**	**15.98**	**15.52**	**15.21**	**14.98**	**14.80**	**14.66**	**14.54**	**14.45**	**14.37**	**14.24**	**14.15**	**14.02**	**13.93**	**13.83**	**13.74**	**13.69**	**13.61**	**13.57**	**13.52**	**13.48**	**13.46**
5	6.61	5.79	5.41	5.19	5.05	4.95	4.88	4.82	4.78	4.74	4.70	4.68	4.64	4.60	4.56	4.53	4.50	4.46	4.44	4.42	4.40	4.38	4.37	4.36
	16.26	**13.27**	**12.06**	**11.39**	**10.97**	**10.67**	**10.45**	**10.27**	**10.15**	**10.05**	**9.96**	**9.89**	**9.77**	**9.68**	**9.55**	**9.47**	**9.38**	**9.29**	**9.24**	**9.17**	**9.13**	**9.07**	**9.04**	**9.02**
6	5.99	5.14	4.76	4.53	4.39	4.28	4.21	4.15	4.10	4.06	4.03	4.00	3.96	3.92	3.87	3.84	3.81	3.77	3.75	3.72	3.71	3.69	3.68	3.67
	13.74	**10.92**	**9.78**	**9.15**	**8.75**	**8.47**	**8.26**	**8.10**	**7.98**	**7.87**	**7.79**	**7.72**	**7.60**	**7.52**	**7.39**	**7.31**	**7.23**	**7.14**	**7.09**	**7.02**	**6.99**	**6.94**	**6.90**	**6.88**
7	5.59	4.74	4.35	4.12	3.97	3.87	3.79	3.73	3.68	3.63	3.60	3.57	3.52	3.49	3.44	3.41	3.38	3.34	3.32	3.29	3.28	3.25	3.24	3.23
	12.25	**9.55**	**8.45**	**7.85**	**7.46**	**7.19**	**7.00**	**6.84**	**6.71**	**6.62**	**6.54**	**6.47**	**6.35**	**6.27**	**6.15**	**6.07**	**5.98**	**5.90**	**5.85**	**5.78**	**5.75**	**5.70**	**5.67**	**5.65**
8	5.32	4.46	4.07	3.84	3.69	3.58	3.50	3.44	3.39	3.34	3.31	3.28	3.23	3.20	3.15	3.12	3.08	3.05	3.03	3.00	2.98	2.96	2.94	2.93
	11.26	**8.65**	**7.59**	**7.01**	**6.63**	**6.37**	**6.19**	**6.03**	**5.91**	**5.82**	**5.74**	**5.67**	**5.56**	**5.48**	**5.36**	**5.28**	**5.20**	**5.11**	**5.06**	**5.00**	**4.96**	**4.91**	**4.88**	**4.86**
9	5.12	4.26	3.86	3.63	3.48	3.37	3.29	3.23	3.18	3.13	3.10	3.07	3.02	2.98	2.93	2.90	2.86	2.82	2.80	2.77	2.76	2.73	2.72	2.71
	10.56	**8.02**	**6.99**	**6.42**	**6.06**	**5.80**	**5.62**	**5.47**	**5.35**	**5.26**	**5.18**	**5.11**	**5.00**	**4.92**	**4.80**	**4.73**	**4.64**	**4.56**	**4.51**	**4.45**	**4.41**	**4.36**	**4.33**	**4.31**
10	4.96	4.10	3.71	3.48	3.33	3.22	3.14	3.07	3.02	2.97	2.94	2.91	2.86	2.82	2.77	2.74	2.70	2.67	2.64	2.61	2.59	2.56	2.55	2.54
	10.04	**7.56**	**6.55**	**5.99**	**5.64**	**5.39**	**5.21**	**5.06**	**4.95**	**4.85**	**4.78**	**4.71**	**4.60**	**4.52**	**4.41**	**4.33**	**4.25**	**4.17**	**4.12**	**4.05**	**4.01**	**3.96**	**3.93**	**3.91**
11	4.84	3.98	3.59	3.36	3.20	3.09	3.01	2.95	2.90	2.86	2.82	2.79	2.74	2.70	2.65	2.61	2.57	2.53	2.50	2.47	2.45	2.42	2.41	2.40
	9.65	**7.20**	**6.22**	**5.67**	**5.32**	**5.07**	**4.88**	**4.74**	**4.63**	**4.54**	**4.46**	**4.40**	**4.29**	**4.21**	**4.10**	**4.02**	**3.94**	**3.86**	**3.80**	**3.74**	**3.70**	**3.66**	**3.62**	**3.60**
12	4.75	3.88	3.49	3.26	3.11	3.00	2.92	2.85	2.80	2.76	2.72	2.69	2.64	2.60	2.54	2.50	2.46	2.42	2.40	2.36	2.35	2.32	2.31	2.30
	9.33	**6.93**	**5.95**	**5.41**	**5.06**	**4.82**	**4.65**	**4.50**	**4.39**	**4.30**	**4.22**	**4.16**	**4.05**	**3.98**	**3.86**	**3.78**	**3.70**	**3.61**	**3.56**	**3.49**	**3.46**	**3.41**	**3.38**	**3.36**
13	4.67	3.80	3.41	3.18	3.02	2.92	2.84	2.77	2.72	2.67	2.63	2.60	2.55	2.51	2.46	2.42	2.38	2.34	2.32	2.28	2.26	2.24	2.22	2.21
	9.07	**6.70**	**5.74**	**5.20**	**4.86**	**4.62**	**4.44**	**4.30**	**4.19**	**4.10**	**4.02**	**3.96**	**3.85**	**3.78**	**3.67**	**3.59**	**3.51**	**3.42**	**3.37**	**3.30**	**3.27**	**3.21**	**3.18**	**3.16**

Source: Reproduced from Snedecor and Cochran, 'Statistical Methods,'' 6th ed., Iowa State University Press, Ames, Iowa, 1967, by permission of the author and publisher.

Table A.2 The 5 (Roman Type) and 1 (Boldface Type) Percent Points for the Distribution of F (*Continued*)

n_2	n_1 degrees of freedom (for greater mean square)																							
	1	2	3	4	5	6	7	8	9	10	11	12	14	16	20	24	30	40	50	75	100	200	500	∞
14	4.60	3.74	3.34	3.11	2.96	2.85	2.77	2.70	2.65	2.60	2.56	2.53	2.48	2.44	2.39	2.35	2.31	2.27	2.24	2.21	2.19	2.16	2.14	2.13
	8.86	**6.51**	**5.56**	**5.03**	**4.69**	**4.46**	**4.28**	**4.14**	**4.03**	**3.94**	**3.86**	**3.80**	**3.70**	**3.62**	**3.51**	**3.43**	**3.34**	**3.26**	**3.21**	**3.14**	**3.11**	**3.06**	**3.02**	**3.00**
15	4.54	3.68	3.29	3.06	2.90	2.79	2.70	2.64	2.59	2.55	2.51	2.48	2.43	2.39	2.33	2.29	2.25	2.21	2.18	2.15	2.12	2.10	2.08	2.07
	8.68	**6.36**	**5.42**	**4.89**	**4.56**	**4.32**	**4.14**	**4.00**	**3.89**	**3.80**	**3.73**	**3.67**	**3.56**	**3.48**	**3.36**	**3.29**	**3.20**	**3.12**	**3.07**	**3.00**	**2.97**	**2.92**	**2.89**	**2.87**
16	4.49	3.63	3.24	3.01	2.85	2.74	2.66	2.59	2.54	2.49	2.45	2.42	2.37	2.33	2.28	2.24	2.20	2.16	2.13	2.09	2.07	2.04	2.02	2.01
	8.53	**6.23**	**5.29**	**4.77**	**4.44**	**4.20**	**4.03**	**3.89**	**3.78**	**3.69**	**3.61**	**3.55**	**3.45**	**3.37**	**3.25**	**3.18**	**3.10**	**3.01**	**2.96**	**2.89**	**2.86**	**2.80**	**2.77**	**2.75**
17	4.45	3.59	3.20	2.96	2.81	2.70	2.62	2.55	2.50	2.45	2.41	2.38	2.33	2.29	2.23	2.19	2.15	2.11	2.08	2.04	2.02	1.99	1.97	1.96
	8.40	**6.11**	**5.18**	**4.67**	**4.34**	**4.10**	**3.93**	**3.79**	**3.68**	**3.59**	**3.52**	**3.45**	**3.35**	**3.27**	**3.16**	**3.08**	**3.00**	**2.92**	**2.86**	**2.79**	**2.76**	**2.70**	**2.67**	**2.65**
18	4.41	3.55	3.16	2.93	2.77	2.66	2.58	2.51	2.46	2.41	2.37	2.34	2.29	2.25	2.19	2.15	2.11	2.07	2.04	2.00	1.98	1.95	1.93	1.92
	8.28	**6.01**	**5.09**	**4.58**	**4.25**	**4.01**	**3.85**	**3.71**	**3.60**	**3.51**	**3.44**	**3.37**	**3.27**	**3.19**	**3.07**	**3.00**	**2.91**	**2.83**	**2.78**	**2.71**	**2.68**	**2.62**	**2.59**	**2.57**
19	4.38	3.52	3.13	2.90	2.74	2.63	2.55	2.48	2.43	2.38	2.34	2.31	2.26	2.21	2.15	2.11	2.07	2.02	2.00	1.96	1.94	1.91	1.90	1.88
	8.18	**5.93**	**5.01**	**4.50**	**4.17**	**3.94**	**3.77**	**3.63**	**3.52**	**3.43**	**3.36**	**3.30**	**3.19**	**3.12**	**3.00**	**2.92**	**2.84**	**2.76**	**2.70**	**2.63**	**2.60**	**2.54**	**2.51**	**2.49**
20	4.35	3.49	3.10	2.87	2.71	2.60	2.52	2.45	2.40	2.35	2.31	2.28	2.23	2.18	2.12	2.08	2.04	1.99	1.96	1.92	1.90	1.87	1.85	1.84
	8.10	**5.85**	**4.94**	**4.43**	**4.10**	**3.87**	**3.71**	**3.56**	**3.45**	**3.37**	**3.30**	**3.23**	**3.13**	**3.05**	**2.94**	**2.86**	**2.77**	**2.69**	**2.63**	**2.56**	**2.53**	**2.47**	**2.44**	**2.42**
21	4.32	3.47	3.07	2.84	2.68	2.57	2.49	2.42	2.37	2.32	2.28	2.25	2.20	2.15	2.09	2.05	2.00	1.96	1.93	1.89	1.87	1.84	1.82	1.81
	8.02	**5.78**	**4.87**	**4.37**	**4.04**	**3.81**	**3.65**	**3.51**	**3.40**	**3.31**	**3.24**	**3.17**	**3.07**	**2.99**	**2.88**	**2.80**	**2.72**	**2.63**	**2.58**	**2.51**	**2.47**	**2.42**	**2.38**	**2.36**
22	4.30	3.44	3.05	2.82	2.66	2.55	2.47	2.40	2.35	2.30	2.26	2.23	2.18	2.13	2.07	2.03	1.98	1.93	1.91	1.87	1.84	1.81	1.80	1.78
	7.94	**5.72**	**4.82**	**4.31**	**3.99**	**3.76**	**3.59**	**3.45**	**3.35**	**3.26**	**3.18**	**3.12**	**3.02**	**2.94**	**2.83**	**2.75**	**2.67**	**2.58**	**2.53**	**2.46**	**2.42**	**2.37**	**2.33**	**2.31**
23	4.28	3.42	3.03	2.80	2.64	2.53	2.45	2.38	2.32	2.28	2.24	2.20	2.14	2.10	2.04	2.00	1.96	1.91	1.88	1.84	1.82	1.79	1.77	1.76
	7.88	**5.66**	**4.76**	**4.26**	**3.94**	**3.71**	**3.54**	**3.41**	**3.30**	**3.21**	**3.14**	**3.07**	**2.97**	**2.89**	**2.78**	**2.70**	**2.62**	**2.53**	**2.48**	**2.41**	**2.37**	**2.32**	**2.28**	**2.26**
24	4.26	3.40	3.01	2.78	2.62	2.51	2.43	2.36	2.30	2.26	2.22	2.18	2.13	2.09	2.02	1.98	1.94	1.89	1.86	1.82	1.80	1.76	1.74	1.73
	7.82	**5.61**	**4.72**	**4.22**	**3.90**	**3.67**	**3.50**	**3.36**	**3.25**	**3.17**	**3.09**	**3.03**	**2.93**	**2.85**	**2.74**	**2.66**	**2.58**	**2.49**	**2.44**	**2.36**	**2.33**	**2.27**	**2.23**	**2.21**
25	4.24	3.38	2.99	2.76	2.60	2.49	2.41	2.34	2.28	2.24	2.20	2.16	2.11	2.06	2.00	1.96	1.92	1.87	1.84	1.80	1.77	1.74	1.72	1.71
	7.77	**5.57**	**4.68**	**4.18**	**3.86**	**3.63**	**3.46**	**3.32**	**3.21**	**3.13**	**3.05**	**2.99**	**2.89**	**2.81**	**2.70**	**2.62**	**2.54**	**2.45**	**2.40**	**2.32**	**2.29**	**2.23**	**2.19**	**2.17**
26	4.22	3.37	2.98	2.74	2.59	2.47	2.39	2.32	2.27	2.22	2.18	2.15	2.10	2.05	1.99	1.95	1.90	1.85	1.82	1.78	1.76	1.72	1.70	1.69
	7.72	**5.53**	**4.64**	**4.14**	**3.82**	**3.59**	**3.42**	**3.29**	**3.17**	**3.09**	**3.02**	**2.96**	**2.86**	**2.77**	**2.66**	**2.58**	**2.50**	**2.41**	**2.36**	**2.28**	**2.25**	**2.19**	**2.15**	**2.13**

Table A.2 The 5 (Roman Type) and 1 (Boldface Type) Percent Points for the Distribution of F (*Continued*)

n_2	n_1 degrees of freedom (for greater mean square)																							
	1	2	3	4	5	6	7	8	9	10	11	12	14	16	20	24	30	40	50	75	100	200	500	∞
27	4.21	3.35	2.96	2.73	2.57	2.46	2.37	2.30	2.25	2.20	2.16	2.13	2.08	2.03	1.97	1.93	1.88	1.84	1.80	1.76	1.74	1.71	1.68	1.67
	7.68	**5.49**	**4.60**	**4.11**	**3.79**	**3.56**	**3.39**	**3.26**	**3.14**	**3.06**	**2.98**	**2.93**	**2.83**	**2.74**	**2.63**	**2.55**	**2.47**	**2.38**	**2.33**	**2.25**	**2.21**	**2.16**	**2.12**	**2.10**
28	4.20	3.34	2.95	2.71	2.56	2.44	2.36	2.29	2.24	2.19	2.15	2.12	2.06	2.02	1.96	1.91	1.87	1.81	1.78	1.75	1.72	1.69	1.67	1.65
	7.64	**5.45**	**4.57**	**4.07**	**3.76**	**3.53**	**3.36**	**3.23**	**3.11**	**3.03**	**2.95**	**2.90**	**2.80**	**2.71**	**2.60**	**2.52**	**2.44**	**2.35**	**2.30**	**2.22**	**2.18**	**2.13**	**2.09**	**2.06**
29	4.18	3.33	2.93	2.70	2.54	2.43	2.35	2.28	2.22	2.18	2.14	2.10	2.05	2.00	1.94	1.90	1.85	1.80	1.77	1.73	1.71	1.68	1.65	1.64
	7.60	**5.42**	**4.54**	**4.04**	**3.73**	**3.50**	**3.33**	**3.20**	**3.08**	**3.00**	**2.92**	**2.87**	**2.77**	**2.68**	**2.57**	**2.49**	**2.41**	**2.32**	**2.27**	**2.19**	**2.15**	**2.10**	**2.06**	**2.03**
30	4.17	3.32	2.92	2.69	2.53	2.42	2.34	2.27	2.21	2.16	2.12	2.09	2.04	1.99	1.93	1.89	1.84	1.79	1.76	1.72	1.69	1.66	1.64	1.62
	7.56	**5.39**	**4.51**	**4.02**	**3.70**	**3.47**	**3.30**	**3.17**	**3.06**	**2.98**	**2.90**	**2.84**	**2.74**	**2.66**	**2.55**	**2.47**	**2.38**	**2.29**	**2.24**	**2.16**	**2.13**	**2.07**	**2.03**	**2.01**
32	4.15	3.30	2.90	2.67	2.51	2.40	2.32	2.25	2.19	2.14	2.10	2.07	2.02	1.97	1.91	1.86	1.82	1.76	1.74	1.69	1.67	1.64	1.61	1.59
	7.50	**5.34**	**4.46**	**3.97**	**3.66**	**3.42**	**3.25**	**3.12**	**3.01**	**2.94**	**2.86**	**2.80**	**2.70**	**2.62**	**2.51**	**2.42**	**2.34**	**2.25**	**2.20**	**2.12**	**2.08**	**2.02**	**1.98**	**1.96**
34	4.13	3.28	2.88	2.65	2.49	2.38	2.30	2.23	2.17	2.12	2.08	2.05	2.00	1.95	1.89	1.84	1.80	1.74	1.71	1.67	1.64	1.61	1.59	1.57
	7.44	**5.29**	**4.42**	**3.93**	**3.61**	**3.38**	**3.21**	**3.08**	**2.97**	**2.89**	**2.82**	**2.76**	**2.66**	**2.58**	**2.47**	**2.38**	**2.30**	**2.21**	**2.15**	**2.08**	**2.04**	**1.98**	**1.94**	**1.91**
36	4.11	3.26	2.86	2.63	2.48	2.36	2.28	2.21	2.15	2.10	2.06	2.03	1.98	1.93	1.87	1.82	1.78	1.72	1.69	1.65	1.62	1.59	1.56	1.55
	7.39	**5.25**	**4.38**	**3.89**	**3.58**	**3.35**	**3.18**	**3.04**	**2.94**	**2.86**	**2.78**	**2.72**	**2.62**	**2.54**	**2.43**	**2.35**	**2.26**	**2.17**	**2.12**	**2.04**	**2.00**	**1.94**	**1.90**	**1.87**
38	4.10	3.25	2.85	2.62	2.46	2.35	2.26	2.19	2.14	2.09	2.05	2.02	1.96	1.92	1.85	1.80	1.76	1.71	1.67	1.63	1.60	1.57	1.54	1.53
	7.35	**5.21**	**4.34**	**3.86**	**3.54**	**3.32**	**3.15**	**3.02**	**2.91**	**2.82**	**2.75**	**2.69**	**2.59**	**2.51**	**2.40**	**2.32**	**2.22**	**2.14**	**2.08**	**2.00**	**1.97**	**1.90**	**1.86**	**1.84**
40	4.08	3.23	2.84	2.61	2.45	2.34	2.25	2.18	2.12	2.07	2.04	2.00	1.95	1.90	1.84	1.79	1.74	1.69	1.66	1.61	1.59	1.55	1.53	1.51
	7.31	**5.18**	**4.31**	**3.83**	**3.51**	**3.29**	**3.12**	**2.99**	**2.88**	**2.80**	**2.73**	**2.66**	**2.56**	**2.49**	**2.37**	**2.29**	**2.20**	**2.11**	**2.05**	**1.97**	**1.94**	**1.88**	**1.84**	**1.81**
42	4.07	3.22	2.83	2.59	2.44	2.32	2.24	2.17	2.11	2.06	2.02	1.99	1.94	1.89	1.82	1.78	1.73	1.68	1.64	1.60	1.57	1.54	1.51	1.49
	7.27	**5.15**	**4.29**	**3.80**	**3.49**	**3.26**	**3.10**	**2.96**	**2.86**	**2.77**	**2.70**	**2.64**	**2.54**	**2.46**	**2.35**	**2.26**	**2.17**	**2.08**	**2.02**	**1.94**	**1.91**	**1.85**	**1.80**	**1.78**
44	4.06	3.21	2.82	2.58	2.43	2.31	2.23	2.16	2.10	2.05	2.01	1.98	1.92	1.88	1.81	1.76	1.72	1.66	1.63	1.58	1.56	1.52	1.50	1.48
	7.24	**5.12**	**4.26**	**3.78**	**3.46**	**3.24**	**3.07**	**2.94**	**2.84**	**2.75**	**2.68**	**2.62**	**2.52**	**2.44**	**2.32**	**2.24**	**2.15**	**2.06**	**2.00**	**1.92**	**1.88**	**1.82**	**1.78**	**1.75**
46	4.05	3.20	2.81	2.57	2.42	2.30	2.22	2.14	2.09	2.04	2.00	1.97	1.91	1.87	1.80	1.75	1.71	1.65	1.62	1.57	1.54	1.51	1.48	1.46
	7.21	**5.10**	**4.24**	**3.76**	**3.44**	**3.22**	**3.05**	**2.92**	**2.82**	**2.73**	**2.66**	**2.60**	**2.50**	**2.42**	**2.30**	**2.22**	**2.13**	**2.04**	**1.98**	**1.90**	**1.86**	**1.80**	**1.76**	**1.72**
48	4.04	3.19	2.80	2.56	2.41	2.30	2.21	2.14	2.08	2.03	1.99	1.96	1.90	1.86	1.79	1.74	1.70	1.64	1.61	1.56	1.53	1.50	1.47	1.45
	7.19	**5.08**	**4.22**	**3.74**	**3.42**	**3.20**	**3.04**	**2.90**	**2.80**	**2.71**	**2.64**	**2.58**	**2.48**	**2.40**	**2.28**	**2.20**	**2.11**	**2.02**	**1.96**	**1.88**	**1.84**	**1.78**	**1.73**	**1.70**

Table A.2 The 5 (Roman Type) and 1 (Boldface Type) Percent Points for the Distribution of F (*Continued*)

n_2	n_1 degrees of freedom (for greater mean square)																						
	1	2	3	4	5	6	7	8	9	10	11	12	14	16	20	24	30	40	50	75	100	200	500	∞
50	4.03	3.18	2.79	2.56	2.40	2.29	2.20	2.13	2.07	2.02	1.98	1.95	1.90	1.85	1.78	1.74	1.69	1.63	1.60	1.55	1.52	1.48	1.46	1.44
	7.17	**5.06**	**4.20**	**3.72**	**3.41**	**3.18**	**3.02**	**2.88**	**2.78**	**2.70**	**2.62**	**2.56**	**2.46**	**2.39**	**2.26**	**2.18**	**2.10**	**2.00**	**1.94**	**1.86**	**1.82**	**1.76**	**1.71**	**1.68**
55	4.02	3.17	2.78	2.54	2.38	2.27	2.18	2.11	2.05	2.00	1.97	1.93	1.88	1.83	1.76	1.72	1.67	1.61	1.58	1.52	1.50	1.46	1.43	1.41
	7.12	**5.01**	**4.16**	**3.68**	**3.37**	**3.15**	**2.98**	**2.85**	**2.75**	**2.66**	**2.59**	**2.53**	**2.43**	**2.35**	**2.23**	**2.15**	**2.06**	**1.96**	**1.90**	**1.82**	**1.78**	**1.71**	**1.66**	**1.64**
60	4.00	3.15	2.76	2.52	2.37	2.25	2.17	2.10	2.04	1.99	1.95	1.92	1.86	1.81	1.75	1.70	1.65	1.59	1.56	1.50	1.48	1.44	1.41	1.39
	7.08	**4.98**	**4.13**	**3.65**	**3.34**	**3.12**	**2.95**	**2.82**	**2.72**	**2.63**	**2.56**	**2.50**	**2.40**	**2.32**	**2.20**	**2.12**	**2.03**	**1.93**	**1.87**	**1.79**	**1.74**	**1.68**	**1.63**	**1.60**
65	3.99	3.14	2.75	2.51	2.36	2.24	2.15	2.08	2.02	1.98	1.94	1.90	1.85	1.80	1.73	1.68	1.63	1.57	1.54	1.49	1.46	1.42	1.39	1.37
	7.04	**4.95**	**4.10**	**3.62**	**3.31**	**3.09**	**2.93**	**2.79**	**2.70**	**2.61**	**2.54**	**2.47**	**2.37**	**2.30**	**2.18**	**2.09**	**2.00**	**1.90**	**1.84**	**1.76**	**1.71**	**1.64**	**1.60**	**1.56**
70	3.98	3.13	2.74	2.50	2.35	2.23	2.14	2.07	2.01	1.97	1.93	1.89	1.84	1.79	1.72	1.67	1.62	1.56	1.53	1.47	1.45	1.40	1.37	1.35
	7.01	**4.92**	**4.08**	**3.60**	**3.29**	**3.07**	**2.91**	**2.77**	**2.67**	**2.59**	**2.51**	**2.45**	**2.35**	**2.28**	**2.15**	**2.07**	**1.98**	**1.88**	**1.82**	**1.74**	**1.69**	**1.62**	**1.56**	**1.53**
80	3.96	3.11	2.72	2.48	2.33	2.21	2.12	2.05	1.99	1.95	1.91	1.88	1.82	1.77	1.70	1.65	1.60	1.54	1.51	1.45	1.42	1.38	1.35	1.32
	6.96	**4.88**	**4.04**	**3.56**	**3.25**	**3.04**	**2.87**	**2.74**	**2.64**	**2.55**	**2.48**	**2.41**	**2.32**	**2.24**	**2.11**	**2.03**	**1.94**	**1.84**	**1.78**	**1.70**	**1.65**	**1.57**	**1.52**	**1.49**
100	3.94	3.09	2.70	2.46	2.30	2.19	2.10	2.03	1.97	1.92	1.88	1.85	1.79	1.75	1.68	1.63	1.57	1.51	1.48	1.42	1.39	1.34	1.30	1.28
	6.90	**4.82**	**3.98**	**3.51**	**3.20**	**2.99**	**2.82**	**2.69**	**2.59**	**2.51**	**2.43**	**2.36**	**2.26**	**2.19**	**2.06**	**1.98**	**1.89**	**1.79**	**1.73**	**1.64**	**1.59**	**1.51**	**1.46**	**1.43**
125	3.92	3.07	2.68	2.44	2.29	2.17	2.08	2.01	1.95	1.90	1.86	1.83	1.77	1.72	1.65	1.60	1.55	1.49	1.45	1.39	1.36	1.31	1.27	1.25
	6.84	**4.78**	**3.94**	**3.47**	**3.17**	**2.95**	**2.79**	**2.65**	**2.56**	**2.47**	**2.40**	**2.33**	**2.23**	**2.15**	**2.03**	**1.94**	**1.85**	**1.75**	**1.68**	**1.59**	**1.54**	**1.46**	**1.40**	**1.37**
150	3.91	3.06	2.67	2.43	2.27	2.16	2.07	2.00	1.94	1.89	1.85	1.82	1.76	1.71	1.64	1.59	1.54	1.47	1.44	1.37	1.34	1.29	1.25	1.22
	6.81	**4.75**	**3.91**	**3.44**	**3.14**	**2.92**	**2.76**	**2.62**	**2.53**	**2.44**	**2.37**	**2.30**	**2.20**	**2.12**	**2.00**	**1.91**	**1.83**	**1.72**	**1.66**	**1.56**	**1.51**	**1.43**	**1.37**	**1.33**
200	3.89	3.04	2.65	2.41	2.26	2.14	2.05	1.98	1.92	1.87	1.83	1.80	1.74	1.69	1.62	1.57	1.52	1.45	1.42	1.35	1.32	1.26	1.22	1.19
	6.76	**4.71**	**3.88**	**3.41**	**3.11**	**2.90**	**2.73**	**2.60**	**2.50**	**2.41**	**2.34**	**2.28**	**2.17**	**2.09**	**1.97**	**1.88**	**1.79**	**1.69**	**1.62**	**1.53**	**1.48**	**1.39**	**1.33**	**1.28**
400	3.86	3.02	2.62	2.39	2.23	2.12	2.03	1.96	1.90	1.85	1.81	1.78	1.72	1.67	1.60	1.54	1.49	1.42	1.38	1.32	1.28	1.22	1.16	1.13
	6.70	**4.66**	**3.83**	**3.36**	**3.06**	**2.85**	**2.69**	**2.55**	**2.46**	**2.37**	**2.29**	**2.23**	**2.12**	**2.04**	**1.92**	**1.84**	**1.74**	**1.64**	**1.57**	**1.47**	**1.42**	**1.32**	**1.24**	**1.19**
1000	3.85	3.00	2.61	2.38	2.22	2.10	2.02	1.95	1.89	1.84	1.80	1.76	1.70	1.65	1.58	1.53	1.47	1.41	1.36	1.30	1.26	1.19	1.13	1.08
	6.66	**4.62**	**3.80**	**3.34**	**3.04**	**2.82**	**2.66**	**2.53**	**2.43**	**2.34**	**2.26**	**2.20**	**2.09**	**2.01**	**1.89**	**1.81**	**1.71**	**1.61**	**1.54**	**1.44**	**1.38**	**1.28**	**1.19**	**1.11**
∞	3.84	2.99	2.60	2.37	2.21	2.09	2.01	1.94	1.88	1.83	1.79	1.75	1.69	1.64	1.57	1.52	1.46	1.40	1.35	1.28	1.24	1.17	1.11	1.00
	6.64	**4.60**	**3.78**	**3.32**	**3.02**	**2.80**	**2.64**	**2.51**	**2.41**	**2.32**	**2.24**	**2.18**	**2.07**	**1.99**	**1.87**	**1.79**	**1.69**	**1.59**	**1.52**	**1.41**	**1.36**	**1.25**	**1.15**	**1.00**

Table A.3 Table of χ^2†

Degrees of Freedom df	P = .99	.98	.95	.90	.80	.70	.50	.30	.20	.10	.05	.02	.01
1	.000157	.000628	.00393	.0158	.0642	.148	.455	1.074	1.642	2.706	3.841	5.412	6.635
2	.0201	.0404	.103	.211	.446	.713	1.386	2.408	3.219	4.605	5.991	7.824	9.210
3	.115	.185	.352	.584	1.005	1.424	2.366	3.665	4.642	6.251	7.815	9.837	11.341
4	.297	.429	.711	1.064	1.649	2.195	3.357	4.878	5.989	7.779	9.488	11.668	13.277
5	.554	.752	1.145	1.610	2.343	3.000	4.351	6.064	7.289	9.236	11.070	13.388	15.086
6	.872	1.134	1.635	2.204	3.070	3.828	5.348	7.231	8.558	10.645	12.592	15.033	16.812
7	1.239	1.564	2.167	2.833	3.822	4.671	6.346	8.383	9.803	12.017	14.067	16.622	18.475
8	1.646	2.032	2.733	3.490	4.594	5.527	7.344	9.524	11.030	13.362	15.507	18.168	20.090
9	2.088	2.532	3.325	4.168	5.380	6.393	8.343	10.656	12.242	14.684	16.919	19.679	21.666
10	2.558	3.059	3.940	4.865	6.179	7.267	9.342	11.781	13.442	15.987	18.307	21.161	23.209
11	3.053	3.609	4.575	5.578	6.989	8.148	10.341	12.899	14.631	17.275	19.675	22.618	24.725
12	3.571	4.178	5.226	6.304	7.807	9.034	11.340	14.011	15.812	18.549	21.026	24.054	26.217
13	4.107	4.765	5.892	7.042	8.634	9.926	12.340	15.119	16.985	19.812	22.362	25.472	27.688
14	4.660	5.368	6.571	7.790	9.467	10.821	13.339	16.222	18.151	21.064	23.685	26.873	29.141
15	5.229	5.985	7.261	8.547	10.307	11.721	14.339	17.322	19.311	22.307	24.996	28.259	30.578
16	5.812	6.614	7.962	9.312	11.152	12.624	15.338	18.418	20.465	23.542	26.296	29.633	32.000
17	6.408	7.255	8.672	10.085	12.002	13.531	16.338	19.511	21.615	24.769	27.587	30.995	33.409
18	7.015	7.906	9.390	10.865	12.857	14.440	17.338	20.601	22.760	25.989	28.869	32.346	34.805
19	7.633	8.567	10.117	11.651	13.716	15.352	18.338	21.689	23.900	27.204	30.144	33.687	36.191
20	8.260	9.237	10.851	12.443	14.578	16.266	19.337	22.775	25.038	28.412	31.410	35.020	37.566
21	8.897	9.915	11.591	13.240	15.445	17.182	20.337	23.858	26.171	29.615	32.671	36.343	38.932
22	9.542	10.600	12.338	14.041	16.314	18.101	21.337	24.939	27.301	30.813	33.924	37.659	40.289
23	10.196	11.293	13.091	14.848	17.187	19.021	22.337	26.018	28.429	32.007	35.172	38.968	41.638
24	10.856	11.992	13.848	15.659	18.062	19.943	23.337	27.096	29.553	33.196	36.415	40.270	42.980
25	11.524	12.697	14.611	16.473	18.940	20.867	24.337	28.172	30.675	34.382	37.652	41.566	44.314
26	12.198	13.409	15.379	17.292	19.820	21.792	25.336	29.246	31.795	35.563	38.885	42.856	45.642
27	12.879	14.125	16.151	18.114	20.703	22.719	26.336	30.319	32.912	36.741	40.113	44.140	46.963
28	13.565	14.847	16.928	18.939	21.588	23.647	27.336	31.391	34.027	37.916	41.337	45.419	48.278
29	14.256	15.574	17.708	19.768	22.475	24.577	28.336	32.461	35.139	39.087	42.557	46.693	49.588
30	14.953	16.306	18.493	20.599	23.364	25.508	29.336	33.530	36.250	40.256	43.773	47.962	50.892

† For larger values of df, the expression $\sqrt{2\chi^2} - \sqrt{2(df) - 1}$ may be used as a normal deviate with unit standard error.

Source: From Fisher, "Statistical Methods for Research Workers," table III, Oliver & Boyd, Edinburgh, by permission of the author and publishers.

Table A.4 Natural Logarithms (Negative Values, Decreasing)

N	0	1	2	3	4	5	6	7	8	9
0.000		-9.21034	-8.51719	-8.11173	-7.82405	-7.60090	-7.41858	-7.26443	-7.13090	-7.01312
0.001	-6.90776	-6.81245	-6.72543	-6.64539	-6.57128	-6.50229	-6.43775	-6.37713	-6.31997	-6.26590
0.002	-6.21461	-6.16582	-6.11930	-6.07485	-6.03229	-5.99146	-5.95224	-5.91450	-5.87814	-5.84304
0.003	-5.80914	-5.77635	-5.74460	-5.71383	-5.68398	-5.65499	-5.62682	-5.59942	-5.57275	-5.54678
0.004	-5.52146	-5.49677	-5.47267	-5.44914	-5.42615	-5.40368	-5.38170	-5.36019	-5.33914	-5.31852
0.005	-5.29832	-5.27851	-5.25910	-5.24005	-5.22136	-5.20301	-5.18499	-5.16729	-5.14990	-5.13280
0.006	-5.11600	-5.09947	-5.08321	-5.06721	-5.05146	-5.03595	-5.02069	-5.00565	-4.99083	-4.97623
0.007	-4.96185	-4.94766	-4.93367	-4.91988	-4.90628	-4.89285	-4.87961	-4.86653	-4.85363	-4.84089
0.008	-4.82831	-4.81589	-4.80362	-4.79150	-4.77952	-4.76769	-4.75599	-4.74443	-4.73300	-4.72170
0.009	-4.71053	-4.69948	-4.68855	-4.67774	-4.66705	-4.65646	-4.64599	-4.63563	-4.62537	-4.61522
0.010	-4.60517	-4.59522	-4.58537	-4.57561	-4.56595	-4.55638	-4.54690	-4.53751	-4.52821	-4.51899
0.011	-4.50986	-4.50081	-4.49184	-4.48295	-4.47414	-4.46541	-4.45675	-4.44817	-4.43966	-4.43122
0.012	-4.42285	-4.41455	-4.40632	-4.39816	-4.39006	-4.38203	-4.37406	-4.36615	-4.35831	-4.35053
0.013	-4.34281	-4.33514	-4.32754	-4.31999	-4.31250	-4.30507	-4.29769	-4.29036	-4.28309	-4.27587
0.014	-4.26870	-4.26158	-4.25451	-4.24750	-4.24053	-4.23361	-4.22673	-4.21991	-4.21313	-4.20639
0.015	-4.19971	-4.19306	-4.18646	-4.17990	-4.17339	-4.16692	-4.16048	-4.15409	-4.14775	-4.14144
0.016	-4.13517	-4.12894	-4.12274	-4.11659	-4.11047	-4.10439	-4.09835	-4.09235	-4.08638	-4.08044
0.017	-4.07454	-4.06868	-4.06285	-4.05705	-4.05129	-4.04555	-4.03986	-4.03419	-4.02856	-4.02295
0.018	-4.01738	-4.01184	-4.00633	-4.00085	-3.99540	-3.98998	-3.98459	-3.97923	-3.97390	-3.96859
0.019	-3.96332	-3.95807	-3.95285	-3.94765	-3.94248	-3.93734	-3.93223	-3.92714	-3.92207	-3.91704
0.020	-3.91202	-3.90704	-3.90207	-3.89713	-3.89222	-3.88733	-3.88246	-3.87762	-3.87280	-3.86801
0.021	-3.86323	-3.85848	-3.85375	-3.84905	-3.84436	-3.83970	-3.83506	-3.83044	-3.82585	-3.82127
0.022	-3.81671	-3.81218	-3.80766	-3.80317	-3.79869	-3.79424	-3.78981	-3.78539	-3.78099	-3.77662
0.023	-3.77226	-3.76792	-3.76360	-3.75930	-3.75502	-3.75075	-3.74651	-3.74228	-3.73807	-3.73388
0.024	-3.72970	-3.72554	-3.72140	-3.71728	-3.71317	-3.70908	-3.70501	-3.70095	-3.69691	-3.69289
0.025	-3.68888	-3.68489	-3.68091	-3.67695	-3.67301	-3.66908	-3.66516	-3.66126	-3.65738	-3.65351
0.026	-3.64966	-3.64582	-3.64200	-3.63819	-3.63439	-3.63061	-3.62684	-3.62309	-3.61935	-3.61563
0.027	-3.61192	-3.60822	-3.60454	-3.60087	-3.59721	-3.59357	-3.58994	-3.58632	-3.58272	-3.57913
0.028	-3.57555	-3.57199	-3.56843	-3.56489	-3.56137	-3.55785	-3.55435	-3.55086	-3.54738	-3.54391
0.029	-3.54046	-3.53702	-3.53359	-3.53017	-3.52676	-3.52337	-3.51998	-3.51661	-3.51325	-3.50990
0.030	-3.50656	-3.50323	-3.49991	-3.49661	-3.49331	-3.49003	-3.48676	-3.48349	-3.48024	-3.47700
0.031	-3.47377	-3.47055	-3.46734	-3.46414	-3.46095	-3.45777	-3.45460	-3.45144	-3.44829	-3.44515
0.032	-3.44202	-3.43890	-3.43579	-3.43269	-3.42960	-3.42652	-3.42344	-3.42038	-3.41733	-3.41428
0.033	-3.41125	-3.40822	-3.40521	-3.40220	-3.39920	-3.39621	-3.39323	-3.39026	-3.38729	-3.38434
0.034	-3.38139	-3.37846	-3.37553	-3.37261	-3.36970	-3.36680	-3.36390	-3.36102	-3.35814	-3.35527
0.035	-3.35241	-3.34955	-3.34671	-3.34387	-3.34104	-3.33822	-3.33541	-3.33260	-3.32981	-3.32702
0.036	-3.32424	-3.32146	-3.31870	-3.31594	-3.31319	-3.31044	-3.30771	-3.30498	-3.30226	-3.29954
0.037	-3.29684	-3.29414	-3.29145	-3.28876	-3.28608	-3.28341	-3.28075	-3.27810	-3.27545	-3.27280
0.038	-3.27017	-3.26754	-3.26492	-3.26231	-3.25970	-3.25710	-3.25450	-3.25192	-3.24934	-3.24676
0.039	-3.24419	-3.24163	-3.23908	-3.23653	-3.23399	-3.23145	-3.22893	-3.22640	-3.22389	-3.22138
0.040	-3.21888	-3.21638	-3.21389	-3.21140	-3.20893	-3.20645	-3.20399	-3.20153	-3.19907	-3.19663
0.041	-3.19418	-3.19175	-3.18932	-3.18689	-3.18447	-3.18206	-3.17966	-3.17725	-3.17486	-3.17247
0.042	-3.17009	-3.16771	-3.16534	-3.16297	-3.16061	-3.15825	-3.15590	-3.15356	-3.15122	-3.14888
0.043	-3.14656	-3.14423	-3.14191	-3.13960	-3.13730	-3.13499	-3.13270	-3.13041	-3.12812	-3.12584
0.044	-3.12357	-3.12130	-3.11903	-3.11677	-3.11452	-3.11227	-3.11002	-3.10778	-3.10555	-3.10332
0.045	-3.10109	-3.09887	-3.09666	-3.09445	-3.09224	-3.09004	-3.08785	-3.08566	-3.08347	-3.08129
0.046	-3.07911	-3.07694	-3.07478	-3.07261	-3.07046	-3.06830	-3.06615	-3.06401	-3.06187	-3.05974
0.047	-3.05761	-3.05548	-3.05336	-3.05125	-3.04913	-3.04703	-3.04492	-3.04282	-3.04073	-3.03864
0.048	-3.03655	-3.03447	-3.03240	-3.03032	-3.02826	-3.02619	-3.02413	-3.02208	-3.02003	-3.01798
0.049	-3.01594	-3.01390	-3.01186	-3.00983	-3.00780	-3.00578	-3.00376	-3.00175	-2.99974	-2.99773

$\log_e 10 = 2.30259$, $\log_e 100 = 4.60517$, $\log_e 1000 = 6.90776$

Table A.4 Natural Logarithms (Negative Values, Decreasing) (*Continued*)

N	0	1	2	3	4	5	6	7	8	9
0.050	-2.99573	-2.99373	-2.99174	-2.98975	-2.98776	-2.98578	-2.98380	-2.98183	-2.97986	-2.97789
0.051	-2.97593	-2.97397	-2.97202	-2.97006	-2.96812	-2.96617	-2.96423	-2.96230	-2.96037	-2.95844
0.052	-2.95651	-2.95459	-2.95267	-2.95076	-2.94885	-2.94694	-2.94504	-2.94314	-2.94124	-2.93935
0.053	-2.93746	-2.93558	-2.93370	-2.93182	-2.92994	-2.92807	-2.92621	-2.92434	-2.92248	-2.92062
0.054	-2.91877	-2.91692	-2.91507	-2.91323	-2.91139	-2.90955	-2.90772	-2.90589	-2.90407	-2.90224
0.055	-2.90042	-2.89861	-2.89679	-2.89498	-2.89318	-2.89137	-2.88957	-2.88778	-2.88598	-2.88419
0.056	-2.88240	-2.88062	-2.87884	-2.87706	-2.87529	-2.87351	-2.87175	-2.86998	-2.86822	-2.86646
0.057	-2.86470	-2.86295	-2.86120	-2.85945	-2.85771	-2.85597	-2.85423	-2.85250	-2.85077	-2.84904
0.058	-2.84731	-2.84559	-2.84387	-2.84215	-2.84044	-2.83873	-2.83702	-2.83532	-2.83361	-2.83191
0.059	-2.83022	-2.82852	-2.82683	-2.82515	-2.82346	-2.82178	-2.82010	-2.81842	-2.81675	-2.81508
0.060	-2.81341	-2.81175	-2.81008	-2.80842	-2.80677	-2.80511	-2.80346	-2.80181	-2.80017	-2.79852
0.061	-2.79688	-2.79524	-2.79361	-2.79198	-2.79035	-2.78872	-2.78709	-2.78547	-2.78385	-2.78224
0.062	-2.78062	-2.77901	-2.77740	-2.77579	-2.77419	-2.77259	-2.77099	-2.76939	-2.76780	-2.76621
0.063	-2.76462	-2.76303	-2.76145	-2.75987	-2.75829	-2.75672	-2.75514	-2.75357	-2.75200	-2.75044
0.064	-2.74887	-2.74731	-2.74575	-2.74420	-2.74264	-2.74109	-2.73954	-2.73799	-2.73645	-2.73491
0.065	-2.73337	-2.73183	-2.73030	-2.72876	-2.72723	-2.72571	-2.72418	-2.72266	-2.72114	-2.71962
0.066	-2.71810	-2.71659	-2.71507	-2.71357	-2.71206	-2.71055	-2.70905	-2.70755	-2.70605	-2.70456
0.067	-2.70306	-2.70157	-2.70008	-2.69860	-2.69711	-2.69563	-2.69415	-2.69267	-2.69119	-2.68972
0.068	-2.68825	-2.68678	-2.68531	-2.68385	-2.68238	-2.68092	-2.67946	-2.67801	-2.67655	-2.67510
0.069	-2.67365	-2.67220	-2.67075	-2.66931	-2.66787	-2.66643	-2.66499	-2.66356	-2.66212	-2.66069
0.070	-2.65926	-2.65783	-2.65641	-2.65498	-2.65356	-2.65214	-2.65073	-2.64931	-2.64790	-2.64649
0.071	-2.64508	-2.64367	-2.64226	-2.64086	-2.63946	-2.63806	-2.63666	-2.63526	-2.63387	-2.63248
0.072	-2.63109	-2.62970	-2.62832	-2.62693	-2.62555	-2.62417	-2.62279	-2.62141	-2.62004	-2.61867
0.073	-2.61730	-2.61593	-2.61456	-2.61319	-2.61183	-2.61047	-2.60911	-2.60775	-2.60640	-2.60504
0.074	-2.60369	-2.60234	-2.60099	-2.59964	-2.59830	-2.59696	-2.59561	-2.59428	-2.59294	-2.59160
0.075	-2.59027	-2.58893	-2.58760	-2.58628	-2.58495	-2.58362	-2.58230	-2.58098	-2.57966	-2.57834
0.076	-2.57702	-2.57571	-2.57439	-2.57308	-2.57177	-2.57046	-2.56916	-2.56785	-2.56655	-2.56525
0.077	-2.56395	-2.56265	-2.56136	-2.56006	-2.55877	-2.55748	-2.55619	-2.55490	-2.55361	-2.55233
0.078	-2.55105	-2.54977	-2.54849	-2.54721	-2.54593	-2.54466	-2.54338	-2.54211	-2.54084	-2.53957
0.079	-2.53831	-2.53704	-2.53578	-2.53452	-2.53326	-2.53200	-2.53074	-2.52949	-2.52823	-2.52698
0.080	-2.52573	-2.52448	-2.52323	-2.52199	-2.52074	-2.51950	-2.51826	-2.51702	-2.51578	-2.51454
0.081	-2.51331	-2.51207	-2.51084	-2.50961	-2.50838	-2.50715	-2.50593	-2.50470	-2.50348	-2.50226
0.082	-2.50104	-2.49982	-2.49860	-2.49738	-2.49617	-2.49496	-2.49375	-2.49254	-2.49133	-2.49012
0.083	-2.48891	-2.48771	-2.48651	-2.48531	-2.48411	-2.48291	-2.48171	-2.48052	-2.47932	-2.47813
0.084	-2.47694	-2.47575	-2.47456	-2.47337	-2.47219	-2.47100	-2.46982	-2.46864	-2.46746	-2.46628
0.085	-2.46510	-2.46393	-2.46275	-2.46158	-2.46041	-2.45924	-2.45807	-2.45690	-2.45574	-2.45457
0.086	-2.45341	-2.45225	-2.45109	-2.44993	-2.44877	-2.44761	-2.44646	-2.44530	-2.44415	-2.44300
0.087	-2.44185	-2.44070	-2.43955	-2.43841	-2.43726	-2.43612	-2.43497	-2.43383	-2.43269	-2.43156
0.088	-2.43042	-2.42928	-2.42815	-2.42702	-2.42588	-2.42475	-2.42362	-2.42250	-2.42137	-2.42024
0.089	-2.41912	-2.41800	-2.41687	-2.41575	-2.41463	-2.41352	-2.41240	-2.41128	-2.41017	-2.40906
0.090	-2.40795	-2.40684	-2.40573	-2.40462	-2.40351	-2.40241	-2.40130	-2.40020	-2.39910	-2.39800
0.091	-2.39690	-2.39580	-2.39470	-2.39360	-2.39251	-2.39142	-2.39032	-2.38923	-2.38814	-2.38705
0.092	-2.38597	-2.38488	-2.38380	-2.38271	-2.38163	-2.38055	-2.37947	-2.37839	-2.37731	-2.37623
0.093	-2.37516	-2.37408	-2.37301	-2.37194	-2.37086	-2.36979	-2.36873	-2.36766	-2.36659	-2.36553
0.094	-2.36446	-2.36340	-2.36234	-2.36127	-2.36021	-2.35916	-2.35810	-2.35704	-2.35599	-2.35493
0.095	-2.35388	-2.35283	-2.35178	-2.35073	-2.34968	-2.34863	-2.34758	-2.34654	-2.34549	-2.34445
0.096	-2.34341	-2.34237	-2.34133	-2.34029	-2.33925	-2.33821	-2.33718	-2.33614	-2.33511	-2.33408
0.097	-2.33304	-2.33201	-2.33098	-2.32996	-2.32893	-2.32790	-2.32688	-2.32585	-2.32483	-2.32381
0.098	-2.32279	-2.32177	-2.32075	-2.31973	-2.31871	-2.31770	-2.31668	-2.31567	-2.31466	-2.31365
0.099	-2.31264	-2.31163	-2.31062	-2.30961	-2.30860	-2.30760	-2.30659	-2.30559	-2.30459	-2.30359

$\log_e 10 = 2.30259$, $\log_e 100 = 4.60517$, $\log_e 1000 = 6.90776$

Table A.4 Natural Logarithms (Negative Values, Decreasing) *(Continued)*

N	0	1	2	3	4	5	6	7	8	9
0.100	-2.30259	-2.29263	-2.28278	-2.27303	-2.26336	-2.25379	-2.24432	-2.23493	-2.22562	-2.21641
0.110	-2.20727	-2.19823	-2.18926	-2.18037	-2.17156	-2.16282	-2.15417	-2.14558	-2.13707	-2.12863
0.120	-2.12026	-2.11196	-2.10373	-2.09557	-2.08747	-2.07944	-2.07147	-2.06357	-2.05573	-2.04794
0.130	-2.04022	-2.03256	-2.02495	-2.01741	-2.00992	-2.00248	-1.99510	-1.98777	-1.98050	-1.97328
0.140	-1.96611	-1.95900	-1.95193	-1.94491	-1.93794	-1.93102	-1.92415	-1.91732	-1.91054	-1.90381
0.150	-1.89712	-1.89048	-1.88387	-1.87732	-1.87080	-1.86433	-1.85790	-1.85151	-1.84516	-1.83885
0.160	-1.83258	-1.82635	-1.82016	-1.81401	-1.80789	-1.80181	-1.79577	-1.78976	-1.78379	-1.77786
0.170	-1.77196	-1.76609	-1.76026	-1.75446	-1.74870	-1.74297	-1.73727	-1.73161	-1.72597	-1.72037
0.180	-1.71480	-1.70926	-1.70375	-1.69827	-1.69282	-1.68740	-1.68201	-1.67665	-1.67131	-1.66601
0.190	-1.66073	-1.65548	-1.65026	-1.64507	-1.63990	-1.63476	-1.62964	-1.62455	-1.61949	-1.61445
0.200	-1.60944	-1.60445	-1.59949	-1.59455	-1.58964	-1.58475	-1.57988	-1.57504	-1.57022	-1.56542
0.210	-1.56065	-1.55590	-1.55117	-1.54646	-1.54178	-1.53712	-1.53248	-1.52786	-1.52326	-1.51868
0.220	-1.51413	-1.50959	-1.50508	-1.50058	-1.49611	-1.49165	-1.48722	-1.48281	-1.47841	-1.47403
0.230	-1.46968	-1.46534	-1.46102	-1.45672	-1.45243	-1.44817	-1.44392	-1.43970	-1.43548	-1.43129
0.240	-1.42712	-1.42296	-1.41882	-1.41469	-1.41059	-1.40650	-1.40242	-1.39837	-1.39433	-1.39030
0.250	-1.38629	-1.38230	-1.37833	-1.37437	-1.37042	-1.36649	-1.36258	-1.35868	-1.35480	-1.35093
0.260	-1.34707	-1.34323	-1.33941	-1.33560	-1.33181	-1.32803	-1.32426	-1.32051	-1.31677	-1.31304
0.270	-1.30933	-1.30564	-1.30195	-1.29828	-1.29463	-1.29098	-1.28735	-1.28374	-1.28013	-1.27654
0.280	-1.27297	-1.26940	-1.26585	-1.26231	-1.25878	-1.25527	-1.25176	-1.24827	-1.24479	-1.24133
0.290	-1.23787	-1.23443	-1.23100	-1.22758	-1.22418	-1.22078	-1.21740	-1.21402	-1.21066	-1.20731
0.300	-1.20397	-1.20065	-1.19733	-1.19402	-1.19073	-1.18744	-1.18417	-1.18091	-1.17766	-1.17441
0.310	-1.17118	-1.16796	-1.16475	-1.16155	-1.15836	-1.15518	-1.15201	-1.14885	-1.14570	-1.14256
0.320	-1.13943	-1.13631	-1.13320	-1.13010	-1.12701	-1.12393	-1.12086	-1.11780	-1.11474	-1.11170
0.330	-1.10866	-1.10564	-1.10262	-1.09961	-1.09661	-1.09362	-1.09064	-1.08767	-1.08471	-1.08176
0.340	-1.07881	-1.07587	-1.07294	-1.07002	-1.06711	-1.06421	-1.06132	-1.05843	-1.05555	-1.05268
0.350	-1.04982	-1.04697	-1.04412	-1.04129	-1.03846	-1.03564	-1.03282	-1.03002	-1.02722	-1.02443
0.360	-1.02165	-1.01888	-1.01611	-1.01335	-1.01060	-1.00786	-1.00512	-1.00239	-0.99967	-0.99696
0.370	-0.99425	-0.99155	-0.98886	-0.98618	-0.98350	-0.98083	-0.97817	-0.97551	-0.97286	-0.97022
0.380	-0.96758	-0.96496	-0.96233	-0.95972	-0.95711	-0.95451	-0.95192	-0.94933	-0.94675	-0.94418
0.390	-0.94161	-0.93905	-0.93649	-0.93395	-0.93140	-0.92887	-0.92634	-0.92382	-0.92130	-0.91879
0.400	-0.91629	-0.91379	-0.91130	-0.90882	-0.90634	-0.90387	-0.90140	-0.89894	-0.89649	-0.89404
0.410	-0.89160	-0.88916	-0.88673	-0.88431	-0.88189	-0.87948	-0.87707	-0.87467	-0.87227	-0.86988
0.420	-0.86750	-0.86512	-0.86275	-0.86038	-0.85802	-0.85567	-0.85332	-0.85097	-0.84863	-0.84630
0.430	-0.84397	-0.84165	-0.83933	-0.83702	-0.83471	-0.83241	-0.83011	-0.82782	-0.82554	-0.82326
0.440	-0.82098	-0.81871	-0.81645	-0.81419	-0.81193	-0.80968	-0.80744	-0.80520	-0.80296	-0.80073
0.450	-0.79851	-0.79629	-0.79407	-0.79186	-0.78966	-0.78746	-0.78526	-0.78307	-0.78089	-0.77871
0.460	-0.77653	-0.77436	-0.77219	-0.77003	-0.76787	-0.76572	-0.76357	-0.76143	-0.75929	-0.75715
0.470	-0.75502	-0.75290	-0.75078	-0.74866	-0.74655	-0.74444	-0.74234	-0.74024	-0.73814	-0.73605
0.480	-0.73397	-0.73189	-0.72981	-0.72774	-0.72567	-0.72361	-0.72155	-0.71949	-0.71744	-0.71539
0.490	-0.71335	-0.71131	-0.70928	-0.70725	-0.70522	-0.70320	-0.70118	-0.69917	-0.69716	-0.69515
0.500	-0.69315	-0.69115	-0.68916	-0.68717	-0.68518	-0.68320	-0.68122	-0.67924	-0.67727	-0.67531
0.510	-0.67334	-0.67139	-0.66943	-0.66748	-0.66553	-0.66359	-0.66165	-0.65971	-0.65778	-0.65585
0.520	-0.65393	-0.65201	-0.65009	-0.64817	-0.64626	-0.64436	-0.64245	-0.64055	-0.63866	-0.63677
0.530	-0.63488	-0.63299	-0.63111	-0.62923	-0.62736	-0.62549	-0.62362	-0.62176	-0.61990	-0.61804
0.540	-0.61619	-0.61434	-0.61249	-0.61065	-0.60881	-0.60697	-0.60514	-0.60331	-0.60148	-0.59966
0.550	-0.59784	-0.59602	-0.59421	-0.59240	-0.59059	-0.58879	-0.58699	-0.58519	-0.58340	-0.58161
0.560	-0.57982	-0.57803	-0.57625	-0.57448	-0.57270	-0.57093	-0.56916	-0.56740	-0.56563	-0.56387
0.570	-0.56212	-0.56037	-0.55862	-0.55687	-0.55513	-0.55339	-0.55165	-0.54991	-0.54818	-0.54645
0.580	-0.54473	-0.54300	-0.54128	-0.53957	-0.53785	-0.53614	-0.53444	-0.53273	-0.53103	-0.52933
0.590	-0.52763	-0.52594	-0.52425	-0.52256	-0.52088	-0.51919	-0.51751	-0.51584	-0.51416	-0.51249

$\log_e 10 = 2.30259$, $\log_e 100 = 4.60517$, $\log_e 1000 = 6.90776$

Table A.4 Natural Logarithms (Negative Values, Decreasing) (*Continued*)

N	0	1	2	3	4	5	6	7	8	9
0.600	-0.51083	-0.50916	-0.50750	-0.50584	-0.50418	-0.50253	-0.50088	-0.49923	-0.49758	-0.49594
0.610	-0.49430	-0.49266	-0.49102	-0.48939	-0.48776	-0.48613	-0.48451	-0.48289	-0.48127	-0.47965
0.620	-0.47804	-0.47642	-0.47482	-0.47321	-0.47161	-0.47000	-0.46841	-0.46681	-0.46522	-0.46362
0.630	-0.46204	-0.46045	-0.45887	-0.45728	-0.45571	-0.45413	-0.45256	-0.45099	-0.44942	-0.44785
0.640	-0.44629	-0.44473	-0.44317	-0.44161	-0.44006	-0.43851	-0.43696	-0.43541	-0.43386	-0.43232
0.650	-0.43078	-0.42925	-0.42771	-0.42618	-0.42465	-0.42312	-0.42159	-0.42007	-0.41855	-0.41703
0.660	-0.41552	-0.41400	-0.41249	-0.41098	-0.40947	-0.40797	-0.40647	-0.40497	-0.40347	-0.40197
0.670	-0.40048	-0.39899	-0.39750	-0.39601	-0.39453	-0.39304	-0.39156	-0.39008	-0.38861	-0.38713
0.680	-0.38566	-0.38419	-0.38273	-0.38126	-0.37980	-0.37834	-0.37688	-0.37542	-0.37397	-0.37251
0.690	-0.37106	-0.36962	-0.36817	-0.36673	-0.36528	-0.36384	-0.36241	-0.36097	-0.35954	-0.35810
0.700	-0.35668	-0.35525	-0.35382	-0.35240	-0.35098	-0.34956	-0.34814	-0.34672	-0.34531	-0.34390
0.710	-0.34249	-0.34108	-0.33968	-0.33827	-0.33687	-0.33547	-0.33408	-0.33268	-0.33129	-0.32989
0.720	-0.32850	-0.32712	-0.32573	-0.32435	-0.32296	-0.32158	-0.32021	-0.31883	-0.31745	-0.31608
0.730	-0.31471	-0.31334	-0.31197	-0.31061	-0.30925	-0.30788	-0.30653	-0.30517	-0.30381	-0.30246
0.740	-0.30111	-0.29975	-0.29841	-0.29706	-0.29571	-0.29437	-0.29303	-0.29169	-0.29035	-0.28902
0.750	-0.28768	-0.28635	-0.28502	-0.28369	-0.28236	-0.28104	-0.27971	-0.27839	-0.27707	-0.27575
0.760	-0.27444	-0.27312	-0.27181	-0.27050	-0.26919	-0.26788	-0.26657	-0.26527	-0.26397	-0.26266
0.770	-0.26136	-0.26007	-0.25877	-0.25748	-0.25618	-0.25489	-0.25360	-0.25232	-0.25103	-0.24974
0.780	-0.24846	-0.24718	-0.24590	-0.24462	-0.24335	-0.24207	-0.24080	-0.23953	-0.23826	-0.23699
0.790	-0.23572	-0.23446	-0.23319	-0.23193	-0.23067	-0.22941	-0.22816	-0.22690	-0.22565	-0.22439
0.800	-0.22314	-0.22189	-0.22065	-0.21940	-0.21816	-0.21691	-0.21567	-0.21443	-0.21319	-0.21196
0.810	-0.21072	-0.20949	-0.20826	-0.20702	-0.20580	-0.20457	-0.20334	-0.20212	-0.20089	-0.19967
0.820	-0.19845	-0.19723	-0.19602	-0.19480	-0.19358	-0.19237	-0.19116	-0.18995	-0.18874	-0.18754
0.830	-0.18633	-0.18513	-0.18392	-0.18272	-0.18152	-0.18032	-0.17913	-0.17793	-0.17674	-0.17554
0.840	-0.17435	-0.17316	-0.17198	-0.17079	-0.16960	-0.16842	-0.16724	-0.16605	-0.16487	-0.16370
0.850	-0.16252	-0.16134	-0.16017	-0.15900	-0.15782	-0.15665	-0.15549	-0.15432	-0.15315	-0.15199
0.860	-0.15082	-0.14966	-0.14850	-0.14734	-0.14618	-0.14503	-0.14387	-0.14272	-0.14156	-0.14041
0.870	-0.13926	-0.13811	-0.13697	-0.13582	-0.13468	-0.13353	-0.13239	-0.13125	-0.13011	-0.12897
0.880	-0.12783	-0.12670	-0.12556	-0.12443	-0.12330	-0.12217	-0.12104	-0.11991	-0.11878	-0.11766
0.890	-0.11653	-0.11541	-0.11429	-0.11317	-0.11205	-0.11093	-0.10982	-0.10870	-0.10759	-0.10647
0.900	-0.10536	-0.10425	-0.10314	-0.10203	-0.10093	-0.09982	-0.09872	-0.09761	-0.09651	-0.09541
0.910	-0.09431	-0.09321	-0.09212	-0.09102	-0.08992	-0.08883	-0.08774	-0.08665	-0.08556	-0.08447
0.920	-0.08338	-0.08230	-0.08121	-0.08013	-0.07904	-0.07796	-0.07688	-0.07580	-0.07472	-0.07365
0.930	-0.07257	-0.07150	-0.07042	-0.06935	-0.06828	-0.06721	-0.06614	-0.06507	-0.06401	-0.06294
0.940	-0.06188	-0.06081	-0.05975	-0.05869	-0.05763	-0.05657	-0.05551	-0.05446	-0.05340	-0.05235
0.950	-0.05129	-0.05024	-0.04919	-0.04814	-0.04709	-0.04604	-0.04500	-0.04395	-0.04291	-0.04186
0.960	-0.04082	-0.03978	-0.03874	-0.03770	-0.03666	-0.03563	-0.03459	-0.03356	-0.03252	-0.03149
0.970	-0.03046	-0.02943	-0.02840	-0.02737	-0.02634	-0.02532	-0.02429	-0.02327	-0.02225	-0.02122
0.980	-0.02020	-0.01918	-0.01816	-0.01715	-0.01613	-0.01511	-0.01410	-0.01309	-0.01207	-0.01106
0.990	-0.01005	-0.00904	-0.00803	-0.00702	-0.00602	-0.00501	-0.00401	-0.00300	-0.00200	-0.00100
1.000	-0.00000	0.00100	0.00200	0.00300	0.00399	0.00499	0.00598	0.00698	0.00797	0.00896
1.010	0.00995	0.01094	0.01193	0.01292	0.01390	0.01489	0.01587	0.01686	0.01784	0.01882
1.020	0.01980	0.02078	0.02176	0.02274	0.02372	0.02469	0.02567	0.02664	0.02761	0.02859
1.030	0.02956	0.03053	0.03150	0.03247	0.03343	0.03440	0.03537	0.03633	0.03730	0.03826
1.040	0.03922	0.04018	0.04114	0.04210	0.04306	0.04402	0.04497	0.04593	0.04688	0.04784
1.050	0.04879	0.04974	0.05069	0.05164	0.05259	0.05354	0.05449	0.05543	0.05638	0.05732
1.060	0.05827	0.05921	0.06015	0.06109	0.06204	0.06297	0.06391	0.06485	0.06579	0.06672
1.070	0.06766	0.06859	0.06953	0.07046	0.07139	0.07232	0.07325	0.07418	0.07511	0.07603
1.080	0.07696	0.07789	0.07881	0.07973	0.08066	0.08158	0.08250	0.08342	0.08434	0.08526
1.090	0.08618	0.08709	0.08801	0.08893	0.08984	0.09075	0.09167	0.09258	0.09349	0.09440

$\log_e 10 = 2.30259$, $\log_e 100 = 4.60517$, $\log_e 1000 = 6.90776$

Table A.4 Natural Logarithms (Negative Values, Decreasing) (*Continued*)

N	0	1	2	3	4	5	6	7	8	9
1.100	0.09531	0.09622	0.09713	0.09803	0.09894	0.09985	0.10075	0.10165	0.10256	0.10346
1.110	0.10436	0.10526	0.10616	0.10706	0.10796	0.10885	0.10975	0.11065	0.11154	0.11244
1.120	0.11333	0.11422	0.11511	0.11600	0.11689	0.11778	0.11867	0.11956	0.12045	0.12133
1.130	0.12222	0.12310	0.12399	0.12487	0.12575	0.12663	0.12751	0.12839	0.12927	0.13015
1.140	0.13103	0.13190	0.13278	0.13366	0.13453	0.13540	0.13628	0.13715	0.13802	0.13889
1.150	0.13976	0.14063	0.14150	0.14237	0.14323	0.14410	0.14497	0.14583	0.14669	0.14756
1.160	0.14842	0.14928	0.15014	0.15100	0.15186	0.15272	0.15358	0.15444	0.15529	0.15615
1.170	0.15700	0.15786	0.15871	0.15956	0.16042	0.16127	0.16212	0.16297	0.16382	0.16467
1.180	0.16551	0.16636	0.16721	0.16805	0.16890	0.16974	0.17059	0.17143	0.17227	0.17311
1.190	0.17395	0.17479	0.17563	0.17647	0.17731	0.17815	0.17898	0.17982	0.18065	0.18149
1.200	0.18232	0.18315	0.18399	0.18482	0.18565	0.18648	0.18731	0.18814	0.18897	0.18979
1.210	0.19062	0.19145	0.19227	0.19310	0.19392	0.19474	0.19557	0.19639	0.19721	0.19803
1.220	0.19885	0.19967	0.20049	0.20131	0.20212	0.20294	0.20376	0.20457	0.20539	0.20620
1.230	0.20701	0.20783	0.20864	0.20945	0.21026	0.21107	0.21188	0.21269	0.21350	0.21430
1.240	0.21511	0.21592	0.21672	0.21753	0.21833	0.21914	0.21994	0.22074	0.22154	0.22234
1.250	0.22314	0.22394	0.22474	0.22554	0.22634	0.22714	0.22793	0.22873	0.22952	0.23032
1.260	0.23111	0.23190	0.23270	0.23349	0.23428	0.23507	0.23586	0.23665	0.23744	0.23823
1.270	0.23902	0.23980	0.24059	0.24138	0.24216	0.24295	0.24373	0.24451	0.24530	0.24608
1.280	0.24686	0.24764	0.24842	0.24920	0.24998	0.25076	0.25154	0.25231	0.25309	0.25387
1.290	0.25464	0.25542	0.25619	0.25696	0.25774	0.25851	0.25928	0.26005	0.26082	0.26159
1.300	0.26236	0.26313	0.26390	0.26467	0.26544	0.26620	0.26697	0.26773	0.26850	0.26926
1.310	0.27003	0.27079	0.27155	0.27231	0.27308	0.27384	0.27460	0.27536	0.27612	0.27687
1.320	0.27763	0.27839	0.27915	0.27990	0.28066	0.28141	0.28217	0.28292	0.28367	0.28443
1.330	0.28518	0.28593	0.28668	0.28743	0.28818	0.28893	0.28968	0.29043	0.29118	0.29192
1.340	0.29267	0.29342	0.29416	0.29491	0.29565	0.29639	0.29714	0.29788	0.29862	0.29936
1.350	0.30010	0.30084	0.30158	0.30232	0.30306	0.30380	0.30454	0.30528	0.30601	0.30675
1.360	0.30748	0.30822	0.30895	0.30969	0.31042	0.31115	0.31189	0.31262	0.31335	0.31408
1.370	0.31481	0.31554	0.31627	0.31700	0.31773	0.31845	0.31918	0.31991	0.32063	0.32136
1.380	0.32208	0.32281	0.32353	0.32425	0.32498	0.32570	0.32642	0.32714	0.32786	0.32858
1.390	0.32930	0.33002	0.33074	0.33146	0.33218	0.33289	0.33361	0.33433	0.33504	0.33576
1.400	0.33647	0.33719	0.33790	0.33861	0.33932	0.34004	0.34075	0.34146	0.34217	0.34288
1.410	0.34359	0.34430	0.34501	0.34571	0.34642	0.34713	0.34784	0.34854	0.34925	0.34995
1.420	0.35066	0.35136	0.35206	0.35277	0.35347	0.35417	0.35487	0.35557	0.35627	0.35697
1.430	0.35767	0.35837	0.35907	0.35977	0.36047	0.36116	0.36186	0.36256	0.36325	0.36395
1.440	0.36464	0.36534	0.36603	0.36672	0.36742	0.36811	0.36880	0.36949	0.37018	0.37087
1.450	0.37156	0.37225	0.37294	0.37363	0.37432	0.37501	0.37569	0.37638	0.37707	0.37775
1.460	0.37844	0.37912	0.37980	0.38049	0.38117	0.38185	0.38254	0.38322	0.38390	0.38458
1.470	0.38526	0.38594	0.38662	0.38730	0.38798	0.38866	0.38934	0.39001	0.39069	0.39137
1.480	0.39204	0.39272	0.39339	0.39407	0.39474	0.39541	0.39609	0.39676	0.39743	0.39810
1.490	0.39878	0.39945	0.40012	0.40079	0.40146	0.40213	0.40279	0.40346	0.40413	0.40480
1.500	0.40546	0.40613	0.40680	0.40746	0.40813	0.40879	0.40946	0.41012	0.41078	0.41145
1.510	0.41211	0.41277	0.41343	0.41409	0.41475	0.41542	0.41607	0.41673	0.41739	0.41805
1.520	0.41871	0.41937	0.42002	0.42068	0.42134	0.42199	0.42265	0.42330	0.42396	0.42461
1.530	0.42527	0.42592	0.42657	0.42723	0.42788	0.42853	0.42918	0.42983	0.43048	0.43113
1.540	0.43178	0.43243	0.43308	0.43373	0.43438	0.43502	0.43567	0.43632	0.43696	0.43761
1.550	0.43825	0.43890	0.43954	0.44019	0.44083	0.44148	0.44212	0.44276	0.44340	0.44404
1.560	0.44469	0.44533	0.44597	0.44661	0.44725	0.44789	0.44852	0.44916	0.44980	0.45044
1.570	0.45108	0.45171	0.45235	0.45298	0.45362	0.45425	0.45489	0.45552	0.45616	0.45679
1.580	0.45742	0.45806	0.45869	0.45932	0.45995	0.46058	0.46121	0.46185	0.46247	0.46310
1.590	0.46373	0.46436	0.46499	0.46562	0.46625	0.46687	0.46750	0.46813	0.46875	0.46938

$\log_e 10 = 2.30259$, $\log_e 100 = 4.60517$, $\log_e 1000 = 6.90776$

Table A.4 Natural Logarithms (Negative Values, Decreasing) *(Continued)*

N	0	1	2	3	4	5	6	7	8	9
1.600	0.47000	0.47063	0.47125	0.47188	0.47250	0.47312	0.47375	0.47437	0.47499	0.47561
1.610	0.47623	0.47685	0.47748	0.47810	0.47872	0.47933	0.47995	0.48057	0.48119	0.48181
1.620	0.48243	0.48304	0.48366	0.48428	0.48489	0.48551	0.48612	0.48674	0.48735	0.48797
1.630	0.48858	0.48919	0.48981	0.49042	0.49103	0.49164	0.49225	0.49286	0.49348	0.49409
1.640	0.49470	0.49531	0.49591	0.49652	0.49713	0.49774	0.49835	0.49896	0.49956	0.50017
1.650	0.50077	0.50138	0.50199	0.50259	0.50320	0.50380	0.50440	0.50501	0.50561	0.50621
1.660	0.50682	0.50742	0.50802	0.50862	0.50922	0.50982	0.51043	0.51103	0.51162	0.51222
1.670	0.51282	0.51342	0.51402	0.51462	0.51522	0.51581	0.51641	0.51701	0.51760	0.51820
1.680	0.51879	0.51939	0.51998	0.52058	0.52117	0.52177	0.52236	0.52295	0.52354	0.52414
1.690	0.52473	0.52532	0.52591	0.52650	0.52709	0.52768	0.52827	0.52886	0.52945	0.53004
1.700	0.53063	0.53122	0.53180	0.53239	0.53298	0.53356	0.53415	0.53474	0.53532	0.53591
1.710	0.53649	0.53708	0.53766	0.53825	0.53883	0.53941	0.54000	0.54058	0.54116	0.54174
1.720	0.54232	0.54291	0.54349	0.54407	0.54465	0.54523	0.54581	0.54639	0.54696	0.54754
1.730	0.54812	0.54870	0.54928	0.54985	0.55043	0.55101	0.55158	0.55216	0.55273	0.55331
1.740	0.55388	0.55446	0.55503	0.55561	0.55618	0.55675	0.55733	0.55790	0.55847	0.55904
1.750	0.55962	0.56019	0.56076	0.56133	0.56190	0.56247	0.56304	0.56361	0.56418	0.56475
1.760	0.56531	0.56588	0.56645	0.56702	0.56758	0.56815	0.56872	0.56928	0.56985	0.57041
1.770	0.57098	0.57154	0.57211	0.57267	0.57324	0.57380	0.57436	0.57493	0.57549	0.57605
1.780	0.57661	0.57717	0.57774	0.57830	0.57886	0.57942	0.57998	0.58054	0.58110	0.58166
1.790	0.58222	0.58277	0.58333	0.58389	0.58445	0.58500	0.58556	0.58612	0.58667	0.58723
1.800	0.58779	0.58834	0.58890	0.58945	0.59001	0.59056	0.59111	0.59167	0.59222	0.59277
1.810	0.59333	0.59388	0.59443	0.59498	0.59553	0.59609	0.59664	0.59719	0.59774	0.59829
1.820	0.59884	0.59939	0.59993	0.60048	0.60103	0.60158	0.60213	0.60267	0.60322	0.60377
1.830	0.60432	0.60486	0.60541	0.60595	0.60650	0.60704	0.60759	0.60813	0.60868	0.60922
1.840	0.60977	0.61031	0.61085	0.61139	0.61194	0.61248	0.61302	0.61356	0.61410	0.61464
1.850	0.61519	0.61573	0.61627	0.61681	0.61735	0.61788	0.61842	0.61896	0.61950	0.62004
1.860	0.62058	0.62111	0.62165	0.62219	0.62272	0.62326	0.62380	0.62433	0.62487	0.62540
1.870	0.62594	0.62647	0.62701	0.62754	0.62807	0.62861	0.62914	0.62967	0.63021	0.63074
1.880	0.63127	0.63180	0.63233	0.63287	0.63340	0.63393	0.63446	0.63499	0.63552	0.63605
1.890	0.63658	0.63711	0.63763	0.63816	0.63869	0.63922	0.63975	0.64027	0.64080	0.64133
1.900	0.64185	0.64238	0.64291	0.64343	0.64396	0.64448	0.64501	0.64553	0.64606	0.64658
1.910	0.64710	0.64763	0.64815	0.64867	0.64919	0.64972	0.65024	0.65076	0.65128	0.65180
1.920	0.65232	0.65285	0.65337	0.65389	0.65441	0.65493	0.65544	0.65596	0.65648	0.65700
1.930	0.65752	0.65804	0.65856	0.65907	0.65959	0.66011	0.66062	0.66114	0.66166	0.66217
1.940	0.66269	0.66320	0.66372	0.66423	0.66475	0.66526	0.66578	0.66629	0.66680	0.66732
1.950	0.66783	0.66834	0.66885	0.66937	0.66988	0.67039	0.67090	0.67141	0.67192	0.67243
1.960	0.67294	0.67345	0.67396	0.67447	0.67498	0.67549	0.67600	0.67651	0.67702	0.67753
1.970	0.67803	0.67854	0.67905	0.67955	0.68006	0.68057	0.68107	0.68158	0.68209	0.68259
1.980	0.68310	0.68360	0.68411	0.68461	0.68511	0.68562	0.68612	0.68663	0.68713	0.68763
1.990	0.68813	0.68864	0.68914	0.68964	0.69014	0.69064	0.69114	0.69165	0.69215	0.69265
2.000	0.69315	0.69365	0.69415	0.69465	0.69514	0.69564	0.69614	0.69664	0.69714	0.69764
2.010	0.69813	0.69863	0.69913	0.69963	0.70012	0.70062	0.70111	0.70161	0.70211	0.70260
2.020	0.70310	0.70359	0.70409	0.70458	0.70508	0.70557	0.70606	0.70656	0.70705	0.70754
2.030	0.70804	0.70853	0.70902	0.70951	0.71000	0.71050	0.71099	0.71148	0.71197	0.71246
2.040	0.71295	0.71344	0.71393	0.71442	0.71491	0.71540	0.71589	0.71637	0.71686	0.71735
2.050	0.71784	0.71833	0.71881	0.71930	0.71979	0.72028	0.72076	0.72125	0.72173	0.72222
2.060	0.72271	0.72319	0.72368	0.72416	0.72465	0.72513	0.72561	0.72610	0.72658	0.72706
2.070	0.72755	0.72803	0.72851	0.72900	0.72948	0.72996	0.73044	0.73092	0.73141	0.73189
2.080	0.73237	0.73285	0.73333	0.73381	0.73429	0.73477	0.73525	0.73573	0.73621	0.73668
2.090	0.73716	0.73764	0.73812	0.73860	0.73908	0.73955	0.74003	0.74051	0.74098	0.74146

$\log_e 10 = 2.30259$, $\log_e 100 = 4.60517$, $\log_e 1000 = 6.90776$

Table A.4 Natural Logarithms (Negative Values, Decreasing) (*Continued*)

N	0	1	2	3	4	5	6	7	8	9
2.10	0.74194	0.74669	0.75142	0.75612	0.76081	0.76547	0.77011	0.77473	0.77932	0.78390
2.20	0.78846	0.79299	0.79751	0.80200	0.80648	0.81093	0.81536	0.81978	0.82418	0.82855
2.30	0.83291	0.83725	0.84157	0.84587	0.85015	0.85442	0.85866	0.86289	0.86710	0.87129
2.40	0.87547	0.87963	0.88377	0.88789	0.89200	0.89609	0.90016	0.90422	0.90826	0.91228
2.50	0.91629	0.92028	0.92426	0.92822	0.93216	0.93609	0.94001	0.94391	0.94779	0.95166
2.60	0.95551	0.95935	0.96317	0.96698	0.97078	0.97456	0.97833	0.98208	0.98582	0.98954
2.70	0.99325	0.99695	1.00063	1.00430	1.00796	1.01160	1.01523	1.01885	1.02245	1.02604
2.80	1.02962	1.03318	1.03674	1.04028	1.04380	1.04732	1.05082	1.05431	1.05779	1.06126
2.90	1.06471	1.06815	1.07158	1.07500	1.07841	1.08181	1.08519	1.08856	1.09192	1.09527
3.00	1.09861	1.10194	1.10526	1.10856	1.11186	1.11514	1.11841	1.12168	1.12493	1.12817
3.10	1.13140	1.13462	1.13783	1.14103	1.14422	1.14740	1.15057	1.15373	1.15688	1.16002
3.20	1.16315	1.16627	1.16938	1.17248	1.17557	1.17865	1.18173	1.18479	1.18784	1.19089
3.30	1.19392	1.19695	1.19996	1.20297	1.20597	1.20896	1.21194	1.21491	1.21788	1.22083
3.40	1.22378	1.22671	1.22964	1.23256	1.23547	1.23837	1.24127	1.24415	1.24703	1.24990
3.50	1.25276	1.25562	1.25846	1.26130	1.26413	1.26695	1.26976	1.27257	1.27536	1.27815
3.60	1.28093	1.28371	1.28647	1.28923	1.29198	1.29473	1.29746	1.30019	1.30291	1.30563
3.70	1.30833	1.31103	1.31372	1.31641	1.31909	1.32176	1.32442	1.32707	1.32972	1.33237
3.80	1.33500	1.33763	1.34025	1.34286	1.34547	1.34807	1.35067	1.35325	1.35584	1.35841
3.90	1.36098	1.36354	1.36609	1.36864	1.37118	1.37372	1.37624	1.37877	1.38128	1.38379
4.00	1.38629	1.38879	1.39128	1.39377	1.39624	1.39872	1.40118	1.40364	1.40610	1.40854
4.10	1.41099	1.41342	1.41585	1.41828	1.42070	1.42311	1.42552	1.42792	1.43031	1.43270
4.20	1.43508	1.43746	1.43984	1.44220	1.44456	1.44692	1.44927	1.45161	1.45395	1.45629
4.30	1.45861	1.46094	1.46326	1.46557	1.46787	1.47018	1.47247	1.47476	1.47705	1.47933
4.40	1.48160	1.48387	1.48614	1.48840	1.49065	1.49290	1.49515	1.49739	1.49962	1.50185
4.50	1.50408	1.50630	1.50851	1.51072	1.51293	1.51513	1.51732	1.51951	1.52170	1.52388
4.60	1.52606	1.52823	1.53039	1.53256	1.53471	1.53687	1.53902	1.54116	1.54330	1.54543
4.70	1.54756	1.54969	1.55181	1.55393	1.55604	1.55814	1.56025	1.56235	1.56444	1.56653
4.80	1.56862	1.57070	1.57277	1.57485	1.57691	1.57898	1.58104	1.58309	1.58515	1.58719
4.90	1.58924	1.59127	1.59331	1.59534	1.59737	1.59939	1.60141	1.60342	1.60543	1.60744
5.00	1.60944	1.61144	1.61343	1.61542	1.61741	1.61939	1.62137	1.62334	1.62531	1.62728

$\log_e 10 = 2.30259$, $\log_e 100 = 4.60517$, $\log_e 1000 = 6.90776$

INDEX